Love and promotion of the values associated with air shows can only be felt by a high analysis of risk reduction, by the work of excellence and the fair vision of flight safety.

Dr. M. Karachalios here shows an essential book of bedside that should be appreciated and understood by all demo pilots, be they beginners or experienced. To recommend without moderation!

Jean-Noël Bouillaguet
Chairman of the French Air Display Association
(L'Association Française du Spectacle Aérien—FSA)

Flying in air shows hasn't made me a better pilot, but rather a better person because there are always people watching me and what I do. Manolis' book vividly captures the transition to excellence that air show performers achieve through consistent training. This book is not only for display pilots but also for anyone seeking high performance in all they do.

Patty Wagstaff
American Aerobatic Pilot, Three-time
US National Aerobatic Champion,
Inductee of the National Aviation Hall of Fame

Excellence in Air Show Performers

This publication reveals to readers the secrets and mindsets of air show performances. Serving as a sequel to its predecessor, *Air Show Performers: Safety, Risk Management, and Psychological Factors* (9781032556147), it builds upon the insights of the first volume and offers an exploration of what distinguishes exceptional performances amid the high-stakes environment of air shows.

From looking at the pathways to excellence to the hidden intricacies of resilient safety strategies, this publication reveals the key approaches to safety and good practice for air show performers. This publication combines academic research with real-world experiences from professionals in the field. It dissects the methodologies and practices essential for shaping elite air show performers and takes a critical look at training regimes. Technical skill, mental resilience, and continuous improvement in performance are crucial for success in this field and this publication explores resilient safety strategies to ensure that pilot and spectator protection remains paramount in every aspect of an air show performance. The reader will not just gain an insight into a pilot's training schedule but also perspectives into mindfulness and the psychological state of those who take part.

Excellence in Air Show Performers: Training for Resilient Safety is a must-read for professionals in health and safety, aviation, and event management seeking to enhance their practices to academics researching the complexities of high-risk environments.

Excellence in Air Show Performers

Training for Resilient Safety

Manolis Karachalios

CRC Press
Taylor & Francis Group
Boca Raton London New York

CRC Press is an imprint of the
Taylor & Francis Group, an **informa** business

Designed cover image: © 2026 HAFGS

First edition published 2026
by CRC Press
2385 NW Executive Center Drive, Suite 320, Boca Raton FL 33431

and by CRC Press
4 Park Square, Milton Park, Abingdon, Oxon, OX14 4RN

CRC Press is an imprint of Taylor & Francis Group, LLC

ISBN: 978-1-032-82405-5 (hbk)
ISBN: 978-1-032-82492-5 (pbk)
ISBN: 978-1-003-50480-1 (ebk)
ISBN: 978-1-003-59213-6 (enhanced ebk)

DOI: 10.1201/9781003504801

Typeset in Times
by Newgen Publishing UK

Dedication

To my beloved wife Mariangela.

Contents

PART III Resilient Safety

PART IV Top-Notch Performers—The Pinnacle of Air Show Excellence

Foreword

EXCELLENCE IN AIRSHOWS

Talking about Excellence in Airshows is a very large subject that we should approach from three different perspectives:

- The Public
- The Organizers
- The Performers

As a performer, having flown more than 3,000 displays in 40 countries, mainly as a team leader but also as an unlimited aerobatics and warbird display pilot…I will, of course, focus on the performer perspective, which inevitably interacts with the 2 other ones.

Safety But Not Only

The first aspect of excellence is often considered to be safety, which is true and is largely covered by the author of this book, but it is only a part of the equation…

I used to say that we are governed by two keywords:

Quality and Safety…which goes together…one cannot exist without the other…

Of course, basically, a flying display must be entertaining; otherwise, it is of no interest neither for the public nor for the organizers…

A display is a compromise between the performance of the aircraft and the skill and training of the pilot (pilots for a team).

The aim is to communicate energy from the cockpit to the spectators…not to fly to please ourselves…or shine our ego.

We are in the entertainment business, which means showmanship is a must.

We offer to the spectators an emotional, sensorial experience beyond the maneuvers…Music, narration, pyro, the rhythm of the display, precision, and elegance of the formations and maneuvers (even the smell of jet fuel for some) are part of this multi-sensorial experience...not to speak about precise aircraft alignment on the ramp, clean aircraft, synchronized start-up for teams.

Performers have to think all the time about how to improve their performance, even in the smallest detail.

The Display But Not Only

A performer who reaches for Excellence is engaged in a commitment far beyond the display...it is a global question of behavior.

Being able to set up a fair proposal to organizers, providing an attractive media package, showing up on time, with the predicted number of people, with no last-minute request to the organizers…and with a smiley face...being available to interact with the press, the public, and the fans, wearing a clean coverall (with appropriate shoes) even during hot days...returning courtesy cars without damage…are all details that the organizers and the public will remember of your venue…

In the Airshows, we meet three categories of performers:

Amateur, Professional, and finally Excellent…

Excellence can be reached by amateurs as well as professionals…it is just a question of attitude and motivation.

Of course, performers expect in return from the organizers to adopt the same attitude…providing them the contractual requests, without starting to think about how to provide promised services only when they see the performer's aircraft landing…

Remember the 5 P rule:

Proper Planning Prevents Poor Performance.

A well-planned event, a well-planned display is the best way to deliver to our customers: the public, and sponsors…an Excellent experience that pleases people and builds a future relationship…

Remember that we can always improve something in our process to reach Excellence.

If I had selected from my long career which venue would be closest to Excellence, it would most probably be the Swiss Air Force Airshow AIR14 in Payerne.

All the above-mentioned points were more than covered with 9 days (including two weekends) monster solo jets and jet teams displays…but let me give you an example of the sense of detail those guys have provided:

When the Swedish pilot landed with his Gripen, he discovered that the GPS of his courtesy car was with his hotel address set, and the GPS language choice was…in Swedish language.

Who got a more Excellent experience?

Jacques Bothelin
French Aerobatic Jet Team Leader,
Honorary Board Member European Airshow Council

Preface

Manolis Karachalios and Daniel Kwasi Adjekum

"*Excellence in Air Show Performers: Training for Resilient Safety*" is crafted as a complementary and expansive sequel to the first book in the series, "*Air Show Performers: Safety, Risk Management, and Psychological Factors*." This second volume aims to dive deeper into the nuanced aspects of air show performances and the stringent training required for resilient safety.

The primary aim of this volume is to provide an exhaustive and in-depth exploration of what constitutes high-caliber performance in the high-risk environment of air shows. Furthermore, this volume looks closer at the training pathways to achieving excellence, analyzing the skills, mental resilience, and continuous improvement processes that set apart elite performers in this unique and challenging field.

A substantial focus is placed on the comprehensive training regimes that underpin these elevated performance levels. The book details the methodologies and practices essential in shaping top-tier air show performers, covering the entire spectrum from initial selection to advanced skill development. This not only extends the discussion from the first volume but also provides a more detailed view of the developmental journey of these performers.

In addition, building on the safety themes explored in the first book, this volume expands into the concept of resilient safety strategies. It aims to dissect how these strategies are formulated to enhance adaptability, balance, and robustness, thereby ensuring the highest levels of safety in air show performances.

By integrating academic research with real-world experiences from seasoned air show professionals, the book bridges the gap between theory and practice. It is both a comprehensive academic resource and a practical guide for industry practitioners.

Ultimately, this book aspires to be an insightful resource for current and future air show professionals, enriching their knowledge and inspiring them to achieve excellence in their careers while retaining safety. It also aims to captivate a broad readership, including aviation professionals, high-performance specialists, academic scholars, and air show enthusiasts, effectively complementing and extending the discourse initiated in the series' first book.

Acknowledgments

I extend my heartfelt gratitude to Dr. Daniel Kwasi Adjekum for his inspirational mentorship in the field of safety science. His invaluable guidance and contributions have profoundly shaped this publication.

Author

Manolis Karachalios, PhD, MBA, ATP, CFI, FRAeS

Manolis Karachalios is a visionary leader in the aviation industry with over 25 years of experience, renowned for his expertise in global operations and strategic management. He has a remarkable track record of leading multinational teams, driving organizational excellence, and pioneering innovative strategies that have significantly enhanced aviation safety and risk management worldwide.

Karachalios earned his PhD in Aerospace Studies from the University of North Dakota and an MBA in Aviation Management from Coventry University. Graduating with honors from the Hellenic Air Force Academy in 2000, he laid a strong foundation for his distinguished career. Notably, he served as the Hellenic Air Force's F-16 Demo Team "ZEUS" Display Pilot from 2010 to 2012, which opened the way for his active involvement within the air show community.

As a Board Member of the European Airshow Council, Karachalios has been instrumental in directing strategic initiatives and fostering safety innovations. His role as Flying Display Director for the Athens Flying Week International Air Show highlights his commitment to ensuring the safe execution of airshow displays. Throughout his career, he has successfully overseen strategic growth and operations in aviation consultancy and safety programs, managing performance standards and mitigating operational risks for pilot training and demonstration teams.

Karachalios has achieved zero safety incidents in major aviation events while significantly expanding his consultancy client base. His expertise in strategic planning, global team leadership, and risk management is widely recognized.

An accomplished author, Karachalios's second book, "*Excellence in Air Show Performers: Training for Resilient Safety*" (2023), reflects his deep insights and extensive experience. He also prepares the annual Des Barker Air Show Safety Report, contributing valuable knowledge to the air show community.

Beyond his professional accomplishments, Dr. Karachalios shares a fulfilling life with his wife, Mariangela, and their two daughters, Maria and Dimitra. His family's support has been the cornerstone of his successful career.

Contact: *mkarachalios@gmail.com*, LinkedIn: *linkedin.com/in/manoliskarahalios*, Athens, Greece.

Contributors

Daniel Kwasi Adjekum
Aviation Department
University of North Dakota
North Dakota, USA

Manolis Karachalios
European Airshow Council

Abbreviations

Abbreviation	Definition
AAM	Advanced Air Mobility
ACC	Air Combat Command
ACE	Aerobatic Competency Evaluator
ANOVA	Analysis of Variance
ADM	Aeronautical Decision Making
AER	Airshow Excellence Rate
AGFI	Adjusted Goodness-of-Fit Index
A-LOC	Almost (G-Induced) Loss of Consciousness
AMT	Airshow Mindfulness Training
AoA	Angle of Attack
APE	Allowable Performance Error
APER	Air Show Performer Excellence Rate
ARL	Air Show Risk Level
ARM	Air Show Resource Management
AsDM	Air Show Decision Making
ASMS	Air Show Safety Management Systems
ASRM	Air Show Safety Risk Management
AVE	Average Variance Extracted
BAAR	Barker Airshow Accident Rate
BADA	British Air Display Association
BC	Bias-Corrected Confidence
CAA	Civil Aviation Authority
CFA	Confirmatory Factor Analysis
CFI	Comparative Fit Index
CMIN	Minimum Discrepancy or Model Chi-Square
COVID-19	Coronavirus Disease 2019
CPA	Causal Path Analysis
CR	Critical Ratios
CST	Critical-Stress Threshold
DA	Density Altitude
EAC	European Airshow Council
EASA	European Union Aviation Safety Agency
EPI	Excellence Performance Indicator

Abbreviation	Definition
ETA	Event Tree Analysis
FAA	Federal Aviation Authority, United States
FAI	Fédération Aéronautique Internationale
FSA	France Spectacle Aérien
FAST	Formation and Safety Team
FCC	Flying Control Committee
FMEA	Failure Mode and Effects Analysis
FMECA	Failure Mode Effects and Criticality Analysis
FRAM	Functional Resonance Analysis Method
FRMS	Fatigue Risk Management Systems
FTA	Fault Tree Analysis
GFI	Goodness-of-Fit Index
GIVD	G-Induced Vestibular Dysfunction
G-LOC	G-Induced Loss Of Consciousness
HA	Hazardous Attitudes
HFTC	Heritage Flight Training Course
HRO	High-Reliability Organization
IAC	International Aerobatic Club
ICAO	International Civil Aviation Organization
ICAS	International Council of Air Shows
IFI	Incremental Fit Index
IFR	Instrument Flight Rules
IMC	Instrument Meteorological Conditions
MAC	Midair Collision
MF	Mindfulness
MI	Modification Indices
ML	Maximum Likelihood Estimation
MLE	Maximum Likelihood Estimates
NFI	Normed Fit Index
PA	Path Analysis
PCFI	Parsimony Comparative Fit Index
PIF	Performance Influencing Factor
PNFI	Parsimony Normed Fit Index
RI	Risk Indicator
RMSEA	Root Means Square Error of Approximation
RP	Risk Perception

Abbreviation	Definition
RPM	Resilient Performance Model
RSC	Resilient Safety Culture
RT	Risk Tolerance
SA	Situational Awareness
SAC	Statement of Aerobatic Competency
SEM	Structural Equation Modelling
SMS	Safety Management System
SOP	Standard Operating Procedures
SPI	Safety Performance Indicator
SRM	Safety Risk Management
STPA	System Theoretic Process Analysis
SVFR	Special Visual Flight Rules
TSM	Time Safety Margin
UPRT	Upset Prevention and Recovery Training
VFR	Visual Flight Rules
VMC	Visual Meteorological Conditions

Prologue

Introduction to Excellence in Air Show Performers

Manolis Karachalios

Contributor
Daniel Kwasi Adjekum

We begin this prologue with a thought-provoking query that might arise in the minds of our readers: why is excellence synonymous with air show performers? The concept of excellence is multifaceted and deeply ingrained in the world of aerial performance. While safety is a fundamental expectation for all aviators, including air show performers, the pursuit of excellence transcends the basic necessity of safety. It is universally acknowledged that air show performers must prioritize safety for themselves and their audience, just like any professional engaging with the public. However, the essence of their craft extends beyond mere safety precautions. The core question is not just about being safe; it is about striving for excellence. This excellence is what sets them apart, captivating and inspiring their audience with every maneuver in the sky.

In the world of air shows, the safe execution of events and precision flying by display pilots is not just expected; it is imperative. The era when air show performers were pioneering adventurers, flying unregulated and without bounds, solely to make a name for themselves, has long passed. In today's world, no reputable air show or aerial display event settles for a pilot whose only consideration is safety devoid of excellence in aerial skills. There needs to be an equitable balance between excellent performance and safety. The ethos of air show performance is deeper, surpassing the baseline of safety to embody a broader representation of the aviation industry and human achievements in aviation and aerospace.

Historically, the narrative of air shows has evolved significantly. In the early 20th century, air shows were largely unregulated events where pilots performed daring stunts to captivate audiences. Most of these aerial antics pushed safe boundaries, especially when most of the aircraft used were of basic design and had rudimentary aviation technology. Some pilots also executed novel and complex aerial maneuvers extemporaneously with untested skills. For instance, the legendary feats of pilots like Lincoln Beachey, known for his death-defying maneuvers in the 1910s, encapsulate the era's spirit of daring and innovation (Karachalios & Adjekum, 2023). However, this unregulated environment also posed significant risks, leading to the evolution of safety protocols and performance standards.

DOI: 10.1201/9781003504801-1

In the contemporary air show circuit, the integration of safety and performance excellence has become paramount. Performers like the US Navy Blue Angels and the Royal Air Force Red Arrows epitomize this blend, executing complex maneuvers with precision while adhering to strict safety guidelines. Their performances are not just about thrilling the audience but also about demonstrating the capabilities of humans and machines in perfect harmony.

The culture of excellence in air shows is comparable to other high-stakes, high-performance, high-resilient environments. For example, Formula 1® drivers such as Lewis Hamilton and Max Verstappen exemplify excellence (van Kesteren & Bergkamp, 2023), where split-second decisions and precise control are critical to both safety and victory. Equally high-performing athletes engaged in high-risk activities are alpine skiers who compete against time in a highly competitive and extreme weather environment. Through their developed mental, physical, and technical skills, they manage to excel in their races (Gilgien et al., 2018; Supej & Holmberg, 2019). Similarly, in the world of professional ballet, dancers like Misty Copeland showcase how years of rigorous training and dedication to a craft can result in performances that combine artistry with physical and mental resilience (DeFrank-Cole & Nicholson, 2016).

Air show performers, much like these elite athletes and artists, train meticulously to perfect their craft. They engage in continuous learning, physical conditioning, and mental preparation to ensure that each performance is executed with the highest level of skill and safety. This relentless pursuit of excellence is what distinguishes them in the competitive field of aviation, serving as an inspiration across various high-performance domains.

The air show industry is a spectacular showcase of human and technological achievements, mirroring the intense and meticulous nature of performances seen in other high-stakes fields like Formula 1® racing, the American NASCAR®, theatrical productions, and circuses. These arenas don't aim for mere adequacy or safety; they strive for peak performance, pushing the limits of human capability and technological innovation.

An air show is more than a display of aerial acrobatics; it's a holistic demonstration of the ultimate levels of human skill, aircraft potential, and the collective achievements of humanity in mastering the skies. Like a Formula 1® driver who must maintain optimal speed, execute precise braking, and achieve flawless pit stops to excel, or a circus performer who performs with the utmost precision and skill before a large audience, air show performers aim for a level of excellence that inherently incorporates safety.

In this high-octane environment, attention to detail is paramount. Every maneuver, every roll, and every dive is a testament to the relentless pursuit of perfection, paralleling the exactitude required in Formula 1 racing and the precision seen in circus acts. This pursuit is not just about avoiding harm but about reaching the zenith of performance, where achieving excellence ensures safety is naturally integrated. In the world of air shows, excellence is not just a goal; it's the standard where safety and spectacular performance are inseparably linked. When excellence is the goal, safety is a given.

For an air show performer, adhering strictly to the criteria of a "safe display" could imply mere survival of the performance or executing maneuvers at higher than intended altitudes, which, while safe, may not align with the anticipated level of excellence. This misalignment can lead to audience disappointment, concern, or even an inability to fully appreciate the intricacies of the performance. For instance, if executing a loop at 500 feet is the standard, as seen in Figure 0.1, performing it safely at 1,000 feet might meet safety criteria but fail to deliver the expected thrill and precision, potentially diluting the spectacle and excitement of the show. Moreover, a performer executing a loop at 1,000 feet may not provide a lasting and positive experience for the audience compared to the more exhilarating performance at 500 feet. A non-exciting performance may wane interest and popularity for such air shows and does not truly reflect the existence of the air show industry.

Consider the Blue Angels, for example, renowned for their tight, precise formations. If they were to opt for a "safer" display, flying 6–20 feet apart, instead of their customary close formation, this would not only alter the visual impact but also affect the pilots' spatial awareness and timing. Such a formation, while seemingly safer due to increased separation, could introduce challenges like misjudged visual cues, inaccurate perception of the lead pilot's maneuvers, and delayed reactions. Ironically, what appears to be a "safer" option might, in practice, increase risk due to these complications.

FIGURE 0.1 Loop aerobatic maneuver. (Copyright: 2017 Mr. Gugger/Shutterstock.com.)

Another pivotal question we endeavor to address in this book is whether excellence serves as the ultimate outcome or the foundational motivation for air show performers. Is it the shining beacon they aim for from the start or the crowning glory of their rigorous training, experience, and mental preparation? Some argue that excellence is the culmination of an arduous journey marked by detailed training, extensive experience, focused attention to detail, and comprehensive mental preparation (Schein, 1990, 2017). From this perspective, excellence is viewed as the end result, achievable after a dedicated period of systematic training and practice, signifying the peak of performance that comes with time and effort. Conversely, others assert that excellence is the initial driving force for air show performers, the core objective guiding every flight and maneuver (Arthur Jr. et al., 2012; Donoso-Morales et al., 2017; Güllich et al., 2022; Hodge et al., 2014; Iso-Ahola, 2024; Iso-Ahola & Dotson, 2016; Lubinski & Benbow, 2000; Yarrow et al., 2009). For these individuals, excellence is not just an outcome but a fundamental goal that shapes their approach to performance, acting as both the stimulus and the guiding light for their endeavors. They are goal-oriented from the outset, with a clear aim to achieve the highest levels of excellence in their performances. Ultimately, it seems that the most effective approach may be a blend of both perspectives. Starting with excellence as the primary motivation sets a high standard and clear direction for performers while recognizing it as the end result underscores the importance of the process and the journey toward achieving it. This synergy ensures that air show performers not only aim for but also attain a consistently high level of performance, where accuracy and safety are inherently intertwined, leading to the ultimate realization of both excellence and safety in their aerial activities.

The approach to excellence by air show performers is underpinned by a proactive mentality that constantly seeks improvement. This mentality leaves no room for error, making it a critical component for air show performers to achieve both accuracy in their performances and safety in all their activities. This "excellence approach" is not only about perfecting their aerial maneuvers but also about embracing a resilient safety culture that has become increasingly prevalent in the aviation industry (Adjekum & Fernandez-Tous, 2020; Schwarz et al., 2016). This shift toward excellence and safety has led to significant enhancements in the flight safety records of this aviation sector (Gill & Shergill, 2004; Krois et al., 2015; Raj, 2019).

The concept of excellence infuses every aspect of an air show performer's life, from ground preparation and flying displays to interviews, communication styles, work ethics, and actual performances. This pervasive culture of excellence is not only an example for the aviation sector but also serves as an inspiration for other high-performance and high-risk fields. The adoption of this culture by air show performers in recent years sets a commendable standard for individuals across various domains, including students, professionals, managers, doctors, and drivers, urging them to strive for excellence in the demanding and high-pressure environment we all navigate today.

The culture of excellence instilled in air show performers offers valuable lessons on managing both physical and psychological challenges, providing a blueprint for overcoming adversity and maintaining focus on long-term objectives. It exemplifies how a steadfast commitment to excellence can inspire individuals to ascend to new heights of performance and achievement. This cultural ethos is not just about achieving success in high-stakes situations; it's about fostering resilience, precision, and a continuous pursuit of improvement that can profoundly influence and inspire across all walks of life, helping people to navigate and excel in the complex, high-demand world in which we live.

DEFINITION OF EXCELLENCE

> *Excellence is the possession chiefly of good qualities in an eminent or unusual degree surpassing merit or skill.*
>
> *Oxford English Dictionary (2024)*

The Oxford English Dictionary defines excellence as "the quality of being outstanding or extremely good" or as "the possession chiefly of good qualities in an eminent or unusual degree surpassing merit or skill." This definition emphasizes superiority in quality and skill, signifying a high standard that distinctly surpasses the norm.

Additionally, in the context of organizational performance, excellence is described as "surpassing merit; thing in which persons, etc. excel," where to excel implies being superior (Gapp, 2004; Sonntag, 2010). This perspective focuses on surpassing established standards to achieve superior outcomes.

Overall, excellence involves a significant degree of quality and superiority, going beyond ordinary standards to achieve exceptional results.

HISTORICAL CONTEXT AND EVOLUTION

Aristotle, the ancient Greek philosopher, had a profound impact on the concept of excellence. He believed that excellence, or *aretē* (αρετή) in Greek, was not just about achieving greatness but about fulfilling one's purpose or function to the best of their ability. His famous quote encapsulates this idea: "*We are what we repeatedly do. Excellence, then, is not an act, but a habit*" (Durant, 1991).

Aristotle's perspective on happiness and excellence is particularly illuminating. He viewed happiness as a form of activity rather than a state, aligning it with the exercise of virtue. According to Aristotle, the highest form of happiness involves a life of intellectual contemplation because this is the highest and most sustainable form of virtue. He famously described happiness as "a working of the soul in the way of excellence" and emphasized that this requires both virtue and learning or discipline (Niu Hai-fang, 2004).

For Aristotle, the pursuit of excellence was about finding the golden mean, or the moderate path between excess and deficiency, which is key to achieving moral virtue and ultimate happiness. This approach was integral to his ethical teachings and was meant to guide individuals toward living a balanced and fulfilling life. This interpretation of excellence as a habitual practice of virtue and the mean between extremes provides a timeless framework that continues to influence contemporary thought on personal and ethical development.

Throughout time, the ethos of excellence has evolved from not just a theory but also a practice that resonates across various sectors beyond aviation, offering a template for high performance and meticulous precision. In the world of space exploration, for example, the same principles of excellence and safety seen in air shows are crucial. The detailed planning, robust teamwork, and precision required for a successful space mission echo the dynamics of an air show performance. The synchronization and flawless execution in space launches, where astronauts and ground control work in perfect harmony, mirror the coordination of air show teams performing complex aerial routines.

The story of the Mars Rover missions, such as the Curiosity Rover's landing on Mars, serves as a testament to the importance of excellence in execution (Bonetti et al., 2013). The precision required to land a rover on the Martian surface, millions of miles away from Earth, underlines the critical nature of accuracy, just as the precision of a pilot in an air show must be faultless to ensure both spectacle and safety (Mandalia & Braun, 2015). This parallel highlights how the pursuit of excellence transcends earthly boundaries and becomes essential in the exploration of space.

In the corporate world, the drive for excellence can be found in the pursuit of innovation and perfection in product development and customer service. From an empirical perspective, researchers such as Hermel and Ramis-Pujol (2003) and Mi Dahlgaard-Park and Dahlgaard (2007) analyzed some of the historical models and frameworks used to quantify excellence by looking at their strengths as well as weaknesses. Models and frameworks identified and discussed in their research were Peters' and Waterman's eight excellence attributes, Peters' and Austin's simplified excellence model, Xerox's excellence models, the European excellence model, Dahlgaard-Park and Dahlgaard's 4P model, and Toyota's 4P model. The authors, in their findings, suggested that excellence models and frameworks were generally inspired by Japanese practices, and they recognize the importance of the soft dimension of organizational realities. The limitations of their findings were that there could be a tendency to interpret these models from reductionist and positivistic views and ignore the human aspect and the soft dimension when organizations try to implement the model in their struggle to achieve excellence. From a more contemporary perspective, companies like Apple (2020), known for their meticulous attention to detail and commitment to delivering high-quality products, embody a similar ethos of excellence, which is reflected in the global appeal and utilization of their products. The design and execution of Apple products, where every element is thought through to provide the user with an optimal experience, reflect the same dedication to excellence and precision that air show performers demonstrate in their routines.

Furthermore, the influence of excellence extends into the realm of education and personal development. The rigorous training and constant striving for improvement that define an air show performer's journey are akin to the lifelong learning and personal growth pursued by individuals in various fields. In a multi-year effort to create cultures of excellence among cohorts of graduate students, Mintrom and Cheng (2014) analyzed subsequent student performance on a significant, related task, and the outcomes indicated that explicitly promoting a culture of excellence among course participants can have a positive and sustained impact on their individual practices. This finding is supported by Slyusarenko et al. (2022), who emphasize the significance of personal development as a criterion for higher education quality, linking it directly to improved educational outcomes and individual growth. Similarly, Bubnys and Kauneckienė (2017) discuss the role of personal and professional development in fostering teacher leadership and innovation within schools, illustrating the parallel between individual growth and professional enhancement.

Finally, de Jong et al. (2022) have developed a conceptual framework to assess excellence in higher education called FACE (Framework for Analyzing Conceptions of Excellence) to provide a reflective tool for analyzing ideas on who is excellent, what is student excellence, why is student excellence important, and how is excellence taught. The main components of the FACE model are based on giftedness, motivation, and excellence in higher education. This model underlines the importance of setting high standards, embracing challenges, and continually pushing the boundaries of what is possible, whether in the classroom, in a professional setting, or in personal endeavors.

In the realm of project management and engineering, operational excellence and risk management strategies are evident in complex engineering projects, with meticulous planning, rigorous safety protocols, and effective team coordination being essential practices. Clark and Fang (2012) illustrate this by analyzing risk interactions in large engineering projects through network theory, underscoring the importance of understanding risk interdependencies and applying comprehensive risk management frameworks. Similarly, Fang et al. (2013) develop an integrated framework for risk response planning under resource constraints, highlighting the critical need for strategic resource allocation in managing interdependent risks in large-scale engineering projects.

The culture of excellence serves as a beacon for various high-risk professions, particularly in the medical field (Sklar, 2015; Lee et al., 2016). Surgeons and healthcare professionals operate in high-stakes environments where precision, teamwork, and a commitment to excellence are crucial for patient safety and successful outcomes. The same principles that guide air show performers in their precise and coordinated displays can be found in the operating room, where surgeons must execute their tasks with the utmost accuracy and care. Studies have shown that teamwork in surgical teams is vital for safe surgery, as it enhances communication and cooperation, which are crucial for managing the complex dynamics of surgical procedures (Aerden et al., 2014; Undre et al., 2007). Similarly, the implementation of structured teamwork and communication protocols in surgery can significantly improve patient outcomes (Makary et al., 2006). Finally, the role of feedback, especially from peers, has been highlighted as important for building excellence in medical care. Rosenthal et al. (2020), as part of a review of a clinical performance improvement tool, highlight how peer feedback is currently used in hospital medicine and present practical steps for hospital medicine programs to implement peer feedback to foster clinical excellence among their clinicians. This finding underscores the essence of peer feedback as an important driver for excellence in high-risk endeavors such as medicine and aerobatic displays.

In the realm of sports, elite athletes often embody the same dedication to excellence as air show performers (Willams & Hodges, 2023; Zentgraf & Raab, 2023). The meticulous preparation, mental fortitude, and continuous striving for improvement seen in Olympic gymnasts, for example, mirror the attributes of top-tier air show pilots (Aquilina, 2013). Both sets of professionals engage in relentless practice and refinement of their skills to achieve peak performance under intense scrutiny and pressure (Swann et al., 2015).

Ultimately, excellence in air show performers transcends the mere execution of complicated maneuvers; it embodies a comprehensive synergy of technical proficiency, relentless preparation, and an unwavering commitment to safety. At the heart of this pursuit is a deep respect for the precision and discipline demanded by the unforgiving nature of flight, where the margin for error is minimal. The performers, often elite pilots from military and civilian backgrounds, engage in continuous, rigorous training to hone their skills, pushing the boundaries of what is possible in the sky. The research underscores the critical importance of psychological and physiological training in enhancing pilot performance, noting the impact of such training on the precision and accuracy of piloting maneuvers, which are vital for both safety and performance in high-stakes environments like air shows (Socha et al., 2020).

STANDARDS AND BENCHMARKS IN AIR SHOW PERFORMANCES

The standards for air show performers differ based on the country and the related aviation authority. Below are discussed what the Federal Aviation Administration (FAA) and UK Civil Aviation Authority (CAA) have regulated to the set of standards as a reference and a case study for other organizations to refer to if needed.

ICAS

The Aerobatic Competency Evaluation Program (International Council of Air Shows, 2019), established by the International Council of Air Shows, sets professional standards and evaluations for civilian air show pilots in North America. It embodies the commitment to maintaining public trust through rigorous competency and safety assessments, ensuring that every performer meets or exceeds the industry-defined thresholds of excellence.

At the heart of achieving excellence in air show performances is a stringent focus on safety, proficiency, and continuous improvement. The program mandates that participants undergo comprehensive evaluations, including both ground and flight assessments, to validate their aerobatic competency. These evaluations are conducted by seasoned Aerobatic Competency Evaluators who themselves embody the highest standards of aerial excellence and safety consciousness.

Moreover, excellence in air show performances is underpinned by a commitment to lifelong learning and adherence to the highest ethical standards. Pilots are encouraged to continuously hone their skills, engage in peer reviews, and partake in professional development activities. This ethos ensures that the air show industry not only preserves its storied tradition of thrilling aerial displays but also advances in line with technological and methodological innovations.

UK CAA

In the United Kingdom, regulatory compliance is paramount (UK CAA, 2024 a, b, c, d, e). All aerial activities intended as exhibitions for public viewing must obtain specific permission from the UK CAA under CAP 403: Flying Displays and Special Events (UK CAA, 2024b). Moreover, pilots participating in these events are mandated to hold a display authorization (DA), ensuring they meet the rigorous safety and skill standards essential for such high-risk performances.

The organizational framework for air shows is meticulously structured to enhance safety and efficiency. The Flying Display Director (FDD) is responsible for the overall management of the air show, ensuring that all activities comply with CAA guidelines and safety protocols. Simultaneously, the event organizer (EO) manages the broader logistical aspects and coordinates with local authorities to safeguard public safety and ensure the event runs smoothly.

Moreover, safety management is a critical focus, with every event requiring comprehensive risk assessments and robust emergency plans to mitigate potential hazards to participants and spectators alike. These preventive measures are crucial in prioritizing the safety and well-being of everyone involved.

Then, operational guidelines are stringently defined to facilitate safe aircraft operations. These guidelines include specific rules regarding minimum separation distances, height restrictions, and protocols for various types of aircraft and special activities such as pyrotechnics and article dropping. Additionally, they encompass regulations for the participation of military and foreign aircraft, ensuring all performances adhere to the highest safety standards.

Training and accreditation are mandatory for both the FDD and participating pilots, incorporating specialized human factors training aimed at managing and mitigating risks associated with human error. This training is vital in maintaining the high safety standards demanded at such events.

Essential components of air show management are proper documentation and effective communication. Required documentation includes detailed maps of the event area, lists of participating aircraft, and comprehensive communication plans that facilitate coordination with emergency services and other stakeholders.

A strong emphasis on human factors highlights the commitment to understanding and managing the human elements involved in air shows. This proactive approach is instrumental in identifying potential risks and developing strategies to address them.

By adhering to these stringent guidelines, air show organizers and participants in the United Kingdom provide a model of excellence that other countries and organizations might find beneficial. This structured approach ensures that each event not only complies with regulatory requirements but also achieves a standard of excellence that enhances overall safety and enjoyment.

Measuring Air Show Excellence

The introduction and application of Excellence Performance Indicators (EPIs) and Safety Performance Indicators (SPIs) are imperative for the air show community. These metrics serve not only as tools for measuring the multifaceted aspects of performance and safety but also as essential foundations for setting benchmarks within the industry. The rationale behind this approach is encapsulated in the adage often attributed to management thinker Peter Drucker (2007): "*If you can't measure it, you can't improve it.*"

EPIs and SPIs provide a structured framework for evaluating the quality of performances, the effectiveness of safety protocols, audience engagement levels, operational efficiency, and numerous other critical factors that contribute to the success of air shows. By establishing clear and quantifiable measures, these indicators allow for the following:

- Objective Assessment: EPIs and SPIs offer objective criteria for assessing performance and safety, moving beyond subjective judgments to provide a standardized basis for evaluation.
- Benchmark Setting: These indicators set industry benchmarks, enabling teams and organizers to gauge their performances against established standards and strive for excellence.
- Identification of Improvement Areas: Through the measurement of specific aspects of air shows, EPIs and SPIs highlight areas needing enhancement, guiding targeted efforts for improvement.
- Continuous Improvement: By tracking changes in these indicators over time, air show participants can embark on continuous improvement cycles, systematically addressing areas for growth and innovation.

A review of existing literature indicates a significant gap in standardized performance evaluations within the air show industry (Barker, 2020; Karachalios & Adjekum, 2023). Previous metrics have predominantly concentrated on safety, often overlooking the broader aspects of quality. This section emphasizes the need for comprehensive evaluation frameworks, with the introduction of the EPIs, which can encompass a wider range of performance criteria and drive continuous improvement in the field of air shows.

The application of EPIs and SPIs in the air show community facilitates a culture of excellence and safety. It enables organizers and performers to make data-driven decisions, enhancing the quality of their offerings and ensuring the well-being of all involved. Furthermore, these metrics foster a competitive yet collaborative environment where participants are motivated to exceed benchmarks and share best practices for the benefit of the entire community. In the words of Lord Kelvin, "*To measure is to know*" (Houser, 2022). By measuring performance and safety through EPIs and SPIs, the air show community sets a baseline from which continuous improvement is not just possible but inevitable, propelling the industry toward higher standards of excellence and safety.

Basic EPIs

The development of the Airshow Excellent Rate (AER) and the Airshow Performer Excellence Rate (APER) involved a comprehensive, multi-staged process aimed at creating a holistic and robust framework for evaluating excellence in air shows. This process was underpinned by collaboration with a wide range of stakeholders, including air show organizers, performers, aviation safety experts, and spectators, to ensure that the EPIs reflect the multifaceted nature of air show performance and excellence.

The indicators are defined as follows:

- The AER formula could be defined as the number of outstanding and exceptional air show events divided by the total number of air show events held during a specific time period, multiplied by 100 to obtain a percentage.

$$AER = \frac{Outstanding\ and\ Exceptional\ Airshow\ Events}{Total\ Number\ of\ Airshow\ Events} \times 100\%$$

- APER is a metric to evaluate individual performers or teams, calculated as follows:

$$APER = \frac{Sum\ of\ Scores}{Maximum\ Possible\ Score} \times 100\%$$

Difference between AER and APER

AER assesses the overall quality and success of air show events. It measures factors such as safety records, audience satisfaction, and innovation within the event, providing an aggregated view of an air show's excellence. AER is crucial for organizers to benchmark their events against industry standards and to identify areas for improvement.

APER, on the other hand, evaluates the individual excellence of performers or teams. It focuses on safety adherence, audience engagement, consistency, and innovation within performances. APER offers performers a tool for self-assessment and a roadmap for professional growth, emphasizing a balanced achievement across crucial performance areas.

Value and Information Gained

Figure 0.2 depicts a comparison of the Excellence Performance Indicators (EPIs) for five hypothetical air shows (AER) and five performers (APER) over the last four years.

From the AER graph of Figure 0.2, readers can gain insights into the performance and improvement trends of different air shows over time. It highlights which air shows consistently excel in quality and which ones have shown significant improvement or decline. This information is invaluable for potential participants, sponsors, and the general audience in making informed decisions.

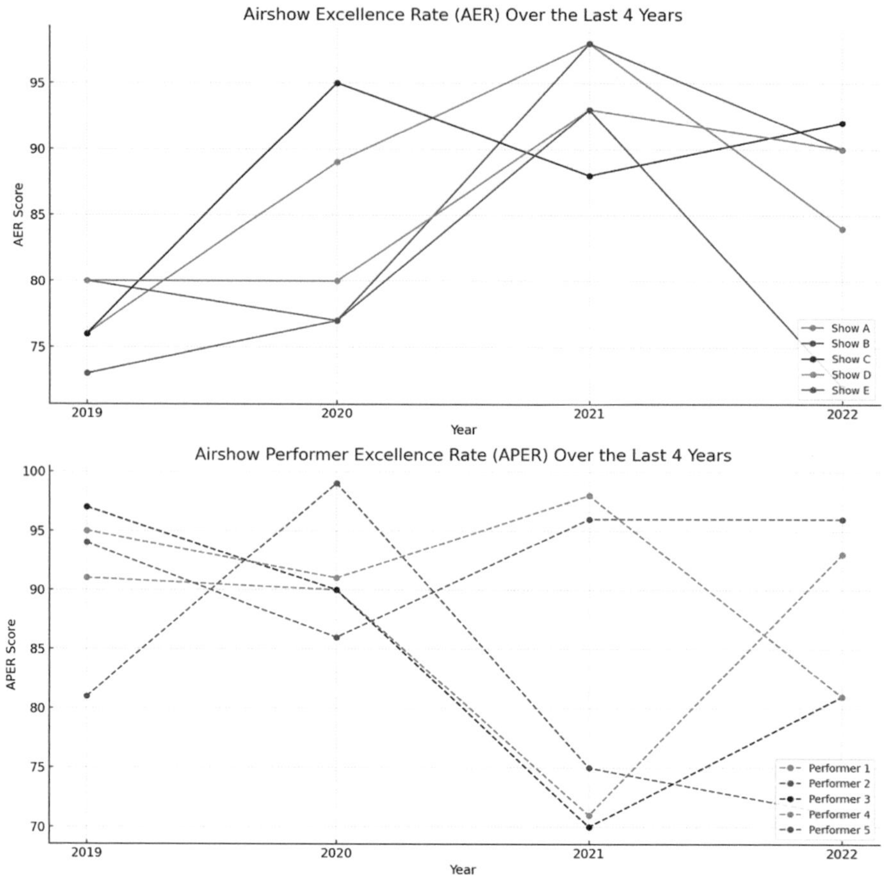

FIGURE 0.2 Comparison of Excellence Performance Indicators (EPIs).

Furthermore, the APER in Figure 0.2 provides a detailed look at the professional development of individual performers or teams. It showcases the consistency of excellence in performances, areas of growth, and, potentially, the impact of targeted efforts to improve specific aspects of their act. Performers can use this information for self-reflection and goal setting, while organizers might use it to scout talent and curate high-quality lineups for their events.

Together, AER and APER offer a comprehensive view of the air show ecosystem's health, driving a culture of excellence that benefits all stakeholders—organizers, performers, and audiences alike.

Airshow Excellence Rate

The AER was introduced in the 2023 Des Barker's annual report (Karachalios, 2024) to capture the quality and excellence of the air show industry. Therefore, the AER would provide an opportunity to acknowledge the outstanding work of air show performers and organizers and recognize the positive impact the air show community has on the aviation industry.

The introduction of the AER could provide a new approach to evaluating air show performance and promote a culture of excellence across our community.

The criteria for what constitute an outstanding air show event could be determined by a panel of experts within the air show community, taking into account factors such as creativity, innovation, and overall excellence. The total number of air show events could be determined by a comprehensive listing of air show events held during the specific period in question. The resulting percentage obtained from the AER formula would provide an overall measure of the level of excellence and high performance achieved by the air show community during that period.

The AER, as presented in the graph (see Figure 0.3), depicts a slight decrease from 99.987% in 2022 to 99.979% in 2023. This marginal decline indicates a small reduction in the overall excellence rate for air shows, which may reflect a minor increase in the number of incidents or accidents or potentially a decrease in adherence to safety standards or effectiveness of safety protocols at air show events.

While the difference between the two percentages is very small (0.008%), in the context of air show safety, even such a slight change can be significant. The goal of achieving the highest possible safety standard means striving for no accidents or incidents, where even a fractional percentage change can represent an important shift.

This small decrease serves as an indicator for the air show community investigating the underlying causes and addressing any potential issues. It may also prompt a reassessment of current safety practices and encourage the implementation of improvements to regain or exceed the previous year's higher excellence rate. The overall high AER across both years, however, does suggest that air shows generally maintain a very high level of safety standards.

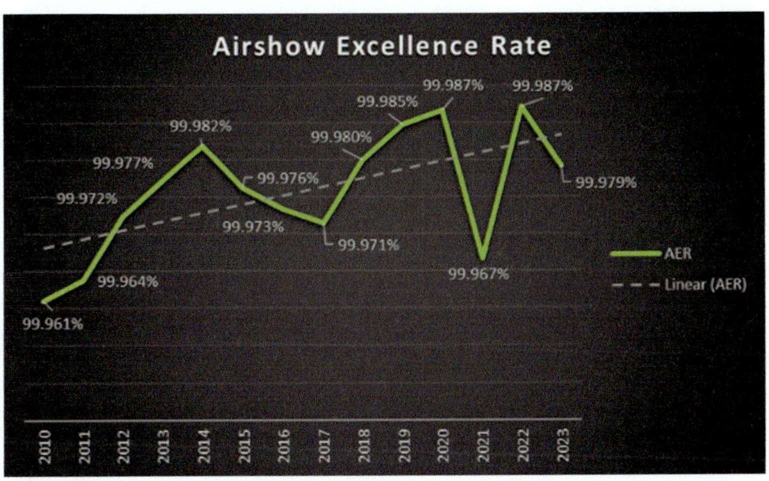

FIGURE 0.3 Airshow Excellence Rate (AER).

FIGURE 0.4 Airshow Excellence Rate (AER), regional comparison for the 2023 air show season.

Regional Comparison for AERs

The bar chart in Figure 0.4 provides a regional comparison of the AER for the year 2023. The AER is a measure that reflects the overall performance quality and safety level of air shows. The data shows the following:

- United States: The AER for the United States stands at 99.989%, which is the highest among the regions compared. This suggests that air shows in the United States are performing exceptionally well, with a near-perfect rate of excellence.
- Europe: Europe's AER is slightly lower at 99.978% but still very high, indicating that European air shows also maintain a very high standard of quality and safety.
- Rest of the World: The AER for the rest of the world is 99.926%, which, while lower than the United States and Europe, still represents a high level of excellence. This rate suggests that there is a strong global standard for air shows, though there may be room for improvement in certain areas to reach the same levels as the United States and Europe.

The regional comparison shows that the air show industry maintains a high standard of excellence across the board, with very small margins differentiating the regions. The differences might be attributed to various factors such as regional safety regulations, cultural differences, performance training quality, investment in air show infrastructure, and possibly the frequency and types of air shows conducted. The high AER values across all regions underline the industry's commitment to delivering quality events that prioritize safety for both the participants and spectators.

Regional Disparities in AER, 2023

The regional disparities in the AER for the year 2023 could reflect not just the operational aspects of air show management but also deeper cultural differences in how safety is perceived and prioritized, the level of oversight by regulatory bodies, and the role of individualism or egoism among performers.

- United States: With the highest AER, this may reflect a culture of stringent safety standards and rigorous oversight. The United States has a mature aviation industry with well-established regulatory frameworks, such as the FAA, which enforces strict air show safety protocols. Additionally, the emphasis on individual achievement in American culture might also contribute to a high degree of professionalism and personal responsibility among performers.
- Europe: Europe's AER at 99.978% indicates a similarly high standard, which could be attributed to the collective approach to safety common in European cultures. The European Union Aviation Safety Agency (EASA) and other national bodies provide comprehensive oversight. This collective approach might balance out individualistic tendencies and promote a safety culture that's based on shared standards and collaborative improvement.
- Rest of the World: The Rest of the World's AER at 99.926%, while still impressive, suggests there might be variations in the enforcement of safety regulations and possibly in cultural attitudes toward risk and safety. In some cultures, there might be a greater emphasis on collective or community success over individual achievement, which could affect the dynamics of safety and performance. Conversely, in some areas, less stringent regulatory oversight or a higher degree of individualism could result in a greater range of safety outcomes.

Cultural factors such as the approach to safety, the level of oversight, the willingness to report, and the balance between collective responsibility and individual recognition can significantly influence the AER (Adjekum, 2014; Meng-Yuan, 2015). In regions where there is less emphasis on individual compliments, the focus may naturally shift toward team success and safety (Noort et al., 2016). However, where individualism and personal achievement are highly valued, this might push performers to take greater risks, which can affect safety rates.

The presence of egoism, if unchecked, can potentially lead to riskier behaviors as performers push boundaries to stand out. It's the role of air show organizers and regulatory bodies to ensure that such tendencies are managed and that the drive for excellence is aligned with the highest safety standards. These cultural nuances are essential for the international air show community to consider as it works toward the goal of achieving an AER of 99.99%, with the ultimate aim of zero fatalities and a Barker Airshow Accident Rate (BAAR) below 1 (Karachalios, 2024).

Airshow Performer Excellence Rate (APER)

An APER could measure various performance criteria, including safety adherence, audience engagement, consistency, and innovation within safety bounds.

For air show performers, especially those looking to gauge and improve the consistency of their flying displays, it's crucial to have tailored metrics that capture the nuances of performance across various dimensions. Here are some additional metrics specifically designed for self-assessment, focusing on consistency along with other critical performance aspects.

Maneuver Consistency Index (MCI)

- Description: This index measures the variability in the execution quality of specific maneuvers over multiple performances. It helps performers understand how consistently they execute each element of their routine.
- Calculation: Compare the scores or subjective assessments of each maneuver across shows, calculating the standard deviation or variance to gauge consistency.

Routine Timing Precision (RTP)

- Description: Accurate timing is essential in synchronized air shows. Routine Timing Precision (RTP) assesses how closely the actual timing of performance matches the planned or ideal timing.
- Calculation: Record the duration of key segments or the entire routine and compare it against pre-set targets to identify deviations.

Environmental Adaptability Score (EAS)

- Description: This score evaluates a performer's ability to adapt maneuvers and routines to varying environmental conditions, such as weather changes, which is crucial for maintaining consistency.
- Calculation: Adjustments made and their effectiveness are rated by coaches or peers post-performance, under different environmental conditions.

Performance Reliability Rating (PRR)

- Description: This rating evaluates the frequency of performance issues or failures, such as incomplete maneuvers or technical problems during displays.
- Calculation: Track and rate each performance for any issues, with higher scores indicating fewer problems and greater reliability.

Feedback Consistency Level (FCL)

- Description: This metric assesses how consistent the audience or expert feedback is across multiple performances. It provides insight into the performer's ability to deliver expected quality consistently.
- Calculation: Analyze feedback from judges, peers, or audiences using a structured questionnaire focusing on key performance aspects, calculating average scores and their variability over time.

Maneuver Consistency Index (MCI)

The Maneuver Consistency Index (MCI) measures the variability in the execution quality of specific maneuvers across different performances. It helps identify which maneuvers are performed with high consistency and which are more variable and may need more focused practice.

Formula

The MCI for each maneuver can be calculated using the variance or standard deviation of the scores assigned to that maneuver across multiple performances. A lower variance or standard deviation indicates higher consistency.

$$MCI = \sqrt{\frac{n\sum\left(x_i - \mu\right)^2}{n}}$$

where

x_i = score of the maneuver in the ith performance
μ = mean score of the maneuver across all performances
n = total number of performances

Steps for Calculation

1. Score Each Maneuver: After each performance, the maneuver is scored (e.g., on a scale of 1–100) based on criteria like precision, timing, and execution.
2. Calculate the Mean Score: Average the scores of each specific maneuver across all evaluated performances.
3. Compute Variance: Calculate the variance by finding the average of the squared differences from the mean.
4. Calculate Standard Deviation (MCI): Take the square root of the variance to get the standard deviation, which is the MCI.

Example Calculation

Suppose a performer executes a specific maneuver (loop) in five different shows, and the scores are as follows: *92*, *88*, *90*, *92*, and *91*.

- Calculate the mean score (μ):

$$\mu = \frac{\left(92 + 88 + 90 + 92 + 91\right)}{5} = 90.6$$

- Calculate the variance:

$$Variance = \frac{\left(92 - 90.6\right)^2 + \left(88 - 90.6\right)^2 + \left(90 - 90.6\right)^2 + \left(92 - 90.6\right)^2 + \left(91 - 90.6\right)^2}{5}$$

- Calculate the MCI (standard deviation):

$$MCI = \sqrt{Variance}$$

For the "Loop" performed across five shows, here are the computed values:

- Mean score (μ): 90.6
- Variance: 2.24
- MCI: Approximately 1.50

Interpretation

The MCI of approximately 1.50 indicates that the scores for this maneuver are quite consistent across performances, with a low degree of variability from the mean score. A lower MCI value reflects higher consistency in execution, suggesting that the performer executes this maneuver with high precision and reliability across different shows.

This methodical approach to calculating the MCI enables performers and their coaches to quantitatively assess performance consistency, providing a clear metric to guide practice and improvement strategies.

Example Visualization: Consistency Heatmap

A heatmap could be an effective visualization tool to display these metrics for a performer across multiple performances or seasons (see Figure 0.5). Each row represents a metric, and each column represents a specific performance, with colors indicating scores (e.g., red for high consistency and blue for low consistency).

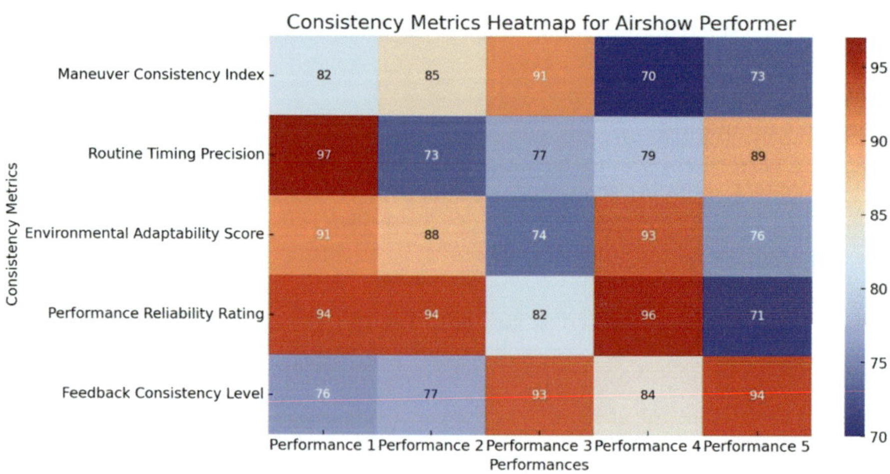

FIGURE 0.5 Consistency heatmap.

The heatmap displayed above visualizes the consistency metrics for an air show performer across five different performances. Each cell represents the score for a particular consistency-related metric during a specific performance, with color intensity indicating the level of consistency (warmer colors represent higher scores, thus greater consistency).

This visualization effectively allows performers and coaches to quickly identify areas of strength and those requiring further improvement, particularly in maintaining consistent performance quality under varying conditions and across multiple shows. Such detailed feedback is essential for targeting specific aspects of training and preparation to enhance overall performance reliability and audience satisfaction.

Excellence Metrics in Diverse Fields

Adopting and adapting metrics from diverse fields, such as sports and surgery, to the air show community can provide innovative ways to evaluate and enhance performance, safety, and audience engagement. Here's how metrics from football, basketball, tennis, surgery, and Formula 1 driving could be integrated into the air show community to augment and expand the EPIs.

Football

- Pass Completion Rate: Analogous to precision in maneuver execution in air shows. Tracking the accuracy of planned vs. executed maneuvers could highlight precision and teamwork.
- Distance Covered: Similar to the extent and variety of aerial coverage during a performance, indicating the dynamic range and energy of the show.

Basketball

- Player Efficiency Rating: A holistic measure of a player's contributions. For air shows, a similar metric could evaluate a performer's overall impact on the show, including safety, engagement, and technical proficiency.
- Turnovers: Reflecting errors or safety breaches in air shows. A lower 'turnover' rate would indicate higher safety and precision.

Tennis

- First Serve Success Rate: Comparable to the success rate of opening maneuvers in air shows, setting the tone for performance quality and audience engagement.
- Unforced Errors: Tracking mistakes not due to environmental conditions but to performer error, providing insight into areas for training and improvement.

Surgery

- Surgical Checklist Compliance: Adherence to pre-flight and safety checklists in air shows could be a critical metric, similar to surgery, ensuring all safety protocols are followed.

- Patient (Audience) Satisfaction Scores: Just as patient feedback is vital in surgery, audience satisfaction could provide direct feedback on the performance's entertainment value and safety perceptions.

Formula 1 Driving

- Pit Stop Time: Reflecting the efficiency of ground crews in air shows. Quick and safe turnaround times, especially during multiple flying displays, could indicate a well-organized and proficient team.
- Lap Time Improvement: The equivalent in air shows could be the improvement in performance timing and execution, showcasing learning and adaptation over time.

Integration into Air Show EPIs

Adopting these metrics into the air show EPIs could involve creating a more nuanced and comprehensive evaluation system that includes the following:

- Technical Precision Metrics: Borrowing from football and basketball to measure the accuracy and efficiency of maneuvers and team coordination.
- Safety and Compliance Metrics: Inspired by surgery and Formula 1, focusing on checklist adherence, error rates, and ground crew efficiency.
- Performance Improvement Metrics: Drawing from tennis and Formula 1, tracking individual and team improvements over time, focusing on reducing unforced errors and enhancing execution speed and precision.
- Engagement and Satisfaction Metrics: Utilizing audience feedback mechanisms similar to those in surgery to gauge the show's impact, ensuring the performances not only dazzle but also connect with the audience on a deeper level.

Enhanced EPIs

Building upon the foundational measures of the AER and the APER, we have expanded our evaluative framework to include comprehensive metrics that draw inspiration from diverse fields such as sports, circus, surgery, and Formula 1 racing, and we propose six additional EPIs:

- Precision Execution Score (PES),
- Safety Compliance Index (SCI),
- Audience Engagement Level (AEL),
- Maneuver Complexity and Execution Score (MCES),
- Performance Improvement Factor (PIF), and
- Innovation within Safety Bounds Index (ISBI).

These indicators are designed to encapsulate a broader spectrum of performance excellence, focusing on precision, safety, engagement, operational efficiency, and continuous improvement.

Safety Compliance Index (SCI)

- Inspiration: Surgery
- Definition: Assesses adherence to established safety protocols and procedures, highlighting the paramount importance of safety in air shows.

Precision Execution Score (PES)

- Inspiration: Football and Basketball
- Definition: Evaluate the accuracy and precision of maneuvers executed by performers, emphasizing technical skill and coordination.

Audience Engagement Level (AEL)

- Inspiration: Circus
- Definition: Measures the level of audience satisfaction and engagement using direct feedback mechanisms to gauge the show's impact.

Maneuver Complexity and Execution Score (MCES)

- Inspiration: Figure skating
- Definition: Reflects the difficulty level of the maneuvers performed as well as the precision and skill with which they are executed, providing a direct measure of aerial performance quality.

Performance Improvement Factor (PIF)

- Inspiration: Tennis and Formula 1
- Definition: Tracks improvement and learning over time, encouraging a culture of continuous development among performers and teams.

Innovation within Safety Bounds Index (ISBI)

- Inspiration: Surgery
- Definition: Evaluates the integration of creative and innovative elements into performances while maintaining strict safety standards. This index highlights how excellence can be achieved through innovation without compromising safety.

Visualization and Analysis

For a pilot-friendly and more intuitive graph, focusing on clarity, simplicity, and direct relevance to performance is key. Pilots benefit from visualizations that quickly convey how their performances align with goals and where improvements can be made. A "Spider Chart" (also known as a Radar Chart) is particularly effective for comparing multiple performance indicators in a way that's easily digestible at a glance. This type of chart can encapsulate various dimensions of performance (EPIs) for a single pilot or team, enabling direct comparison to ideal benchmarks or between different pilots/teams.

The Spider Chart (Radar Chart) in Figure 0.6 illustrates a hypothetical pilot's performance across five EPIs: PES, SCI, AEL, Maneuver Complexity and Execution

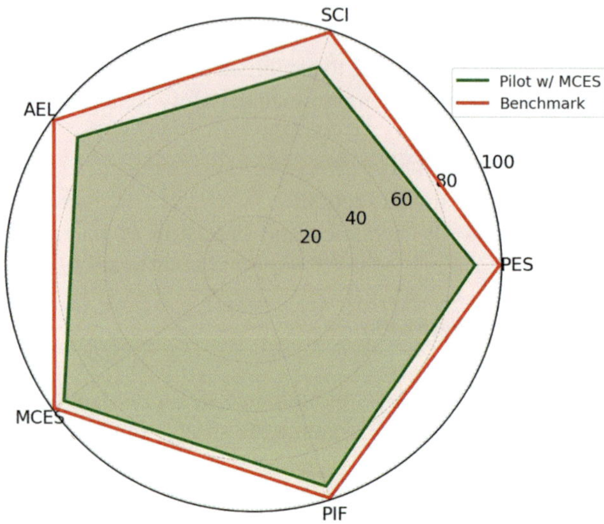

FIGURE 0.6 Pilot performance EPIs with MCES.

Score (MCES), and PIF. Each axis represents one of the EPIs, with scores plotted against a benchmark or maximum possible score (shown in red).

This visualization provides an intuitive way to assess performance:

- Immediate Visual Feedback: The chart offers immediate visual feedback on how the pilot's performance (in green) measures up against the benchmarks (in red) across multiple areas.
- Identify Strengths and Areas for Improvement: It's easy to see which areas exceed expectations and which ones might need more focus. For example, this pilot excels in the PIF but has room for improvement in the SCI.
- Balanced Performance Overview: The Radar Chart format helps in understanding the balance of the pilot's performance across different critical aspects, encouraging a more holistic approach to improvement.

Such a graph is not only pilot-friendly but also actionable, guiding focused improvements and facilitating discussions around performance and development strategies.

The above visualization, by focusing on performer-related metrics like MCES, becomes an even more valuable tool for pilots and teams, offering targeted insights into areas of strength and potential growth. It underscores the importance of not only maintaining high safety standards and engaging the audience but also mastering the technical and artistic elements of air show performances.

Enhanced APER

Excellence in air show performances can be characterized not only by complexity and innovation but also by the simplicity, safety, and accuracy of the display; it's important to adjust our evaluation framework to reflect these values. This perspective shifts the emphasis toward precision, reliability, and adherence to safety standards, which are crucial aspects of air show displays, especially when considering the skills and responsibilities of display pilots.

Evaluating excellence in air show performers requires a nuanced approach to ensure a fair, comprehensive, and constructive evaluation is being conducted. With the above in mind, a formula for the APER that integrates key parameters, as suggested in the enhanced EPIs section, critical to air show performances has been developed.

The Criteria

1. SCI: The paramount importance of safety cannot be overstated. A pilot who prioritizes safety, adheres to all regulations, and maintains a spotless safety record should be rated highly. Safety is not just a necessity but the foundation upon which all performances are built.
2. PES: Excellence is demonstrated through the precision and accuracy of maneuvers. A flawlessly executed simple routine can be more impressive than a complex display flawed by errors. This criterion underscores the importance of meticulous control and skill.
3. MCI: The ability to deliver high-quality performances consistently, show after show, is a mark of true excellence. Consistency reflects a pilot's dedication, thorough preparation, and professional integrity.
4. AEL: The ability to engage and captivate the audience, even with simpler routines, is crucial. Pilots who tell a story or convey the joy of flight through their maneuvers can create a lasting impression.
5. ISBI: Innovation doesn't necessarily mean taking higher risks. Pilots who introduce new elements or ideas into their routines while staying within safety margins contribute significantly to the overall excellence of the show.

The Weighted Formula

To quantify these criteria, we have assigned specific weights to each parameter based on its importance. Here is the breakdown:

1. SCI: 30%
2. PES: 25%
3. MCI: 20%
4. AEL: 15%
5. ISBI: 10%

The total score for each display pilot is calculated using the following formula:

$$APER\,Score = (0.30 \times SCI) + (0.25 \times PES) + (0.20 \times MCI) + (0.15 \times AEL) + (0.10 \times ISBI)$$

Range for Excellence Evaluation Criteria

To provide a clear and standardized measure, we define the following ranges for each criterion on a scale of 1–10 (see Table 0.1).

Given this scoring range, the total APER score for each pilot will fall within a defined range. Here is the interpretation of the final score:

- 9.0–10: Excellent.
- 8.0–8.9: Good.
- < 8.0: Needs Improvement.

Sample Scorecard

To illustrate how this formula works in practice, let us consider a sample scorecard, as shown in Table 0.2.

Let's take a closer look at the calculation for Display Pilot ID 001:

1. Safety (SCI): 9.5
2. Precision and accuracy (PES): 8.5
3. Consistency (MCI): 9.0
4. Audience Engagement (AEL): 9.0
5. Innovation within Safety Bounds (ISBI): 9.0

The final APER score is calculated as follows:

$$APER = (0.30 \times 9.5) + (0.25 \times 8.5) + (0.20 \times 9.0) + (0.15 \times 9.0) + (0.10 \times 9.0) = 9.0$$

This weighted formula ensures that each parameter is given due consideration, not only to a flight display director or an organizer, to evaluate the levels of excellence of air show performers objectively, but also to allow display pilots themselves to fairly and comprehensively self-evaluate their performance levels.

Adjusting the Excellence Levels Framework

Given these criteria, the excellence levels framework could be adjusted to equally value simplicity, safety, and accuracy. This adjustment means redefining what constitutes a "legendary" performance to include those pilots who may not perform the most complex maneuvers but excel in precision, safety, and audience connection, such as Kent Pietsch's[1] elegant performances, who would have constantly achieved an *Excellence Score* of *10*.

Example of Adjusted Framework Application

Under this revised framework, a pilot who performs a simpler routine with exceptional precision, safety, and audience engagement could score as high as one performing inconsistently more complex maneuvers. The scoring system would need to reflect these values, perhaps through adjusted weights or specific criteria that score safety and precision on par with complexity.

TABLE 0.1
Excellence Evaluation Criteria

Category	9–10	7–8	5–6	3–4	1–2
Safety (SCI)	Exceptional adherence to safety, no incidents.	Very good safety record, minor issues.	Adequate safety, some concerns.	Below average, notable safety issues.	Poor safety record, significant safety breaches.
Precision and Accuracy (PES)	Flawless execution of maneuvers.	High precision, minor errors.	Average accuracy, noticeable errors.	Below average, frequent errors.	Poor precision, numerous errors.
Consistency (MCI)	Consistently high-quality performances.	Generally consistent, occasional variances.	Average consistency, some fluctuations.	Below average, inconsistent performances.	Highly inconsistent, significant variances.
Audience Engagement (AEL)	Captivates and engages the audience thoroughly.	Good engagement, audience generally interested.	Average engagement, mixed audience reactions.	Below average, struggles to engage the audience.	Poor engagement, audience disengaged.
Innovation (ISBI)	Highly innovative while maintaining safety.	Good innovation, mostly within safety bounds.	Average innovation, occasionally pushing safety limits.	Below average, often risks safety.	Poor innovation, frequently endangers safety.

TABLE 0.2
Excellence Scorecard

Display Pilot ID	SCI (30%)	PES (25%)	MCI (20%)	AEL (15%)	ISBI (10%)	APER
001	9.5	8.5	9.0	9.0	9.0	9.0
002	8.0	7.0	8.0	8.0	7.5	7.7
003	9.0	8.0	8.0	7.0	7.0	8.1
004	10.0	7.5	9.5	8.5	8.0	8.9
005	9.0	9.5	9.0	9.0	8.5	9.1

Visualization Adjustment

In a visualization of this adjusted framework, display pilots would be categorized not just by the technical difficulty of their performances but also by how well they meet the excellence evaluation criteria of safety, precision, consistency, audience engagement, and innovation. This approach ensures that excellence is broadly and inclusively defined, recognizing the diverse skills and priorities that pilots bring to their performances, while at the same time keeping air bosses alert about areas of improvement of the performers attending their aerial event.

Impact on Performers and Air Shows

The introduction of EPIs, including the AER and the APER, constitutes a significant advancement within the air show community. This concept delineates a rigorous, quantifiable framework intended to elevate standards of performance and safety, fundamentally transforming the modalities through which excellence is gauged and attained in air shows.

The deployment of this innovative framework could commence with training sessions for all pertinent stakeholders—comprising organizers, performers, and air bosses—to ensure a comprehensive understanding of the importance and application of the EPIs. Central to this initiative, the AER serves as a paramount metric that evaluates the overall quality and safety of air shows, compelling organizers to prioritize these elements at every stage of event planning and execution.

In parallel, the APER concentrates on individual performers and teams, assessing their consistency, engagement, and innovative contributions within their performances. This bifurcated approach guarantees that both the collective event and individual contributions adhere to elevated standards.

Supplementary metrics such as the SCI and the AEL provide additional evaluative dimensions, focusing acutely on adherence to safety protocols and audience satisfaction, respectively. Furthermore, the MCES augments operational excellence by scrutinizing the precision and efficiency of flying displays.

To ascertain the enduring efficacy and pertinence of the proposed EPIs, AER, and APER, a perpetual study and refinement strategy is imperative. This entails the establishment of a robust feedback loop wherein data from each air show is constantly collected and analyzed. Such a data-driven approach facilitates ongoing modifications and enhancements to the indicators, ensuring their alignment with the evolving challenges and expectations endemic to the air show industry.

Routine analyses and feedback from performers and organizers are essential in this iterative process, as they provide critical insights that can inform the recalibration of EPIs. This continuous refinement process not only refines the accuracy of assessments but also cultivates a culture of proactive improvement within the community.

All in all, the introduction of this novel framework of EPIs could revolutionize the international air show community by establishing new benchmarks for excellence, safety, and public engagement. By embracing this proposal, air shows globally can ensure adherence to the most stringent standards, delivering exhilarating yet safe experiences for all attendees. This commitment to excellence and continual enhancement could solidify the air show community's status as a vanguard in performance and safety, inspiring future generations of performers and organizers.

NOTE

1 Kent Pietsch, an aerobatic pilot since 1973, has performed at over 400 shows across the US. Known for his unique routines, Kent executes three main acts: a dead-stick landing from 6,000 feet, a rooftop landing on a moving RV, and a comedy act featuring a detached aileron. He flies an Interstate Cadet and prioritizes audience entertainment, emphasizing interaction with fans (International Council of Air Shows Foundation, 2022; Kent Pietsch, n.d.).

REFERENCES

Adjekum, D. K. (2014). Safety culture perceptions in a collegiate aviation program: A systematic assessment. *Journal of Aviation Technology and Engineering, 3*(2), Article 8. https://doi.org/10.7771/2159-6670.1086

Adjekum, D. K., & Fernandez-Tous, M. (2020). Assessing the relationship between organizational management factors and a resilient safety culture in a collegiate aviation program with safety management systems (SMS). *Safety Science, 131*, 104909. https://doi.org/10.1016/j.ssci.2020.104909

Aerden, D., Smets, D., Poelaert, J., Oste, J., & van den Brande, P. (2014). Fighting human error: What surgeons can learn from aviators. *Acta Chirurgica Belgica, 114*(4), 228–232. https://doi.org/10.1080/00015458.2014.11681017

Apple Inc. (2020). *Apple environmental progress report*. Apple.

Aquilina, D. (2013). A study of the relationship between elite athletes' educational development and sporting performance. *The International Journal of the History of Sport, 30*, 374–392.

Arthur Jr., W., Glaze, R. M., Bhupatkar, A., Villado, A. J., Bennett Jr., W., & Rowe, L. J. (2012). Team task analysis: Differentiating between tasks using team relatedness and team workflow as metrics of team task interdependence. *Human Factors, 54*(2), 277–295. https://doi.org/10.1177/0018720811435234

Barker, D. (2020). *Anatomy of airshow accidents* (1st ed., Vol. 1). International Council of Air Shows, International Air Show Safety Team.

Bonetti, D., Zaiacomo, G. D., & Ramos, R. (2013). High-precision landing guidance for Mars: Design and system MDO. *Journal of Guidance Control and Dynamics, 23*, 620–628.

Bubnys, R., & Kauneckienė, N. (2017). Personal professional development as a premise for teacher's leadership. *Social Welfare: Interdisciplinary Approach*, *7*(2), 71–85.

Clark, R., & Fang, C. (2012). Network theory-based analysis of risk interactions in large engineering projects. *Reliability Engineering & System Safety, 106*, 1–10.

de Jong, N. A., Boon, M., van Gorp, B., Büttner, S. A., Kamans, E., & Wolfensberger, M. V. C. (2022). Framework for analyzing conceptions of excellence in higher education: a reflective tool. *Higher Education Research & Development, 41*(5), 1468–1482. https://doi.org/10.1080/07294360.2021.1915964

DeFrank-Cole, L., & Nicholson, R. K. (2016). The slow-changing face of leadership in ballet: An interdisciplinary approach to analysing women's roles. *Leadership and the Humanities, 4*(2), 73–91. https://doi.org/10.4337/lath.2016.02.01

Donoso-Morales, D., Bloom, G. A., & Caron, J. G. (2017). Creating and sustaining a culture of excellence: insights from accomplished university team-sport coaches. *Research Quarterly for Exercise and Sport, 88*(4), 503–512. https://doi.org/10.1080/02701367.2017.1370531

Drucker, P. (2007). *The practice of management.* Routledge. https://doi.org/10.4324/9780080942360

Durant, W. (1991). *The story of philosophy: The lives and opinions of the world's greatest philosophers.* Simon & Schuster/ Pocket Books.

Fang, C., Marle, F., Xie, M., & Zio, E. (2013). An integrated framework for risk response planning under resource constraints in large engineering projects. *IEEE Transactions on Engineering Management, 60*(3), 627–639.

Gapp, R. (2004). The risk to organisational excellence by processes that limit managerial knowledge and perception. *Journal of Manufacturing Technology Management, 15*, 387–393. https://doi.org/10.1108/17410380410540372

Gilgien, M., Reid, R., Raschner, C., Supej, M., & Holmberg, H.-C. (2018). The training of Olympic alpine ski racers. *Frontiers in Physiology, 9*. https://doi.org/10.3389/fphys.2018.01772

Gill, G. K., & Shergill, G. S. (2004). Perceptions of safety management and safety culture in the aviation industry in New Zealand. *Journal of Air Transport Management, 10*(4), 231–237. https://doi.org/10.1016/j.jairtraman.2004.02.002

Güllich, A., Macnamara, B. N., & Hambrick, D. Z. (2022). What makes a champion? Early multidisciplinary practice, not early specialization, predicts world-class performance. *Perspectives on Psychological Science: A Journal of the Association for Psychological Science, 17*(1), 6–29. https://doi.org/10.1177/1745691620974772

Hermel, P., & Ramis-Pujol, J. (2003). An evolution of excellence: Some main trends. *The TQM Magazine, 15*(4), 230–243. https://doi.org/10.1108/09544780310486146

Hodge, K., Henryh, G., & Smit, W. (2014). A case study of excellence in elite sport: motivational climate in a world champion team. *The Sport Psychologist, 28*(1), 60–74. https://doi.org/10.1123/tsp.2013-0037

Houser, K. W. (2022). *To measure is to know … or not.* Leukos. www.tandfonline.com/doi/abs/10.1080/15502724.2022.2029086

International Council of Air Shows Foundation. (2022). Kent Pietsch. https://airshowfoundation.org/kent-pietsch/

International Council of Air Shows. (2019). *ACE manual—Revision 9.* International Council of Air Shows.

Iso-Ahola, S. E. (2024). A theory of the skill-performance relationship. *Frontiers in psychology, 15*, 1296014. https://doi.org/10.3389/fpsyg.2024.1296014

Iso-Ahola, S. E., & Dotson, C. O. (2016). Psychological momentum-a key to continued success. *Frontiers in Psychology, 7*, 1328. https://doi.org/10.3389/fpsyg.2016.01328

Karachalios, M. (2024). *Des Barker report 2023, airshow incident and accident review* (p. 12). European Airshow Council.

Karachalios, M., & Adjekum, D. K. (2023). *Air show performers: Safety, risk management, and psychological factors*. CRC Press. https://doi.org/10.1201/9781003431879

Krois, P., Armenis, D., Joly, R., Kirwan, B., Marrison, C., May, N., Piccione, D., & Schwarz, M. (2015). *Toward a human performance standard of excellence in air traffic management*. www.semanticscholar.org/paper/Toward-a-Human-Performance-Standard-of-Excellence-Krois-Armenis/c393c7a69dce945db437350480a7adab763e073a

Lee, K., Wright, S. M. & Wolfe, L. (2016). The clinically excellent primary care physician: Examples from the published literature. *BMC Family Practice, 17*, 169. https://doi.org/10.1186/s12875-016-0569-x

Lubinski, D., & Benbow, C. P. (2000). States of excellence. *The American Psychologist, 55*(1), 137–150. https://doi.org/10.1037//0003-066x.55.1.137

Makary, M. A., Makary, M., Sexton, J. B., Freischlag, J., Holzmueller, C., Millman, E., & Pronovost, P. (2006). Operating room teamwork among physicians and nurses: Teamwork in the eye of the beholder. *Journal of the American College of Surgeons, 202*(5), 746–752.

Mandalia, A., & Braun, R. (2015). Supersonic retropropulsion thrust vectoring for Mars precision landing. *Journal of Spacecraft and Rockets, 52*, 827–835.

Meng-Yuan, L. (2015). Safety culture in commercial aviation: Differences in perspective between Chinese and Western pilots. *Safety Science, 79*, 193–205. https://doi.org/10.1016/j.ssci.2015.05.011

Mi Dahlgaard-Park, S., & Dahlgaard, J. J. (2007). Excellence – 25 years evolution. *Journal of Management History, 13*(4), 371–393. https://doi.org/10.1108/17511340710819606

Mintrom, M., & Cheng, M. (2014). Creating cultures of excellence: Strategies and outcomes. *Cogent Education, 1*(1). https://doi.org/10.1080/2331186X.2014.934084

Niu, H.-f. (2004). The commentary on Aristotle's theory of happiness. *Journal of Gui Yang Teacher's College (Social-Science Edition), 19*(4), 10–14.

Noort, M. C., Reader, T. W., Shorrock, S., & Kirwan, B. (2016). The relationship between national culture and safety culture: Implications for international safety culture assessments. *Journal of Occupational and Organizational Psychology, 89*, 515–538. https://doi.org/10.1111/joop.12139

Oxford English Dictionary. (2024). Excellence, n. Oxford University Press. https://doi.org/10.1093/OED/7301319564

Pietsch, K. (n.d.). *Kent Pietsch airshows*. Retrieved July 24, 2024, from https://kentpietschairshows.com/pietsch/About.html

Raj, R. (2019). Safety culture app: An Innovation in safety performance system at the aviation industry. *International Journal of Engineering and Advanced Technology*. https://doi.org/10.35940/ijeat.b3620.129219

Rosenthal, M. A., Sharpe, B. A., & Haber, L. A. (2020). Using peer feedback to promote clinical excellence in hospital medicine. *Journal of General Internal Medicine, 35*(12), 3644–3649. https://doi.org/10.1007/s11606-020-06235-w

Schein, E. H. (1990). Organizational culture. *American Psychologist, 45*(2), 109–119. https://doi.org/10.1037/0003-066X.45.2.109

Schein, E. H. (2017). *Organizational culture and leadership*. John Wiley and Sons, Inc. http://archive.org/details/organizationalcu0000sche_d2h5

Schwarz, M., Kallus, K. W., & Gaisbachgrabner, K. (2016). Safety culture, resilient behavior, and stress in air traffic management. *Aviation Psychology and Applied Human Factors, 6*(1), 12–23. https://doi.org/10.1027/2192-0923/a000091

Sklar, D. P. (2015). Building a culture of excellence: Learning from our successes. *Academic Medicine, 90*(7), 835–837. https://doi.org/10.1097/ACM.0000000000000756

Slyusarenko, O., Talanova, Z., Debych, M., Levkulych, V., & Shchegliuk, D. (2022). *Personal development as a key criterion for higher education quality*. Wisdom. https://doi.org/10.24234/wisdom.v4i3.894

Socha, V., Socha, L., Socha, L., Hanakova, L., Karapetyan, L., Valenta, V., Kusmirek, S., Pilmannová, T., & Olexa, P. (2020). Effect of psychological training on pilot's performance. *Transportation Research Procedia, 51*, 252–260.

Sonntag, V. (2010). Excellence in spinal neurosurgery. *Clinical Neurosurgery*. www.semanticscholar.org/paper/Excellence-in-spinal-neurosurgery.-Sonntag/5487053c6b0f44c004b8ec24025a8cfdc26d6db6

Supej, M., & Holmberg, H. C. (2019). Recent kinematic and kinetic advances in Olympic alpine skiing: Pyeongchang and beyond. *Frontiers in Physiology, 10*. https://doi.org/10.3389/fphys.2019.00111

Swann, C., Moran, A., & Piggott, D. (2015). Defining elite athletes: Issues in the study of expert performance in sport psychology. *Psychology of Sport and Exercise, 16*, 3–14.

UK Civil Aviation Authority (UK CAA). (2024a). *Air displays*. Retrieved July 24, 2024, from www.caa.co.uk/Safety-initiatives-and-resources/How-we-regulate/Safety-Plan/Mitigating-key-safety-risks/Air-displays

UK Civil Aviation Authority (UK CAA). (2024b). *CAP 403: Flying displays and special events: safety and administrative requirements and guidance* (21st ed.). UK Civil Aviation Authority. www.caa.co.uk/our-work/publications/documents/content/cap-403/

UK Civil Aviation Authority (UK CAA). (2024c). *Display symposia events*. www.caa.co.uk/general-aviation/flying-displays-and-special-events/display-symposia-events/

UK Civil Aviation Authority (UK CAA). (2024d). *Human factors in air displays*. www.caa.co.uk/safety-initiatives-and-resources/working-with-industry/human-factors/human-factors-in-air-displays/

UK Civil Aviation Authority (UK CAA). (2024e). *The CAA behavioural and attitudinal fitness assessment*. www.caa.co.uk/general-aviation/flying-displays-and-special-events/the-caa-behavioural-and-attitudinal-fitness-assessment/

Undre, S., Sevdalis, N., Healey, A., Darzi, A., & Vincent, C. (2007). Observational teamwork assessment for surgery (OTAS): Refinement and application in urological surgery. *World Journal of Surgery, 31*(6), 1373–1381.

van Kesteren, E.-J., & Bergkamp, T. (2023). Bayesian analysis of Formula One race results: Disentangling driver skill and constructor advantage. *Journal of Quantitative Analysis in Sports, 19*(4), 273–293. https://doi.org/10.1515/jqas-2022-0021

Williams, A. M., & Hodges, N. J. (2023). Effective practice and instruction: A skill acquisition framework for excellence. *Journal of Sports Sciences*, *41*(9), 833–849. https://doi.org/10.1080/02640414.2023.2240630

Yarrow, K., Brown, P., & Krakauer, J. W. (2009). Inside the brain of an elite athlete: The neural processes that support high achievement in sports. *Nature Reviews Neuroscience, 10*(8), 585–596. https://doi.org/10.1038/nrn2672

Zentgraf, K., & Raab, M. (2023). Excellence and expert performance in sports: what do we know and where are we going? *International Journal of Sport and Exercise Psychology, 21*(5), 766–786. https://doi.org/10.1080/1612197X.2023.2229362

Part I

Pilot Selection

The Foundation of Excellence

Manolis Karachalios

Contributor
Daniel Kwasi Adjekum

FIGURE PI.1 Steps to air show excellence: Pilot selection.

DOI: 10.1201/9781003504801-2

INTRODUCTION

> *To get to the top ranks of World Aerobatic Competition or air show flying, you're looking at approximately ten years of practice and training. Maybe more. There is simply no easy way to get to the top except for old-fashioned blood, sweat, and tears.*
>
> *Goulian (2012)*

The quote is from Michael Goulian, one of North America's most decorated aerobatic pilots and recognized aviation professionals. Michael has devoted himself to excellence in the air and epitomizes the values of hard work, dedication, and an unwavering focus on the pursuit of perfection. What goes into selecting individuals like Michael for an air show performance?

Selecting the *right* individual for an air show performance is foundational to achieving excellence and safety in these spectacular events. This meticulous process is central to the industry, not just for the safety of the performers and spectators but also for the overall reputation and viability of air shows. The process goes beyond assessing technical skills to evaluate the pilot's mental and emotional fitness, including identifying overt and concealed hazardous attitudes that can jeopardize safety. This is especially true when we are exploring excellence in air shows, which is not merely a function of flying skills; it encompasses a broader spectrum of attributes, including the pilot's ability to manage risk and exhibit the right attitude.

In the first book of this series (Karachalios & Adjekum, 2023), it was emphasized the necessity of a thorough selection process that gauges flying abilities and scrutinizes the candidate's psychological disposition. This is critical in ensuring that air show pilots can perform under the unique pressures of the air show environment while adhering to stringent safety standards. Hazardous attitudes, as identified by the Federal Aviation Administration (FAA) (2020, Chapter 1), such as macho, invulnerability, impulsivity, resignation, and anti-authority, can significantly compromise safety. These attitudes must be meticulously screened during the selection process to ensure that pilots have the technical capabilities and the right mindset for safe and successful performances.

This notion was expanded by the concept of concealed hazardous attitudes, which may not be immediately apparent but can be just as detrimental to safety (Karachalios, 2022). These underlying attitudes can include overconfidence, complacency, or an undue reliance on past success, which might lead to disregarding standard safety protocols or underestimating risks. The importance of disciplined behavior in aviation cannot be overstated, as its non-existence can lead to neglect of pre-flight checks, disregard for standard operating procedures (SOPs), and an overall increase in risk-taking behavior.

Emotional distractibility can impair a pilot's focus, leading to oversight of critical details and mismanagement of unexpected situations. Then, egocentric and entitled attitudes can result in a failure to collaborate effectively with the team, potentially ignoring valuable input or underestimating the complexity of the aerial maneuvers.

In addition, disorganization or sloppiness in attitude can manifest in erratic flying patterns, miscommunication, and overall degradation of the safety and smooth execution of the air show.

It is evident that the ability of mentors, instructors, and air show evaluators to identify concealed attitudes early in the selection process is crucial for cultivating a resilient safety culture within the air show community. Several strategies can be implemented to enhance this capability:

- Behavioral Interviews and Psychological Testing: Conducting structured behavioral interviews that dive into a candidate's history and how they have managed specific scenarios—like a sudden aircraft malfunction or responding to a teammate's mistake—can provide deep insights into their judgment and decision making skills. Psychological tests can assess risk tolerance, stress resilience, and obedience to authority. For instance, using the Minnesota Multiphasic Personality Inventory (MMPI) (Butcher, 2010) can help identify traits not conducive to safe flying practices, such as impulsivity or overconfidence.
- Simulated Scenarios: Flight simulators create high-pressure environments that mimic routine and critical situations, allowing evaluators to observe a candidate's real-time problem-solving and emotional control. Flight simulations also allow for significant benefits, such as improved safety and reduced cost of a pilot training process (Myers et al., 2018; Vidakovic et al., 2021). For example, a simulation might involve a sudden engine failure during a critical phase of flight, assessing the pilot's ability to maintain composure and follow emergency protocols accurately.
- 360-Degree Feedback: This involves collecting feedback from various individuals who have interacted with the candidate, including peers, subordinates, and other trainers. Such a holistic view can uncover inconsistencies in a candidate's behavior that might not be evident in a one-on-one interview or self-reported assessments. Nominally, two types of 360-degree feedback are expected and unexpected results. Expected information from the raters is defined under two headings: "developmental areas" and "strengths." Developmental areas include the attitudes the raters find improvement necessary. Strengths represent the appraisee's powerful sides, which the rater evaluates. When the ratee receives feedback on how the rater sees them and is surprised by these results, it is called unexpected information (Kuzulu & Iyem, 2016). For instance, a candidate might present a calm and controlled demeanor during formal assessments but might be known among peers for taking unnecessary risks during less formal flying sessions.
- Continuous Training and Evaluation: Pilots should engage in ongoing training that refines their technical skills and reinforces the importance of safety and ethical conduct. An example is competency-based training, which has become the main educational approach in training professional pilots. This type of training creates a systematic approach to educating competent pilots who can operate an airplane safely and efficiently through continuous monitoring and evaluation of performance (Kearns et al., 2017; Tuhkala et al., 2024).

This training can be augmented by workshops on ethical decision making in aviation, and case study discussions on past air show incidents can help reinforce the cultural norms and safety standards expected. For example, a yearly seminar could focus on analyzing a recent near-miss incident at an air show, discussing what was learned and how similar situations can be avoided.

- Peer Safety Reporting Systems: Implementing a system where peers can report concerns anonymously can help catch potential issues before they escalate. Such a system encourages a proactive approach to safety, where all team members are vigilant and responsible for maintaining safety standards. Such reporting systems can also be confidential but have similar incentives by enabling peers to voluntarily report potential violations and inadvertent errors during flight operations without fearing that their actions may have adverse consequences for their person. Examples of large-scale safety reporting systems are the Aviation Safety Reporting System (ASRS) in the United States, which collects, analyzes, and responds to voluntarily submitted aviation safety incident reports in order to lessen the likelihood of aviation accidents, and the European Aviation Safety Agency (EASA) confidential safety reporting system (Aviation Safety Reporting System (ASRS), 2024; European Aviation Safety Agency (EASA), 2024). Any identifiable information should not be disclosed to anyone outside the confidential safety reporting team, including regulators, organizations, and companies. An example could be a mobile app or a secure online form where team members can report any safety concerns directly to a safety officer without fear of reprisal.
- Cultural Fit Assessments: Cultural fit refers to how aligned an air show candidate is with an air show organization's culture. That means the candidate's goals, values, and belief system connect with the organization's own goals and values (Scott, 2024; Tholen, 2024). Evaluating how well candidates mesh with the air show community's values and operational ethos is crucial. This might include assessing their ability to work collaboratively under stress or their willingness to participate in community outreach programs. Target-oriented selection of prospective air show candidates is the first important premise in meeting the safety criteria of air show organizers and aviation authorities while ensuring that successful candidates fit well into the organization's culture (Zinn et al., 2019). For instance, part of the selection process could involve candidates participating in team-building exercises or community engagement activities to see how well they integrate with existing teams and represent the organization publicly.
- Selection Process Utility: As part of the air show pilot candidate's selection process assessment, measuring the individual's utility within the air show industry is essential. In personnel selection, this is known as "utility." Utility in this context refers to the usefulness of the pilot selection procedure in achieving organizational goals. An air show performer candidate should perform the target job of demonstrating aerial display proficiency safely at or above some level of acceptable industry performance standards. Selecting air show pilot candidates with better airmanship can increase the margin of safety; conversely, failing to eliminate air show pilot candidates with poor

airmanship can increase the risks to safety and property. Selection process utility eliminates those individuals who do not have a reasonable likelihood of performing at or above that level (Broach et al., 2019).

Through these strategies, air show organizations can more effectively identify pilots whose attitudes and behaviors align with stringent safety and performance standards, ensuring a positive safety culture permeating every aspect of their operations. This proactive approach not only enhances the safety of the events but also contributes to the long-term reputation and success of the air show community.

Moreover, the selection process is vital in identifying and mitigating risk factors related to flying skills and pilot attitudes. This comprehensive evaluation ensures that only the most suitable candidates, who exhibit exceptional flying abilities and a strong safety-conscious mindset and are willing to listen and accept constructive criticism for their performance improvement, are chosen to represent the pinnacle of air show performance.

Interestingly, research on pilot selection methods began shortly before World War I and was accomplished by medical personnel, physiologists, and psychologists. In their research on the history of pilot selection, Carretta and King (2019) state that Italy was the first country with a pilot selection research program before World War I, and Germany developed special medical standards for pilots in 1910, followed by the United States in 1912. They further state that the French and British efforts in pilot selection were less comprehensive. The British were mostly interested in studying the effects of altitude flying, simple motor coordination tests, and "staleness" in pilots rather than in the examination of aptitude in pilot candidates. The authors conclude that a study by O. Selz in Germany in the early 1900s suggested that airplane accidents were caused mostly by pilot mental and physical failures.

Incorporating findings from aviation psychology research in the late 1990s and early 2000s, such as those by Jensen (1995) on pilot judgment and decision making, further enriches the selection criteria. Jensen's study offers critical insights into the cognitive processes that underpin pilot performance, emphasizing rational and motivational judgment aspects and the pivotal question of whether these can be effectively taught through Crew Resource Management (CRM). Jensen also highlights the importance of managing stress and risk within aviation, which are crucial considerations in selecting pilots who can perform under the high-stakes conditions of an air show. The discussion extends to how CRM can bolster pilot judgment, thereby enhancing flight safety—a paramount concern during air show performances.

This framework is relevant and essential for setting the criteria for selecting air show pilots, ensuring they can maintain composure and make sound decisions during performances. Furthermore, Jensen's (1995) exploration into different flying contexts, such as ballooning and gliding, provides additional perspectives that are beneficial for understanding the diverse skill sets required for various aerial exhibition roles. This comprehensive approach to pilot judgment and CRM forms a foundational element in our endeavor to select the right candidates and ensure that those chosen to exhibit the highest standards of cognitive aptitude, decision making proficiency, and stress management capabilities critical for success in the demanding arena of air shows.

Prior studies on critical examination of pilot selection batteries by Damos (1996) and the determinants of elite pilot performance by O'Hare (1997) have also been useful.

The taxonomic approach to human factors in accident analysis, as introduced by O'Hare (2000), is pivotal for evaluating the complex interplay of skills, attitudes, and behaviors that define pilot performance and safety in aviation. This framework is particularly valuable when assessing potential air show pilots. The taxonomy helps identify whether errors during flight arise from a lack of skill or the misapplication of skills under stress or unusual conditions. Such distinctions are crucial for choosing pilots with the necessary technical abilities and who excel in applying these skills in an air show's dynamic and demanding environment. Moreover, we can examine how pilots have made decisions and responded in past critical situations to gauge their attitudes toward risk, safety, and compliance.

This analysis is essential for selecting pilots whose personal attitudes align with the stringent safety standards required at air shows. Finally, the framework categorizes pilot behaviors, which are influenced by individual capabilities, external pressures, and organizational culture. Understanding these behaviors is key to identifying patterns that either enhance safety or pose risks. This knowledge allows for developing selection criteria focused on safety and performance excellence, ensuring that only the most qualified pilots participate in the high-pressure, visually spectacular context of air shows.

Another taxonomy that provides an understanding of the interplay between air show pilots, environmental, supervisory, and organizational factors related to human performance and its effects on errors is the Human Factors Analysis and Classification System (HFACS). HFACS is a theoretically based tool for investigating and analyzing human error associated with accidents and incidents. Previous research has shown that HFACS can be reliably used to identify human factors trends associated with military and general aviation accidents (Shappell et al., 2007). Reports from HFACS analysis of accidents and incidents provide essential areas of focus when accessing potential candidates undergoing assessment to be considered as air show pilots.

In terms of the different statistical considerations in the selection process, Weissmuller and Damos (2014), in a review of various statistical approaches to pilot selection, suggest that pilot selection systems traditionally use one of three statistical approaches to model candidate performance: multiple linear regression, linear discriminant analysis, and logistic regression. They highlight the accuracy of selection decisions and the limitations of these approaches and recommend a selection system using a progressive process rather than the traditional single- or multi-stage process.

The progressive process strives to maximize the available data to ensure a cost-effective selection system that can identify the best applicants at all stages of the selection process. Nevertheless, any progressive process is complex and requires substantial effort and implementation planning. These features might limit its adoption, especially by organizations that hire few pilots and have limited historical pilot selection and training databases.

In addition to technical and psychological assessments, the selection process must consider the candidate's fit within the air show community's unique culture and operational ethos, ensuring they can contribute positively to the collective performance and safety ethos (Butler, 1988). Kristof-Brown et al. (2023) suggest that the selection process must specify the kinds of fit that will benefit both performance and safety goals and use validated selection methods to assess those forms of fit. Butler's analysis provides a crucial framework that is highly relevant to the context of air shows, where the selection of pilots must be particularly stringent due to the high-risk nature of the performances.

In civil commercial aviation, an airline such as British Airways' approach to selecting candidates with a commercial pilot's license reflects a need for foundational skills upon which specialized training can build. The process comprehensively evaluates a candidate's abilities beyond technical skills. Non-technical skills, such as leadership, communication, problem-solving, situational awareness, workload management, teamwork, and decision making, are crucial for safe and efficient operations. A candidate must demonstrate proficiency in these core non-technical competencies through their ability to collaborate, communicate, and adapt in a high-pressure environment (Skyborne Airline Academy, n.d.). Regarding the personality characteristics that predict training success for pilot candidates, Breuer et al. (2023) suggest that it has a small significant positive relationship with conscientiousness and self-confidence and a small negative association with neuroticism.

For air shows, this implies that candidates should not only meet general commercial flying standards but also exhibit advanced aeronautical skills and a heightened awareness of safety in dynamic environments. The training philosophy highlighted by Butler (1988), which focuses on ground-based simulations that replicate both routine and emergency scenarios, is particularly applicable to air show pilot training. These simulations are instrumental in preparing pilots for the unique challenges posed by air shows, where executing precise maneuvers in close formation and managing unexpected situations are routine.

Moreover, the concept of shared responsibility and error detection among crew members is a critical safety principle that must be ingrained in air show teams. Ensuring that all team members can identify and correct potential errors in real time enhances safety and the overall success of the performance. Thus, the insights from Butler's study are directly applied to establishing a selection and training process for air show pilots that prioritizes advanced skill sets, team cohesion, and an uncompromising commitment to safety. This rigorous approach ensures that selected pilots can perform complex aerial routines and are ambassadors of safety and professionalism, reflecting the high standards of their teams and the air show community at large.

The selection of air show pilots is a multifaceted process that goes beyond mere flying skills to include a thorough evaluation of the candidates' psychological and emotional attributes. By meticulously assessing these qualities and uncovering any potentially hazardous attitudes, the air show community can guarantee that only the most capable and dependable pilots are chosen to fly. This careful selection process is essential for maintaining the tradition of excellence and safety that defines the air show experience.

REFERENCES

Aviation Safety Reporting System (ASRS) (2024). *Confidential and voluntary safety reporting*. https://asrs.arc.nasa.gov/overview/summary.html

Breuer, S., Ortner, T.M., Gruber, F.M., Hofstetter, D., Scherndl, T. (2023). Aviation and personality: Do measures of personality predict pilot training success? Updated meta-analyses. *Personality and Individual Differences, 202*, 111918. https://doi.org/10.1016/j.paid.2022.111918

Broach, D., Schroeder, D., & Gildea, K. (2019). *Best practices in pilot selection* (DOT/FAA/AM-19/6; p. 73). Federal Aviation Administration (FAA). https://libraryonline.erau.edu/online-full-text/faa-aviation-medicine-reports/AM19-06.pdf

Butcher, J. N. (2010). *Minnesota multiphasic personality inventory*. In *The Corsini encyclopedia of psychology* (pp. 1–3). John Wiley & Sons, Ltd. https://doi.org/10.1002/9780470479216.corpsy0573

Butler, A. (1988). *Safety aspects of personnel selection and training of pilots*. Baillière's Clinical Anaesthesiology.

Carretta, R.T. & King, R. E. (2019). *History of pilot selection*. In *Pilot selection: Psychological principles and practice* (1st ed., pp. 9–21). CRC. https://doi.org/10.4324/9780429492105

Damos D. L. (1996). *Pilot selection batteries: Shortcomings and perspectives. The International Journal of Aviation Psychology, 6*(2), 199–209. https://doi.org/10.1207/s15327108ijap0602_6

European Aviation Safety Agency (EASA) (2024). *Confidential safety reporting system*. www.easa.europa.eu/en/confidential-safety-reporting

Federal Aviation Administration (FAA). (2020). *Aviation instructor's handbook (FAA-H-8083-9)*. Federal Aviation Administration (FAA). www.faa.gov/regulations_policies/handbooks_manuals/aviation/aviation_instructors_handbook

Goulian, M. (2012). *I want to do what you do*. Goulian Aerosports. https://mikegoulian.com/news-category/i-want-to-do-what-you-do

Jensen, R. S. (1995). *Pilot judgment and crew resource management*. Avebury Aviation; Ashgate Publishing Co.

Karachalios, M. (2022). *An evaluation of the relationships between resilient safety culture, safety risk parameters, and mindfulness in the international air show community* [PhD dissertation, University of North Dakota]. https://commons.und.edu/theses/4267/

Karachalios, M., & Adjekum, D. K. (2023). *Air show performers: Safety, risk management, and psychological factors*. CRC Press. https://doi.org/10.1201/9781003431879

Kearns, S. K., Mavin, T. J., & Hodge, S. (2017). *Competency-based education in aviation: Exploring alternate training pathways*. Routledge.

Kristof-Brown, A., Schneider, B., & Su, R. (2023). Person-organization fit theory and research: Conundrums, conclusions, and calls to action. *Personnel Psychology*, 76, 375–412. https://doi.org/10.1111/peps.12581

Kuzulu, E., & Iyem, C. (2016). Is 360 degree feedback appraisal an effective way of performance evaluation? *International Journal of Academic Research in Business and Social Sciences, 6*. https://doi.org/10.6007/IJARBSS/v6-i5/2124

Myers, P. L., Starr, A. W., & Mullins, K. (2018). Flight simulator fidelity, training transfer, and the role of instructors in optimizing learning. *International Journal of Aviation, Aeronautics, and Aerospace, 5*(1). https://doi.org/10.15394/ijaaa.2018.1203

O'Hare, D. (1997). Cognitive ability determinants of elite pilot performance. *Human Factors, 39*(4), 540–552. https://doi.org/10.1518/001872097778668004

O'Hare, D. (2000). The 'Wheel of Misfortune': A taxonomic approach to human factors in accident investigation and analysis in aviation and other complex systems. *Ergonomics, 43*(12), 2001–2019. https://doi.org/10.1080/00140130050201445

Scott, C. (2024). 25 cultural fit interview questions plus sample answers you could expect. Academy to innovate HR blog. Retrieved June 06, 2024. www.aihr.com/blog/cultural-fit-interview-questions/

Shappell, S., Detwiler, C., Holcomb, K., Hackworth, C., Boquet, A., & Wiegmann, D. A. (2007). Human error and commercial aviation accidents: An analysis using the human factors analysis and classification system. *Human Factors, 49*(2), 227–242. https://doi.org/10.1518/001872007X312469

Skyborne Airline Academy (n.d.). *British Airways Speedbird Pilot Academy*. Retrieved June 06, 2024. https://skyborne.com/uk/british-airways-speedbird-faq/

Tholen, G. (2024). Matching candidates to culture: How assessments of organisational fit shape the hiring process. *Work, Employment and Society, 38*(3), 705–722. https://doi.org/10.1177/09500170231155294

Tuhkala, A., Heilala, V., Lämsä, J., Helovuo, A., Tynkkynen, I., Lampi, E., Sipiläinen, K., Hämäläinen, R., & Kärkkäinen, T. (2024). The interconnection between evaluated and self-assessed performance in full flight simulator training. *Vocations and Learning, 17*(2), 253–276. https://doi.org/10.1007/s12186-023-09339-6

Vidakovic, J., Lazarevic, M., Kvrgic, V., et al. (2021). Flight simulation training devices: Application, classification, and research. *International Journal of Aeronautical and Space Sciences, 22*, 874–885. https://doi.org/10.1007/s42405-021-00358-y

Weissmuller, J., & Damos, D. (2014). Improving the pilot selection system: statistical approaches and selection processes. *The International Journal of Aviation Psychology, 24*. https://doi.org/10.1080/10508414.2014.892764

Zinn, F., Goerke, P., & Marggraf-Micheel, C. (2019). Selecting for cockpit crew. In *Pilot Selection: Psychological Principles and Practice* (1st ed., pp. 21–34). CRC. https://doi.org/10.4324/9780429492105

1 Qualifications, Criteria, and Predictors

Manolis Karachalios

Contributor
Daniel Kwasi Adjekum

INTRODUCTION

The importance of establishing accurate and relevant *qualifications*, *predictors*, and *criteria* for air show performers cannot be overstressed, as these elements are crucial in ensuring the safety and success of aviation events. Various research studies (ALMamari & Traynor, 2019; Broach et al., 2019; Avers et al., 2016) have demonstrated that precise selection criteria directly correlate with pilots' operational safety and performance efficiency. This introductory Chapter explores how these factors form an integral part of the meticulous process of selecting the crème de la crème of air show performers.

Qualifications encapsulate the hard skills and formal training necessary for air show performers. These include advanced aerobatic certifications, extensive flight hours in specific aircraft types, and a deep understanding of aviation safety regulations.

Criteria for selection merge these elements, forming a holistic view of a pilot's capabilities. This includes evaluating their adaptability to different air show formats, be it single-ship precision routines or multi-ship formations, which require impeccable timing and coordination with other pilots. The criteria also consider the pilot's background, distinguishing between the regimented training typical of military aviators and the more varied experiences of civilian pilots.

Predictors extend beyond tangible qualifications, emphasizing the soft skills and inherent attributes that indicate a pilot's potential success in the air show arena. Psychological resilience, situational awareness, and decision making under stress are key predictors.

DOI: 10.1201/9781003504801-3

HISTORICAL EVOLUTION

The evolution of qualifications, criteria, and predictors for air show performers encapsulates a transformative journey, mirroring the advancements in aviation technology and safety paradigms. The book "Air Show Performers: Safety, Risk Management, and Psychological Factors" (Karachalios & Adjekum, 2023) has provided a detailed account of this progression, highlighting the pivotal shifts in the air show landscape.

The Inception Era (1900s–1930s)

During the Inception Era (1900s–1930s), the world of aviation was a burgeoning field, marked by a spirit of exploration and innovation. This period in aviation history is particularly notable for the way in which it merged with the realms of spectacle and discovery, drawing crowds eager to witness the boundaries of technology and human courage being pushed. Early air show performers were often self-taught aviators, many of whom had backgrounds in engineering or mechanics, which enabled them to build and modify their own aircraft to perform unique maneuvers that captivated audiences.

A lack of formal pilot training programs or certifications characterized the era. Aviators like the Wright brothers, who are often credited with conducting the first formal air shows, epitomized this period of pioneering flight. Their exhibitions in the United States were not only demonstrations of their inventions but also served to validate the potential of controlled, powered flight. These shows were fundamentally experimental, with pilots assessing the capabilities of their aircraft in real time, often in front of live audiences.

Another quintessential example from this era is Lincoln Beachey (Figure 1.1), known as the "Man Who Owns the Sky." Beachey was a daredevil who pushed his aircraft to its limits, performing death-defying loops and spirals that had never been seen before. His performances helped popularize air shows as a form of entertainment while simultaneously demonstrating the capabilities of early aircraft (Onkst, n.d.).

Barnstorming, which became popular in the 1920s, was another hallmark of this era. Barnstormers were itinerant pilots who traveled across the country, performing aerial stunts such as wing-walking, parachute jumping, and mid-air plane transfers. These stunt pilots, like the famous Pancho Barnes and Clyde Pangborn, would often land in farmers' fields and offer short rides to locals for a small fee or perform spontaneous air shows that challenged the conventional notions of flight and aerobatic capability.

The air races of the 1930s also contributed to the development of pilot skills and aircraft technology. Events like the Schneider Trophy, which was contested by seaplanes and focused on speed and agility, not only pushed technological advancements but also set the stage for the development of faster, more agile aircraft that would later be pivotal in World War II.

FIGURE 1.1 Lincoln Beachey ready to fly over Niagara Falls on June 27, 1911. (Photo credit: 2002 National Air and Space Museum, Smithsonian Institution (SI Neg. 93-9680).)

The Military Influence (1940s–1960s)

The Military Influence era from the 1940s to the 1960s brought about a significant transformation in the criteria and qualifications for air show performers in the United States, particularly with the establishment of the US Navy Flight Demonstration Squadron, the Blue Angels, the US Air Force's (USAF's) Air Demonstration Team, the Thunderbirds, and in Greece the All Aces Team ("*Καρέ των Άσσων*")[1] for the Hellenic Air Force (see Figure 1.2). This period marked a shift from the earlier, more informal nature of air shows to a model characterized by strict selection criteria, rigorous standards, and disciplined training methodologies (Guyton, 1950) akin to military aviation.

Establishment of Aerobatic Teams and Their Standards

The formation of prestigious aerobatic teams, such as the US Navy's Blue Angels in 1946 (U.S. Navy, 2024) and the USAF's Thunderbirds in 1953 (U.S. Air Force,

FIGURE 1.2 Flypast over the Acropolis, Athens: All Aces Team ("*Καρέ των Άσσων*") in Formation. (Copyright: Hellenic Air Force.)

2024), exemplified this shift. These teams were not merely about spectacular flying but also about demonstrating the capabilities of well-trained, highly disciplined military aviators (Garretson, n.d.). The selection of pilots for these teams was stringent, focusing on extensive combat flying experience (Armstrong & Voris, 2016; International Council of Air Shows (ICAS), 2019a; U.S. Air Force, 2003), advanced aerobatic and formation training, and a proven record of adherence to military discipline (Armstrong & Voris, 2016; Garretson, n.d.; Wilcox, 2004). These qualifications ensured that only the top echelons of military pilots were chosen, those who could reliably perform complex maneuvers at high speeds and in tight formations.

Criteria for Pilot Selection

The criteria for selecting pilots during this era were rigorous (U.S. Navy, 2024; U.S. Air Force, 2024; Armstrong & Voris, 2016; International Council of Air Shows, 2019a; U.S. Air Force, 2003). Candidates were typically required to have several hundred hours of jet aircraft experience, a spotless flying record, and recommendations from commanding officers. This focus on proven ability and existing military credentials ensured that the performances not only thrilled audiences but also adhered to the highest safety standards. Pilots' ability to execute precise maneuvers in close formation was a fundamental criterion, reflecting their individual skill and capacity to function seamlessly within a team.

Technological Competence

As aircraft technology advanced rapidly during the 1940s–1960s, technological competence became an essential qualification for air show pilots. Mastery of the latest aviation technology and aircraft was imperative, as aerobatic teams often used cutting-edge military jets to perform their routines. The ability to safely and confidently manage these powerful and complex machines was a critical criterion that directly impacted the selection and training processes.

In 1949, the Blue Angels transitioned from the F8F Bearcat to the jet-powered F9F-2 Panther. The Panther's jet propulsion and straight-wing design introduced new flight dynamics, requiring pilots to adapt to faster speeds and different handling characteristics. This shift allowed for smoother and more stable flight, enabling tighter and more precise formations but demanding higher skill levels, quicker reaction times, and enhanced formation coordination to execute complex maneuvers safely (U.S. Navy, 2024).

Similarly, in 1956, the Thunderbirds adopted the North American F-100 Super Sabre, an aircraft that represented a significant leap in aviation technology as the first USAF aircraft capable of supersonic speed in level flight. The transition to flying the F-100 required pilots to adapt to new flight characteristics and more efficient energy management to maintain in formation while performing advanced aerobatic maneuvers (National Museum of the United States Air Force, n.d.; U.S. Air Force, 2003, 2024).

These technological advancements not only enhanced the performance capabilities of air show teams but also set higher standards for pilot proficiency and safety.

Military Discipline and Training

Military discipline and training were key predictors of success in air show performances. The structured environment of military training ensured that pilots developed strong situational awareness, reflexive responsiveness to changing conditions, and an ingrained adherence to safety protocols. These attributes were crucial for the success of the air show routines and for minimizing risks during performances, which now featured increasingly complex aerobatics.

The Safety Paradigm Shift (1970s–1990s)

The Safety Paradigm Shift era (1970s–1990s) marked a critical juncture in the evolution of air show safety, qualifications, and performance criteria. During this period, the focus on safety became paramount, catalyzing significant changes in how pilots and performances were evaluated and regulated. This era's transformative nature can be traced to a series of safety events that underscored the vital need for robust safety measures, culminating in institutional changes that set new standards for air show safety.

The Ramstein Air Show Disaster as a Turning Point

The Ramstein air show disaster in 1988, which resulted in 70 deaths and hundreds of injuries, was a stark reminder of the risks associated with aerial displays (Vanderpool,

2022). It served as a wake-up call to the air show community, prompting an immediate and thorough review of safety protocols. This tragic event led to the implementation of more stringent regulations concerning the pilots and the design and execution of air show routines. For example, new rules were established regarding the minimum safe distances between performing aircraft and spectators and the angles and types of maneuvers that could be safely performed in proximity to an audience.

Standardization of Qualifications and Performance Criteria

The push toward standardized qualifications was primarily driven by the realization that inconsistent safety practices could have catastrophic outcomes. This led to a reassessment of what qualifications were necessary for pilots performing in air shows. The emphasis shifted toward flying skills, situational awareness, adherence to safety protocols, and the ability to perform under tightly controlled conditions. Regulatory bodies like the Federal Aviation Administration (FAA), with the assistance of the International Council of Air Shows (ICAS), began implementing standardized evaluations that pilots must pass to qualify for participation in air shows. These evaluations included rigorous assessments of a pilot's ability to execute complex maneuvers safely and consistently.

Role of Air Show Associations

Air show associations such as ICAS became instrumental in enforcing these new standards. They played a crucial role in setting the criteria and monitoring compliance at air shows around North America. These bodies also facilitated ongoing education and training for air show organizers and performers, ensuring that safety remained at the forefront of every event. The participation criteria grew to include technical flying ability and a demonstrated understanding of and commitment to these enhanced safety protocols.

Predictors of Success under New Safety Standards

Under the new safety paradigms, the predictors of a successful air show performance shifted significantly. Previously, the emphasis might have been on the spectacular nature of the stunts performed. However, in the post-1980s, the ability to execute these stunts within the strict confines of the new safety regulations became the new benchmark of success. Compliance with safety measures, a pilot's record in adhering to safety protocols, and the ability to maintain control of the aircraft under stringent conditions became key indicators of a pilot's suitability for performing in air shows.

Examples of Enhanced Safety Measures

Air show routines were redesigned to minimize risks in response to the new safety standards. For instance, maneuvers that involved direct flight toward the audience, previously a popular spectacle, were banned or modified to ensure that aircraft would always be oriented away from spectators during critical maneuvers (see Figure 1.3). Pilots also began to engage in more comprehensive pre-show briefings and debriefings, where they would go over every aspect of the display routine with safety officers and other pilots to ensure full compliance and understanding of safety measures.

FIGURE 1.3 The Blue Angels pulling for a loop at Pt Mugu, October 1978. (Copyright: Mark Carlisle.)

The Contemporary Framework (2000s to Present)

In the 21st century, the landscape for air show performers has transformed significantly compared to previous decades, driven by technological advances, heightened safety regulations, and a more rigorous approach to training and performance evaluation.

The *qualifications* framework for air show performers has experienced considerable refinement and evolution, spurred by technological advancements and a deeper comprehension of human factors influencing performance. Today's performers are expected to demonstrate advanced technical skills, such as precision maneuvering and mastery over a diverse array of aerobatic techniques, and the ability to operate high-performance aircraft in extremely demanding scenarios. These rigorous expectations ensure that only the most skilled and versatile pilots take to the skies at air shows.

Achieving certifications in aerobatic competency has become a central requirement, serving as a formal endorsement of a pilot's capabilities and safety awareness. These certifications are often rigorous, involving both theoretical and practical examinations to assess a pilot's knowledge and skill in executing complex aerobatic maneuvers. Additionally, accumulating extensive flight hours in aerobatics is crucial. These hours are a testament to a pilot's experience and resilience, providing them with the exposure needed to refine their skills and adapt to various aerial challenges.

This enhanced qualifications framework is about ensuring that pilots can perform impressive maneuvers and guaranteeing their ability to do so safely and consistently. The inclusion of human factors into the training regimen—such as stress management, situational awareness, and decision making under pressure—further amplifies the qualifications. Pilots are trained to maintain high-performance levels even when faced with unexpected conditions or emergencies, emphasizing the critical integration of psychological resilience into their skill set.

Moreover, the advancements in simulation technology have allowed these *qualifications* to be evaluated in safe, controlled environments before pilots execute maneuvers in actual air shows. Simulators can mimic a wide range of conditions pilots might face, from mechanical failures to sudden weather changes, providing an invaluable tool for enhancing safety and skill. This rigorous and holistic approach to qualifying air show performers ensures that they are adept at handling the physical demands of their shows and are prepared mentally and emotionally to tackle the challenges of this high-stakes field.

The *criteria* for selecting air show performers have become more stringent over time, reflecting a comprehensive approach that integrates both the skillful mastery of flight and the mental acumen required to thrive under intense pressures. With the inherent risks associated with performing complex aerobatics at air shows, the psychological well-being of the performers is not just an additional consideration—it is central to their ability to perform safely and effectively.

Rigorous psychological evaluations are a critical part of the selection process. These evaluations are designed to assess a pilot's stress tolerance and decision making capabilities—two indispensable traits for any performer tasked with executing high-stakes maneuvers in front of large audiences. Managing stress effectively and making split-second decisions can often mean the difference between a flawless performance and one that could lead to serious safety incidents.

Beyond psychological resilience, the selection criteria heavily emphasize safety compliance and the ability to manage risks proactively. The air show industry's shift toward a zero-accident culture has heightened the importance of these traits. Performers are expected to regularly participate in safety training sessions that keep them abreast of the latest safety protocols and technological advancements and reinforce their proactive risk management skills. This involves detailed planning and constant evaluation of potential hazards that could arise during performances, with strategies ready to be implemented on the fly to mitigate these risks effectively.

In addition to technical skills, the selection criteria consider several non-technical (human) skills that are crucial for pilot selection and will be analyzed in detail later in the Chapter. These include passion for aviation, motivation, use of knowledge, self-discipline, awareness of personal limitations, information management and situational awareness, emotional awareness, emotional resilience, adaptability, and continuous improvement. Leadership and team cohesion, interpersonal communication and teamwork, and addressing and transforming hazardous attitudes are also vital. While Crew Resource Management (CRM) or Threat and Error Management (TEM) courses are not currently required for air show pilot selection; however, pilots with a commercial background often fulfill these requirements.

These selection criteria ensure that performers not only possess exceptional flying skills but are also equipped mentally and emotionally to manage the rigors of their profession. The modern *criteria* for selecting air show performers are defined by a dual focus on technical proficiency and psychological endurance, supported by a strong commitment to safety at every level. This holistic approach helps maintain the excitement and allure of air shows while ensuring the highest levels of safety and professionalism, thereby protecting both performers and spectators.

Predictors of success for air show performers have undergone a significant transformation in the 21st century. Whereas physical stamina and manual flying skills were once the primary markers of a pilot's prowess, today's indicators of success are far more multifaceted. Modern air show performers must demonstrate a keen ability to adapt to rapidly changing flying conditions, a commitment to continuous learning through data-driven feedback, and a capacity to engage effectively with audiences through social media, without letting this engagement distract from their primary duties.

Advanced technologies such as high-fidelity simulators, virtual reality, and advanced data analytics play a critical role in training and performance assessment in this new era. Simulators and virtual reality (VR) tools allow pilots to rehearse complex maneuvers and safely handle various simulated emergency scenarios. Advanced data analytics software enables detailed analysis of performance and flight parameters, including speed, altitude, G-forces, roll rate, heading, flight path, and ground track, providing actionable insights that drive continuous improvement. According to the FAA, this technology helps in aircraft and pilot performance monitoring, while enhancing overall training effectiveness (FAASTeam, 2021; William, 2020). Additionally, Artificial Intelligence (AI) models complex maneuvers and predicts performance based on a set of predictive metrics, further improving training outcomes (Jiang et al., 2023).

Moreover, a successful performer's skill set now includes savvy engagement with audiences through social media and online platforms. This engagement is essential for promoting and sustaining interest in their performances. However, it must be managed judiciously to support their flying responsibilities rather than detract from them. The ability to strike this balance is crucial, as digital distraction can adversely affect performance quality and safety.

This holistic approach to performance broadens air show performers' competencies and significantly enhances safety and audience interaction. These changes mark a profound shift from the previous, more manually focused and less regulated practices, steering the profession toward higher standards of professionalism and precision. This evolution reflects broader trends in the integration of technology and digital communication within specialized fields, highlighting how modern display pilots (DPs) must be as adept with digital tools and public engagement as they are with aircraft controls.

QUALIFICATIONS, CRITERIA, AND PREDICTORS

In the current landscape of air shows, the qualifications, criteria, and predictors for selecting performers are defined by a sophisticated interplay of hard and soft skills, military and civilian backgrounds, and the operational demands of single-ship versus multi-ship demonstrations. The methods used to select pilot applicants are identified as predictors, while measures of work performance are labeled criteria. When a psychological test is used to select air show pilots, the test is a predictor.

An adequately designed assessment procedure may help reduce the chances of accidents, decrease training costs, and minimize costly dropouts (Bartram, 2018). To assess how well the test suits this purpose, it has to be conducted a validation study in which test results for applicants are compared to actual work performance or academic results. Both work performance and assessment grades are examples of criteria (Martinussen & Hunter, 2018).

As part of any comprehensive air show pilot selection process, it is essential to start with a thorough review of the job to determine what skills, abilities, and qualities are important for the candidate to possess. Such a systematic review is called a job analysis, which involves a detailed survey of the tasks involved in the job. A job analysis mainly consists of a job description and a person specification. A job description accounts for the activities or tasks to be performed; a person specification lists the skills, expertise, knowledge, and other physical and personal qualities the person must have to perform these activities (Wilson, 2007).

Various tests are utilized in pilot selection. These include semi-structured interviews and a work sample test, where the candidate performs a similar task or the same task as the person would do as part of the job. The task is scored using standardized rules for how the performance should be rated. The idea is that the behavior will predict similar behavior later. In their study, Hausknecht et al. (2004) suggested that interviews and work samples were perceived more favorably than cognitive ability tests, which were perceived more favorably than personality inventories, honesty tests, biodata, and graphology.

Other tests, such as assessment center methods, involve the candidate receiving tasks similar or relevant to the job sought. These often involve simulated scenarios in which small groups try to resolve a problem together. This makes it possible to study how candidates interact with each other, their leadership abilities, communication skills, and team dynamics, to name a few desired skills. Several trained observers observe the applicants, who usually use standardized forms to rate their performance (Joiner, 2002).

A number of psychological tests may also be used as predictors, including ability and aptitude tests and personality tests (Campbell et al., 2009; Cox et al., 2017). Interestingly, most of the major new developments in pilot aptitude testing have occurred in the military setting and have involved computer-based assessment. Effective selection procedures must integrate the assessment of ability and aptitude with other assessments: physical and medical, temperament, social skills assessment, and teamwork (Bartram, 2018). Past flying experience is also sometimes used for selection purposes (Johnston & Catano, 2013). If a person already has some work experience, it would be reasonable to obtain references from former employers and for pilots' experience regarding the total number of flying hours, certificates, and type of aircraft flown.

Harris (2011) emphasizes the importance of pilot selection, acknowledging it as a critical factor in ensuring the safety and efficiency of flight operations. Then, discusses how commercial aviation has evolved into a primarily cognitive endeavor, requiring pilots to have more than technical flying skills. He suggests that pilot selection should not only focus on an individual's ability to control the aircraft but also on their decision making, situational awareness, and ability to manage stress and fatigue. Moreover, Harris advocates for a holistic approach in pilot selection, integrating technical and soft skills to identify candidates best suited for the multifaceted demands of modern flight decks.

This includes assessing a pilot's ability to manage complex systems, communicate effectively, and work collaboratively in a crew environment. The approach also considers the impact of stress, fatigue, and other environmental factors on a pilot's performance. Furthermore, Harris's work reflects on the continuous development and updating of selection processes to adapt to the changing dynamics of the aviation industry. He underscores the need for pilot selection criteria to evolve with advancements in aircraft technology and the increasing complexity of airspace management. As Harris outlines, the selection process must be *rigorous* and *comprehensive*, ensuring that pilots are technically proficient and equipped with the necessary *soft skills* to manage flight operations' cognitive and interpersonal aspects.

In the context of air show performers, the relevance of personality traits and non-personality-related tests differs somewhat from that of airline pilots, reflecting the unique demands of each aviation sector. Chidester et al. (1991) found that better-performing pilots in commercial aviation scored higher on traits like mastery and expressivity and lower on hostility and aggression. These traits indicate an individual's ability to control situations effectively and communicate clearly without succumbing to negative emotions that could impair judgment or teamwork.

Ji et al. (2018), in a study of proactive personality and situational judgment of civil flying cadets from the Civil Aviation Flight University of China, found that proactive personality has both a direct and an indirect effect mediated by risk perception on flying cadets' situational judgment. Furthemore, the direct effect was found to be moderated by cognitive flexibility which weakened the relationship between proactive personality and situational judgment among flying cadets.

In the high-stakes, fast-paced environment of air shows, performers also require mastery to execute precise and complex maneuvers, and expressivity to engage with the audience and fellow performers. For air show performers, mastery could translate into the ability to flawlessly execute complex aerobatic maneuvers, while expressivity may relate to the performer's ability to connect with the audience during the show. This is supported by Lambirth et al. (2003), who suggested that military pilots in fast-paced flight environments often appeared calm, less inhibited, and more willing to tolerate risk than a normative sample.

Lower levels of hostility and aggression are preferred traits in pilots, suggesting a calm and composed demeanor essential for managing the high-stress flying environment. This is even more critical for air show performers, who often operate under the pressure of performing daring stunts in front of live audiences. The ability to remain calm and focused during intense aerobatic sequences is vital for ensuring the performance's safety and the wow factor that audiences expect. This is supported by the findings of Meško et al. (2013), who also suggest that high-performance military pilots had better emotional control, higher emotional stability, extroversion, and impulse control than the general population and control groups.

These traits of high-performance pilots, such as fighter pilots and air show pilots, contrast with those of commercial airline pilots. Where mastery might focus on safely navigating a range of flight conditions, and expressivity could relate to clear communication with the crew and passengers, air show performers need to demonstrate their skills in a more public and visually impactful manner.

In their literature review of pilot personalities, Chaparro et al. (2020) suggested that pilots tend to exhibit personality traits lower in neuroticism, higher in extraversion, equivalent in openness, lower in agreeableness, and higher in conscientiousness. However, the trends are not as ubiquitous when different pilot categories are examined. For instance, commercial pilot research consistently shows pilots have higher levels of conscientiousness than the general population; however, the results for military and student pilots are not equivocal.

Non-personality-related tests, such as verbal and numerical reasoning, show strong correlations with performance in airline pilots, emphasizing the importance of flight administration and planning (Harris, 2011). Also, high-performance aerobatic pilots are sometimes required to operate at their physiological limits and to react quickly and autonomously to emergencies when flying high-speed jets; civil pilots often work in highly predictable environments. Civil pilots' main objectives include operating the aircraft preferably in the exact same manner for every flight, abiding by routine operating procedures, and analyzing potential emergency situations with the airline crew before taking action (Ernsting et al., 2006).

Just like desired attributes for high-performance military pilots, high-level cognitive abilities are essential for air show performers, and they should be balanced with spatial and situational awareness and motor skills that enable them to execute aerobatic maneuvers safely and accurately (Carretta, 2011).

Cook's (2009) findings on the importance of cognitive skills in personnel selection have significant implications for air show performers. While commercial pilots need strong verbal and numerical reasoning for flight planning and administration, air show performers require quick cognitive processing for real-time decision making during performances. The ability to rapidly assess and respond to changing conditions, such as weather or aircraft behavior, is crucial for executing a safe and successful air show. The cognitive demands in air shows are more immediate and perceptual, focusing on real-time decision making and physical execution of flight routines rather than the prolonged planning and administrative tasks common in commercial aviation.

Johnston (1996) advocates for a psychological evaluation to reduce attrition in training, a principle applicable to air show performers. Psychological resilience and stability are vital in enduring the rigorous training and the mental demands of performing complex aerial stunts in front of live audiences. A psychological evaluation can help identify individuals with the mental fortitude and stress-coping mechanisms necessary for this high-pressure role.

Besco (1994) and Bartram and Baxter (1996) highlight the contentious role of personality assessments in pilot selection, with some arguing against their utility and others emphasizing the growing importance of CRM-related qualities. While technical flying skills are paramount for air show performers, the "*management role*" akin to CRM in commercial aviation is replaced by an "*entertainment role*." This role places a premium on qualities like showmanship, the ability to connect with the audience, and the mental agility to adapt to varying performance conditions. Therefore, for air show performers, alongside technical skills, personality traits that facilitate public engagement, creativity in performance, and the ability to maintain composure under the spotlight are crucial.

Qualifications

Qualifications serve as the primary threshold for assessing a pilot's foundational skills and knowledge, setting the stage for their effectiveness in the demanding environment of air displays. According to both the UK National Air Traffic Services (NATS) (Cox, n.d.) and academic research, such as Bor et al. (2019), in aviation psychology, these qualifications are not just formalities but are crucial for equipping pilots with the essential competencies required for managing the complexities of modern air operations. This includes a comprehensive understanding of aircraft systems, aerodynamics, aerobatic techniques, and emergency procedures—vital for safe and efficient display flying management.

These qualifications typically stem from a blend of theoretical knowledge obtained through structured educational programs and practical experience gained through hands-on flight training. For example, pilots must accrue a specified number of flight hours and pass rigorous tests to assess their ability to apply theoretical knowledge in real-world scenarios. This dual approach ensures that pilots are knowledgeable and proficient in applying their expertise under the dynamic pressures and unpredictability of flying.

In addition to technical proficiency, NATS highlights the critical roles of professionalism and motivation in the success of DPs. In this context, *professionalism* involves strict adherence to ethical standards, meticulous preparation, and the precise execution of flying tasks with a commitment to upholding safety and operational standards at all times. *Motivation* fuels pilots to pursue excellence, driven by a profound passion for aviation and a desire to contribute positively to the air show industry. It propels them to continuously seek improvement, embrace challenges, and maintain high energy and a positive outlook—qualities essential for the high-pressure environment of air shows.

Together, these qualifications, professionalism, and motivation form a robust framework that assesses a pilot's readiness for air displays and ensures they embody the spirit and integrity required to elevate air show standards and enhance spectator safety.

Criteria

Criteria encompass a wider range of considerations, including a pilot's experience in aerobatic flying, teamwork abilities, and adherence to safety protocols. Safety reports often cite the lack of comprehensive criteria in pilot selection as a contributing factor in air show accidents. Historical analyses (Barker, 2020) provide a sobering reminder of the critical need for thorough and holistic evaluation processes that align with the air show community's safety culture and performance standards.

Flight Discipline

The tragic crash of *Czar 52*, a B-52 at Fairchild Air Force Base (see Figure 1.4), serves as a potent case study highlighting the critical importance of appointing the right individuals to key positions within military and aviation settings (Kern, 1995). This case, thoroughly documented and analyzed by Kern, emphasizes how the lack of appropriate leadership and oversight can lead to fatal outcomes. *Czar 52's* pilot was a highly skilled aviator, yet his repeated violations of air discipline and disregard for safety protocols were well-known issues—a perfect example of normalization of deviance—that were insufficiently addressed by his superiors over the years. Despite numerous warning signs and incidents that underscored the Captain's increasingly risky behavior, he remained in a position where he could make critical decisions, ultimately leading to the crash that killed all crew members on board.

This case emphasizes the necessity of filling key positions with individuals who possess the requisite skills and demonstrate a robust commitment to safety, discipline, and adherence to regulations. *Czar-52* captain's case shows that even skilled individuals may not be suitable for roles where their decisions can have life-or-death

FIGURE 1.4 A B-52 Stratofortress performs a pass over the Langkawi International Maritime and Aerospace Exhibition 2019 in Padang Mat Sirat, Malaysia, March 26, 2019. U.S. Navy Mass Communication. (Photo by Specialist 1st Class Leonard Adams.)

consequences if they do not align with the organization's essential values and safety standards.

The lessons from the *Czar 52* incident stress the need for rigorous selection processes, *continuous* evaluation of personnel suitability, and the courage to make difficult decisions regarding personnel changes when safety is at stake. These steps are crucial in preventing similar tragedies and ensuring that leadership positions enhance, rather than compromise, the safety and integrity of operations.

Health Condition

Then, the tragic accident at the Concord International Air Festival in 1993 highlights the crucial importance of maintaining stringent health criteria for DPs. As detailed by Barker (2020), this incident highlights how even seemingly minor medical conditions can lead to catastrophic outcomes. Shelly, who had been suffering from flu-like symptoms and fatigue prior to the show, experienced in-flight incapacitation, which likely contributed to the loss of control and subsequent crash of his aircraft during a performance.

The need for rigorous medical standards for pilots, especially those involved in the high-risk realm of air show performances, cannot be overstressed. Medical exams for pilots often reveal conditions that, while not directly causing accidents, are significant enough to potentially impair a pilot's ability to perform optimally. Such was the case with Shelly, whose pre-existing heart disease and reported flu symptoms may have led to sudden incapacitation at a critical moment of the flight.

To prevent such incidents, it is essential that pilots undergo thorough medical evaluations that go beyond basic fitness checks. These evaluations should include cardiovascular assessments, particularly for pilots with a history of heart conditions, as cardiovascular issues can lead to sudden incapacitation. Additionally, pilots should be assessed for their resilience to physical and mental stress, especially in the context of the demanding performances required at air shows (Davenport et al., 2021; Federal Aviation Administration (FAA), 1984; Maculewicz et al., 2022).

DPs should be required to report any health changes immediately—even seemingly minor issues like flu symptoms or unusual fatigue—as these can be early indicators of more severe health problems. The oversight authorities should ensure that pilots have a current and compliant medical certificate and affirm their physical condition before each performance as part of the standard pre-flight checks.

For example, in Australia, the Civil Aviation Safety Authority (CASA) requires air show organizers to ensure that pilots participating in the display have the required medicals, licenses, and ratings to participate. An aerobatic pilot must hold a current medical certificate, or medical exemption, as required for the flight crew license or pilot authorization held, and a pilot new to aerobatics should have a check-up with an Aviation Medical Examiner before aerobatic training (Civil Aviation Safety Authority [CASA], 2025).

All aerobatic pilots should know that lack of recent practice, illness, and poor physical condition can reduce tolerance to accelerative forces and increase the possibility of a dangerous loss of consciousness. Moreover, the culture within the air show community should encourage transparency and caution regarding health issues.

FIGURE 1.5 Jacques Bothelin with the author preparing for a media flight in Athens, Greece, 2011. (Copyright: Hellenic Air Force.)

Pilots must be supported in making decisions prioritizing safety over performance, including the difficult decision to withdraw from a scheduled flight due to health concerns.

Approachability and Leadership Skills for Air Show Performers

Air show performers must exhibit approachability and strong leadership skills, especially when conducting media flights. Approachability involves engaging positively with audiences, media, and stakeholders, fostering a welcoming and professional image, which helps build public relations and enhance the overall experience for spectators.

In the photograph shown in Figure 1.5, Jacques Bothelin, the charismatic leader of the Breitling Jet Team, stands in animated conversation with the author, the then Zeus Demo Pilot for the Hellenic Air Force. This image vividly illustrates the critical blend of approachability and leadership required in air show performers. Bothelin's *approachable* demeanor exemplifies how connecting with audiences in engaging dialogues enhances the public's experience, making air shows more than just aerial displays but personal and memorable events.

Leadership, as demonstrated by Bothelin, is essential for formation display flying. This includes ensuring safety while delivering a thrilling experience, requiring clear communication, confidence, and quick decision making. Bothelin's leadership in guiding the Breitling Jet Team for several years exemplifies the balance of performance excellence and public relations, acting as an aviation ambassador to the world.

Predictors

While basic *qualifications* form the bedrock of pilot certification, advanced *predictors* play a critical role in fine-tuning the selection process for air show pilots. They provide a nuanced view of a pilot's capabilities, ensuring that only those with the right skills, temperament, and cognitive abilities are chosen for these high-stakes aviation events. *Predictors* provide deeper insights into a pilot's potential to excel in air shows' specialized and demanding context. Notably, Ree and Carreta (1996) underscore the significance of predictive validity in cognitive and psychomotor tests for evaluating a pilot's aptitude for effectively operating an aircraft. These predictors are designed to assess a comprehensive spectrum of attributes that are crucial for success in high-performance aviation environments.

These attributes include spatial orientation, which is the ability to understand and position oneself within three-dimensional space—a skill essential for performing complex aerial maneuvers. Risk management skills enable pilots to effectively evaluate and mitigate potential hazards, maintaining safety despite air shows' dynamic and sometimes unpredictable nature. Additionally, adaptability to high-stress conditions is vital, as it allows pilots to maintain composure and decision making quality under pressure, ensuring both performance and safety are upheld during critical moments. For example, CASA requires air DPs to conduct a risk assessment of the area to be flown over—the point on the ground or water vertically below the aircraft is not within 150 meter of a person, vessel, vehicle, structure, or livestock (CASA, 2022).

Furthermore, these predictive assessments often include tests on reaction time, decision making under duress, and problem-solving in rapidly changing scenarios. Such evaluations are crucial in determining a pilot's ability to process information quickly and respond precisely—skills that are imperative during air shows where split-second decisions can be the difference between a successful performance and a potential mishap.

Beyond individual performance, these predictors can also gauge a pilot's potential for teamwork and communication. Air shows frequently involve multiple aircraft and require precise coordination with other pilots and ground crews. Hence, a pilot's ability to communicate effectively, follow complex directives, and operate in sync with a team becomes as important as their individual flying skills.

In expanding the role of predictors in pilot selection for air shows, aviation psychologists and trainers can tailor training programs more effectively. By understanding the specific strengths and areas for improvement in pilot candidates, these programs can focus on developing skills that enhance the safety and spectacle of air shows.

Advancing Excellence: Enhancing Qualifications, Criteria, and Predictors for Display Pilots

Role of Simulator Training

Simulator training has become indispensable in the qualification, selection criteria, and predictive success factors for DPs in the 21st century. This training tool offers diverse benefits, ranging from enhancing basic flying skills to mastering complex aerobatics under various conditions (Allerton, 2010; Bell & Waag, 1998; Vidakovic et al., 2021). It enhances their qualifications, serves as a critical criterion for their selection, and is a reliable predictor of their success, ensuring that pilots are skilled, safe, and adept at leveraging the best technology offers.

Simulators allow pilots to practice and refine their flying skills without the risk of performing high-stakes maneuvers in an actual aircraft. This includes everything from basic flight operations to advanced aerobatic skills. The ability to repeat specific maneuvers in a simulator helps pilots achieve a level of precision and reliability that is difficult to attain through traditional flight training alone. Research suggests that the level of practice and training in a flight simulator significantly influences flying skills (Haslbeck et al., 2014).

Then, DPs can experience a wide range of scenarios in a simulator that might be impractical to recreate in real life. This includes severe weather conditions, system failures, and emergency response situations. This exposure is critical in preparing pilots for any unforeseen circumstances they might encounter during a show, thus broadening their qualifications.

Criteria for Selection Informed by Simulator Performance

Simulator performance is an excellent metric for evaluating a pilot's readiness and capability (Hebbar & Pashikar, 2017). During the selection process for display teams or air show pilots, evaluators can use simulators to assess candidates under uniform conditions, providing an objective basis for comparison. Key criteria derived from simulator assessments are the ability to manage stress, make quick decisions, and maintain control over the aircraft in simulated high-pressure environments.

Regular simulator training ensures that pilots keep their flying skills and knowledge of aircraft systems up to date. Competence in a simulator reflects a pilot's ability to continuously learn and adapt to new technologies and techniques, which is a crucial criterion for selection in competitive display environments.

Predictors of Success Enhanced by Simulator Use

Error Reduction

The intensive use of simulators in training significantly reduces errors in actual flights. By practicing maneuvers repeatedly in a simulator, pilots develop higher retentive memory and cognitive patterns that enhance their performance reliability, strongly predicting success in air show environments (McCarthy & Agnarsson, 2018).

Psychological Resilience

Simulator training helps build psychological resilience by exposing pilots to stressful situations in a controlled environment. Pilots learn to manage their emotions and decision making processes effectively under pressure. This resilience is a critical predictor of a pilot's ability to perform consistently well in routine displays and situations where quick, calm reactions are necessary. Research suggests that flight crews that are highly responsive early in simulated scenarios and exhibit resilient behaviors (i.e., monitor, anticipate, respond, learn) are better able to address potential threats well before they emerge to become significant hazards (Stephens et al., 2021).

Adaptation to Advanced Technologies

Proficiency in using simulators also indicates a pilot's adaptability to advanced aviation technologies. As simulators incorporate the latest advancements in aviation tech, pilots who train with these tools are better prepared to transition these skills to real-world applications, which is a valuable predictor of their ability to handle new aircraft and systems in the evolving field of aerial display.

Role of Video Recording

The integration of video recording, especially with the expanded use of onboard cameras mounted either in the cockpit or at the aircraft to provide an external view of the flying display, view, and data analysis, has revolutionized the selection process, training, and performance evaluation of DPs in the 21st century (Kikkawa & Mavin, 2018). These technologies are now central to defining the qualifications, criteria, and predictors of success for pilots in air shows and display events. Here is an expanded look at how these tools are used:

Performance Review and Enhancement

Pilots and their coaches extensively use video recordings of performances to debrief after training sessions and actual performances. These recordings allow pilots to visually review their maneuvers, alignment, timing, and spatial orientation during complex aerobatics, providing a valuable tool for critiquing and improving each aspect of their flying.

Technical Skill Assessment

Video footage of previous flights can be invaluable during the selection process for new DPs. It allows selection committees to assess a candidate's skills in a real-world context, observing how they handle the aircraft during high-pressure situations and execute precision maneuvers.

Role of Flight Data Analysis

Modern aircraft are equipped with systems that can collect a wide array of data during flight, including speed, altitude, G-forces, and engine parameters. By analyzing this data, coaches and pilots can gain insights into the performance of both the aircraft and the pilot, identifying areas for improvement, such as flight path optimization and energy management (Wang et al., 2017).

Data analysis tools can predict outcomes based on historical performance data. This predictive capability is instrumental in assessing potential risks and preparing pilots for various scenarios they might encounter during a show (Evdokimenkov et al., 2020; Lee, 2006). This technology enhances safety by simulating potential problems and training pilots on the appropriate responses.

Influence on Qualifications, Criteria, and Predictors

In the contemporary landscape of aerial display aviation, video recording, and data analysis integration have significantly influenced the qualifications, criteria, and predictors of success for DPs. These technological tools have revolutionized how pilot qualifications are assessed, shifting from subjective interpretations to a more objective, data-driven framework. Selection panels can make informed, impartial decisions by utilizing quantifiable performance metrics derived from detailed data and video recordings. This approach allows for precise evaluations of pilots' abilities to execute maneuvers with exactitude, maintain consistency across performances, and adhere to stringent safety protocols.

The essential skill set for DPs now prominently includes proficiency in analyzing flight data and video debriefs, a crucial criterion in the pilot selection process. For example, pilots often must review video footage to assess and improve their performance critically. A pilot might analyze a sequence to perfect the timing and precision of a loop or roll, using slow-motion playback to identify second errors in execution or to verify alignment during formations.

Furthermore, the ability to adeptly manipulate these technological tools is pivotal for continuous professional development and strongly correlates with a pilot's potential success in this field. Pilots who can seamlessly incorporate feedback from data analysis and video reviews into their training routines demonstrate a higher capacity for adaptability and learning. Such capabilities are particularly beneficial for identifying and correcting subtle flaws in technique, thereby enhancing overall performance quality. For instance, pilots might use telemetry data to adjust their gate[2] entry speed or altitude during specific maneuvers, optimizing safety and spectacle.

This focused reliance on technology does not merely refine current practices but sets new standards in pilot training and performance evaluation. It ensures that DPs are equipped to meet the evolving demands of modern aerial displays and are at the forefront of adopting innovations that enhance safety, precision, and engagement in their performances. The role of these technologies in shaping the qualifications, criteria, and predictors of success in aerial display aviation underscores a shift toward more rigorous, quantifiable, and technology-driven standards in the industry.

Impact of Social Media

The impact of social media on the qualifications, criteria, and predictors of success for DPs is a modern development that reflects the broader influence of digital culture in specialized professions. Social media has reshaped how pilots are evaluated and selected, engage with audiences, and manage their professional reputations.

While social media might not directly affect the technical qualifications required of DPs, it profoundly influences how they are perceived, selected, and evaluated in terms of engagement, professionalism, and adaptability. As the aviation industry evolves, social media's role in shaping DPs' careers is likely to become even more significant (International Council of Air Shows (ICAS), 2022a).

Here is an overview of the effects of social media in this context:

Public Engagement and Reputation

Qualifications and Criteria

Social media proficiency has become an unofficial part of the qualifications for DPs. Those who can effectively use social media may benefit from increased visibility and support, potentially influencing selection decisions. For teams and sponsors, a pilot with a solid social media following can be seen as more valuable due to their ability to engage with fans and generate interest in events. ICAS conducted two national social media marketing programs in the mid-summer and early fall of 2022 to raise air show awareness throughout the United States and Canada. In addition, ICAS and the Foundation collaborated to conduct a series of local social media campaigns focused on promoting ticket sales to specific shows in specific communities (ICAS, 2022b).

Predictors of Success

The ability to maintain a positive and professional online presence can be a predictor of a pilot's success in terms of marketability and public relations. Pilots who actively engage with their audience, share insights into their training, and promote aviation safety and education are often viewed more favorably.

Enhanced Visibility and Scrutiny

Qualifications and Criteria

The visibility provided by social media means that every aspect of a pilot's performance and behavior is more publicly scrutinized. This transparency can pressure pilots to maintain high standards of professionalism both online and offline. As such, social media can indirectly raise the behavioral and ethical standards expected of DPs.

Predictors of Success

Pilots who effectively manage their online reputation and use social media to showcase their skills and achievements may find greater opportunities within the industry. Consistent, positive social media interactions can build a pilot's brand and foster loyalty among fans, which can be crucial for career advancement.

Feedback and Community Interaction

Qualifications and Criteria

Social media platforms provide direct feedback from spectators and peers, which can be invaluable for self-improvement and professional development. Pilots who are responsive to feedback and actively participate in online aviation communities may demonstrate a commitment to excellence and continuous learning—critical qualities in the selection process.

Predictors of Success

The ability to engage constructively with feedback and use social media as a learning tool can indicate a pilot's dedication to personal and professional growth. Such engagement can enhance pilots' skills and knowledge, aiding their performance in air shows.

Crisis Management and Communication Skills

Qualifications and Criteria

Managing crises effectively has become a critical skill for DPs in the digital age. Social media can amplify any incident, and pilots must be adept at navigating these situations. Therefore, effective communication during crises can be a significant criterion in evaluating a pilot's qualifications and having a guide to planning an effective response to accidents, major incidents, and other reputation threats is vital (International Air Transport Association (IATA), 2018).

Predictors of Success

Pilots who manage social media crises with transparency, professionalism, and tact will likely successfully sustain their reputations and careers. This capability strongly predicts a pilot's ability to manage public perceptions and maintain confidence among sponsors and fans.

HARD SKILLS REGULATED

In the domain of air show performances, the hard skills required for performers are meticulously regulated and required to ensure the utmost safety and precision. As mandated by national civil aviation safety regulatory bodies like the FAA in the United States and the UK Civil Aviation Authority (UK CAA) under regulations such as CAP 403, these skills delineate the technical proficiency needed. The hard skills encompass a broad range of competencies, including intricate maneuver execution, comprehensive aircraft handling, and a profound understanding of aerodynamics and safety protocols.

Air show performers must master complex aerobatic maneuvers, demonstrating control and precision that reflect extensive training and experience. This mastery is essential for the spectacle of performance and meeting the stringent safety standards set forth by regulatory bodies (UK Civil Aviation Authority (UK CAA), 2025b). The performers must also have in-depth knowledge of their aircraft's mechanics and capabilities, ensuring they can handle the aircraft safely under various aerobatic conditions. Additionally, they must be adept at interpreting and adapting to changing weather conditions, which is crucial for maintaining safety during outdoor air show events (FAA, 2020).

In addition to the regulations set by civil aviation authorities, military organizations often have their own distinct set of requirements for air show performers. These guidelines may differ regarding the complexity and nature of the performances, reflecting the unique operational and safety frameworks within military aviation contexts. For instance, military air show acts might necessitate additional competencies related to tactical maneuvers, formation flying, and the operation of military-specific aircraft systems (Military Aviation Authority (MAA), UK, 2024).

The hard skills required for air show performers are also significantly influenced by the complexity of the air show act itself. Acts such as formation flying or wing walking demand specialized skills beyond solo aerobatic flying. Formation flying, for example, requires pilots to possess exceptional spatial awareness, precision timing, and the ability to maintain strict formation positions while executing complex aerobatic maneuvers. On the other hand, wing walking necessitates by the pilot a thorough understanding of weight and balance principles and the ability to coordinate closely with the wing walker to ensure safety throughout the performance.

Based on Professional Background

Military Air Show Performers

Military air show performers typically originate from highly disciplined military backgrounds, where they have undergone rigorous training programs specific to the armed forces. These performers are usually seasoned pilots with extensive experience, flying advanced military aircraft under challenging conditions. Their backgrounds often include completing elite flight schools, participating in demanding operational missions, and having a deep familiarity with military protocols and procedures (U.S. Marine Corps, 2022).

For example, as of 2022, pilot applicants for the US Navy Flight Demonstration Squadron, the Blue Angels, should be career-oriented, active-duty commissioned officers. The F/A-18 Hornet demonstration pilot applicants must be carrier-qualified tactical jet pilots with approximately 1,250 hours of total jet flight time by 30 September of the year applying and have completed an operational fleet tour. For the Red Arrows of the UK Royal Air Force (RAF), demonstration pilots must have a minimum of 1,500 flying hours, have completed a frontline operational tour, and be assessed as above average in their flying role to meet these criteria to apply for selection into the team. A shortlist of up to nine applicants is examined during a thorough selection week. They are put through a grueling flying test, formal interview, media test, and peer assessments (Royal Air Force, 2024).

This extensive training ensures they possess the technical flying skills required for complex aerobatics and the discipline and precision that are hallmarks of military operations. Furthermore, their experience in structured military environments instills a strong sense of teamwork and responsibility—qualities that are essential for the synchronized and safe execution of elaborate aerial maneuvers during shows. As representatives of their respective military branches, these performers also embody the professionalism and excellence that the armed forces aim to showcase to the public and the international community.

Recruitment and Strategic Representation in Military Air Display Teams

The selection of military DPs for either single-ship displays or demonstration teams involves a comprehensive set of considerations that extend beyond the individual's flying acumen. This meticulous selection process aligns with the armed forces' broader objectives and strategic needs, ensuring that the chosen pilots represent the best of military values and operational excellence.

CAREER PROGRESSION AND TENURE

Military authorities carefully consider the impact of a pilot's tenure on a display team on their long-term career. Typically, appointments to these high-profile positions are for a limited duration, usually two to three years, to prevent stalling the pilot's career progression within the armed forces. This approach allows pilots to contribute to the team's success while ensuring that their career development remains on track, with opportunities for promotion and advancement unaffected by their temporary assignment. For example, the tenure as a team member of the Red Arrows is three years, and once they have finished their tour with the team, they will return to operational, training, staff duties, or other roles in the RAF.

OPERATIONAL INTEGRITY

The selection process for display team pilots must align with the service's operational demands, ensuring that pilots' participation does not compromise their unit's operational readiness or capability. A critical aspect is maintaining a balance between the prestige and recruitment advantages of display teams and the need to fulfill primary defense and combat roles. For example, the Hellenic Air Force carefully selects pilots for its Zeus Demo Team, ensuring that their involvement enhances public relations without detracting from their essential air defense duties.

Similarly, the USAF Thunderbirds have consistently used combat-capable fighter pilots for their performances, except during the years immediately following the 1973 oil embargo by Arab members of OPEC (Gary, 2024).

INSTRUCTIONAL CAPABILITY

A vital aspect of a DP's role is the ability to mentor and train successors. Candidates for these positions are often selected based on their instructional qualifications, focusing on their potential to impart skills and knowledge to the next generation of DPs. This ensures the sustainability of the team's quality and the continuous flow of expertise.

RECRUITMENT AND PUBLIC RELATIONS TOOL

Display teams are robust military recruitment and public relations tools, showcasing the armed forces' precision, discipline, and professionalism. For example, the Red Arrows are touted as representing the speed, agility, and precision of the RAF and are the public face of the service. They assist in recruiting to the Armed Forces, act as ambassadors for the United Kingdom at home and overseas, and promote the best of Britain. In selecting pilots for these teams, the military often considers how well they embody the force's values and whether they can effectively promote the service branch. This extends to encouraging diversity and inclusiveness, reflecting the modern, multifaceted nature of the armed forces.

PROMOTING DIVERSITY AND INCLUSIVITY

Some armed forces worldwide are increasingly recognizing the importance of diversity in enhancing team performance and public perception. Selecting pilots representing different genders, ethnicities, and geographic regions helps present an inclusive image of the armed forces. It also plays a crucial role in attracting a diverse pool of potential recruits, demonstrating that the military is an institution where various backgrounds and perspectives are valued and can thrive.

In 1994, the selection of Captain Donnie Cochran as the leader for the Blue Angels was historic as he was the first African American Blue Angels commanding officer (U.S. Navy, n.d.). On March 11, 2023, Lieutenant Amanda Lee became the first female fighter jet pilot in the Blue Angels squadron's 76-year history. More than 12,000 attendees witnessed her historic flight. She is the first female fighter jet pilot to fly with the squadron and the second female pilot to fly for the team. Marine Major. Katie Higgins broke that barrier in 2014, flying the team's C-130 "Fat Albert" for two seasons (de Herrera, 2023).

In essence, selecting military DPs is a strategic decision encompassing a broad spectrum of factors. These pilots are ambassadors of aerial excellence and key figures in the armed forces narrative, exemplifying leadership, operational expertise, instructional capability, and the values of diversity and inclusivity. Their role, therefore, is pivotal in shaping the public image of the military and inspiring future generations to explore a career in aviation and defense.

Pilot Selection in USAF's ACC Single-Ship Demonstration Team

The selection and training protocol for single-ship demonstration pilots in the Air Combat Command (ACC) provides a prime example of the rigorous qualifications, criteria, and predictors used in military air show performer selection. As outlined in the ACC guidelines (U.S. Air Force (USAF), 2022), the process begins with wing commanders selecting new demonstration pilots in August. This timing ensures selected pilots can commence their dedicated training by September, well ahead of their first air show season the following year.

This procedure highlights the military's commitment to meticulous preparation and specialization. Serving as a demonstration pilot is considered a two-year full-time assignment, emphasizing the extensive commitment required to achieve excellence in this role. Pilots are relieved of all other duties during this period, allowing them to focus entirely on their demonstration responsibilities. This exclusivity is crucial, as it ensures that the pilots can concentrate on refining their flying skills and perfecting their performance routines without the distractions of additional military tasks.

This focused training and operational model strongly predicts success in military air show performances. By allocating a specific timeline for training and ensuring that pilots have no other duties, the ACC maximizes the likelihood of peak performance levels. Such dedicated preparation not only enhances the safety and execution of complex aerial maneuvers but also significantly contributes to the overall spectacle and precision that are hallmarks of military air shows.

Thus, the ACC's selection and training guidelines reflect a comprehensive approach designed to cultivate elite air show performers who are thoroughly prepared to meet the high expectations and stringent safety standards of military aerial demonstrations.

Civilian Air Show Performers

Civilian air show performers come from diverse backgrounds, often characterized by a blend of commercial, general aviation, and, sometimes, amateur flying experiences. Unlike their military counterparts, civilian performers may not have a formalized training pathway; instead, their skills are typically honed through civilian flying schools, aerobatics training programs, and self-guided practice. Many begin their careers as private pilots, commercial aviators, or enthusiasts passionate about aerobatics and choose to develop their skills to perform at air shows.

Their performances often showcase high creativity and individuality, reflecting their varied experiences and personal styles. Civilian performers tend to have greater flexibility in choreographing their routines, allowing them to innovate and entertain with unique maneuvers that might not strictly adhere to the structured formations typical of military displays. They are adept at engaging audiences with thrilling stunts highlighting their flying prowess and showmanship.

Moreover, civilian air show performers often participate in competitions and flying clubs, providing networking opportunities and skill enhancement outside of formal military structures.

This environment fosters a community-driven approach to learning and performance, emphasizing peer-to-peer interactions and shared experiences. They also have to manage the entrepreneurial aspects of their careers, such as marketing, show bookings, and sponsor relations, making them well-rounded in aerial and business skills. Their role is pivotal in popularizing aviation among the general public, making air shows accessible and enjoyable for diverse audiences.

Depending on the country and the regulations set forth by the respective Civil Aviation Authority (CAA), civilian air show performers' qualifications and the specific skills required to perform can vary significantly. Each country's CAA establishes its own standards and criteria that govern the certification, safety regulations, and performance requirements for civilian air show pilots. These regulations are designed to ensure the safety of both the performers and the public, but the depth and nature of the skills evaluated can differ markedly from one jurisdiction to another.

For instance, some countries may strongly emphasize specific aerobatic certifications and minimum performance hours before a pilot can participate in air shows. Others might focus more on the pilot's overall flying experience and record of safety without mandating specialized aerobatic qualifications. Additionally, the rigorousness of medical examinations, the pilot's proficiency in maneuvering aircraft under various conditions, and even the types of aircraft permitted for use in performances can all vary.

This variability means that civilian air show performers must be highly adaptable and knowledgeable about the specific regulations and expectations of the CAAs in the countries where they perform. It also implies that while the core skills of flying and aerobatics are universally valued, the precise qualifications and evaluative criteria are tailored to meet local regulatory standards and safety concerns. As a result, civilian performers must continuously update their skills and certifications to align with the evolving requirements of different CAAs, ensuring they remain compliant and competitive in the international air show circuit.

United Kingdom

The UK CAA behavioral and attitudinal fitness assessment (UK Civil Aviation Authority, 2024) critically evaluates potential air show performers on essential aspects such as compliance history, attitude, and risk management. Reflecting on past regulatory adherence and enforcement actions, this process helps identify candidates who not only meet the technical qualifications but also possess the right behavioral traits and attitude toward safety and professionalism. Such an assessment ensures that selected performers align with the stringent safety and ethical standards required in the air show domain.

Moreover, the UK CAA outlines a comprehensive set of qualifications, criteria, and predictors for air show performers under the CAP 403 (UK Civil Aviation Authority, 2025a), known as the "*Flying Displays and Special Events: A Guide to Safety and Administrative Arrangements*." This document is the cornerstone for ensuring that all air show operations within the United Kingdom meet stringent safety and performance standards. Here is a synopsis of the key elements derived from CAP 403:

QUALIFICATIONS

CAP 403 mandates that all DPs possess a valid Display Authorization (DA), which is a testament to their proficiency and safety in aerial displays. To obtain a DA, pilots must undergo a rigorous assessment, including a practical flying test and an evaluation of their knowledge of display flying. This ensures that only pilots with the requisite skills and experience are authorized to perform, maintaining high safety standards at air shows.

EXPERIENCE

CAP 1724 (UK Civil Aviation Authority, 2025b) suggests that the Display Authorization Examiner (DAE) should focus on the experience of a display flying applicant, among other areas of interest. Specifically, the minimum experience required for a DA application varies by aircraft type, as depicted in Table 1.1:

- For pilots of aircraft with piston engines of 800 horsepower or greater, mass of 2,730 kilogram or greater, jet-powered aircraft, helicopters, and gyroplanes, a minimum of 500 hours of total flight time is required, with at least 300 hours as pilot-in-command. This category demands the highest level of experience due to the complexity and risk associated with operating high-power and

TABLE 1.1
Minimum Experience Required for DA Application

Aircraft Type	Total Flight Time (hours)	Pilot-in-Command Time (hours)
Piston Engines (800 horsepower or Greater), Mass (2,730 kilogram or greater), Jet-Powered, Helicopter, Gyroplane	500	300
Fixed-Wing Aircraft (Not in the above Category)	200	100
Microlight Aircraft	100	50
Gliders	100	50
Powered Parachute, Paramotors, Powered Hang Gliders, Hang Gliders, Paragliders	50	25

high-mass aircraft. Pilots must demonstrate substantial flight time to ensure they possess the necessary skills and judgment for display flying.
- For pilots of fixed-wing aircraft that do not fall within the above categories, a minimum of 200 hours of total flight time is required, with at least 100 hours as pilot-in-command. This requirement is lower than for high-power or high-mass aircraft, reflecting the balance between ensuring sufficient experience and not imposing excessively high barriers for pilots of less complex aircraft.
- For microlight aircraft pilots, a minimum of 100 hours of total flight time is required, with at least 50 hours as pilot-in-command. Microlight aircraft are typically simpler and lighter, so the experience requirements are lower, focusing on ensuring pilots have enough flight time to handle the specific characteristics of microlight aircraft confidently.
- For glider pilots, a minimum of 100 hours of total flight time is required, with at least 50 hours as pilot-in-command. Similar to microlight aircraft, gliders require specific skills, especially in unpowered flight, and the experience requirement ensures pilots have enough command experience to manage the unique challenges of gliding.
- For pilots of powered parachutes, paramotors, powered hang gliders, hang gliders, and paragliders, a minimum of 50 hours of total flight time is required, with at least 25 hours as pilot-in-command. These are among the simplest forms of aircraft, and the experience requirements reflect their lower complexity and risk, emphasizing ensuring pilots have sufficient time to master control of these lighter and less complex aircraft.

The CAP 1724 guidelines for DA applicants emphasize the importance of flight experience tailored to the specific aircraft type. The varying requirements reflect each category's complexity, risk, and skill levels. By setting these minimum experience thresholds, CAP 1724 aims to ensure that DPs have the necessary expertise to perform safely and effectively, thus enhancing overall air show safety and performance quality.

MEDICAL FITNESS

The medical certificate ensures the pilot is physically and mentally fit to perform the demanding tasks of display flying. The standards for these certificates are stringent and include comprehensive medical examinations that assess various health aspects, such as vision, hearing, cardiovascular health, neurological and psychological conditions, and general physical fitness.

For a UK or EU medical certificate, the examination must be conducted by a certified Aeromedical Examiner (AME) who follows the protocols set by aviation regulatory authorities. The AME assesses the pilot's health against the regulatory standards designed to minimize the risk of in-flight medical emergencies that could compromise safety.

The International Civil Aviation Organization (ICAO) medical certificate, recognized internationally, must also meet or exceed these standards. This ensures that pilots from different countries are held to consistent and high levels of health and safety requirements. The ICAO certificate includes similar comprehensive medical evaluations and is accepted globally, facilitating the participation of international pilots in display flying events.

CAP 1724 requires that the DA for pilots of all registered aircraft be only valid if the pilot holds a medical certificate that meets specific standards. This ensures that all pilots authorized for display flying are medically fit and capable of managing such activities' unique stresses and demands. This requirement helps maintain high safety standards and minimizes the risk of health-related incidents during air shows and other display events. Transport Canada (2015) also requires a pilot who is part of an air show in Canada to have a medical certificate appropriate to the aircraft to be operated in the special aviation competition.

CRITERIA

The criteria for participation in air shows in the United Kingdom are clearly defined in CAP 403 and 1724 and include the need for thorough preparation and planning. DPs must submit detailed display profiles, which include descriptions of the sequences of maneuvers, the heights at which they will be performed, and the mitigation measures for any associated risks.

Additionally, the aircraft used in displays must meet specific technical standards and be suitable for the planned maneuvers. Ensuring compliance with these criteria is imperative to guarantee that both the pilot and the aircraft are adequately prepared for the unique demands of display flying.

PREDICTORS OF SUCCESS

According to various regulatory guidelines such as the UK CAP 403, predictors of a successful air show performer include a strong record of compliance with safety standards, continuous professional development, and an ability to adapt to varying conditions and regulatory changes. Successful DPs often exhibit high situational awareness, exceptional decision making skills under pressure, and a proactive approach to safety management.

SAFETY AND ADMINISTRATIVE ARRANGEMENTS

The UK CAP 403 and 1724 emphasize the administrative arrangements that must be in place for air shows, including insurance, emergency planning, and coordination with local authorities. These arrangements ensure that all potential risks are managed appropriately and that there are clear protocols for addressing any incidents that may occur.

These regulatory documents establish a rigorous framework that air show performers must navigate to participate in displays. This framework ensures that only the most skilled and prepared performers participate in these events, thereby upholding the safety and quality of air shows across the United Kingdom. By adhering to these detailed regulations, pilots contribute to the ongoing success and sustainability of air show events, ensuring they are safe and enjoyable for all participants and spectators.

United States

In the United States, the FAA regulatory requirements for individualized aerobatics flying are not explicitly defined in terms of license or certification apart from ensuring that the aircraft being used is certified for aerobatic flight and the context of the flight conditions within a set of restrictions, determines whether or not aerobatics can be performed (eCFR 14 Part 91.303, 2024). However, for pilots to perform aerobatics, dynamic maneuvering, or certain other flight operations at an aviation event (air show), the FAA, per Order 8900.1, Volume 5, Chapter 9, Section 1, requires additional qualifications and endorsements (FAA, 2023).

The FAA depends on industry-designated evaluators under an FAA-accepted ACE (aerobatic competency) Program for evaluations of civil pilot certification and qualifications to perform at air shows by making appropriate recommendations to the FAA for the issuance, renewal, reevaluation, and rescission of a statement of aerobatic competency (SAC). By FAA policy, a pilot who wishes to perform aerobatics or certain other flight operations at an aviation event (air show) issued FAA Form 7711-1, Certificate of Waiver (CoW) or Authorization, must possess a valid SAC (FAA Form 8710-7) or TC Form 26-0307 (FAA, 2023).

The SAC, a crucial document for air show performers, is issued by the FAA in accordance with Order 8900.1, Volume 5, Chapter 9, Section 1. It is issued to a pilot performing aerobatics, dynamic maneuvering, or certain other flight operations at an aviation event (air show) after demonstrating competence to an industry Aerobatic Competency Evaluator authorized under an FAA-accepted industry program. The recognized industry organizations (RIOs), such as the ICAS and the Experimental Aircraft Association (EAA) Warbirds of America (WOA), are the only RIOs that have FAA-accepted ACE Programs, and their designated Aerobatic Competency Evaluators are recognized under each individual program. This thorough process ensures the highest safety and quality standards in air shows.

ICAS recommendations guide most civilian air show performers in the United States. According to the ICAS *Air Show Manual, 7th Edition*, it is suggested that all civilian performers should possess at least a commercial pilot's license and a current SAC to participate in air shows. This set of requirements ensures that all performers have a foundational level of expertise and the specific skills necessary to execute aerobatic maneuvers in a public display setting safely.

COMMERCIAL PILOT'S LICENSE

After passing the applicable knowledge and practical tests, a person obtains an FAA-issued commercial pilot certificate under 14 CFR Part 61.133 (Office of the Federal Register (OFR) of the National Archives and Records Administration (NARA) and US Government Publishing Office (GPO, 2024). It permits the holder to carry people or property for compensation or hire. This license is more advanced than a basic private pilot's license, requiring more extensive flight training, a higher level of theoretical knowledge, and the ability to operate aircraft under varying conditions. For air show performers, this license ensures that they have a comprehensive understanding of aircraft operations and can handle complex flight operations.

STATEMENT OF AEROBATIC COMPETENCY

Certification is crucial for air show performers as it assesses a pilot's ability to conduct aerobatics safely. The process includes both ground and practical evaluations by qualified Aerobatic Competency Evaluators (International Council of Air Shows, 2019b).

The ground evaluation session is equally, if not more important than the practical evaluation session. It provides a critical opportunity to review not only the display profile selected but also all knowledge areas unique to air show flying. For some pilots, this may be their only formal review of essential topics such as air show aerodynamics, the effects of density altitude, physiology, and energy management throughout the display. Therefore, the Aerobatic Competency Evaluator must allocate ample time and emphasize the importance of this phase to ensure that each applicant fully understands and remembers the key concepts and principles discussed. Pilots must demonstrate an understanding of the relationship between turn radius, true airspeed, and G-forces. They must also comprehend the effects of personal physiological factors like stress and fatigue. Additionally, pilots must demonstrate knowledge of the kinetic and potential energy relationships in air show flying, as well as the effect of density altitude on true airspeed and engine performance.

The practical evaluation session is essential for assessing a pilot's ability to perform aerobatic maneuvers safely and proficiently. For sport aerobatics, aircraft are evaluated in one of four performance classes: Level 4 (800 feet AGL minimum), Level 3 (500 feet AGL minimum), Level 2 (250 feet AGL minimum), and Level 1 (Unrestricted). The evaluation should match the aircraft's capabilities and exclude inappropriate maneuvers. Typical aircraft, such as the Extra, Edge, Pitts, MX-2, and Yak-55, must demonstrate inverted flight, advanced maneuvers like torque rolls and inverted flat spins, and gyroscopic maneuvers. The Aerobatic Competency Evaluator may add necessary maneuvers for a thorough evaluation, ensuring they are safe and appropriate. If familiar with the applicant, the Aerobatic Competency Evaluator may omit certain maneuvers for higher-level evaluations. The practical evaluation also includes crosswind correction and the execution of a full aerobatic sequence.

This comprehensive certification ensures that pilots and aircraft are well-prepared for the unique demands of display flying, maintaining high safety standards and operational integrity.

These requirements are put in place to safeguard the performers and protect public spectators and property at air shows. By adhering to these guidelines, performers demonstrate their commitment to safety and professionalism, which are crucial to conducting successful and safe air shows. Additionally, compliance with ICAS standards can help performers secure future air show participation, as event organizers and sponsors often prefer or require performers who meet these recognized standards to ensure the highest levels of safety and professionalism at their events.

Moreover, these certifications must be regularly updated and maintained, requiring performers to continuously hone their skills and stay abreast of best practices in aerobatics and air show performances. This continuous professional development helps to maintain high safety standards and enhances the overall quality and appeal of air shows in the United States.

FAST

The Formation and Safety Team (FAST) is a worldwide educational organization that teaches safe formation flying in restored, vintage military and civilian aircraft. FAST comprises 16 signatory organizations that aim to support education in the restoration, maintenance, and flight of their members' aircraft (FAST, 2024a)). The "FAST Airshow Operations" document (Formation and Safety Team [FAST], 2021) establishes a detailed framework for pilots participating in air shows under a CoW. This framework emphasizes rigorous safety standards, procedural knowledge, and formation flying competency, particularly for non-aerobatic maneuvers. The document outlines several vital components essential for ensuring that pilots comply with the highest safety standards.

As a RIO, FAST provides formation credentials for pilots operating within the air show environment where a CoW is in effect. The FAST Airshow Operations document was written to help provide FAST pilots with the knowledge and background information needed to operate safely and legally in CoW airspace (Formation and Safety Team (FAST, 2021). Each FAST pilot operating under a CoW must have been evaluated and demonstrated both knowledge of air show procedures and in-flight formation competency to meet the standards and be issued a formation credential (FAST Card). The complete lists of evaluation metrics and Practical Test Standards for Wingman, Flight Lead, and Check Pilots are included in the FAST Foundation and Principles and FAST Practical Test Guides.

Although it is a comprehensive learning tool, it is not intended to be the only source of information. Much of the material in the FAST Airshow Operations document is taken from FAA Order 8900.1A, Flight Standards Information Management System (FAA, 2022a).

AUTHORIZATION AND QUALIFICATIONS Pilots holding FAST cards are authorized to perform specific non-aerobatic maneuvers, following strict training and evaluation criteria detailed in the document. The FAST authorizations are evaluated and administered by FAST. There are two additional levels of qualification available to form air show pilots. These levels are the Formation Dynamic Maneuvering (FDM) endorsement and the SAC card (FAA Form 8710-7 or TC Form 26-0307) with a Formation endorsement (an aerobatic qualification).

The FDM qualification/endorsement allows for maneuvers that separate aircraft from the formation inside the Flying Display Area, including solo or formation opposing or crossing passes, "bomb-bursts," or extended trail maneuvering. The FDM card program is administered by the ICAS, while both ICAS and EAA-WOA administer the aerobatic SAC card program. The FAA issues credentials for these levels.

Pilots aspiring to engage in more dynamic aerial displays require an advanced qualification like the SAC, which the FAA requires to perform aerobatics or certain other flight operations at an aviation event (air show) issued FAA Form 7711-1,

CoW or Authorization. This additional certification allows for more complex aerobatic performances within designated flying display areas, ensuring pilots meet the demanding skills required for such maneuvers.

DOCUMENTATION AND COMPLIANCE The FAST air show operating policies and procedures (Formation and Safety Team (FAST, 2023)) mandate that all participating pilots and aircraft meet stringent documentation requirements to confirm compliance with airworthiness standards. Even though the FAST program is based on voluntary compliance and the integrity of all member formation pilots, the FAST national guideline, combined with the signatory training manual and standard operating procedures (SOPs), is considered the minimum requirement to qualify for a FAST formation credential.

All organizations that are signatories to FAST are free to write their own policy manuals if they want to make any of the current policies more restrictive or include additional policies omitted from FAST guidelines, but they must ensure that less restrictive policies are written or followed. Signatory organizations sponsoring or promoting flying events will require all participants to provide credentials of formation flight competence and currency in accordance with FAST guidelines created to show compliance and will accept the FAST approvals. These documents reinforce the importance of adherence to regulatory guidelines.

OPERATIONAL PROTOCOLS Detailed operational protocols, including comprehensive flight preparation procedures, briefing requirements, communication protocols, and established emergency procedures, are outlined in the FAST operating policies and procedures and the air show operations manual. These protocols are designed to enhance operational safety and efficiency, providing clear guidelines for pilots during the high-pressure scenarios typical at air shows.

MANEUVERING AND SAFETY Specific guidelines for maneuvering, energy management, and low-altitude flying are provided to emphasize safety. These guidelines help pilots maintain safe distances from the crowd and adhere to altitude limits, which is crucial for preventing accidents and ensuring the safety of both performers and spectators.

WEATHER AND ENVIRONMENTAL CONSIDERATIONS The document specifies the environmental conditions under which performances can safely proceed, including visibility, ceiling, and wind limits. These considerations ensure that air show operations are conducted safely, accounting for variable and often unpredictable weather conditions.

In addition to these operational standards, the qualifications and criteria for air show performers, as outlined in the Airshow Operations Appendix F (FAA, 2022a; Formation and Safety Team (FAST), 2021, 2023, 2024b), are rigorously structured to ensure pilots possess the requisite expertise and situational awareness. Pilots must hold a FAST card for non-aerobatic fly-bys, demonstrating their knowledge of air show procedures and competency in in-flight formation. More advanced aerobatic and dynamic maneuvers require a SAC card issued by the ICAS, which confirms that a pilot has met the exacting standards necessary for such performances.

AEROBATIC COMPETENCY EVALUATION (ACE) PROGRAM

The "Applicant Procedures 6.1 Preparation" section of the Aerobatic Competency Evaluation (ACE) Manual (International Council of Air Shows, 2019b) outlines the detailed steps and responsibilities that an applicant must undertake before undergoing an evaluation for aerobatic competency within the ACE framework. Additionally, the "Pilot Credentials" section specifies the essential documentation and qualifications required for an air show performer to be eligible for the evaluation process.

In air show performance, initial applicants and those seeking recurrent status must adhere to the meticulously defined criteria and qualifications outlined in the ACE Manual. This ensures that all performers meet high safety, proficiency, and performance standards.

A summary of the qualifications and criteria for air show performers is listed below:

1. Practical Experience and Preparation
   To qualify, applicants need a robust background in aerobatics, documented through experiences such as participation in competitions, instructional roles, or relevant military experience. They must also provide a comprehensive account of their preparation for the air show evaluation, demonstrating readiness and a commitment to high safety and performance standards.
2. Professional Credentials
   Performers must present a comprehensive set of credentials, including a valid pilot certificate with the appropriate ratings, a current medical certification, any previous SAC, a recent flight review, and a detailed performance sequence/profile. These documents establish the pilot's legal and medical fitness to perform aerobatics at public events.
3. Aircraft and Airspace Credentials
   Documentation confirming the aircraft's registration and airworthiness is required, alongside proof of the aircraft's capability to perform proposed aerobatic maneuvers. Performers must also secure and provide evidence of the appropriate airspace/waiver for the evaluation, ensuring the location and conditions are suitable for the aerobatic activities planned.
4. Knowledge and Skills Evaluation
   Applicants undergo an exhaustive evaluation of their air show flying knowledge, including aerodynamics, the impact of density altitude on aircraft performance, and specifics of air show accidents. The skills assessment covers their ability to execute the maneuvers in their performance sequence/profile efficiently, manage energy during flights, and adhere to safety protocols.
5. Proficiency and Currency
   It is crucial for air show performers to maintain proficiency and currency. They must possess a current and valid SAC card, demonstrating proficiency in the required aircraft category, level, and endorsements for their performance. Currency is shown by performing in the specific aircraft category and

endorsements within 45 days before the air show, ensuring readiness and safety.

6. Evaluation by an Authorized Evaluator
   Coordinating with a qualified Aerobatic Competency Evaluator is essential. Applicants must work with the Aerobatic Competency Evaluator to secure authorized/waivered airspace and set a date and location for the evaluation, guaranteeing that the assessment occurs under standardized and controlled conditions.
7. Endorsement Requirements
   Different endorsements within the SAC card, such as rolls only, solo aerobatics, and formation aerobatics, require specific practice sessions and evaluations. Each endorsement has a defined qualification period, ensuring that performers regularly refine their skills and are consistently assessed for their aerobatic competencies.

This structured and comprehensive approach to qualifying air show performers guarantees that only the most capable and thoroughly prepared individuals participate in these high-risk, public-facing events, thus maintaining the safety and integrity of the air show industry.

Other Countries

Similar to the United Kingdom and the United States, air show participation criteria in various countries require detailed preparation and planning. For instance, pilots must submit comprehensive display plans detailing maneuvers, heights, and risk mitigation measures (CASA, 2025; FAA, 2023; SACAA, 2021; Transport Canada, 2015). Additionally, aircraft must meet specific technical standards for planned maneuvers. Moreover, guidelines such as CASA AC 91-21, CAR 623.06, and the SACAA SAE Handbook emphasize safety compliance, professional development, and adaptability. Furthermore, administrative requirements like insurance, emergency planning, and local coordination ensure the safety and quality of air shows in Australia, Canada, and South Africa.

The Canadian requirements for operating an aircraft in an air show are thorough, necessitating the pilot to hold a license and medical certificate appropriate to the aircraft to be operated in the air show (Transport Canada, 2015). To perform solo air show aerobatic maneuvers, a pilot must possess one of the following documents: a TC SAC (form 26-0307) (Federal Aviation Administration (FAA), 2023, p. 13), an FAA Statement of Acrobatic Competency (form 8710-7) (Federal Aviation Administration (FAA), 2023, p. 10), or an aerobatic competency certificate that meets the previous competency requirements and is recognized by TC.

To join an aerobatics team in Canada, a pilot must meet all the previously mentioned requirements. Additionally, within the preceding 12 months, they must have performed with the other members of the formation in eight aerobatic performances or completed a minimum of 25 aerobatic practice sessions with the other members of the formation. The pilot is also expected to attend a briefing or review led by the formation leader or another designated formation member, and a representative of the certificate holder, preferably the person in charge of flight operations, must be present.

The topics covered in the briefing or review include safety and operational matters such as the designation of the formation leader and alternate leaders, the selection of maneuvers to be performed and their sequence, formation positions, alternate positions in case of aborts, radio procedures, and call signs, emergency procedures, expected speeds, and power settings, and visual signals. This detailed process ensures that all team members are well-prepared and understand their roles and responsibilities.

In South Africa, in order for any civilian pilot to fly at a special air event or air show, the pilot must hold a valid DA issued by the South African Civil Aviation Authority (SACAA, 2021). A DAE must evaluate the pilot applicant's display competency and submit a report for SACAA. A DAE must have received a recommendation from an organization associated with a particular display discipline or from another DAE and recommended by the Display Authorization Committee (DAC) to SACAA.

The DAE must have personal knowledge or equivalent experience acceptable to the DAC, hold a valid pilot's license with normally a minimum of 1,000 hours as a pilot in command, or equivalent experience acceptable to the DAC. The DAE should be an active DP who has participated in at least four displays or aerobatic competitions within the previous 12 months at SACAA-approved competitions unless proof of a recent military rating can be provided. Apart from having the DA, recency rules require a pilot who intends to be part of an air show to have flown at least four displays in a calendar year assigned by a flying display director (FDD).

In Australia, the CASA AC 91-21 Part 5.8 requires aerobatic performers to hold a current medical certificate or medical exemption, as required for the flight crew license or pilot authorization held. The display organizer is responsible for reviewing the DP and essential crew medical certificates before participating in a display. This information needs to be acknowledged by the Display Organizers as being present and suitable, and CASA may request this information from time to time for further review (Civil Aviation Safety Authority [CASA], 2025).

Enhancing Air Show Safety and Performance: An Analysis of Regulatory Organizations

As established by regulators such as the FAA, UK CAA, TC, CASA, and SACAA, civil air show performers' qualifications, criteria, and predictors provide robust frameworks designed to ensure safety and professionalism within the air show industry. However, there are areas where these authorities could enhance their guidelines better to address emerging challenges and technological advancements in aviation.

Regular, in-depth skill assessments can provide a more continuous overview of a pilot's capabilities and readiness for aerobatic performance, going beyond the existing biennial flight review requirement. Research suggests that more frequent, task-specific evaluations can improve performance and safety outcomes in high-risk industries (Smith & Jones, 2020). Given the high-stress nature of aerobatic performances, mental health evaluations should also be considered a critical component of a pilot's overall fitness. Studies have shown that mental health significantly impacts decision making and performance, especially in high-pressure scenarios typical of air shows (Doe, 2019).

Establishing a public performance record would enhance transparency and allow for better monitoring of safety and performance trends within the air show community (Anderson & Lee, 2021). This initiative could lead to more informed decision making by event organizers and regulatory bodies, potentially lowering the risk of accidents.

Continuous education tailored to the unique demands of air show flying can further enhance pilots' skills and knowledge. This could include advanced aerodynamic training, emergency procedures, and the latest aircraft design technology and performance advancements. Regular workshops, seminars, and online courses could be instrumental in keeping pilots abreast of the latest developments in the field (Brown, 2018). Regular skill assessments, mental health evaluations, public performance records, and continuous education could further improve air show performers' safety and performance standards.

In summary, while the various regulatory bodies' current qualifications and criteria for air show performers are robust, there is room for enhancement. The November 12, 2022, fatal mid-air collision between a Bell P-63F Kingcobra and a Boeing B-17G Flying Fortress participating in the air show, "Wings over Dallas," (WOD) at Dallas Executive Airport, Dallas, Texas (RBD) is evident of such a need for continuous safety monitoring and systems improvements. Both airplanes were destroyed in the post-impact fire that ensued in that accident.

The P-63F pilot and all five crew members aboard the B-17G were fatally injured. No ground injuries were reported. Both airplanes were owned by the American Airpower Heritage Flying Museum and operated by volunteers of the Commemorative Air Force (CAF). The flight was operated under Title 14 Code of Federal Regulations (CFR) Part 91 with a CoW for the air show. Both flights originated at the RBD with an intended destination of the RBD (National Transportation Safety Board (NTSB), 2024).

Financial Criteria

In the world of civilian air show performances, financial considerations play a crucial role in the qualifications, criteria, and predictors for selecting performers. These economic factors are not only pivotal in ensuring the feasibility of air shows but also serve as a lens through which civil aviation authorities assess the viability and sustainability of air show performers.

The financial implications of organizing and conducting flying displays by civilian performers are considerable and warrant a detailed discussion that, unfortunately, falls beyond the scope of this text. Organizing an air show involves a complex array of expenses, including but not limited to aircraft maintenance and operation costs, fuel, insurance premiums, fees for aerobatic certifications, and payments to meet stringent safety standards set by Civil Aviation Authorities.

Civilian performers often bear the brunt of these costs themselves, unlike their military counterparts, who are supported by government funding. The expense of maintaining high-performance aircraft in optimal condition for displays and the necessary modifications and repairs can be substantial. Additionally, the cost of travel, accommodation for the team, and equipment transport to various show locations significantly add to the financial load.

Marketing and promotional activities, which are crucial for attracting sponsorships and audience attendance, also require substantial financial investment. Civilian air show performers must often engage in extensive self-promotion and business networking to secure the funding needed to cover these costs, relying on ticket sales, merchandise, and sponsorships to help finance their operations. For example, top-notch aerobatic performer Michael Goulian has earned sponsorship deals with Whelen Engineering, a leader in automotive and mass notification lighting, Pentagon Federal Credit Union (PenFed), Cirrus Aircraft, BOSE, which manufactures advanced active noise-reducing headsets specifically for pilots, and Hartzell Propeller, a global leader in advanced technology propeller design and manufacturing (Goulian, 2024).

Given these myriad costs, the economic viability of performing in air shows can be challenging and requires careful financial planning and management. This aspect of air show performances, while critical to the sustainability of civilian performers, involves financial strategies and considerations that are complex enough to merit a separate, in-depth analysis. Thus, while this text acknowledges the significant financial burdens civilian air show performers face, a comprehensive exploration of these economic factors is beyond its current purview.

Financial Stability as a Qualification

Civil aviation authorities increasingly view financial stability as a key qualification for civilian air show performers. This is because financial solvency ensures that performers can maintain their aircraft in optimal condition, cover insurance costs, and manage other expenses related to air show participation. Feng and Wang (2000) highlight the significance of financial performance in the aviation sector, underscoring the need for comprehensive financial evaluation.

Kalemba and Campa-Planas (2019) suggested a significant relationship between safety and an airline's revenue flow, and Fardnia et al. (2020) document an inverse relationship between air carriers' profitability and accident propensity. They found that other financial variables, such as liquidity, asset utilization, and financial leverage, also appear to affect an airline's safety record and that flight equipment maintenance and overhaul expenditures are negatively related to accident rates. These factors demonstrate a vital need for financial diligence in the interest of air show safety.

Economic Viability as a Selection Criterion

The economic viability of performers is becoming a critical criterion in their selection for air shows. Authorities may consider performers' ability to secure sponsorships, manage operational costs effectively, and generate revenue through merchandise sales and appearances as indicators of their economic viability and long-term sustainability. Wang and Sarkis (2017) discuss the impact of corporate social responsibility on financial performance, hinting at the broader implications of financial health and sustainability in the aviation industry.

Financial Management as a Predictor of Success

The ability to manage financial resources effectively is a strong predictor of a civilian air show performer's success. Those who demonstrate acumen in budgeting, financial planning, and cost control are more likely to deliver successful air show performances and ensure their continuity in the circuit. As per the study by Lakshmiprabha and Amuthan (2023), understanding the financial performance of aviation entities can provide insights into their operational success, which can be extrapolated to air show performers.

Future Considerations for Civil Aviation Authorities

Civil aviation authorities should consider integrating financial assessments as part of the licensing and approval process for air show performers. This could include reviewing performers' financial statements, funding sources, and sponsorship agreements to ensure they have a solid financial foundation that supports their air show activities. Moreover, as Fardnia et al. (2020) suggest, financial health can correlate with safety and operational efficiency, making it a critical factor for consideration.

In conclusion, financial considerations are becoming increasingly central to the qualifications, criteria, and predictors for civilian air show performers. Civil aviation authorities, recognizing the integral role of financial health in ensuring the success and sustainability of air shows, may need to place greater emphasis on economic factors in their assessment and approval processes for air show participants in the future.

Comparison between Military and Civilian Demonstration Teams

When comparing military demonstration teams, such as the Red Arrows, Thunderbirds, or Blue Angels, with civilian air show performers, several distinct differences in selection processes, performance expectations, and public engagement responsibilities become apparent.

SELECTION PROCESSES AND CRITERIA

The selection criteria for military demonstration teams are exceptionally stringent, reflecting the elite status of these groups. Candidates for these teams must demonstrate not only top-tier flying skills but also extensive experience of piloting high-performance military aircraft. The selection process often includes rigorous physical and psychological testing to ensure that pilots can manage the intense pressures and physical demands of aerobatic maneuvers at high speeds. Additionally, military pilots must show an ability to work within highly precise and coordinated team environments where flawless synchronization is critical.

In contrast, civilian air show performers may come from a variety of backgrounds, including commercial and general aviation, and their selection criteria can vary widely depending on the team or the type of performances they deliver. While civilian performers also require excellent flying skills, the emphasis on military precision and teamwork is typically less intense than in military squads. Civilian performers often have more flexibility in choreographing routines and may perform solo or in less rigidly structured groups.

PERFORMANCE EXPECTATIONS

Military demonstration teams are renowned for their precise, choreographed routines, often involving complex formation flying and synchronized maneuvers. These performances are designed to entertain and demonstrate the high levels of skill and advanced capabilities of military aviators and their equipment. Every maneuver is meticulously planned and practiced, ensuring safety and precision, reflecting the discipline and excellence expected of military personnel.

Civilian air show performers, while also highly skilled, might focus more on individual flair and creativity in their routines. They have the liberty to innovate with their flying styles and may not require the same level of strict formation and timing as military teams. Their performances might include a wider variety of stunts and can be tailored to different themes or audiences.

PUBLIC AND MEDIA ENGAGEMENT

Military demonstration teams often serve as ambassadors for their respective armed forces, and as such, pilots are expected to excel in public and media engagements. This aspect of their role includes conducting interviews, making public appearances, and participating in community outreach programs, all aimed at fostering good public relations and enhancing the military's visibility. For example, the Red Arrows has a public and media engagement team comprising a Chief of Staff, a Public Relations officer, and photographers. The DPs also receive media relations training.

Civilian performers, on the other hand, while they do engage with audiences and media, typically carry less responsibility for representing national or military institutions. Their interactions focus more on entertainment, promoting their personal or team brands, and connecting with aviation enthusiasts.

While both types of performers share the common goals of skill demonstration and audience engagement, their context and the expectations they must meet are shaped by fundamentally different objectives. Military teams are integral to national representation and military discipline, whereas civilian performers emphasize entertainment and personal or team branding. To provide a clearer understanding of the differences between military demonstration teams and civilian air show performers, let us explore their selection processes, performance expectations, and public engagement responsibilities in more detail through Table 1.2, which underlines the structural and operational distinctions between military and civilian performers in the air show industry.

Based on Complexity

The ACE Manual (International Council of Air Shows, 2019b, Appendix D) outlines the Practical Flight Evaluation Standards, focusing on the essential aspects of aerobatic flight evaluations for air show performers. It establishes a framework to ensure pilots demonstrate the necessary skills and competencies in a structured, safe

TABLE 1.2
Hard Skills Comparison for Military and Civilian Air Show Performers

Aspect	Military Demonstration Teams	Civilian Air Show Performers
Selection Process	Highly rigorous with strict criteria. Requires top-tier flying skills, extensive military flying experience, and rigorous physical and psychological tests. Teamwork within precise formations is critical.	It varies widely. Generally, it requires excellent flying skills but less emphasis on military precision. There is more flexibility in team structure and performance style.
Performance Expectations	Performances are highly choreographed, showcasing precise formation flying and synchronized maneuvers. The aim is to demonstrate military precision and capabilities.	Performances may focus on individual creativity and flair, with a broader range of stunts and less rigid choreography.
Public Engagement	Serve as ambassadors for their respective armed forces, engaging extensively in public and media relations to enhance military visibility and public relations.	Engage primarily in entertainment, focusing on promoting personal or team brands and connecting with aviation enthusiasts.
Safety and Discipline	Adherence to strict safety protocols reflecting military discipline. Every maneuver and operation is meticulously planned and reviewed.	Adhere to strict safety standards, but with potentially more latitude in execution and innovation in routines.
Training Regime	Intensive and continuous, often part of ongoing military training programs. It focuses on maintaining peak physical and mental conditioning.	Training varies by team and individual, often dictated by personal goals and the nature of the performances.
Role and Representation	Symbolic role representing military excellence and national pride. Highly visible in national events and international shows.	Primarily entertainers and performers without the institutional representation and responsibilities of military pilots.

environment. Here is an expanded overview with examples and imagery descriptions of the maneuvers discussed:

Evaluation Preparedness

Pilots must assess their readiness for the evaluation and only proceed when confident in performing the required maneuvers. For instance, if a pilot is scheduled to demonstrate a complex maneuver like a Lomcevak, see Figure 1.6, a tumbling maneuver involving complex aircraft rotations, they should only proceed if they have thoroughly

FIGURE 1.6 Frecce Tricolori's solo performing a Lomcevak maneuver. (Copyright: 2014, Alessandro Scagliusi/ Shutterstock.com.)

practiced and mastered the technique in various conditions to ensure safety and precision during the evaluation.

Aircraft Suitability

The standards acknowledge that not all maneuvers are suitable for every aircraft type, necessitating adjustments or omissions based on the aircraft's capabilities. For example, an aircraft without an inverted fuel and oil system may be unable to perform sustained inverted flight maneuvers safely. In such cases, evaluators might omit these maneuvers from the evaluation or substitute them with alternatives that evaluate similar skill sets but are safe for the specific aircraft.

Maneuver Requirements

The manual lists maneuvers that applicants must demonstrate, including standard aerobatic figures like loops, rolls, and hammerheads. For instance, a pilot might be required to perform a "hammerhead" or stall turn (see Figure 1.7[3]), where the aircraft climbs vertically until nearly stalled, then pivots around its vertical axis (the nose drops and the tail swings) before descending along the same vertical line. This maneuver tests the pilot's ability to manage energy and control the aircraft during high-angle, low-speed flight.

Evaluation Criteria and Flexibility

Evaluators have the discretion to modify the evaluation based on the aircraft's performance class and capabilities. For example, while evaluating a high-performance

FIGURE 1.7 Hammerhead turn. (Copyright: 2017, Juanita Bairley/Shutterstock.com.)

monoplane like the Extra 300, the Aerobatic Competency Evaluator might request advanced maneuvers like torque rolls or tailslides that showcase the aircraft's agility and the pilot's control during extreme aerobatic flight. Conversely, for a vintage biplane like a Stearman, the evaluation might focus on fundamental aerobatic maneuvers, such as barrel rolls and slow rolls, which align with the aircraft's operational envelope.

These examples and depictions highlight the Practical Flight Evaluation Standards' emphasis on safety, competency, and the adaptability of the evaluation process to suit the specific aircraft and the pilot's expertise in aerobatic flying.

Aerobatic Flying

To excel as air show performers, individuals must demonstrate a multifaceted set of skills and attributes encompassing physical fitness and technical competence.

Research across various domains has shed light on the intricate demands of aerobatic flying. More specifically, the intense cardiovascular demands of aerobatics necessitate rigorous selection and monitoring of pilots to prevent sudden mid-air incapacitations, ensuring aviation safety (FAA, 1984; Perrier et al., 2005; Nicol et al., 2018).

Moreover, the physiological responses of pilots during aerobatic flights, including heart rate and variability, indicate the high levels of physical and mental workload experienced (Guo et al., 2012). Fresnel et al. (2021) looked at the relationships between sudden accelerative G-force changes on cerebral oxygenation levels in aerobatic pilots and found a significant and positive correlation between changes in accelerative G-force exposures and oxygen hemoglobin concentration, underscoring the need for a high level of physical fitness for aerobatics.

Single-Ship Performance

In the field of air shows, there exists a dazzling dichotomy between formation display teams and single-ship display teams. Both entities epitomize the pinnacle of piloting dexterity, yet they diverge in their expression of aerobatic mastery.

Formation display teams are a harmonious assembly of pilots, each an expert in their craft, woven together to create an aerial ballet. The flight experience within these teams is formidable, often surpassing thousands of flight hours, a testament to their advanced capabilities. Qualifications among these aviators are not left to chance; flight leadership and instructional prowess are common prerequisites. In their operational composition, these teams exhibit a hierarchy that includes designated leaders and wingmen, each crucial to orchestrating their airborne symphony. Their operational history often paints a canvas of diverse experiences, including combat and quick reaction alert responsibilities, underlining their multifaceted expertise.

The meticulous selection process for formation teams underscores an emphasis on peerless flight skills married with a collaborative spirit. Training for these aviators is relentless, centered on the art of formation flying—where precision is not merely desired but demanded. Their performances on the world stage are more than displays; they are a representation of national pride and aeronautical heritage.

In stark contrast stand the single-ship display teams, where the limelight shines on a solitary figure against the vast theatre of the sky. These pilots are the custodians of an immense repository of flight hours, each one a narrative of their journey. Their qualifications, too, are notable, often encompassing instructor or test pilot experience. The single-ship pilot is the quintessential hero in this narrative, their lone aircraft a brush with which they paint their aerial stunts.

The selection process here is tailored to identify not just a pilot but a performer capable of enthralling an audience, marrying technical prowess with showmanship. The training is singularly focused, honing the skills needed to command the skies alone. When they take to the air, it is a powerful exhibition of human and machine in perfect unison, a solo act that has the power to inspire the next generation of aviators.

Both types of teams fly with a weight of responsibility, with formation teams perfecting synchronicity and single-ship pilots mastering the nuances of their craft.

The aircraft they command are extensions of their being, chosen for their performance and emblematic presence. As these pilots ascend through their careers, they often transition from leadership roles within their squadrons to the echelons of the aerobatic elite, each step furtherance of their legacy.

Whether as part of a meticulously coordinated team or a solo envoy of the skies, these pilots embody the spirit of aviation at its most exhilarating peak.

Multi-Ship and Multi-Crew Performance

The qualifications, criteria, and predictors for complex air show performances, such as multi-ship flying, multi-crew performance, wing walking, and drone show performance, vary significantly depending on the nature and complexity of the act.

The essential qualifications in multi-ship and multi-crew performances extend beyond individual flying skills. Non-technical skills such as teamwork, communication, and precise coordination are also vital (DiazGranados et al., 2023; Hein et al., 2021). Foushee and Helmreich (1988) and Kanki et al. (2010) emphasize the importance of group interaction and crew effectiveness, where successful missions require technical proficiency and effective group function. Kanki (2010) further states that communication as a CRM skill serves as a social function when it helps to form team relationships and creates a work atmosphere that affects how crews perform their duties.

The selection criteria often focus on the pilots' ability to operate in sync with others, maintaining formation while executing complex maneuvers. Both Hagen (2018) and Hedge et al. (2000) highlight the relevance of CRM skills in selecting pilots, suggesting that those with strong CRM capabilities are more likely to excel in multi-crew environments.

Formation Display Teams

In a Chapter dedicated to the quintessence of aerial excellence, the rigorous selection and composition of some of the world's premier military aerobatic teams—the UK RAF's Red Arrows and the US Navy's Blue Angels (see Figure 1.8)—emphasize the extraordinary blend of skill, experience, and character required for admission to these elite squadrons.

The Red Arrows, representing the pinnacle of the RAF's public display, select team members with a rich tapestry of backgrounds. From "*Creamies*," an affectionate term for top-tier pilots who return as instructors, to the seasoned combat veterans and DPs, the team members bring with them an extensive record of operational experience, including Quick Reaction Alert duties and multinational operations.

Educational diversity is also prominent, with degrees spanning air transport management, aerospace engineering, and natural sciences. The selection process for these pilots is multi-layered, involving rigorous flying tests, interviews, media assessments, and peer evaluations, ensuring that those who don the prestigious red suits are top-tier pilots and embody the ethos of the RAF and can articulate it on a global stage.

Across the Atlantic, the Blue Angels are no less formidable in their criteria. The Blue Angels pilots are renowned for their tactical expertise, having accrued over 1,250 tactical jet flying hours and carrier qualifications, reflecting their naval heritage.

FIGURE 1.8 The Red Arrows and Blue Angels in a spectacular joint formation flight during Leeuwarden air show, June 2006. (Copyright: Alan Meloy.)

RED ARROWS

The selection criteria for the Red Arrows, the RAF aerobatic display team, clearly illustrate the rigorous standards required for air show performers at the highest level.

The selection criteria for the Red Arrows, the RAF's aerobatic team, are meticulously designed to identify candidates who demonstrate the highest levels of flying skill and operational competence. A prerequisite of 1,500 flying hours ensures that each candidate has a wealth of experience managing various aircraft under diverse conditions, which speaks to their technical proficiency and depth of flying experience.

Additionally, candidates must have completed a frontline operational tour, confirming they possess real-world, operational flying experience that often comes under challenging and high-stress conditions—essential for the precision required in air show flying. Moreover, a requirement to be assessed as above average in their flying role further filters the applicant pool to include only the top performers from the RAF. This criterion emphasizes the need for exceptional, consistent performance and skill, ensuring that only the most capable and distinguished pilots are considered for the prestigious Red Arrows team.

Selection Week Activities

During the selection week for the Red Arrows (Royal Air Force, 2024), candidates face an intensive series of tests and evaluations designed to ascertain their fitness for one of the most prestigious roles in military aviation. The week begins with a rigorous flying test, where candidates' skills are thoroughly assessed in complex aerobatic maneuvers. This test is crucial as it gauges their technical proficiency and ability to handle an aircraft during demanding routines and their composure under intense pressure—a vital trait for any pilot performing in high-stakes air show environments.

Following the flying portion, candidates undergo a formal interview. This stage of the selection process is comprehensive, assessing candidates beyond their flying capabilities, including their interpersonal and communication skills, leadership potential, and ability to mesh well with the existing team dynamics. These factors are critical in maintaining the Red Arrows' reputation for precision and excellence, both in the air and as a cohesive unit.

In addition to technical and personal assessments, candidates are also put through a media test. This aspect of the selection process is indispensable given the Red Arrows' role as ambassadors of the RAF. Effective media interaction requires candidates to articulate clearly and confidently, manage public relations adeptly, and engage with both national and international audiences, reflecting the team's public-facing nature.

Lastly, peer assessments play a significant role during the Selection Week. These evaluations involve current team members observing and providing feedback on the candidates' ability to integrate into the team. This step ensures that new members not only possess outstanding flying skills but also embody the team spirit, collaborative ethos, and mutual respect that define the Red Arrows. This aspect of the selection process highlights the importance of teamwork and the capacity for seamless collaboration within a high-performing group, ensuring the team operates as a well-coordinated whole during training and performances.

These varied components of Selection Week are meticulously designed to select the best candidates who will uphold the Red Arrows' illustrious tradition of aerial excellence and team unity. Each test and evaluation contribute to a holistic view of each candidate's capabilities and potential role within the team, ensuring that those chosen are truly the best fit for this prestigious Team.

Team Leader Selection

The selection of the Team Leader, or *the Boss*, is a process marked by careful consideration and specificity, reflecting the critical role this position plays in the overall success and management of the team. Given the complexities and high stakes of aerial display performances, the leader must possess not only superior flying skills but also comprehensive leadership abilities, excellent communication skills, and a profound commitment to the safety and coordination of the team.

To ensure that *the Boss* can effectively manage these responsibilities, candidates are typically chosen from among pilots who have prior experience with the team. This requirement guarantees that the selected individual thoroughly understands the team's dynamics, the precise nature of the routines, and the specific operational demands involved. Familiarity with the team's culture and existing members is crucial, as this allows for smoother communication, quicker decision making, and more effective leadership.

Moreover, having prior experience with the team ensures that the new leader is already well-versed in the unique challenges and pressures of performing in air shows under the public eye. They understand the importance of maintaining a positive team atmosphere and the nuances of managing both the technical and interpersonal aspects of team operations. This experience also means that the individual has observed and possibly participated in critical decision making processes and is, therefore, better equipped to handle the responsibilities of planning, executing, and revising complex display routines.

In addition to experience, the selection process also emphasizes the importance of demonstrated leadership during their tenure with the team. Potential leaders are evaluated based on their ability to inspire confidence, foster teamwork, and manage the logistics and safety protocols critical to successful air shows. Their leadership style, effectiveness in communication, and ability to maintain high morale and motivation among team members are closely scrutinized.

This thorough approach ensures that the selected *Boss* is an expert in aerobatic skills, a capable steward of the team's legacy, and an effective conductor of its daily operations and strategic visions. The continuity provided by selecting from within the team helps preserve the established practices and traditions of excellence while also infusing the leadership role with fresh perspectives and renewed vigor, which is vital for the ongoing success and evolution of the display team.

US NAVY, BLUE ANGELS

The qualifications, predictors, and criteria for a US Navy pilot to become a Blue Angel, based on the profiles of current team members, include a combination of academic achievements, extensive flight experience, leadership roles, and personal decorations.

Flight Leader

To become the flight leader of the Blue Angels, the elite flight demonstration squadron of the US Navy, a candidate must meet a highly selective set of qualifications, demonstrate key predictors of success, and fulfill stringent criteria. The qualifications for this prestigious position typically include extensive naval aviation experience, demonstrated leadership capabilities within military aviation units, and a proven record of exceptional performance in the air and command/administrative capacities.

Key predictors of success for a potential leader of the Blue Angels include advanced aerial maneuver skills, the ability to manage high-pressure situations with composure, and strong communication and team management abilities. Furthermore, the criteria for selection would also emphasize impeccable safety records, a deep understanding of flight dynamics and aerobatics, and the charisma to represent the Blue Angels and the Navy effectively at public events. The selection process prioritizes outstanding flying skills and leadership qualities that inspire confidence and foster team cohesion, ensuring the squadron maintains its renowned standards of excellence and precision. More specifically, it encompasses:

1. Strong Educational Background: A degree in aerospace engineering or a related field provides a foundational understanding of aerodynamics,

mechanics, and aviation systems, essential for mastering complex flight maneuvers and understanding the technical aspects of various aircraft.

2. Naval Aviator Designation: Achieving the status of a Naval Aviator indicates the successful completion of rigorous training and assessment, ensuring the individual possesses the basic flying skills and aeronautical knowledge necessary for advanced aviation roles.
3. Varied Operational Roles: Experience in diverse operational positions, such as Weapons Training Officer or Safety Officer, equips a pilot with a broad perspective on aviation operations, enhancing decision making skills and operational safety awareness.
4. Combat and Deployment Experience: Participation in combat deployments tests a pilot's ability to perform under pressure and in challenging environments, proving their competence in real-world military operations and ability to adapt and make rapid decisions.
5. Leadership and Awards: Recognition through leadership awards signifies a pilot's ability to lead and inspire, which is crucial for the Blue Angels' team-oriented and high-visibility environment.
6. Advanced Tactical Training: Completing courses like the TOP GUN illustrates a pilot's advanced tactical skills, aerial combat acumen, and a deep understanding of fighter jet capabilities, which are vital for executing the precise and coordinated maneuvers of the Blue Angels.
7. Instruction and Mentorship Role: Experience as an instructor pilot shows the ability to effectively teach and communicate complex concepts, an essential trait for Blue Angels pilots who often engage in public outreach and mentorship.
8. Senior Command Experience: Holding senior command positions demonstrates leadership, responsibility, and the ability to manage complex operations, aligning with the Blue Angels' ethos of excellence and precision.
9. Extensive Flight Hours and Carrier Landings: A high number of flight hours and carrier-arrested landings indicate vast experience in varied flight conditions and carrier operations, ensuring the pilot's adeptness in handling the high-performance aircraft used by the Blue Angels.

These qualifications affect a pilot's performance and skills by ensuring they have the necessary technical knowledge, practical experience, leadership ability, and mental fortitude to perform the demanding and precise aerobatic maneuvers for which the Blue Angels are known. The combination of educational background, flight experience, combat exposure, tactical training, and leadership roles prepares a pilot to meet the high standards and unique challenges of flying with and leading the Blue Angels.

COMPARISON AMONGST DEMO TEAMS

To provide a comprehensive comparison between the selection criteria for the United Kingdom's Red Arrows and the US Navy's Blue Angels, Table 1.3 summarizes the key components of each team's selection process:

TABLE 1.3
Primary Criteria for Pilot Selection of Military Demonstration Teams

Criteria/ Component	Red Arrows (UK)	Blue Angels (US)
Minimum Flying Hours	At least 1,500 flying hours are required.	Over 1,250 tactical jet flying hours are needed.
Operational Experience	Completion of a frontline operational tour is necessary.	Must be a carrier-qualified, active-duty naval aviator or naval flight officer.
Performance Assessment	Must be assessed as above-average in their flying role.	Must have an above-average performance rating.
Selection Process	Involves a flying test, formal interview, media test, and peer assessments.	Involves a formal application, interviews, and demonstration flights with the team.
Team Composition	Up to three new pilots are chosen annually. Team Leader selected from former pilots.	Six jet pilots are selected annually, with specific roles like Flight Leader and Opposing Solo.
Public and Media Engagement	Media test to assess public engagement abilities.	Demonstrated ability to represent the Navy and engage with the public and media.
Teamwork and Dynamics	Peer assessments to gauge teamwork capabilities.	Strong emphasis on teamwork and the ability to integrate within the team.
Leadership Experience	Team Leader must have completed a prior three-year tour with the team.	Leadership qualities are crucial, often with prior command experience required.
Academic Background	Degrees in various fields, including management and engineering.	Bachelor's degree in fields like aerospace engineering, history, or similar.

The table highlights the rigorous and demanding selection criteria for both the Red Arrows and the Blue Angels, emphasizing the elite nature of these aerobatic teams. While there are similarities in the emphasis on extensive flying experience, operational expertise, and public engagement abilities, specific requirements and selection processes reflect each team's unique operational and cultural contexts.

STANDARDS FOR HERITAGE FLIGHT DISPLAYS

The qualifications, criteria, and predictors for DPs within the USAF Heritage Flight Foundation (AFHFF) are meticulously tailored to the complexity and demands of heritage flight displays. These standards ensure that pilots exhibit exceptional flying skills and a profound respect for the historical significance of the aircraft they operate.

Qualifications
Pilots aiming to participate in heritage flights must have extensive experience with vintage military aircraft. This includes a minimum of either 500 hours flying USAF fighter-type warbirds or a combination of 50 hours in such warbirds plus 2,000 hours in active-duty USAF fighter-type aircraft. This experience is vital as it ensures that pilots are deeply familiar with historic warplanes' unique handling characteristics and performance constraints.

Criteria
Pilots must also hold specific certifications affirming their expertise in general and specialized flying disciplines. These include a current FAST Training certificate with a four-ship stamp or equivalent, proving their capability to lead complex multi-aircraft formations. Additionally, pilots must possess a current Class II Medical Certificate, a Commercial Pilot Certificate, and a Level II SAC card or equivalent, which indicate their health, general flying proficiency, and aerobatic skills, respectively.

Moreover, pilots must be under age 65 by the final day of the annual Heritage Flight Training Course (HFTC) and have access to or own an appropriately aged warbird that has passed a rigorous annual inspection. This age criterion balances the need for experienced pilots with considerations for optimal health and reaction times essential for performing precise maneuvers.

Predictors of Success
The ongoing proficiency and capability to perform safely in air shows are further ensured through regular assessments. Pilots must undergo annual re-certifications and continuous training, which include updating their skills according to the latest flying standards and safety protocols. The ability to meet these stringent renewal criteria strongly predicts a pilot's ongoing ability to safely and effectively perform complex aerial maneuvers.

Operational Requirements
Pilots looking to fly a new warbird type must demonstrate specific readiness through additional qualifications. Depending on the aircraft's familiarity, they must log at least 50 hours of flight time in the new aircraft type, plus an extra 5–10 hours in the specific or a similar warbird. Before participating in their first public heritage flight with the new aircraft, pilots must complete a detailed training profile with an experienced HF pilot, ensuring they are fully prepared to integrate the new aircraft into their performance repertoire.

These comprehensive qualifications and criteria are crucial for maintaining the high standards of performance and safety expected at air shows, particularly those involving historically significant aircraft. They ensure that each pilot is technically proficient and capable of honoring the heritage and legacy of the aircraft they fly, thereby enhancing the public's appreciation of these flying monuments to aviation history.

All these measures are relevant in the wake of the recent accident during the Wings Over Dallas 2022 air disaster (NTSB, 2024).

Comparison of Single-Ship versus Multi-Ship Diplay Teams

Table 1.4 encapsulates the shared characteristics and expectations of formation and single-ship display teams, highlighting the specialized nature of aerobatic flying within military aviation. Both types of teams require a profound depth of experience and skill, with formation teams emphasizing the collective artistry of precision flying, while single-ship display teams focus on the pilot's individual acumen and the technical prowess of the aircraft.

Additionally, Table 1.5 contrasts the qualifications and criteria for formation display team pilots with those for single-ship display teams, using the Patrouille de France and Hellenic Air Force *Zeus* Demo Team as representative examples:

TABLE 1.4
Commonalities within Military Single-Ship versus Formation Display Teams

Commonalities	Single-Ship Display Teams	Formation Display Teams
Flight Experience	Pilots usually have a high number of flight hours in their respective aircraft, demonstrating their expertise.	Pilots typically have extensive flight hours, often exceeding 1,500 hours, reflecting their advanced flying skills.
Pilot Qualification	Pilots are highly skilled in their craft, and their qualifications often include being instructors or test pilots.	Members are often required to have specific qualifications, such as being a flight leader or instructor.
Team Composition	Composed of a single pilot who represents the capabilities of the aircraft and the air force with dynamic maneuvers.	Composed of multiple pilots with roles such as leader, right and left wing, slot, and soloists for larger formations.
Operational Experience	Typically, they have operational experience, including combat missions or specialized roles like QRA.	Many have operational flying experience, which may include combat or QRA (Quick Reaction Alert) duties.
Selection Process	Selection often includes an evaluation of solo flying skills, showmanship, and the ability to engage with the public.	Selection involves assessment by the team or higher command, with an emphasis on flight skills and the ability to work in a team.
Training	Focused training on mastering the aircraft for complex solo aerobatics and display routines.	Intensive training focused on formation flying, aerobatics, and maintaining tight formations at various speeds and altitudes.

TABLE 1.4 (Continued)
Commonalities within Military Single-Ship versus Formation Display Teams

Commonalities	Single-Ship Display Teams	Formation Display Teams
Public Engagement	Solo pilots perform at air shows highlighting individual skill and aircraft maneuverability, often inspiring future pilots.	Teams perform at air shows worldwide, showcasing precision and teamwork, and serve as ambassadors for their respective air forces.
Roles and Responsibilities	Responsibilities are centered around individual aircraft handling, safety, and executing a choreographed routine.	Responsibilities include perfecting formation positions, maintaining safety within the team, and executing coordinated maneuvers.
Aircraft Types	Fly a specific aircraft type, often the main or most iconic fighter or aerobatic aircraft of the air force.	Typically, they fly the same type of aircraft within the team for uniformity in performance and visuals.
Career Path	Pilots usually ascend through the ranks via instructional roles, operational missions, and exhibiting exemplary flying skills.	Pilots often have a background in instruction, flight safety, or other leadership roles within their squadrons before joining.

TABLE 1.5
Comparison of Pilot Selection Criteria between Patrouille de France and Hellenic Air Force "Zeus" Demo Team

Criteria/ Component	Formation Display Teams (Patrouille de France)	Single-Ship Display Teams (Hellenic AF Zeus)
Team Composition	Aerobatic team with precise formations.	Solo pilot performing individual maneuvers.
Flight Hours	A minimum of 1,500 flight hours is required.	More than 1,600 flying hours, with 1,200 on F-16.
Qualifications	"Chef de Patrouille" (Patrol Leader) qualification.	Qualified in F-16.
Selection Process	Determined by commanding structure, with evaluation.	Part of the Zeus Solo Display Team.
Operational Roles	Emphasis on teamwork and coordination in formations.	Focus on individual piloting prowess and display skills.
Training	Advanced aerobatic and tactical proficiency.	Extensive experience in operational and demo roles.
Public Role	Represent France in aerobatic displays worldwide.	Engages the public through F-16 demonstrations.
Additional Roles	"Charognard" for new graduates	Display flying as a side job, emphasizing versatility.

Table 1.5 delineates the distinction between the collaborative nature of formation display teams, where synchronization and teamwork are paramount, and the singular focus of single-ship display teams, which spotlight the pilot's individual skills and ability to engage the audience. Formation teams like the Patrouille de France demand a high level of group discipline and aerial choreography, whereas single-ship teams, represented by the demo pilot of the Hellenic Air Force, demonstrate the agility and capability of a single aircraft and the pilot's mastery.

Wing Walking Performers

Wing walking, a specialty performance, requires not only the pilot's flying expertise but also the wing walker's skill and courage. The qualifications for pilots in these acts include precise low-speed control and the ability to perform under variable conditions, ensuring the safety of the wing walker.

The criteria for wing walkers encompass not only physical fitness and acrobatic ability but also immense trust in the pilot and the aircraft, as their performance is highly reliant on the pilot's expertise and the aircraft's stability.

Drone Show Performance

Drone shows, which involve the coordination of multiple unmanned aerial vehicles (UAVs) in complex choreographies, require operators with a deep understanding of drone technology, programming, and synchronized control. Waibel et al. (2017) discuss the creative potential of drone show systems, which necessitates operators to choreograph drones in 3D space safely and reliably.

The selection for drone show performers is based on technical prowess in UAV operation, programming skills for choreography, and a strong safety orientation. According to Hassanalian and Abdelkefi (2017), the design and operational challenges of drones, including stability and trajectory correction, are critical considerations in drone show performances.

Advanced Air Mobility (AAM) Shows

Advanced Air Mobility (AAM) encompasses the use of advanced, often autonomous, aerial vehicles for urban air mobility (UAM) and beyond. Pilots or operators in this category need to be versed in cutting-edge aviation technology, including electric vertical takeoff and landing (eVTOL) aircraft operation, autonomous navigation systems, and integration with urban airspace management (Al-Rubaye et al., 2023). AAM shows would require performers to understand and interact with complex air traffic management ecosystems and to demonstrate the capabilities of AAM vehicles in a safe, controlled manner.

HUMAN SKILLS, BOTH REGULATED AND EMPIRICAL

According to the guidance set out by the UK CAA titled *Human Factors in Flying Displays – Practical Guidance for Display Pilots & Flying Display Directors* (Cox, n.d.), *professionalism* and *motivation* stand out as the most crucial personal qualities for individuals involved in flying displays, particularly for DPs. In a presentation at the National Training Aircraft Symposium (NTAS), Velázquez (2017) highlights that professionalism is an intrinsic part of CRM predicated on compliance with checklists, using SOPs, conducting appropriate pre-flight actions, and engaging in proper planning. Being a great DP is not only about technical expertise but also significantly depends on the individual's personal drive and professional demeanor.

Professionalism, in this context, refers to the adherence to an ethical standard that includes both the meticulous preparation and the execution of flying tasks with precision and responsibility. It encompasses a comprehensive understanding of the aviation industry's regulatory framework and a commitment to uphold safety and operational standards. This quality ensures that each decision and action taken by DPs not only enhances the quality of the display but also secures the safety of participants and spectators.

Realizing a need for enhanced professionalism among pilots due to a series of fatal aircraft mishaps bordering on unprofessional behaviors observed among the flight crews, the FAA issued the Pilot Professional Development Final Rule on February 20, 2020. This rule requires mentoring training for pilots in command in areas such as professionalism (U.S. Department of Transportation, Federal Aviation Administration, 2020). Even though the rule is basically for commercial airline operators, it is relevant to the air show industry, and key guidelines can be applicable.

Unlike personality, skills, and attitudes may be changed or enhanced, and in the context of air display, a DP's attitude and motivation to display is vital. Findings from the *Human Factors in Air Displays: Transfer of Behaviours and Error Path Study* suggest that motivational factors can impact DPs to pursue excellence in their highly demanding roles (UK CAA, 2018). It is often driven by a profound passion for aviation and a desire to contribute to the success of the air show industry. Motivated individuals are likely to engage in continuous learning and improvement practices, seek feedback actively, and embrace challenges as opportunities to innovate and excel. This inner drive helps maintain high energy levels and a positive outlook, which is crucial for the high-pressure environment of air shows.

Expanding the "*Human Factors Guidance for Display Pilots (DPs) to Safeguard Displays*" (Cox, n.d.) with specific examples and a deeper exploration of the crucial aspects that ensure DPs manage their displays safely and proficiently:

Passion for Aviation

A deep-seated passion for aviation fuels the relentless pursuit of excellence among air show performers. This passion underpins a commitment to continuous learning, driving performers to refine their skills, innovate in their routines, and adapt to technological advancements in the industry. The empirical evidence suggests that an enthusiastic engagement with one's craft is linked to enhanced performance and innovation (Balcar, 2014).

Motivation

As part of assessing the necessary soft skills for DPs, the UK CAA's DAEs use open questioning to establish the motivation of the potential applicant for display flying.

This involves understanding the applicant's reasons for wanting to participate in display flying and ensuring that their motivations align with the principles of safety and professionalism.

One of the key aspects evaluated is whether the applicant genuinely desires to proficiently display the aircraft rather than seeking personal attention or engaging in risky behavior to impress an audience. A DP must prioritize showcasing the aircraft's capabilities with precision and control, demonstrating technical skill and adherence to safety protocols.

To gauge the applicant's motivation accurately, a DAE could ask open-ended questions encouraging the applicant to elaborate on their goals and attitudes toward display flying. Questions might include:

- *"What inspired you to pursue display flying?"*
- *"How do you approach the balance between performance and safety in your flying?"*
- *"Can you describe a situation where you had to prioritize safety overachieving a more dramatic display?"*

Through these questions, the DAE can assess whether the applicant is motivated by a genuine passion for aviation and a commitment to safe and skilled flying or is driven by the desire to show off, which could lead to unnecessary risks.

It is essential for DPs to possess a mindset focused on continuous improvement and responsible flying. They should demonstrate *humility*, a *willingness to learn*, and an *understanding* of their significant responsibility in ensuring the safety of themselves, their aircraft, and their audience. DPs should also exhibit strong situational awareness, the ability to make sound judgments under pressure, and effective communication skills to coordinate with their team and event organizers.

Use of Knowledge and Skill

DPs must harness their extensive knowledge and refined skills to design and execute flying displays that are both impressive and safe. This involves a comprehensive understanding of the aircraft's capabilities and limitations; for instance, a pilot flying a vintage biplane must know the specific stress limits of the aircraft's frame and the optimal speeds for various maneuvers. Pilots should continuously update their training to include the latest aviation safety protocols and technological advancements. By doing so, they ensure that their performance not only adheres to safety standards but also demonstrates the peak capabilities of their aircraft under controlled conditions.

For example, a pilot planning a loop maneuver at a show should calculate the minimum altitude needed and check the aircraft's ability to withstand the g-forces involved, rehearsing under various weather conditions to ensure safety margins are respected.

Self-Discipline

The necessity for strict self-discipline manifests in rigorous adherence to safety protocols and personal performance standards. This means conducting detailed pre-flight checks, sticking to the flight routine as practiced, and making conservative choices when faced with adverse conditions. The FAA Risk Management Handbook-FAA-H-8083-2A (FAA, 2022b) states that discipline stems from good training and habit patterns and implicitly affects aeronautical decision making.

A disciplined pilot will be guarded and inoculated against hazardous attitudes and operational pitfalls. In general, a disciplined pilot will always do the right thing. For example, if unexpected crosswinds are encountered during a display, a disciplined pilot would adjust their flight path to maintain a safe distance from the crowd and other aircraft rather than sticking rigidly to the planned routine. Also, during a display, when a pilot notices a slight engine irregularity, adhering to their disciplined training, they will choose to abbreviate their performance and land safely rather than risk a full routine that might exacerbate the issue.

Awareness of Personal Limitations

Pilots must be keenly aware of their physical and mental state and how it may affect their performance. A pilot must also understand the limits of their expertise and know that sometimes, competing task demands, task ambiguity, and organizational pressures can adversely affect cognitive abilities, making them vulnerable to errors (Dismukes et al., 2007). For example, recognizing signs of fatigue or stress is crucial, as these can impair reaction times and decision making abilities.

The FAA Risk Management Handbook provides guidance for such personal limitations' awareness through the IMSAFE approach (FAA, 2022b). A pilot might decide to cancel or adjust their participation in a display after a restless night or during periods of personal stress, knowing that their performance could be compromised. In another example, a pilot feels the onset of a cold and recognizes that their reaction time might be slower and their decision making impaired. They choose to discuss adjusting their role in the display with the event coordinator, perhaps flying fewer demanding maneuvers or handing the display over to an alternate pilot.

Information Management and Situation Awareness

Maintaining situational awareness involves managing a wide array of information effectively. This includes monitoring weather changes, maintaining communication with ATC or event coordinators, and being aware of all operational aspects of their aircraft. Pilots must be adept at processing this information quickly and adjusting their actions accordingly (FAA, 2022a; Kalagher et al., 2021; Wickens et al., 2023).

For example, during a display, a pilot receives real-time updates about a sudden change in weather conditions. Using this information, they quickly adapt their routine to avoid areas of turbulence, effectively communicating their changes to the FDD and ensuring that their new flight path is coordinated with other performers.

Emotional Intelligence (EI)

Emotional intelligence (EI) is paramount, equipping performers with the ability to manage stress, recognize their own emotional states, and empathize with team members and the audience, as suggested by Fallon et al. (2014), who found that in stressful situations, higher levels of EI improve tactical decision making in aid of safety and time-sensitivity. Dugger and McCrory (2021) also suggest a positive association between global EI score and safety among army aviators. Beling and Wild (2024), in a study of the relationship between EI and decision making among pilots, found a correlation between individual EI dimensions and decision making scenarios.

Wang et al. (2021) also assessed the impact of EI, safety training, and situational awareness among pilots and found that EI was significantly related to situational awareness only when safety training inadequacy was more salient. The more inadequate safety training was, the greater the indirect effect of EI on safety performance via situational awareness. These extant studies suggest that EI can play a crucial in air shows' dynamic and often unpredictable environment, where split-second decisions can have critical outcomes (Korniienko & Barchi, 2023).

Emotional Resilience

Emotional resilience fortifies performers against the stresses and strains of high-risk maneuvers, enabling them to maintain focus and composure during their routines. These attributes are essential not only for the safety and success of each performance but also for the continuous growth and development of performers in the face of challenges (Dean & East, 2019).

Adaptability and Continuous Improvement

The realm of air shows is ever-changing, with technological advancements and evolving audience expectations shaping its landscape. Adaptability emerges as a vital soft skill, enabling performers to navigate new challenges and integrate innovative practices into their routines. This adaptability is intertwined with a commitment to continuous improvement, a pursuit that sees performers consistently refining their skills and presentations to captivate audiences anew with each performance (Rovenska et al., 2023).

Leadership and Team Cohesion

In multi-ship and multi-crew displays, leadership transcends the conventional, morphing into a shared responsibility where each member's contributions are pivotal to the collective performance. This shared leadership fosters team cohesion, ensuring that each maneuver is a testament to trust, precision, and collective excellence. The fabric of teamwork in such settings is strengthened by each member's ability to communicate effectively, synchronize actions, and support one another through every twist and turn of the aerial ballet (Rofymenko et al., 2023).

Interpersonal Communication and Teamwork

Communicating effectively with team members, ground crews, and FDDs ensures synchronized and safe performances. This extends to multi-ship displays where the complexity of coordination is heightened, necessitating not just clear communication but also unwavering trust and understanding among team members. The essence of successful air show performances lies in the seamless integration of individual actions into a collective, choreographed aerial spectacle (Hendarman & Cantner, 2018).

Addressing and Transforming Hazardous Attitudes

The FAA's emphasis on mitigating hazardous attitudes in air show performers is a call to transform potential vulnerabilities into strengths (FAA, 2022b). Recognizing and addressing attitudes like overconfidence or complacency is not merely about risk mitigation but cultivating a mindset prioritizing safety, excellence, and the well-being of all involved. This transformation is an ongoing process, requiring performers to engage in *self-reflection* and *continuous learning*, thus elevating professionalism and safety standards in air show performances (Karachalios, 2022).

By focusing on these enhanced aspects of human factors, DPs meet the rigorous demands of their roles and model a commitment to safety and professionalism. These principles help pilots develop robust habits that support the dynamic and visually stunning environment of air show flying while maintaining the highest safety standards.

Ultimately, the empirical examination of soft skills within the air show industry reveals their indispensable role in shaping the performers' character, performance, and safety outcomes. These skills, interwoven with technical capabilities, form the foundation upon which the awe-inspiring performances are built, captivating audiences while upholding the highest standards of safety and professionalism.

Comparison between Display Pilots and Commercial Pilots

The selection criteria for both air show performers and commercial pilots demand a high level of technical flying expertise, but the specific skills emphasized in each role are distinct due to the unique demands of their respective duties. Air show performers must have an extraordinary blend of technical *aerobatic proficiency*, *psychological* resilience, and *showmanship* to captivate audiences with thrilling maneuvers under high-pressure situations. This skill-set extends beyond flying; it includes engaging with the public and performing with a level of *flair* and *precision* that turns aerobatic displays into spectacular events.

Conversely, commercial pilots are selected for their comprehensive knowledge of aircraft systems, proficiency in navigating complex flight paths, and ability to maintain calm under pressure, ensuring the safety, efficiency, and comfort of all passengers. Their role demands a focus on operational safety, effective communication with crew and air traffic control, and adherence to strict regulations that govern commercial flights.

Given the breadth of skills required for air show flying, including mastering intricate aerobatics and maintaining sharp mental focus, most DPs possess the foundational flying skills and discipline that could qualify them for commercial piloting. Many DPs might indeed hold commercial licenses and have the capability to transition to commercial flying if they choose.

However, the reverse scenario—commercial pilots moving into air show performance—is less straightforward. While commercial pilots have the necessary flying skills, not all may possess the specific aerobatic training or the flair for showmanship required for air show performances. Transitioning to the specialized and high-risk world of air shows would require additional training in aerobatics and possibly *a shift in mindset* to handle the performative aspects of the role.

FINAL CONSIDERATIONS FOR QUALIFICATIONS, CRITERIA, AND PREDICTORS OF AIR SHOW PERFORMERS

The role of an air show performer extends well beyond the confines of aviation skills. It enters the realm of performance art, requiring a blend of technical prowess, psychological resilience, and a personality that resonates with the public. The consensus among veteran demonstration pilots (Barker, 2020) suggests that while exceptional flying skills are fundamental, the quintessential attributes of a successful DP hinge on *maturity*—defined as *discipline*, *perfectionism*, and *positive assertiveness* in flying.

Maturity in this context transcends mere age or flight hours; it encompasses a pilot's capacity to manage the limelight with grace and confidence without feeling compelled to continually prove their competence. It involves understanding and respecting the limitations of their aircraft, adhering strictly to air show regulations, and handling feedback constructively, especially when it comes from colleagues. This type of feedback is not personal but an honest effort by fellow professionals to ensure a safe and visually appealing display.

The responsibilities of a DPs stretch beyond the cockpit. They encompass a significant public relations component, where the pilot must maintain a persona that aligns with the values of the sponsoring organization and effectively engages the audience. The pilot must manage the physical and mental demands of being the face of the performance, which can be as taxing as it is exhilarating.

In the military, the selection process for DPs typically involves senior squadron leaders selecting candidates. The higher command then approves these candidates based on stringent criteria, including required flight hours and specific experience. The process often culminates in a competitive evaluation to assess candidates' aptitude for managing the unique pressures of display flying. However, flying skills alone do not suffice; a pilot's personality and ability to fulfill the role of an ambassador are equally important and scrutinized.

However, what needs to be remembered while considering the necessary qualifications for someone to become an air show DP is that at the heart of every successful DP lies a deep-rooted *passion for aviation* and an unwavering *love for flying*. These pilots see their roles not just as jobs but as integral parts of their identities, deeply connected to the *aviation community* and the *air show family* to which they proudly belong. This profound connection fuels their desire to excel, not just for their own satisfaction but also in the eyes of their peers, aiming to be second to none in every performance.

For these aviators, flying at air shows is not about the sacrifices; instead, it is viewed through a lens of optimism and an enduring commitment to the community.

Their enthusiasm for aviation extends beyond personal achievement, fostering a spirit of camaraderie and mutual support that is palpable in every maneuver they expertly execute. This optimistic outlook is contagious, enhancing their own performance and enriching the entire air show experience for both performers and spectators alike.

DPs thrive under the pressure of performing complex maneuvers in front of large crowds, driven by a genuine love for the art of flight and an earnest desire to contribute positively to the air show circuit. They are not deterred by the challenges and demands of their roles. Instead, they embrace these aspects as opportunities to showcase their skills and celebrate aviation's heritage and future with fellow enthusiasts.

In this vibrant community, DPs find more than just a professional platform; they find a *family*. This sense of belonging strengthens their commitment and fuels their ambition to perform flawlessly, not only to uphold the high standards set by their predecessors but also to inspire future generations of pilots. Their presence at air shows does more than entertain; it invigorates the aviation community, fostering a shared passion for flying that transcends the events themselves.

Undoubtedly, the intense passion for aviation is a crucial component of the qualifications, criteria, and predictors of success for aspiring DPs. This fervent enthusiasm is what motivates them to pursue and achieve excellence, maintain an optimistic attitude, and engage actively within the air show community. These elements are not merely desirable traits but are essential qualifications that set apart successful air show performers from their peers.

This deep-rooted passion is instrumental in display pilots to constantly refine their skills and push the boundaries of their flying capabilities. It fuels their dedication to the art of aerobatic display, ensuring that they remain committed to continuous improvement and innovation. Such *commitment* is critical, as it translates into meticulous preparation and a compelling presence in the air, traits that are indispensable in the high-stakes environment of air shows.

Moreover, the DPs' enthusiastic participation and optimistic spirit significantly enhance the dynamics within the air show community. Their energy and dedication are a beacon, inspiring fellow pilots and captivating audiences. This influence is a vital predictor of a pilot's success in the field, as the ability to uplift and energize the air show family contributes to these events' overall success and sustainability.

While technical skills and physical capabilities are fundamental, the passion for aviation truly distinguishes a successful DP. This passion is a core aspect of the qualifications and is a critical criterion for selection and continued success in the air show circuit. This unceasing enthusiasm lights up the runways and skies, guiding and inspiring all who share in the airborne spectacle, making it an integral part of what defines and predicts a successful air show performer.

NOTES

1 The "All Aces" recounts the formation of Greece's first aerobatic squadron in 1954, initiated by Major K. Kokkas despite initial reluctance from authorities. Their stunning performance for the Minister of Defense secured official approval, leading to four years of impressive demonstrations in Greece and Europe. Their skills earned international acclaim,

including second place at a NATO air show in Belgium. In 1958, they retired, paving the way for new pilots and modern aircraft (HAF, n.d.).

2 A "gate" in aerial display contexts is a designated point in the sky where aircraft must enter or exit during their routines, adhering to predefined airspeed, altitude, and velocity vector specifications. This setup helps ensure the safety and visual effectiveness of maneuvers by guiding pilots in maintaining precise control over their aircraft throughout the performance.

3 For illustrating the maneuvers in Figures 1.6 and 1.7, a visual depiction of the smoke trail is selected instead of a line-graph to showcase the crowd's perspective of the show. The book's focus is not on providing specific flight parameters or how a maneuver should be flown in a "vacuum" but rather on illustrating the dynamic maneuvering of the aircraft. It highlights the wind's effect on the flight path (through the shifting smoke trail) and ultimately the planning and execution of any air show maneuver: to entertain the audience. Additionally, Chapter 5 provides an analysis of the Aresti system for more details in designing and deciphering aerobatic figures and flight profiles.

REFERENCES

Allerton, D. (2010). The impact of flight simulation in aerospace. *Aeronautical Journal, 114*, 747–756. https://doi.org/10.1017/S0001924000004231

ALMamari, K., & Traynor, A. (2019). Multiple test batteries as predictors for pilot performance: A meta-analytic investigation. *International Journal of Selection and Assessment, 27*, 337–356. https://doi.org/10.1111/ijsa.12258

Al-Rubaye, S., Tsourdos, A., & Namuduri, K. (2023). Advanced air mobility operation and infrastructure for sustainable connected eVTOL vehicle. *Drones, 7*(5), Article 319. https://doi.org/10.3390/drones7050319

Anderson, K., & Lee, D. (2021). The impact of public performance records on safety in aviation. *Journal of Aviation Safety Research, 15*(3), 45–60.

Armstrong, J., & Voris, R. M. (2016). *From POW to Blue Angel: The story of commander Dusty Rhodes* (Reprint edition). University of Oklahoma Press.

Avers, K. B., King, R. E., & Bartholomew, C. L. (2016). *The handbook of pilot selection* (2nd ed.). CRC Press.

Balcar, J. (2014). Soft skills and their wage returns: Overview of empirical literature. *Review of Economic Perspectives, 14*(1), 3–15.

Barker, D. (2020). *Anatomy of airshow accidents* (1st ed., Vol. 1). International Council of Air Shows, International Air Show Safety Team.

Bartram, D. (2018). Aptitude testing and selection in aviation. In *Aviation instruction and training* (pp. 34–51). Routledge.

Bartram, D., & Baxter, J. (1996). The role of the management in pilot performance. *Aviation Psychology Journal, 6*(2), 123–134.

Beling, C., & Wild, G. (2024). The association between emotional intelligence and decision making for pilots. *Journal of Air Transport Management, 114*, 102506. https://doi.org/10.1016/j.jairtraman.2023.102506

Bell, H. H., & Waag, W. L. (1998). Evaluating the effectiveness of flight simulators for training combat skills: a review. *The International Journal of Aviation Psychology, 8*(3), 223–242. https://doi.org/10.1207/s15327108ijap0803_4

Besco, R. (1994). The role of personality assessments in pilot selection. *Journal of Aviation Psychology, 4*(1), 29–35.

Bor, R., Eriksen, C., Hubbard, T., & King, R. (2019). *Pilot selection: Psychological principles and practice*. CRC Press.

Broach, D., Schroeder, D., & Gildea, K. (2019). *Best practices in pilot selection* (DOT/FAA/AM-19/6; p. 73). Federal Aviation Administration (FAA). https://libraryonline.erau.edu/online-full-text/faa-aviation-medicine-reports/AM19-06.pdf

Brown, T. (2018). The importance of continuous education in aviation. *Aerospace Education Quarterly, 22*(4), 25–35.

Campbell, J. S., Castaneda, M., & Pulos, S. (2009). Meta-analysis of personality assessments as predictors of military aviation training success. *The International Journal of Aviation Psychology, 20*(1), 92–109. https://doi.org/10.1080/10508410903415872

Carretta, T. R. (2011). Pilot candidate selection method: Still an effective predictor of US Air Force pilot training performance. *Aviation Psychology and Applied Human Factors, 1*(1), 3–8. https://doi.org/10.1027/2192-0923/a00002

Chaparro, M. E., Carroll, M., & Malmquist, S. (2020). *Personality trends in the pilot population. Collegiate Aviation Review International, 38*(2), Article 100219. https://doi.org/10.22488/okstate.20.100219

Chidester, T. R., Helmreich, R. L., Gregorich, S. E., & Geis, C. E. (1991). Pilot personality and crew coordination: Implications for training and selection. *International Journal of Aviation Psychology, 1*(1), 25–44.

Civil Aviation Safety Authority. (2025). *Advisory circular AC 91-21 v2.3: Air displays* (File ref D25/74814). www.casa.gov.au/sites/default/files/2022-11/advisory-circular-91-21-air-displays.pdf

Cook, M. (2009). *Personnel selection: Adding value through people*. Wiley-Blackwell.

Cox, M. (n.d.). *Human factors in flying displays – Practical guidance for display pilots & flying display directors* (p. 27). National Air Transport Services (NATS). Retrieved July 24, 2024, from www.caa.co.uk/media/pmqpx4gl/human-factors-in-flying-displays-practical-guidance-nats.pdf

Cox, B. D., Schmidt, L. L., Slack, K. J., & Foster, T. C. (2017). Assessment and selection of military aviators and astronauts. In *Aeromedical psychology* (pp. 17–36). CRC Press.

Davenport, E., Palileo, E., & Gore, S. (2021). Cardiovascular screening for pilots, aircrew, and high performance & spaceflight passengers. *Reach, 21*, 100040.

de Herrera, A. (2023). Female Navy pilot makes historic debut with the Blue Angels. *Navy Times*. www.navytimes.com/news/your-navy/2023/03/26/female-navy-pilot-makes-historic-debut-with-the-blue-angels/

Dean, S. A., & East, J. I. (2019). Soft skills needed for the 21st-century workforce. *International Journal of Applied Management and Technology, 18*(1), 17–32. https://doi.org/10.5590/IJAMT.2019.18.1.02

DiazGranados, D., Woods, J. J., Wilder, A., & Curtis, M.T. (2023). Chapter 8 – Team dynamics in the air: A review of team research relevant to Aviation. In J. R. Keebler, E. H. Lazzara, K. A. Wilson, & E. L. Blickensderfer (Eds.), *Human Factors in Aviation and Aerospace* (3rd ed., pp. 199–215). Academic Press. https://doi.org/10.1016/B978-0-12-420139-2.00003-4

Dismukes, R.K., Berman, B.A., & Loukopoulos, L. (2007). *The limits of expertise: Rethinking pilot error and the causes of airline accidents* (1st ed.). Routledge. https://doi.org/10.4324/9781315238654

Doe, J. (2019). Mental health and its impact on decision making in high-risk industries. *Journal of Occupational Psychology, 13*(2), 89–102.

Dugger, Z. T., & McCrory, B. (2021). Emotional intelligence and safety citizenship among Army aviators. *International Journal of Aviation, Aeronautics, and Aerospace, 8*(1), Article 5. https://doi.org/10.15394/ijaaa.2021.1552

Electronic Code of Federal Regulations (e-CFR) Title 14—Aeronautics and Space (2024, May 31). Chapter I—Federal Aviation Administration, Department Of Transportation, subchapter F—Air traffic and general operating rules part 91—general operating and flight rules subpart D—Special Flight Operations § 91.303. Retrieved October 06, 2024 from www.ecfr.gov/current/title-14/chapter-I/subchapter-F/part-91/subpart-D/section-91.303

Ernsting, J., Nicholson, A., & Rainford, D. (2006). *Ernsting's aviation medicine*. Hodder Arnold.

Evdokimenkov, V., Kim, R., & Popov, S. (2020). Risk management by trend analysis of flight information. *Russian Engineering Research, 40*, 160–163. https://doi.org/10.3103/s1068798x20020136 Viper Demo Team (af.mil).

FAASTeam. (2021). *Aircraft performance monitoring*. Federal Aviation Administration (FAA). www.faasafety.gov/files/events/GL/GL09/2021/GL09109193/Aircraft_Performance_Monitoring.pdf

Fallon, C. K., Panganiban, A. R., Wohleber, R., Matthews, G., Kustubayeva, A. M., & Roberts, R. (2014). Emotional intelligence, cognitive ability and information search in tactical decision making. *Personality and Individual Differences, 65*, 24–29.

Fardnia, P., Kaspereit, T., Walker, T., & Xu, S. (2020). Financial performance and safety in the aviation industry. *International Journal of Managerial Finance*. https://doi.org/10.1108/ijmf-03-2019-0095

Federal Aviation Administration (FAA). (1984). AC 91-61 – A hazard in aerobatics: Effects of g-forces of pilots. Accessed 06/11/2024. www.faa.gov/regulations_policies/advisory_circulars/index.cfm/go/document.information/documentID/22429

Federal Aviation Administration (FAA). (2020). AC 121-43. Guidelines for developing and implementing mentoring training for pilots in command. Accessed 06/11/2024. (PIC). www.faa.gov/documentLibrary/media/Advisory_Circular/AC_121-43.pdf

Federal Aviation Administration (FAA). (2022a). *Order 8900.1A – Flight standards information management system.* Federal Aviation Administration. www.faa.gov/regulations_policies/orders_notices/index.cfm/go/document.information/documentID/1041430

Federal Aviation Administration (FAA). (2022b). *Risk management handbook.* Accessed 06/13/2024. www.faa.gov/regulationspolicies/handbooksmanuals/risk-management-handbook-faa-h-8083-2a

Federal Aviation Administration (FAA). (2023). *Order 8900.1, Volume 5, Chapter 9, Section 1: Issue/Renew/Reevaluate/Rescind a Statement of Aerobatic Competency (SAC).* Dynamic Regulatory System. https://drs.faa.gov/browse/excelExternalWindow/DRSDOCID100875355920230912130420.0001

Feng, C., & Wang, R. (2000). Performance evaluation for airlines including the consideration of financial ratios. *Journal of Air Transport Management, 6*, 133–142. https://doi.org/10.1016/S0969-6997(00)00003-X

Formation And Safety Team (FAST). (2021). *FAST air show operations* (Revision 2). https://assets.nicepagecdn.com/272793ad/6169412/files/FASTAirshowOpsREV2.pdf

Formation and Safety Team (FAST) (2023). *Operating policies and procedures version 23.3.* Retrieved November 06, 2024. https://flyfast.org/sites/all/docs/FF&P%20Ver%2023.3,%20Feb%2028,%202023.pdf

Formation And Safety Team (FAST). (2024a). *Formation and Safety Team (FAST).* https://flyfast.org/

Formation And Safety Team (FAST). (2024b). *Formation and Safety Team (FAST), forms and documents. Formation and Safety Team (FAST).* https://flyfast.org/Forms.html

Foushee, H. C., & Helmreich, R. L. (1988). Group interaction and flight crew performance. In E. L. Wiener & D. C. Nagel (Eds.), *Human factors in aviation* (pp. 189–227). Academic Press. https://doi.org/10.1016/b978-0-08-057090-7.50013-8

Fresnel, E., Dray, G., Pla, S., Jean, P., Belda, G., & Perrey, S. (2021). Cerebral oxygenation responses to aerobatic flight. *Aerospace Medicine and Human Performance, 92*(10), 838–842. https://doi.org/10.3357/AMHP.5896.2021

Garretson, M. (n.d.). *Blue Angels Decades 1946–1955: Vol. One*. Retrieved July 20, 2024, https://blueangelsdecades.com/home

Gary, D. (2024). *Thunderbirds*. Encyclopedia Britannica. www.britannica.com/topic/Thunderbirds

Goulian, M. (2024). *Goulian aerosports*. Goulian Aerosports. https://mikegoulian.com/

Guo, H., Zhou, Y., & Jing, B. (2012). Heart rate and heart rate variability characteristics of the pilots in aerobatic flight of high performance fighter aircraft. *Chinese Journal of Aerospace Medicine, 23*, 170–173.

Guyton, B. T. (1950). United States Navy proceedings. *United States Navy Proceedings, 76*(4), 566. www.usni.org/magazines/proceedings/1950/april/blue-angels

Hagen, J. U. (2018). Crew resource management revisited. In J. Hagen (Ed.), *How could this happen?* Palgrave Macmillan. https://doi.org/10.1007/978-3-319-76403-0_14

Harris, D. (2011). *Human performance on the flight deck*. Ashgate Publishing Ltd.

Haslbeck, A., Kirchner, P., Schubert, E., & Bengler, K. (2014). A flight simulator study to evaluate manual flying skills of airline pilots. *Proceedings of the Human Factors and Ergonomics Society Annual Meeting, 58*(1), 11–15. https://doi.org/10.1177/1541931214581003

Hassanalian, M., & Abdelkefi, A. (2017). Classifications, applications, and design challenges of drones: A review. *Progress in Aerospace Sciences, 91*, 99–131.

Hausknecht, J. P., Day, D. V., & Thomas, S. C. (2004). Applicant reactions to selection procedures: An updated model and meta-analysis. *Personnel Psychology, 57*(3), 639–683. https://doi.org/10.1111/j.1744-6570.2004.00003.x

Hebbar, P.A., Pashilkar, A.A. (2017). Pilot performance evaluation of simulated flight approach and landing manoeuvres using quantitative assessment tools. *Sādhanā, 42*, 405–415 https://doi.org/10.1007/s12046-017-0613-0

Hedge, J., Bruskiewicz, K., Borman, W. C., Hanson, M., Logan, K. K., & Siem, F. M. (2000). Selecting pilots with crew resource management skills. *The International Journal of Aviation Psychology, 10*(4), 377–392.

Hein, M. B., Littlepage, G. E., & Georgiou, A. M. (2021). What we know about teamwork and multiteam coordination in aviation: Teamwork processes in aviation. *21st International Symposium on Aviation Psychology*, 420–425. https://corescholar.libraries.wright.edu/isap_2021/70

Hellenic Air Force. (n.d.). *Το Καρέ των Άσσων* [Full of Aces]. Retrieved August 4, 2024, from www.haf.gr/news/publications/acrobatic-team-history/kare-twn-asswn/

Hendarman, A. F., & Cantner, U. (2018). Soft skills, hard skills, and individual innovativeness. *Eurasian Business Review, 8*(2), 139–169.

International Air Transport Association (IATA) (2018). Guidance document crisis communication and reputation management in the digital age: A guide to best practice for the aviation industry.

International Council of Air Shows (ICAS). (2019a). *Roy M. Butch Voris*. International Council of Air Shows Foundation. https://airshowfoundation.org/roy-m-butch-voris/

International Council of Air Shows (ICAS). (2019b, January 7). *ACE manual—Revision 9*. International Council of Air Shows.

International Council of Air Shows (ICAS). (2022a). Social media continues to highlight public engagement opportunities. *Air Shows Magazine*. https://airshowdigest.aero/2022/02/24/social-media-continues-to-highlight-public-engagement-opportunities-2/

International Council of Air Shows (ICAS) (2022b). SASP to expand to 2022. https://airshows.aero/News/ID/596

Ji, M., Xu, Q., Xu, S., Du, Q., & Li, D. (2018). Proactive personality and situational judgment among civil flying cadets: The roles of risk perception and cognitive flexibility. *Transportation Research Part F: Traffic Psychology and Behaviour, 59*, 179–187

Jiang, Y., Tran, T. H., & Williams, L. (2023). Machine learning and mixed reality for smart aviation: Applications and challenges. *Journal of Air Transport Management, 111*, 102437. https://doi.org/10.1016/j.jairtraman.2023.102437

Johnston, N. (1996). The role of psychological evaluation in pilot training. *Aerospace Medicine and Human Performance, 67*(10), 972–977.

Johnston, P. J., & Catano, V. M. (2013). Investigating the validity of previous flying experience, both actual and simulated, in predicting initial and advanced military pilot training performance. *The International Journal of Aviation Psychology, 23*(3), 227–244. https://doi.org/10.1080/10508414.2013.799352

Joiner, D. (2002). Assessment Centers: What's New?. *Public Personnel Management, 31*, 179–185. https://doi.org/10.1177/009102600203100204

Kalagher, H. & de Voogt, A. & Boulter, C. (2021). Situational awareness and general aviation accidents. *Aviation Psychology and Applied Human Factors, 11*. https://doi.org/10.1027/2192-0923/a000207

Kalemba, N., & Campa-Planas, F. (2019). Safety and the economic and financial performance in the airline industry: is there any relationship? *Aviation, 23*(1), 7–14. https://doi.org/10.3846/aviation.2019.9744

Kanki, B. (2010). Communication and crew resource management. *Crew Resource Management*, 111–145. https://doi.org/10.1016/B978-0-12-374946-8.10004-4

Kanki, B., Helmreich, R., & Anca, J. (Eds.). (2010). *Crew resource management* (ch. 12, 2nd ed.). Wiley. https://doi.org/10.1016/C2009-0-02518-5

Karachalios, E. M. (2022). *An evaluation of the relationships between resilient safety culture, safety risk parameters, and mindfulness in the international air show community*. The University of North Dakota.

Karachalios, M., & Adjekum, D. K. (2023). *Air show performers: Safety, risk management, and psychological factors*. CRC Press. https://doi.org/10.1201/9781003431879

Kern, T. (1995). *Darker shades of blue: A case study of failed leadership*. https://docslib.org/doc/11768753/darker-shades-of-blue-a-case-study-of-failed-leadership

Kikkawa, Y., & Mavin, T. J. (2018). Integrated digitised video recordings in postflight-simulator training: A matter of reflection. In C. Harteis (Ed.), *The impact of digitalization in the workplace. Professional and practice-based learning* (Vol. 21). Springer. https://doi.org/10.1007/978-3-319-63257-5_8

Korniienko, I., & Barchi, B. (2023). *The concept of the soft skills and emotional intelligence development*. Psychology Travelogs.

Lakshmiprabha, L., & Amuthan, T. B. (2023). A study on financial performance of Air India. *i-manager's Journal on Economics & Commerce, 3*(2), 36–42. https://doi.org/10.26634/jecom.3.2.19893

Lambirth, T. T., Dolgin, D. L., Rentmeister-Bryant, H. K., & Moore, J. L. (2003). Selected personality characteristics of student naval aviators and student naval flight officers. *International Journal of Aviation Psychology, 13*, 415–427. https://doi.org/10.1207/s15327108ijap1304_06

Lee, W. (2006). Risk assessment modeling in aviation safety management. *Journal of Air Transport Management, 12*, 267–273. https://doi.org/10.1016/J.JAIRTRAMAN.2006.07.007

Maculewicz E, Pabin A, Kowalczuk K, Dziuda Ł, Białek A. (2022). Endogenous risk factors of cardiovascular diseases (CVDs) in military professionals with a special emphasis on military pilots. *Journal of Clinical Medicine, 11*(15), 4314. https://doi.org/10.3390/jcm11154314

Martinussen, M., & Hunter, D. R. (2018). *Aviation psychology and human factors* (2nd ed.). CRC Press.

McCarthy, P., & Agnarsson, A. (2018). Pilot performance assessment in simulators: Exploring alternative assessment methods. In D. Harris (Ed.), *Engineering Psychology and Cognitive Ergonomics*. EPCE 2018. Lecture Notes in Computer Science() (Vol. 10906). Springer. https://doi.org/10.1007/978-3-319-91122-9_50

Meško, M., Karpljuk, D., Štok, Z. M., Videmšek, M., Bertoncel, T., Bertoncelj, A., & Podbregar, I. (2013). Motor abilities and psychological characteristics of Slovene military pilots. *The International Journal of Aviation Psychology, 23*(4), 306–318. https://doi.org/10.1080/10508414.2013.833750

Military Aviation Authority (MAA), UK. (2024). *Regulatory Article (RA) 2335: Flying displays, display flying, role demonstrations and flypasts (Version 9).* Ministry of Defense UK. www.gov.uk/government/publications/regulatory-article-ra-2335-flying-displays-and-special-events

National Museum of the United States Air Force. (n.d.). *North American F-100D Super Sabre*. www.nationalmuseum.af.mil/Visit/Museum-Exhibits/Fact-Sheets/Display/Article/195798/north-american-f-100d-super-sabre/

National Transportation Safety Board (NTSB). (2024). *Project summary: Aviation investigation – 49 docket items – cen23ma034*. Accessed 06/11/2024. https://data.ntsb.gov/Docket?ProjectID=106276

Nicol, E., Rienks, R., Gray, G., Guettler, N., Manen, O., Syburra, T., D'arcy, J., Bron, D., & Davenport, E. (2018). An introduction to aviation cardiology. *Heart, 105*, s3–s8. https://doi.org/10.1136/heartjnl-2018-313019

Office of the Federal Register (OFR) of the National Archives and Records Administration (NARA), & U.S. Government Publishing Office (GPO). (2024). *14 CFR 61.133, Commercial pilot privileges and limitations.* FederalRegister.Gov. www.ecfr.gov/current/title-14/part-61/section-61.133

Onkst, D. H. (n.d.). *Early exhibition aviators. U.S. Centennial of Flight Commission.* Retrieved May 7, 2023, from www.centennialofflight.net/essay/Explorers_Record_Setters_and_Daredevils/early_exhibition/EX7.htm

Perrier, E., Leduc, P., Manen, O., Lerecouvreux, M., Deroche, J., Paris, J., Doireau, P., Quiniou, G., Geffroy, S., & Carlioz, R. (2005). The heart and aerobatics. *Archives des maladies du coeur et des vaisseaux, 98*(1), 47–52.

Ree, M. J., & Carretta, T. R. (1996). Central role of g in military pilot selection. *The International Journal of Aviation Psychology*, *6*(2), 111–123. https://doi.org/10.1207/s15327108ijap0602_1

Rofymenko, O., Savielieva, O., Prokop, Y., Loginova, N., & Dyka, A. (2023). Soft skills for software developers. *Cybersecurity: Education, Science, Technique, 3*(19), 6–19. https://doi.org/10.28925/2663-4023.2023.19.619

Rovenska, V., Smyrnova, I., & Latysheva, O. (2023). *Development of soft skills in the context of performance management skills.* Problems of Systemic Approach in the Economy.

Royal Air Force (2024). The Red Arrows. Retrieved October 06, 2024. www.raf.mod.uk/display-teams/red-arrows/the-team/

Smith, L., & Jones, R. (2020). Evaluating performance metrics in high-risk professions. *Safety Science Monitor, 28*(1), 14–29.

South African Civil Aviation Authority (SACAA). (2021). *Special Air Events Handbook.* Chapter 6 – Pilot display competency. Retrieved October 06, 2024. www.airshowsa.org.za/airshows/docs/SAE%20handbook%20240221%20N%20.pdf

Stephens, C. L., Prinzel, L. J., Kiggins, D., Ballard, K., & Holbrook, J. (2021). Evaluating the use of high fidelity simulator research methods to study airline flight crew resilience. *21st International Symposium on Aviation Psychology*, 140–145.

Transport Canada. (2015). Standard 623 – Special Flight Operations – Canadian Aviation Regulations (CARs). Airshows, Part 623.06. Retrieved October 06, 2024. https://tc.canada.ca/en/corporate-services/acts-regulations/list-regulations/canadian-aviation-regulations-sor-96-433/standards/standard-623-special-flight-operations-canadian-aviation-regulations-cars#623d1_06

U.S. Air Force (USAF). (2003). Thunderbirds at 50. *Air Force Magazine*. www.airandspaceforces.com/PDF/MagazineArchive/Magazine%20Documents/2003/June%202003/0603Tbirds.pdf

U.S. Air Force (USAF). (2022). *Air Force Manual 11-246: Air Force fighter type aircraft demonstrations (A-10, F-16, F-22, F-35)* (Vol. 1). USAF. https://static.e-publishing.af.mil/production/1/af_a3/publication/afman11-246v1/afman11-246.pdf

U.S. Air Force (USAF). (2024). Thunderbirds History. www.airforce.com/thunderbirds/history

U.S. Department of Transportation, Federal Aviation Administration. (2020). *Pilot professional development* (Final rule). *Federal Register, 85*(37), 10896–10935. www.federalregister.gov/documents/2020/02/25/2020-01111/pilot-professional-development

U.S. Marine Corps. (2022). *Marine officer applications being accepted by Navy flight demonstration squadron (Blue Angels).* United States Marine Corps Flagship.

U.S. Navy. (2024). U.S. Navy Blue Angels, History. www.blueangels.navy.mil/history/

UK Civil Aviation Authority (UK CAA). (2018). *CAP 1694-Human factors in air displays: Transfer of behaviours and error path study*. Health and Safety Laboratory. Project number: PE11472. Retrieved November 06, 2024. www.caa.co.uk/our-work/publications/documents/content/cap1694/

UK Civil Aviation Authority (UK CAA). (2024). The CAA behavioural and attitudinal fitness assessment. www.caa.co.uk/general-aviation/flying-displays-and-special-events/the-caa-behavioural-and-attitudinal-fitness-assessment/

UK Civil Aviation Authority. (2025a, February 4). *CAP 403: Flying displays and special events – Safety and administrative requirements and guidance* (Edition 22). www.caa.co.uk/our-work/publications/documents/content/cap-403/

UK Civil Aviation Authority. (2025b, March 10). *CAP 1724: Flying display pilot authorisation and evaluation: Requirements and guidance* (Edition 7). www.caa.co.uk/our-work/publications/documents/content/cap1724/

U.S. Navy. (n.d.). *History: Significant events. Blue Angels*. Retrieved April 18, 2025, from www.blueangels.navy.mil/history/events.htm

Vanderpool, R. (2022). *Looking back: Ramstein air show crash*. Office of Special Investigations. www.osi.af.mil/News/Features/Display/Article/3129659/looking-back-ramstein-air-show-crash/

Velázquez, J. (2017). Pilot professionalism and attitude management: A new CRM generation? [PowerPoint presentation]. *National Training Aircraft Symposium (NTAS), 26*, Slides 16–18. https://commons.erau.edu/cgi/viewcontent.cgi?article=1071&context=ntas

Vidaković, J., Lazarevic, M., Kvrgic, V., Maksimović, I., & Rakic, A. (2021). Flight simulation training devices: Application, classification, and research. *International Journal of Aeronautical and Space Sciences, 22*, 874–885. https://doi.org/10.1007/s42405-021-00358-y

Waibel, M., Keays, B., & Augugliaro, F. (2017). *Drone shows: Creative potential and best practices* (Report No. 50527). ETH Zürich. https://doi.org/10.3929/ethz-a-010831954

Wang, L., Ren, Y., Sun, H., & Dong, C. (2017). A landing operation performance evaluation system based on flight data. In D. Harris (Ed.), *Engineering Psychology and Cognitive Ergonomics: Cognition and Design* (Lecture Notes in Computer Science, Vol. 10276, pp. 297–305). Springer. https://doi.org/10.1007/978-3-319-58475-1_22

Wang, Z., Jiang, Z., & Blackman, A. (2021). Linking emotional intelligence to safety performance: The roles of situational awareness and safety training. *Journal of Safety Research, 78*, 210–220.

Wang, Z., & Sarkis, J. (2017). Corporate social responsibility governance, outcomes, and financial performance. *Journal of Cleaner Production, 162*, 1607–1616. https://doi.org/10.1016/J.JCLEPRO.2017.06.142

Wickens, C., Vidulich, M., & Tsang, P.. (2023). Information processing in aviation. https://doi.org/10.1016/B978-0-12-420139-2.00002-2

Wilcox, R. K. (2004). *First Blue: The story of World War II ace Butch Voris and the creation of the Blue Angels* (1st ed.). Thomas Dunne Books.

William, J. (2020). *The missing link: An inside look at flight data monitoring options for GA* [FAA Safety Briefing]. Cleared for Takeoff. https://medium.com/faa/the-missing-link-abdc1fda5de6

Wilson, M. (2007). A history of job analysis. In, L. Koppes (Ed.), *Historical perspectives in industrial and organizational psychology* (pp. 219–241). Lawrence Erlbaum Associates.

2 Demographics of Air Show Performers

Manolis Karachalios

Contributor
Daniel Kwasi Adjekum

INTRODUCTION

The world of air show performers is a rich tapestry woven from diverse threads, each representing the unique backgrounds, experiences, and journeys of the individuals who dare to dance in the sky. This Chapter examines the demographics of air show performers, exploring the wide variety of ages, genders, nationalities, and backgrounds that converge in the thrilling arena of aerobatic displays.

Air show performers come from a broad spectrum of life paths, each bringing a distinct flavor to the performances they execute with precision and flair. From seasoned veterans with decades of experience to young, dynamic pilots fresh from aerobatic academies, the diversity among air show performers enriches the field and enhances the spectacle for audiences worldwide.

Historically, air shows have been dominated by a fairly homogenous group, typically composed of male pilots with military backgrounds. However, the landscape is rapidly changing. Today, the arena is filled with individuals from various walks of life. Female pilots, pilots from different ethnicities, and even those with non-traditional paths into aviation are increasingly taking to the skies. This shift reflects broader societal changes and signifies a deliberate effort within the industry to promote inclusivity and embrace a wider array of talents.

Young, vigorous pilots bring fresh energy and a modern approach to entertainment via their performances. These newcomers often integrate cutting-edge technology and contemporary flying styles, captivating younger audiences and inspiring the next generation of aviators. Their presence ensures that air shows remain relevant and evolve with the times.

On the other end of the spectrum are the seasoned veterans. These experienced pilots, many of whom have backgrounds in military aviation, bring a wealth of knowledge and a deep understanding of showmanship. Their performances are often characterized by precision and a mastery of complex maneuvers honed over years of practice. The blend of youthful innovation and veteran expertise creates a dynamic, multifaceted air show experience that appeals to a broad audience.

DOI: 10.1201/9781003504801-4

It is observed that the demographics of air show performers are evolving toward greater diversity and inclusivity. Therefore, in Chapter 2, these changes will be investigated in depth, analyzing the historical evolution, future trends, and specific initiatives like those of the Air Combat Command (ACC) demonstration teams, highlighting the industry's proactive steps to embrace and celebrate diversity in all its forms. As the air show community continues to grow and diversify, it stands as a testament to the power of inclusion and the richness that diverse perspectives bring to the art of aerial performance.

HISTORICAL EVOLUTION

The historical evolution of air show performers from the early 1900s to 2024 mirrors broader societal trends, showcasing significant shifts in demographics such as age, gender, background, financial status, marital status, education, and country of origin. This overview provides insights into how these factors have influenced the air show industry over time.

Early 20th Century (early 1900s–1930)

In the early 20th century, air show performers were predominantly young, adventurous men from Europe or the United States. This period marked the nascent stage of aviation, primarily centered in these regions due to early technological advancements and societal fascination with flight. These performers often had military or engineering backgrounds and were financially backed by government or private investors passionate about aviation. The first official air show at the Rheims Aviation Meeting in 1909 (National Air and Space Museum (NASM), 2024) featured international pilots showcasing their skills and aircraft, setting the stage for future air displays.

Mid-20th Century (1931–1960)

During the mid-20th century, the demographics of air show performers began to diversify slightly. Performers typically ranged in age from their 30s to 40s, reflecting a mix of seasoned veterans and younger pilots. While the industry remained predominantly male, with a strong presence in Europe and the United States, it also began to expand to other regions, such as Canada and Australia, as aviation technology proliferated globally. Post-World War I air shows often featured former military pilots who demonstrated their skills in barnstorming acts and traveling air circuses, contributing to the spread of aviation culture.

Late 20th Century (1961–1990)

The latter part of the 20th century saw further diversification in the demographics of air show performers. Ages ranged broadly, with young pilots in their 20s joining experienced aviators in their 50s and 60s. This period marked the entry of more female performers, though they remained a minority. The geographical spread of performers widened, with increasing participation from countries across all continents, reflecting the globalization of the air show industry.

Notable events such as the Cleveland National Air Show and the Sun 'n Fun Fly-In in Florida contributed to this trend. The Cleveland National Air Show in 1976 showcased a mix of modern military jets and historical warplanes, drawing large crowds and diverse performers like the Blue Angels and Bob Hoover (Cleveland National Air Show, 2022). The Sun 'n Fun Fly-In, which began in 1974, quickly grew into one of the largest aviation events in the world, attracting pilots and enthusiasts from a wide range of backgrounds and promoting general aviation (Sun 'n Fun, n.d.).

Early 21st Century to Present (1991–2024)

In recent decades, the age range of performers has continued to broaden, with participants ranging from young adults to those in their 70s, demonstrating the enduring appeal of air shows across generations. Gender diversity has improved significantly, with more women taking center stage as performers. The industry has seen a substantial increase in global participation, with performers coming from diverse countries, including emerging aviation markets in Asia, Africa, and South America.

A prime example of these efforts is the Royal International Air Tattoo (RIAT), held annually in the United Kingdom. RIAT has become one of the world's largest military air shows, showcasing a diverse array of international performers and aircraft. The event actively promotes inclusivity and diversity by featuring pilots and teams from various countries, reflecting the global nature of modern aviation. RIAT's commitment to diversity is evident in its diverse lineup and its efforts to engage with various participants and audiences (Royal International Air Tattoo, 2024).

Throughout these periods, the financial status, marital status, and educational background of performers have also evolved. Financially, the range has expanded from mostly sponsored or military-funded aviators to include independently wealthy performers and those supported by commercial endorsements. Marital status varied widely, with the modern era seeing performers from all walks of life. Educationally, the field has grown from primarily self-taught aviators to include highly educated individuals with specialized degrees in aviation and related fields.

The demographics of air show performers from the 1900s to 2024 illustrate a dynamic and evolving landscape, reflecting changes in societal norms, advancements in aviation technology, and the global expansion of the air show industry. The diversity in age, gender, background, and national origin of performers today speaks to the universal appeal and inclusivity of air shows as a global phenomenon.

Today's Air Show Performers' Demographic

The map illustrated in Figure 2.1 shows the distribution of air show performers per country in 2024. The United States and the United Kingdom dominate the scene, each boasting a high number of performers. European countries like France and Italy have moderate representation. In contrast, countries in Asia, Africa, and South America have fewer performers, reflecting varying levels of investment and interest in aviation

FIGURE 2.1 Estimated number of air show performers per country in 2024. (Created by the author. Map copyright: Bing.)

entertainment. This demographic distribution suggests strong aviation traditions in the United States and United Kingdom and indicates potential growth areas in other regions.

Beyond geographic distribution, the demographics of air show performers provide a fascinating insight into the complex interplay between personal attributes and professional capabilities. Extensive research has underscored the significance of experience in shaping a performer's risk perception, revealing that seasoned pilots or performers often exhibit a heightened awareness and understanding of the risks involved in their aerial endeavors (Crundall et al., 2013; Ferraro et al., 2015; Joseph & Reddy, 2013; Winter et al., 2019; You et al., 2013).

Experience in flying or related activities correlates positively with an individual's ability to perceive and navigate risks effectively. This nuanced understanding of risk is further nuanced by age, with studies indicating an inverse relationship between a performer's age and risk tolerance (Gibson et al., 2013; Hallahan et al., 2004). This relationship suggests that younger performers may exhibit higher risk tolerance, potentially influencing the dynamics of air show performances.

Marital status, too, emerges as a significant factor in this context, affecting risk tolerance levels among performers. This demographic variable, along with others like educational background, paints a comprehensive picture of the factors influencing risk perception in aviation. For instance, Adjekum et al. (2015) highlighted the correlation between educational level and safety perception among aviation students, while Chionis and Karanikas (2018) observed that aviation professionals with a bachelor's degree or lesser qualifications tend to be more risk-averse compared to their postgraduate counterparts.

However, Barker (2003) introduces a thought-provoking perspective by noting a paradox within the air show industry: extensive flying experience and proven expertise, often hailed as protective factors in general aviation, do not unequivocally translate to safer air show performances. This observation suggests that the relationship between a performer's demographic characteristics and their performance safety is not linear but complex and multifaceted.

The Study

To deepen our understanding of air show performers, examining the intricate relationships between their total display experience, military versus civilian flying background, age, educational level, marital status, and their approach to risk and safety in performances became crucial. By unraveling these dynamics, we can better comprehend the diverse tapestry of factors that define and influence the demographics of air show performers. That was one of the initiatives behind the comprehensive study conducted on air show performance in 2022 (Karachalios, 2022). The research provides an insightful analysis of the diverse characteristics and backgrounds of individuals within the air show industry.

The study showcases a broad age distribution among performers, with notable concentrations in the 35–44 and 55–64 age groups, suggesting a blend of youthful enthusiasm and seasoned experience crucial for the rigors of air show performances.

It was observed that the air show industry is predominantly male, pointing to a need for increased gender diversity within the field. Furthermore, the global representation of performers from over 20 countries highlights the international appeal and inclusivity of air shows, manifesting a culturally diverse participant pool.

Marital status primarily reflected a married cohort and the educational background revealed a substantial number of performers with higher education degrees, indicating a highly educated community within the air show sector. These demographic details are essential for understanding air show performers' social and intellectual profiles.

Regarding professional roles, most respondents were active air show performers, many with over a decade of experience, illustrating a profound dedication to the art of aerobatics. The study also found a balanced mix of civilian and military aerobatic backgrounds among the performers, enriching the community with various skills and experiences.

FUTURE DEMOGRAPHIC TRENDS FOR AIR SHOW PERFORMERS

The future of air show performer demographics will likely be shaped by broader societal and technological changes, reflecting shifts in age, gender diversity, educational backgrounds, and international participation. Based on the author's (Karachalios 2022) study, financial constraints significantly impact civilian pilots aspiring to join the air show community. The high costs associated with training, aircraft maintenance, and operational expenses can pose substantial barriers to entry, potentially leading to a decline in civilian single-ship air show performers as the profession becomes financially non-viable.

Despite these challenges, no clear trend indicates that military-background performers will dominate the air show scene in the future. Instead, broader industry trends and evolving support mechanisms likely influence a more nuanced demographic shift. As the aviation industry evolves, younger individuals with a passion for flying and aerobatics are expected to enter the field, attracted by advancements in flight technology and broader career opportunities within the sector. It is also plausible that increased access to aviation education and training programs will drive a gradual shift toward a younger demographic of air show performers.

Performers will likely come from a wider range of educational backgrounds, reflecting modern air shows' interdisciplinary nature, combining entertainment, technology, and aviation elements. This diversification will enrich the skill set within the air show community.

The international representation among air show performers is expected to grow, with increasing cross-border mobility and collaboration within the global aviation community. This trend will contribute to a more culturally diverse and globally interconnected group of performers.

Advancements in aviation technology will not only affect the types of aircraft used in air shows but also influence the demographic profile of performers, with a possible increase in performers skilled in operating advanced, high-tech aircraft.

The financial landscape for air show performers may see positive changes, with increased sponsorship opportunities and funding models making air show participation more accessible to a broader demographic.

Recruitment Strategies

Implementing outreach programs in schools and aviation academies is crucial to attracting younger talent. These programs should highlight the excitement and career potential of air show performers. Establishing mentorship programs where experienced performers can guide and inspire newcomers is equally important, as these mentorships facilitate knowledge transfer and foster a learning community.

Gender, racial, and ethnic diversities can be promoted through targeted recruitment drives and scholarships designed to encourage more inclusivity for people of other gender orientations, ethnicities, and racial backgrounds to pursue careers in air show performance. There may be some challenges to these efforts from a sociocultural perspective in some countries and communities, but it is worth the effort. Publicizing the achievements of especially females and people of color in the air show community will provide role models and further encourage women to enter the field.

Addressing financial stability and developing sponsorship programs and financial support systems can alleviate the economic barriers aspiring performers face. Encouraging partnerships between air show organizations and aviation companies can provide additional financial and logistical support.

To accommodate performers' personal lives, creating family-friendly policies and support networks within the air show community can enhance work–life balance. Offering flexible scheduling and remote participation options in the training and planning phases can accommodate performers with family commitments.

Education and background diversity can be enhanced by facilitating continuous education and training opportunities for air show performers. This ensures they can develop their skills and adapt to advancements in aviation technology. Encouraging diversity in recruitment by valuing varied educational and professional backgrounds will recognize the broad range of skills and perspectives they bring to the industry.

Promoting international exchange programs and collaborative events can attract performers from various countries, enhancing the global appeal and diversity of air show performances. Leveraging digital platforms to reach potential performers worldwide by offering virtual training sessions and interactive air show experiences can further support this goal.

Methods for Retention

Offering clear career progression pathways within the industry is essential to retain air show performers. Regularly assessing and recognizing the achievements of performers through awards, titles, or public recognition can incentivize and acknowledge their contributions.

Fostering a strong sense of community among air show performers through regular social events, forums, and team-building activities can enhance their sense of belonging. Providing mental health and well-being support, recognizing the high-stress nature of the profession, and offering resources to assist performers in managing these challenges are also crucial.

Prioritizing safety in all aspects of air show performances ensures performers feel secure and supported. Investing in state-of-the-art training facilities and simulation technology allows performers to practice and perfect their routines in a safe environment.

By implementing these strategies, the air show industry can attract and retain a diverse and talented pool of new performers through supportive policies, community engagement, and continuous development opportunities. While military-trained pilots will continue to have a significant presence, increased support and accessibility for civilians, combined with a growing appreciation for diverse skill sets and backgrounds, will likely lead to a more heterogeneous demographic profile in the air show community.

DIVERSITY, EQUITY, AND INCLUSIVITY IN THE AIR SHOW ARENA

The future of diversity, equity, and inclusivity in the air show arena is poised for significant transformation. As the air show industry continues to evolve, the drive toward greater diversity is expected to intensify, influenced by global trends for equity, societal demands for inclusivity, and diversity's inherent benefits to the sector.

Historical Perspective on Diversity

In the early 20th century, the air show arena was a stark reflection of the societal norms and constraints of the time. It was predominantly male-dominated, with limited participation from diverse ethnic and socioeconomic backgrounds. Opportunities in aviation and aerobatics were largely restricted to a narrow demographic group, primarily white males with military affiliations. This historical context highlights the significant progress made in the industry's diversity landscape.

Current Diversity Landscape

The contemporary air show industry has significantly shifted toward inclusivity and diversity. Initiatives aimed at promoting gender, racial, and ethnic diversities have seen an increase in females, people of color, and the LGBTQ+ community both in the cockpit and in leadership roles within the industry. Moreover, there has been a conscious effort to embrace cultural diversity, with performers from various ethnic backgrounds adding vibrancy and global appeal to air shows.

Future Directions in Diversity, Equity, and Inclusion

The air show industry is on a trajectory of continuous expansion and inclusivity. The public's support and financial backing of aviation organizations with diverse members are key to driving this change. These groups not only establish best practices but also raise awareness. Leading aviation organizations advocating for Diversity, Equity, and Inclusion (DEI), such as the Organization of Black Aerospace Professionals (OBAP), Women in Aviation International (WAI), The Ninety-Nines, National Gay Pilots Association (NGPA), Professional Asian Pilots Association (PAPA), and Latino Pilots Association (LPA), are instrumental in nurturing the next generation of air show pilots.

Anticipated trends are described in the following sections.

Expanding Global Representation

In the future, we will likely see a broader global representation among air show performers. As air shows gain popularity in different parts of the world, performers from various countries will participate, bringing diverse flying styles, cultural narratives, and aviation heritage to the forefront. This global exchange will enrich the air show industry, offering audiences a more varied and enriching experience.

Addressing Economic Barriers

Economic factors remain a significant hurdle for many aspiring performers. However, the industry will likely witness innovative financial models and partnerships facilitating greater participation. Sponsorships, scholarships, and corporate partnerships can provide the necessary funding to support talented individuals who might otherwise be unable to afford the costs of becoming an air show performer. These initiatives will not only diversify the pool of performers but also ensure that financial constraints do not hinder talent from shining in the air show arena.

Promoting Inclusivity Through Policy and Practice

On the policy front, air show organizations and governing bodies must continue advancing inclusive practices and policies. This includes creating environments that respect and value diversity in all its forms, from gender and ethnicity to socioeconomic background and beyond. Training programs, community outreach, and awareness campaigns will be essential in fostering an inclusive culture that attracts and retains diverse talent in the air show industry.

The future of air show performer demographics is set to reflect a more inclusive and diverse composition driven by societal changes, technological advancements, and strategic initiatives within the industry. This progression toward greater diversity enriches the air show community and ensures its sustainability and relevance in a rapidly changing world. As air shows continue to captivate audiences globally, the performers' diverse backgrounds and experiences will testify to the industry's evolution and embrace a broader spectrum of talents and stories.

CASE STUDY: DIVERSITY, EQUITY, AND INCLUSIVITY IN USAF'S AIR COMBAT COMMAND (ACC) DEMONSTRATION TEAMS

The air show industry, particularly among display pilots and demo teams, has traditionally lacked diversity, often dominated by male pilots from similar ethnic and socioeconomic backgrounds. However, recent efforts have significantly advanced the principles of diversity and inclusivity within this niche, recognizing that diverse teams enhance performance quality and audience engagement.

The USAF's ACC demonstration teams exemplify successful diversity and inclusivity initiatives in the air show sector. The ACC teams, comprising the F-35A Lightning II, F-16 Viper, F-22 Raptor, and A-10C Thunderbolt II Demonstration Teams, have implemented several strategies to promote diversity and inclusivity.

A significant milestone for the ACC teams is the inclusion of female pilots in leadership roles. In 2024, Capt. Melanie Kluesner became the new certified pilot and commander of the F-35A Lightning II Demonstration Team, marking the second consecutive year that a woman has led the F-35 Demo Team (see Figure 2.2). This transition, following Maj. Kristin Wolfe's tenure, demonstrates ACC's commitment to diversity and inclusivity (Air Force Heritage Flight, 2024).

It is interesting that it took almost 15 years to catch up to the standard set by the Patrouille de France (PAF), which is the precision aerobatics demonstration unit of the French Air and Space Force. On 25 November 2009, for the first time in the world, an aerobatic team was commanded by a woman when the Commander of the French Air Force, Virginie Guyot, became the leader of the Patrouille de France (European Airshows, 2024).

Other notable female pilots in ACC demonstration teams include Capt. Aimee Fiedler, 2023 F-16 Viper Demonstration Team commander, and Capt. Lindsay Johnson, commander of the A-10C Thunderbolt II Demonstration Team. These women lead their respective teams and serve as role models, encouraging more women to pursue careers in aviation and air show performance (Aerotech News, 2023; Air Combat Command (ACC), 2023).

In 2020, Captain Remoshay Nelson, US Air Force Public Affairs Officer, became the first black woman to be a United States Air Force Thunderbird. (eurweb.com, 2020). As of 2024, the Thunderbird team leadership consists of eight Caucasian male officers who are demonstration pilots, and four officers serve in critical roles, from

FIGURE 2.2 Capt. Melanie Kluesner, the 2024 F-35A Lightning II Demonstration Team pilot and commander, with her predecessor, Maj. Kristin Wolfe, together at the 2024 Heritage Flight Training Course. (Copyright: U.S. Air Force, photo by 1st Lt. Nathan Poblete.)

medical support to public affairs, of which three are women. More than 120 enlisted members represent nearly 30 different career fields, which is very diverse, with about 24 women and a higher non-Caucasian representation (USAF, 2024).

Impact and Outcomes

Including diverse pilots in the ACC demonstration teams has yielded several positive outcomes. Including qualified and competent women in leadership roles who have been exceptional in their performance has also served as inspiring role models, encouraging many young women to pursue careers in aviation and air show performance. This progression contributes to gradually balancing gender representation within the military aerobatic community.

Enhanced team performance has also been a direct result of these diversity efforts. Diverse teams bring various perspectives and approaches, leading to more innovative and engaging performances that captivate audiences. The unique backgrounds and experiences of the team members enrich the performance quality, making the shows more dynamic and appealing.

Broader audience engagement has been another significant outcome. The diverse composition of the teams resonates with a wider audience, fostering greater community engagement and support for the air shows. This inclusivity helps to build a more connected and supportive fan base, enhancing the overall success and reach of the air shows.

The ACC demonstration teams have set a benchmark for diversity and inclusivity in the air show industry. Their efforts have enriched the performances and established a global standard for other military air show teams, demonstrating the potential for positive change when organizations commit to fostering a diverse and inclusive environment. Despite all these efforts, more work must be done to ensure equity based on the current composition of Thunderbirds flight demonstration team pilots.

REFERENCES

Adjekum, D., Keller, J., Walala, M., Young, J., Christensen, C., & DeMik, R. (2015). Cross-sectional assessment of safety culture perceptions and safety behavior in collegiate aviation programs in the United States. *International Journal of Aviation, Aeronautics, and Aerospace*. https://doi.org/10.15394/ijaaa.2015.1074

Aerotech News. (2023). Meet the 2023 ACC demonstration team pilots. Retrieved July 21, 2024, from www.aerotechnews.com/

Air Combat Command (ACC). (2023). ACC announces 2023 demonstration team pilots. Retrieved July 21, 2024, from www.acc.af.mil/

Air Force Heritage Flight. (2024). Melanie "MACH" Kluesner becomes the F-35 Demo Team's new pilot and commander. Retrieved July 21, 2024, from www.airforceheritageflight.org/

Barker, D. (2003). *Zero error margin: Airshow display flying analysed*. Freeworld Publications.

Chionis, D., & Karanikas, N. (2018). Differences in risk perception factors and behaviours amongst and within professionals and trainees in the aviation engineering domain. *Aerospace, 5*, Article 2. https://doi.org/10.3390/aerospace5020062

Cleveland National Air Show. (2022). 1976 – Cleveland national air show. Retrieved July 21, 2024, from www.clevelandairshow.com/

Crundall, D., van Loon, E., Stedmon, A. W., & Crundall, E. (2013). Motorcycling experience and hazard perception. *Accident Analysis & Prevention, 50*, 456–464. https://doi.org/10.1016/j.aap.2012.05.021

European Airshows. (n.d.). *Patrouille de France*. Retrieved April 20, 2025, from www.europeanairshows.co.uk/performers/patrouille-de-france

Eurweb.com. (2020). Remoshay Nelson: Howard University graduate becomes first black, female Air Force Thunderbird. https://eurweb.com/2020/howard-graduate-makes-history-with-air-force-thunderbirds/

Ferraro, R., VanDyke, D., Zander, M., Anderson, K., & Kuehlen, B. (2015). Risk perception in aviation students: Weather matters. *International Journal of Aviation, Aeronautics, and Aerospace, 2*(1). https://doi.org/10.15394/ijaaa.2015.1044

Gibson, R. J., Michayluk, D., & Van de Venter, G. (2013). Financial risk tolerance: An analysis of unexplored factors. *Financial Services Review, 22*, 23–50. https://opus.lib.uts.edu.au/handle/10453/23532

Hallahan, T. A., Faff, R. W., & McKenzie, M. D. (2004). An empirical investigation of personal financial risk tolerance. *Financial Services Review, 13*, 57–78.

Joseph, C., & Reddy, S. (2013). Risk perception and safety attitudes in Indian army aviators. *The International Journal of Aviation Psychology, 23*(1), 49–62. https://doi.org/10.1080/10508414.2013.746531

Karachalios, M. (2022). *An evaluation of the relationships between resilient safety culture, safety risk parameters, and mindfulness in the international air show community* [PhD dissertation, University of North Dakota]. https://commons.und.edu/theses/4267/

National Air and Space Museum (NASM). (2024). NASM, 1909: The first major international flying meet. https://airandspace.si.edu/stories/editorial/1909-first-major-international-flying-meet

Royal International Air Tattoo. (2024). About RIAT. www.airtattoo.com/

Sun 'n Fun. (n.d.). *About Sun 'n Fun*. Retrieved April 20, 2025, from https://flysnf.org/

USAF. (2024). Thunderbirds. Meet our team. www.airforce.com/thunderbirds/our-team

Winter, S. R., Truong, D., & Keebler, J. R. (2019). The flight risk perception scale (FRPS): A modified risk perception scale for measuring risk of pilots in aviation. *International Journal of Aviation Research, 11*(1), 57–72.

You, X., Ji, M., & Han, H. (2013). The effects of risk perception and flight experience on airline pilots' locus of control with regard to safety operation behaviors. *Accident Analysis & Prevention, 57*, 131–139. https://doi.org/10.1016/j.aap.2013.03.036

FURTHER READINGS

Fighter Jets World. (2020). F-35 Demo pilot explains how to perform inverted-to-inverted pass maneuver. https://fighterjetsworld.com/air/fighter-jets-videos/f-35-demo-pilot-explains-how-to-perform-inverted-to-inverted-pass-maneuver/22501/

Hirschman, D. (2019). Join the tumbling mirth. *AOPA*. www.aopa.org/news-and-media/all-news/2019/january/pilot/join-the-tumbling-mirth

Magazine, S., & Gary, D. (2009). Tumbling with the stars. *Smithsonian Magazine*. www.smithsonianmag.com/air-space-magazine/tumbling-with-the-stars-136694285/

3 Demography, Risk Factors, and Resilient Safety
An Empirical Analysis

Manolis Karachalios

Contributor
Daniel Kwasi Adjekum

INTRODUCTION

This Chapter addresses the crucial need to understand how demographic factors influence safety and risk management within the air show community. Given the high-stakes and dynamic nature of air shows, thoroughly examining the relationships between pilot demographics, the inherent risks of aerial performances, and the elements contributing to a resilient safety culture (RSC) is essential. This empirical analysis aims to fill the knowledge gap and provide a comprehensive, evidence-based understanding of these dynamics.

The diversity within the air show community—including variations in age, experience, educational background, and marital status—significantly impacts individual approaches to risk and safety. Therefore, in Chapter 3, these demographic variables are explored, offering insights into their influence on behavior and decision making in high-pressure environments.

Understanding how age affects performance and risk tolerance (RT) is essential for developing tailored safety protocols and training programs. Research indicates that age-related factors can positively and negatively impact risk perception (RP) and management. Younger pilots may demonstrate higher risk-taking behaviors, while older pilots might leverage their extensive experience to mitigate risks more effectively (EASA, 2022; Reader, 2016).

Experience in civilian or military aviation is another critical factor influencing safety practices, and this Chapter compares the safety behaviors of pilots with different types and levels of experience, highlighting best practices and areas for improvement. Pilots with military backgrounds might have different risk management strategies than those with civilian experience (Pamplona & Jorge, 2020; SKYbrary, 2021; Sicard et al., 2003; Wang & Zhang, 2020).

DOI: 10.1201/9781003504801-5

Educational attainment may be correlated with enhanced risk management and strategic thinking abilities. More specifically, higher education levels can lead to better comprehension of complex safety protocols and more effective implementation of safety measures. With this in mind, this Chapter analyzes how varying levels of education impact pilots' approaches to safety and decision making processes (Chan & Li, 2023).

Though seemingly personal, marital status can influence an aviator's focus and RP. The responsibilities and commitments associated with marriage might affect how pilots perceive and manage risks, and in this Chapter these influences will be probed to provide a more holistic understanding of safety behaviors (EASA, 2022).

The analysis in Chapter 3 is grounded in a wealth of data sources, including personal anecdotes, industry studies, and empirical research.

For instance, studies have shown that cultural and social factors significantly impact safety culture in aviation. Effective communication and teamwork, essential in high-risk aviation environments like air shows, can be influenced by sociocultural dynamics among crew members (Cookson, 2019; Liao, 2015; Matyukhina et al., 2021; Mjøs, 2004).

Furthermore, safety culture is not static; it evolves with continuous assessment and improvement. Implementing safety management systems (SMSs) incorporating demographic insights can lead to more effective safety protocols and training programs (Foster & Adjekum, 2022; Kaspers et al., 2017). Consequently, evidence-based recommendations for enhancing safety culture by integrating demographic considerations into safety management practices will be provided in this Chapter (Foster & Adjekum, 2023; Chan & Li, 2023).

DEMOGRAPHY

The demographics of air show performers, which have been analyzed and discussed in detail in Chapter 2, provide a foundational context for understanding their influence on safety and risk management. The air show community is characterized by significant diversity, encompassing variations in age, experience, educational background, marital status, and specific air show flying experience. These demographic factors play a crucial role in shaping individual approaches to risk and safety. The relevance of these factors was highlighted during the study conducted by the author in 2022 (Karachalios 2022).

AGE

Age is a critical factor that influences performance and RT. Research indicates that age-related factors can positively and negatively impact risk perception and management. Younger pilots may exhibit higher risk-taking behaviors due to factors such as greater physical agility and a propensity for thrill-seeking. Conversely, older pilots might leverage their extensive experience and accumulated wisdom to mitigate risks more effectively.

Wang et al. (2023), in a study on the effects of age on pilot performance, found that flight parameter exceedance rates increased rapidly among pilots aged 21–40, regardless of their flight exposure levels. The authors also found that flight exposure mediated the association between age and exceedance rates among pilots aged 21–35; age moderated the association between flight exposure and exceedance rates among pilots aged 26–35. Age directly affected exceedance rates among pilots aged 41–45 and 56–60.

From a cognitive-performance perspective, Causse et al. (2019) found that the performance of older relative to younger pilots was impaired in two executive tasks from the standardized CANTAB battery, namely Spatial Working Memory (SWM) and One Touch Stockings of Cambridge (OTS) tasks among private pilots, with the most significant impairment observed for the highest-load SWM task.

Luciani et al. (2022), in a sociodemographic and work-related information study that assessed emotional dysregulation, coping styles, and risk attitudes, found that older age and emotional dysregulation were associated with higher risk attitudes in pilots. Understanding these age-related influences is essential for developing tailored safety protocols and training programs (EASA, 2022; Reader, 2016).

Experience

Experience, whether gained in civilian or military aviation, significantly affects safety practices. Li et al. (2003) found that flight experience, as measured by total flight time at baseline, showed a significant protective effect against the risk of air crash involvement. Pilots with military backgrounds often bring different risk management strategies compared to those with purely civilian experience (Kaminska et al., 2021). Military training typically emphasizes discipline, adherence to protocols, and rigorous risk assessment, which can enhance safety practices in high-pressure situations. This section compares the safety behaviors of pilots with different types and levels of experience, highlighting best practices and areas for improvement.

Air Show Flying Experience

Specific air show flying experience is a unique factor that influences how performers manage risks and ensure safety. Pilots who have experience in air show flying often have a deeper understanding of the specific challenges and risks associated with aerial performances. This experience allows them to develop specialized skills and strategies to effectively manage the dynamic environment of air shows (Karachalios & Adjekum, 2023).

Educational Background

Educational attainment is often correlated with enhanced risk management and strategic thinking abilities. Higher education levels can lead to a better comprehension of complex safety protocols and more effective implementation of safety measures. Pilots with advanced degrees or specialized training are typically better equipped to understand and manage the multifaceted risks associated with air show performances (Chan & Li, 2023).

Marital Status

Although a personal detail, marital status can influence an aviator's focus and RP. The responsibilities and commitments associated with marriage might affect how pilots perceive and manage risks. For instance, married pilots might exhibit more cautious behavior due to their responsibilities toward their families. Some studies on people who are engaged in such high-risk activities support this assertion.

Riviere et al. (2011) found that single soldiers had higher risk-taking behavior than married soldiers. Research by Roussanov and Savor (2014) among corporate investment managers suggested that single managers had relatively higher RT in making business decisions than married managers. This Chapter investigates these influences to provide a more holistic understanding of safety behaviors (EASA, 2022).

By synthesizing data from various sources, including personal anecdotes, industry studies, and empirical research, this Chapter offers a robust and nuanced understanding of how demographic factors intersect with risk and safety in air shows. The insights gained from this analysis are crucial for developing effective SMSs and training programs that incorporate demographic considerations to enhance overall safety and performance in the air show industry.

RISK FACTORS

Understanding the various risk factors associated with air show performances is crucial for enhancing safety and developing effective risk management strategies. This section investigates the components of risk assessment, RP, and RT, providing a comprehensive analysis of how these factors influence air show performers' behavior and decision making processes.

Risk Assessment

Risk assessment and management are fundamental to a pilot's decision making process (Martinussen & Hunter, 2018). Inadequate risk assessment can lead to poor decision making, potentially resulting in fatal aviation accidents (Air Accidents Investigation Branch [AAIB], 2017; Brugnara et al., 2022; Jensen & Benel, 1977; Kelly & Efthymiou, 2021). O'Hare (1990) found that pilots' unreasonable risk estimations could contribute to decisions to continue flying into deteriorating weather conditions.

Risk assessment is a systematic process to identify, evaluate, and prioritize potential hazards that could negatively impact safety during air show performances (ICAO, 2018). Risk assessment is continuous because the air show performance activities are constantly changing due to many factors, including updated technology, newer aircraft designs, changes in processes and procedures, human performance limitations, regulatory oversight factors, and updated performance standards.

New hazards can be introduced, and some hazards and associated safety risks may change over time. In addition, the effectiveness of implemented risk mitigation strategies must be monitored to determine if further action is required. This process involves several key steps, including hazard identification, risk analysis, and risk evaluation:

1. Hazard Identification: Recognizing potential dangers that could lead to accidents or incidents, such as mechanical failures, adverse weather conditions, pilot error, and unexpected obstacles in the performance area.
2. Risk Analysis: The likelihood and potential consequences of identified hazards are assessed by evaluating the probability of occurrence and the severity of potential outcomes. Traditional safety risk analysis tools such as Event Tree Analysis (ETA), Fault Tree Analysis (FTA), Failure Mode and Effects Analysis (FMEA)/Failure Mode Effects and Criticality Analysis (FMECA), and Risk Matrices are commonly used (Anes et al., 2022; Banghart et al., 2018; Carlson, 2012; Signoret & Leroy, 2021).

   Newer approaches, such as the Functional Resonance Analysis Method (FRAM) and System Theoretic Process Analysis (STPA), are also used to describe the dynamic and non-linear risk behaviors in sociotechnical systems (Abraha & Liyanage, 2014; Leveson, 2012; Zhang et al., 2022).
3. Risk Evaluation: Comparing analyzed risks against established criteria to determine their acceptability. This evaluation helps prioritize risks and decide on appropriate mitigation measures. Effective risk assessment is fundamental to developing safety protocols and emergency response plans.

The SHOWCENTER model can be applied by air show performers in a dynamic risk assessment, as discussed in the first book of this series (Karachalios & Adjekum, 2023).

Risk Perception

Sjoberg et al. (2004) define RP as a person's subjective evaluation of the likelihood of a specific type of accident occurring and their concern about its potential consequences. Extant research has been conducted on various aspects of RP, such as its impact on pilot decision making (Orasanu et al., 2002). Hunter (2002) elaborates on this by describing RP as an individual's awareness of the risks inherent in a situation and that it can be influenced by both the situation's nature and the individual pilot's characteristics.

Martinussen and Hunter (2018) consider RP a cognitive process that accurately assesses internal and external conditions. Siegrist and Árvai (2020) identify three dormant perspectives under RPs, namely the characteristics of hazards, the characteristics of risk perceivers, and the application of heuristics to inform risk judgments. As part of a review, Ebrahim et al. (2022) suggest that understanding how pilots perceive risk allows for an improved understanding of pilots' propensity to engage in risk and the design of training initiatives to mitigate risky behavior.

Research has consistently shown a positive relationship between experience and RP. For instance, studies indicate that younger drivers tend to have more distorted views of driving hazards than older drivers (Rhodes & Pivik, 2011; Tränkle et al., 1990; White et al., 2011). Hunter's extensive work on RP and tolerance (2002, 2006) among general aviation pilots includes developing scales to measure how these factors correlate with pilots' involvement in hazardous events. His findings suggest that RP is a more crucial predictor of such events than RT. Similarly, research conducted in Australia by Drinkwater and Molesworth (2010) identified a significant negative correlation between RP and acceptance among general aviation pilots who opted to fly in risky scenarios.

In a study of the effects of RP and flight experience on Chinese airline pilots' locus of control with regard to safety operation behaviors, You et al. (2013a) suggested that RP seemed to mediate the relationship between locus of control and safety operation behaviors and that higher RP and an internal locus of control increased pilots' likelihood of engaging in safety-oriented behaviors, such as enhanced situational awareness and more effective decision making.

In a study of Indian army helicopter pilots, Joseph and Reddy (2013) found that lower levels of RP were linked to higher tendencies for risk-taking and higher risk attitude scores. Chionis et al. (2022) analyzed a representative sample of 140 aviation safety investigation reports to identify the degree to which RP and communication and their factors have been influential overall and across various flight operation stages of investigated events and found that environmental factors affected RPs and communication frequently.

The importance of RP extends beyond aviation to other industries. Şimşekoğlu et al. (2013) found that poor RP and high RT are associated with car accidents. High-Reliability Organizations (HROs) in sectors like petrochemicals (Kao et al., 2008) and offshore petroleum (Cairns et al., 2008) also recognize the critical role of RP in maintaining safe operations.

In the context of air shows, pilots' RPs are influenced by various factors, including personal experience, training, cultural background, and psychological traits.

- Personal Experience: As You et al. (2013b) suggested, pilots with extensive experience, such as in air show flying may perceive risks differently than less experienced pilots. Experienced pilots might be more adept at recognizing and mitigating hazards due to their familiarity with the specific challenges of air show performances.
- Training and Education: Chen and Chen (2014) and Adjekum et al. (2015) suggest that comprehensive training programs and higher educational attainment can enhance pilots' ability to perceive risks accurately. Well-trained pilots are likely to have a better understanding of potential hazards and effective risk management strategies.

- Cultural Influences: Vredenburgh and Cohen (1995) suggest that cultural factors influence RPs. Renn and Rohrmann (2000) suggest that cultural factors influence RP directly and indirectly by assigning significance to individual risk characteristics. Cultural background can significantly impact how risks are perceived. For example, pilots from cultures with a high priority for safety in flight operations and higher risk aversion may be more cautious in their approach to air show performances than those from cultures with higher RT.
- Psychological Traits: Individual psychological factors, such as risk-taking propensity, stress tolerance, and confidence levels, also play a critical role in shaping RP (Breakwell, 2007; Ebrahim et al., 2021; Saivasan & Lokhande, 2022). Pilots with high-stress tolerance and confidence may perceive risks as more manageable.

Understanding RP is essential for developing tailored training programs and safety interventions that address air show performers' specific needs and characteristics.

Risk Tolerance

Hunter (2002) pointed out that while RP and RT are related, they are distinct concepts. RT refers to the level of risk an individual is prepared to accept to achieve a particular objective (Hunter, 2002). It is influenced by a person's overall aversion to risk and the importance they place on a specific goal (Martinussen & Hunter, 2018). In aviation, the significance of specific goals can determine the extent of risk a pilot is willing to take.

The concept of RT is not limited to aviation but is also relevant in the finance sector. Research on financial RT demonstrates the balance between the risks individuals are willing to take and the rewards they seek. Studies by Gibson et al. (2013) and Hallahan et al. (2004) indicate that RT tends to decrease with age. Additionally, Hallahan et al. found that marital status, number of dependents, income, and overall wealth are significant factors influencing RT levels.

Numerous studies have identified a strong link between RT and behaviors that can lead to hazards, such as driving violations and workplace injuries (Arnett et al., 1997; Christian et al., 2009; Paul & Maiti, 2007). However, Hunter (2002) observed no significant correlation between RT and hazardous events in aviation.

In the realm of air shows, the variability of performers' RT is a significant factor. This variability is influenced by their individual attitudes, the organizational culture they are part of, and the regulatory frameworks governing their activities. This diversity and complexity of RT among air show performers underscores the need for a nuanced understanding of this concept in the aviation safety domain.

- Personal Attitudes: Individual attitudes toward risk can range from highly risk-averse to highly risk-tolerant. Pilots' attitudes are shaped by personal values, past experiences, and individual goals (Molesworth & Chang, 2009; Nuñez et al., 2019; Scharf & Cross, 2019). Ji et al. (2011) found that RT indirectly affects safety operation behavior among Chinese airline pilots by influencing hazardous attitudes (HA). Pilots with a higher tolerance for risk may be more willing to engage in complex and daring maneuvers, while those with lower tolerance may prefer more conservative performances.

- Organizational Culture: The culture of the organization or team that pilots belong to can also influence RT. Organizations with a strong safety culture emphasize the importance of risk management and encourage behaviors that prioritize safety over performance (Adjekum & Fernandez-Tous, 2020; Pidgeon & O'Leary, 2017). Bhandari and Hallowell (2022) suggest that a safety climate can reduce RT and promote more risk-averse decision making at work across diverse cultures. They further suggest that multi-national organizations can foster a consistent safety culture where operational rules are followed and risks are avoided. Conversely, organizations prioritizing competitive success and crowd-pleasing performances may inadvertently foster higher RT among their pilots.
- Regulatory Frameworks: Regulatory guidelines and industry standards are crucial in setting the boundaries for acceptable risk levels (ICAO, 2018; Civil Aviation Safety Authority [CASA], 2025; Insua et al., 2016; United Kingdom Civil Aviation Authority [UK CAA], 2025). Compliance with these regulations ensures that pilots operate within safe limits, thereby reducing the likelihood of accidents and incidents.

Organizations can create an environment that supports safe and effective air show performances by understanding and managing RT. This involves fostering a safety culture, providing ongoing training and education, and ensuring adherence to regulatory standards.

Hazardous Attitudes

HA have been extensively researched as a potential factor influencing decision making and increasing the likelihood of accidents (Martinussen & Hunter, 2018). According to the FAA, five key attitudes that can lead to hazardous events are macho, antiauthority, invulnerability, impulsivity, and resignation (Federal Aviation Administration [FAA], 2023). It is essential to promptly identify, manage, and correct these HA to reduce flight risk.

The FAA offers specific antidotes for each hazardous attitude among pilots (FAA, 2023, pp. 2–5). Table 3.1 illustrates these HA and their corresponding antidotes.

Several factors can influence a pilot's safety behavior, impacting the safety performance of an aviation organization. A sense of calling, which refers to employees' intrinsic belief that their work is highly significant, can positively affect pilot safety behavior (Xu et al., 2022). Personality traits such as agreeableness, neuroticism, extraversion, and openness to experience have been consistently linked to safety behavior (Doerr, 2020).

TABLE 3.1
Hazardous Attitudes and Corresponding Antidotes

Hazardous Attitude	Antidotes
Macho	Taking chances is foolish.
Antiauthority	Follow the rules. They are usually right.
Invulnerability	It could happen to me.
Impulsivity	Not so fast. Think first.
Resignation	I am not helpless. I can make a difference.

Source: FAA (2016, pp. 2–5).

RT and RP also play roles in shaping pilot safety behavior; improved RP can reduce the negative impact of RT on safety practices (Ji et al., 2011). Additionally, an internal locus of control, the belief in one's influence over outcomes, directly affects a pilot's safety behavior (You et al., 2013a). Mindfulness (MF) has been proposed as a way to enhance safety behavior, with flight experience further strengthening this effect (Ji et al., 2011).

However, personality theories have limitations, especially when considering cultural differences. Western European and North American cultures dominate commercial aviation, which affects the applicability of these theories. Cross-cultural variations, such as those between Chinese personality dimensions and Western scientific literature, should be anticipated (Hofstede, 1984, 2001). Concepts like power distance may not adequately explain Chinese authoritarianism.

Unsafe attitudes can lead to unsafe behavior, impairing display pilot performance. Barker (2020a) defined a rogue display pilot as "an unprincipled pilot living apart from the display community and having destructive tendencies." These pilots prioritize their ego, pushing boundaries with aggression and arrogance, often linked to the Dunning-Kruger effect (Dunning, 2011). Moreover, HA may explain the differences in attitudes among air show performers and highlight endogenous personality traits, such as egotism (Karachalios, 2022). However, such attitudes endanger not only the pilots themselves but also other pilots and spectators. The B-52 accident at Fairchild Air Base in 1994 is a notable example of a rogue aviator incident (Barker, 2003, 2020a; Diehl, 2002; Kern, 1995; Thompson, 1995; USAF, 1994).

To mitigate such egocentric attitudes, effective leadership by air bosses is crucial during air show planning and execution (Barker, 2020a; Chen & Chen, 2014; Schopf et al., 2021). Roger Beazley, former flying display director of the Farnborough International Airshow, advocated for aircrew safety briefings to address egocentric attitudes and ensure open, private discussions on previous issues (Barker, 2020a, p. 50).

Display pilots may also act unsafely due to the latent factor of social facilitation bias without prior indications of HA (Karachalios, 2022; Papadakis, 2008). Jarvis (2009) noted that the pressure from spectators can be so intense that it prevents a display pilot from recognizing a hazardous situation as it develops. Barker (2020a) attributed the higher accident rates during actual air shows compared to practice sessions to this spectator-induced pressure, among other factors.

RESILIENCE IN AVIATION

In the dynamic world of aviation, where unpredictability is a constant, resilience is the cornerstone for safeguarding the aircraft and its pilot. Hollnagel et al. (2011) define resilience as "the intrinsic ability to adapt to unexpected conditions," highlighting it as the aviation industry's proactive defense against unforeseen challenges. This concept goes beyond mere recovery; as Kouzes and Posner (2017) explain, it involves "bouncing forward" and being ready to face future challenges with renewed strength.

Resilience in aviation is about thriving amid disruption and mastering the art of adaptation when faced with adverse conditions. For pilots, especially those in the high-stakes environment of air shows, it involves anticipating potential disturbances and having the capability to withstand and emerge stronger from them. This quality is crucial for maintaining safety and excellence in a demanding field. Recognizing its importance, the aviation sector has integrated resilience training for airport and air carrier employees to help them cope with traumatic events (National Academies of Sciences, Engineering, and Medicine, 2009).

Significance of Resilience in Air Show Performance

Resilience is crucial in the aviation industry, particularly during large or small crises. An individual's capacity for resilience shapes their response to challenges, providing the emotional stability and clarity of thought needed to act confidently under pressure. In a study of resilience among Chinese airline pilots, Zhao et al. (2024) identified two factors that underlie resilience. *Decisiveness* indicates a pilot's high confidence in their abilities and problem-solving skills and a more resolute and decisive approach when dealing with challenges.

The other factor was *adaptability*, which centers on a positive attitude toward coping with uncertainty and change, demonstrating the pilot's emphasis on coping with change, maintaining close relationships, and adopting an optimistic and persevering attitude when faced with uncertainty. Air show performers need to demonstrate decisiveness and adaptability in an environment where safety events can mar a routine, well-laid-out performance, and there will be a need for temporary holds and re-engagement.

Developing, Sustaining, and Enhancing Resilience

Individual Resilience

Individual resilience is characterized by positivity and a forward-looking mindset. According to Fredrickson's (2001) broaden-and-build theory, positive emotions can enhance cognitive, social, and psychological resources. Conner's (1993) model of resilience highlights essential traits such as positivity, focus, flexibility, organization, and proactiveness—critical qualities for pilots.

Mowbray (2011) further explores personal resilience, proposing a seven-element model to strengthen it. The American Psychological Association (2002) suggests practical ways to build resilience, including developing strong, supportive relationships and maintaining a positive perspective on life's challenges.

Martin (2019) suggests that training to enhance non-technical competencies or Crew Resource Management (CRM) skills, such as situational awareness, decision making, leadership, teamwork, workload management, stress management, and fatigue management, can build individual resilience among air show performers during critical events. Zhao et al. (2024) recommend early identification of individuals with lower resilience and enabling the evaluation of intervention efficacy, which can positively impact pilot psychological health and aviation safety in the context of grief and trauma following distressing events.

Organizational Resilience

For organizations, resilience must be embedded in their culture. A resilient organization anticipates, copes, and adapts—essential capabilities for HROs such as aviation. Heese et al. (2014) suggested four dimensions underlying behavior toward organizational resilience: goal-directed/proactive solutions, flexibility, improvisation, and availability of resources.

Serfontein and Govender (2021) suggest that for aviation organizations such as the air show community, a harmonized application of organizational control systems, such as safety assurance and effective organizational governance, stimulates organizational resilience in a socio-technical system through the autonomous advancement of organizational resilience. Furthermore, the robustness of organizational control systems activates an organization's capacity to adapt sustainably whilst maintaining stakeholder value within complex socio-technical systems.

Creating an organizational resilience that is aligned with the safety culture involves a sophisticated blend of safety practices, learning opportunities, continuous improvement, and cost-effectiveness (Shirali et al., 2016a, b). This multifaceted approach integrates people, technology, and administration to enhance the organization's ability to adapt and evolve sustainably. Bell (2020) argues that combining such holistic and traditional strategies can enable organizations to thrive in volatile environments.

Safety Culture and Resilient Safety: A Symbiotic Relationship

Safety culture and resilient safety are fundamentally interconnected and essential to maintaining the integrity of aviation operations (Adjekum & Fernadez-Tous, 2020). Safety culture embodies a collective commitment to prioritizing safety above other competing goals and involves shared values, beliefs, and practices that emphasize the importance of safety in every aspect of operations. (Cooper, 2000; Key et al., 2023; Schwarz et al., 2016). On the other hand, resilient safety introduces a dynamic element, emphasizing adaptability and continuous learning.

To cultivate an RSC, organizations must develop the capability to respond to, monitor, anticipate, and learn from every situation (Hollnagel, 2010). This involves creating an environment where safety is ingrained in the organizational ethos, enabling the team to manage unexpected events effectively and maintain safe operations under pressure.

An RSC is not static; it continuously evolves through learning and adaptation (Akselsson et al., 2009). It integrates robust safety practices, provides opportunities for learning and improvement, and seeks cost-effective solutions to enhance overall safety performance (Shirali et al., 2016a, b). This multifaceted approach combines the efforts of individuals, technology, and management to create an organization that can adapt and thrive in the face of challenges.

In the context of the air show community, resilient safety is paramount. It ensures that pilots and organizations survive challenges and emerge stronger. This Chapter aims to chart a path for air show performers and the industry at large, promoting a future where safety and performance are seamlessly integrated, and the industry is prepared to meet today's and tomorrow's demands. A detailed analysis of the resilient safety of air show performers is covered in Part III of the current publication.

CORRELATION HEATMAP FOR RISK FACTORS

This section explores the relationships between RSC, safety risk parameters, and MF within the international air show community, see Table 3.2, as derived from the study by the author (Karachalios, 2022).

The research question that guides the analysis is "*What are the strengths of relationships between risk perception, risk tolerance, hazardous attitudes, and resilient safety culture when mediated by individual mindfulness among members of the international air show community?*" which aims to explore the intricate connections between various risk and psychological factors that influence safety in the high-pressure context of air shows. RP and RT are crucial for understanding how pilots evaluate and react to potential dangers. Previous studies have demonstrated a negative association between RP and risk-taking behaviors (Drinkwater & Molesworth, 2010; Joseph & Reddy, 2013; You et al., 2013a), but the specific dynamics within the air show environment remain underexplored. HA, such as macho behavior and invulnerability, significantly impact decision making and risk management (Martinussen & Hunter, 2018), and identifying these attitudes is essential for enhancing safety. Additionally, fostering an RSC is vital for minimizing accidents and ensuring pilots can effectively respond to unforeseen circumstances. This question seeks to understand how these factors interact and contribute to an RSC. Individual MF, which has been shown to enhance decision making and reduce anxiety (Gautam & Mathur, 2018; Li et al., 2020), is examined as a mediating variable to understand its influence on the relationship between the other variables.

A key component of this analysis is the correlation heatmap (see Figure 3.1), which visually represents the associations among the research variables. This heatmap was created by triangulating quantitative and qualitative data, resulting in a mixed data set that offers robust and meaningful insights. The heatmap displays 30 correlations

TABLE 3.2
Research Variables

Research Variable	Abbreviation	Definition
Risk Perception	RP	A person's subjective evaluation of the likelihood of a specific type of accident occurring and their concern about its potential consequences (Sjoberg et al., 2004).
Risk Tolerance	RT	The level of risk an individual is prepared to accept to achieve a particular objective (Hunter, 2002).
Hazardous Attitude	HA	Attitudes that can lead to hazardous events, such as macho, antiauthority, invulnerability, impulsivity, and resignation (FAA, 2016).
Mindfulness	MF	Paying attention in a particular way: On purpose, in the present moment, and nonjudgmentally (Kabat-Zinn, 1994, p. 4).
Resilient Safety Culture	RSC	A safety culture that focuses on resilience, learning, continuous enhancements, and cost-effectiveness (Shirali et al., 2016a).

Source: Karachalios (2022).

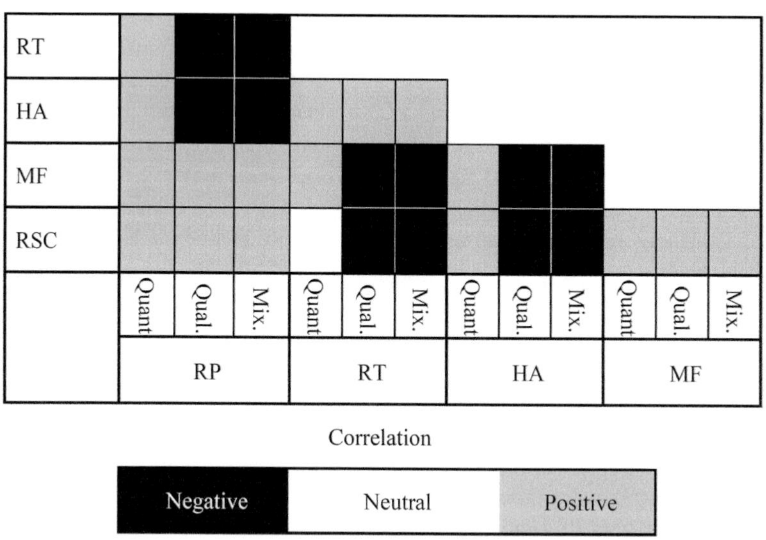

FIGURE 3.1 Correlation heatmap: Joint display of findings for research variables with total triangulated association.

between the research variables in a matrix of color-coded tiles, where each tile indicates a correlation derived from quantitative (Quant), qualitative (Qual), and mixed-methods (Mix) data (Heyvaert et al., 2011; Younas et al., 2021).

Each research variable was analyzed through the lens of both quantitative and qualitative data. This comprehensive approach ensures a thorough understanding of how these variables interact and influence each other within the context of RSC in air shows.

The provided heatmap is a correlation matrix displaying the relationships between the various risk factors and RSC within the context of air show performers.

A breakdown of what the colors represent and a general analysis of each correlation is as follows:

- Black tiles represent a negative correlation between the variables. This suggests that as one variable increases, the other tends to decrease.
- Grey tiles represent a positive correlation, indicating that as one variable increases, the other also tends to increase.
- White tiles signify a neutral relationship, meaning there is no significant correlation between the variables.

Summary of the Findings Leading to the Correlation Heatmap for Risk Factors

The research involved several steps in validating the constructs, discriminating validity, and assessing the reliability of the scale items. The subsequent phase focused on evaluating the goodness-of-fit for all measurement models using Structural Equation Modeling (SEM). Key fit indices reported included χ^2, NFI, IFI, and CFI. Due to the model's complexity, RMSEA fit indices were above 0.10 and thus not used for determining an acceptable fit. Additionally, χ^2, being sensitive to sample size, was not considered relevant.

An initial assessment was conducted on the hypothesized fully mediated measurement model that depicted the relationships among the variables. The model produced satisfactory goodness-of-fit indices across all fit indices. Subsequently, a competing partially mediated measurement model was examined based on recommendations from the IBM SPSS Amos 28 Graphics® modification indices function. In this model, the pathway between RP and RSC was removed. Another round of analysis indicated no significant improvement in fit indices compared to the initial model.

A third competing measurement model was derived by removing the pathway between RT and MF from the second model. While this model showed a better fit in terms of the χ^2 index, other indices were lower compared to the second model.

Ultimately, Model I (Figure 3.2) was selected for hypothesis testing due to its higher NFI, IFI, and CFI indices compared to the other competing models. This decision was based on theoretical considerations and hypothesis testing requirements.

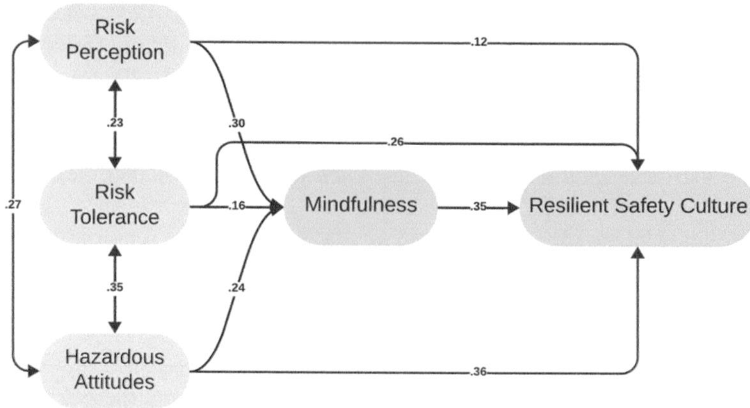

FIGURE 3.2 Model I – Fully mediated measurement model showing paths for all variables.

Mediation Effect of Mindfulness (MF)

The mediation effect of MF on the relationships between RP, RT, HA, and RSC was examined using a bootstrapping method. The results of the regression analyses indicated the following:

1. RP to MF and RSC:
   - RP significantly predicted MF ($b = .40$, $t = 5.27$, $p < .001$).
   - Controlling for MF, RP also significantly predicted RSC ($b = .10$, $t = 2.25$, $p < .05$).
   - A significant indirect positive relationship between RP and RSC was mediated by MF ($a*b = .22$, Bootstrap $CI_{95} = .12$ and .31), accounting for approximately 50% of the total effect.
2. RT to MF and RSC:
   - RT significantly predicted MF ($b = .47$, $t = 3.90$, $p < .001$).
   - A significant indirect positive relationship between RT and RSC was mediated by MF ($a*b = .42$, Bootstrap $CI_{95} = .02$ and .28), accounting for approximately 31% of the total effect.
3. HA to MF and RSC:
   - HA significantly predicted MF ($b = .48$, $t = 4.87$, $p < .001$).
   - Controlling for MF, HA also significantly predicted RSC ($b = .30$, $t = 5.32$, $p < .001$).
   - A significant indirect positive relationship between HA and RSC was mediated by MF ($a*b = .42$, Bootstrap $CI_{95} = .03$ and .25), accounting for approximately 29% of the total effect.

Summary of Hypothesis Testing

The purpose of the study was to examine the effect of risk factors, HA, and MF on the RSC within the international air show community. The SEM analysis tested 12 hypotheses, as shown in Table 3.3.

TABLE 3.3
Summary of Hypothesis Testing

#	Relationship	Statistic	*p*-value	Significance
1	RP and RT	$r = .23$	$< .01$	Statistically significant
2	RP and MF	$\beta = .30$	$< .001$	Statistically significant
3	RP and RSC	$\beta = .11$	$< .05$	Statistically significant
4	RP, MF, and RSC	$\beta = .11$	$< .05$	MF significantly mediates
5	RT and MF	$\beta = .16$	$< .05$	Statistically significant
6	RT, MF, and RSC	$\beta = .42$	$< .001$	MF significantly mediates
7	HA and MF	$\beta = .24$	$< .05$	Statistically significant
8	HA and RSC	$\beta = .30$	$< .001$	Statistically significant
9	HA, MF, and RSC	$\beta = .12$	$< .001$	MF significantly mediates
10	HA and RT	$r = .35$	$< .001$	Statistically significant
11	HA and RP	$\beta = .27$	$< .01$	Statistically significant
12	MF and RSC	$\beta = .39$	$< .001$	Statistically significant

Interpretation of Heatmap Analysis for Air Show Performers

In the context of air show performers, the heatmap analysis provides crucial insights that can inform pilot selection, training, and performance enhancement strategies. Below is an interpretation with examples and relevant academic references:

Risk Perception (RP) and Risk Tolerance (RT)

A negative correlation between RP and RT suggests that pilots who have a high level of RP, meaning they can accurately understand and evaluate the risks associated with aerobatics, are less likely to engage in overly risky maneuvers. For instance, Wiegmann and Shappell (2003) demonstrated that pilots with heightened RP levels are less prone to committing errors that could lead to accidents. This insight is vital for pilot selection and training programs, emphasizing the need to enhance pilots' risk assessment capabilities.

Risk Perception (RP) and Mindfulness (MF)

The positive correlation between RP and MF indicates that pilots who practice MF are more attuned to the nuances of flight dynamics and environmental factors. As shown in studies by Kabat-Zinn (1994), MF training can significantly enhance pilots' focus and situational awareness, which is crucial for the precision required in air shows. For example, incorporating MF techniques into training can improve pilots' ability to stay present and make more informed decisions during complex maneuvers.

Risk Perception (RP) and Resilient Safety Culture (RSC)

The positive relationship between RP and RSC suggests that pilots are encouraged to be vigilant and aware of potential risks in an RSC. Grote (2012) discusses how a resilient organization fosters an environment where risk assessment is a part of the daily routine. This correlation underscores the importance of cultivating a safety culture that values continuous risk assessment and proactive safety measures.

Risk Tolerance (RT), Hazardous Attitudes (HA), and Mindfulness (MF)

The negative correlation between RT and HA implies that a higher willingness to accept risks is often associated with fewer HA. However, the neutral relationship between RT and MF indicates that a pilot's MF practice might not directly influence their risk acceptance. Turner (1995) supports this by suggesting that pilots' risk decisions are multifaceted and not solely affected by their MF state. This highlights the need for a balanced training approach that addresses risk management and MF without assuming one directly influences the other.

Mindfulness (MF) and Hazardous Attitudes (HA)

The negative correlation between MF and HA reflects that more mindful pilots are less likely to display HA such as impulsivity or invulnerability (FAA, 2023). Hunter (2002) emphasizes the importance of self-awareness in aviation safety, showing that mindful pilots tend to make safer decisions and are better at managing stress and distractions, which can prevent hazardous behaviors.

Mindfulness (MF) and Resilient Safety Culture (RSC)

The positive correlation between MF and RSC suggests that a pilot's MF contributes to organizational resilience. Weick and Sutcliffe (2007) argue that MF in organizations enhances the capacity to deal with unexpected events, a vital aspect of an RSC. Training programs that include MF practices can strengthen the overall safety culture by improving individual and collective resilience.

Resilient Safety Culture (RSC) and Hazardous Attitudes (HA)

The negative correlation between RSC and HA across all data sets aligns with Reason's (1997) findings, which argue that a strong safety culture can mitigate HA and behaviors within an aviation organization. This indicates that fostering an RSC is essential for reducing risky behaviors and attitudes among pilots, leading to safer air show performances.

The heatmap analysis offers a comprehensive understanding of how various factors interact within the context of air show performers. We can develop more effective pilot selection, training, and performance improvement strategies by examining these correlations. Integrating RP, MF, and an RSC plays a crucial role in enhancing overall safety and performance in the high-risk environment of air shows.

Effects on the Selection and Training of Air Show Performers

Continuing the analysis, we examine how these correlations influence the selection and training of air show performers.

Integrating Mindfulness Into Training

The positive correlation between MF and RP suggests that integrating MF practices into training could enhance pilots' ability to anticipate and respond to potential hazards. MF training can be designed based on models like those proposed by Langer (1989), focusing on increasing pilots' attention to the present moment and their receptiveness to context and perspective changes during flight. This approach can lead to better situational awareness and improved decision making skills, which are essential for safe aerobatic performances.

Addressing Hazardous Attitudes

The apparent negative correlation between HA and both MF and RSC indicates that these attitudes can be reduced through cultural and individual interventions. Training programs could benefit from incorporating psychological tools to reshape attitudes, as Helmreich and Merritt (1998) discussed, which focuses on modifying pilots' behavioral norms and attitudes. Such programs might include cognitive-behavioral techniques to address and mitigate impulsivity, invulnerability, and other HA.

Promoting a Resilient Safety Culture

An RSC is positively associated with RP and negatively with HA. This suggests that fostering an RSC should be a strategic objective within aerobatic teams. Fostering air show operations that allow for situational adaptability, institutional learning, continuous improvements, and cost-effectiveness in operations, as suggested by Shirali et al. (2016a, b), is vital. As per Rochlin (1999) and Dekker (2012), creating an environment where pilots feel comfortable reporting errors without fear of reprisal can significantly contribute to safety. Encouraging open communication and continuous feedback can help identify potential risks and address them proactively.

Experience and Risk Tolerance

While not directly correlated in the heatmap, experience plays a crucial role in shaping RT. Experienced pilots tend to have a well-calibrated sense of risk, which can be further refined by sharing experiences and learning from past incidents, as documented in case studies by Dismukes et al. (2007). Mentorship programs where experienced pilots guide newer ones can instill a balanced approach to risk-taking and enhance overall safety (FAA, 2020a).

Developing Risk Management Skills

Pilots should receive training emphasizing active risk management skills, enabling them to make informed decisions under pressure (FAA, 2022). This aligns with the findings of Wiegmann and Shappell (2003), who discuss the Human Factors Analysis and Classification System (HFACS), which could be adapted to analyze and mitigate risks in aerobatic flying. Incorporating HFACS into training can help pilots recognize and address human error factors before incidents occur.

System Theoretic approach to Threat and Error Management (TEM) training as part of a more extensive risk assessment of air show pre-planning can also be useful (Koglbauer, 2016). Other tools, such as FMEA, FMECA, and Scenario-based failure analysis, may also be helpful. From a tactical perspective, using tools such as the Flight Risk Assessment Tool (FRAT) should allow operators and pilots to see the risk profile of an air show flight in its planning stages (FAA, n.d.).

FRAT allows operators to determine an acceptable level of risk for their flights based on the type of operation, environment, aircraft used, crew training, and overall operating experience. When the risk for a flight exceeds the acceptable level, the hazards associated with that risk should be further evaluated and the risk reduced. A higher-risk flight should not be operated if the hazards cannot be mitigated to an acceptable level.

Enhancing Decision Making Abilities

With RP positively correlated with resilience and MF, it is evident that improving decision making abilities is crucial. Programs similar to those used in military pilot training, which teach OODA loop principles (Observe, Orient, Decide, Act), could be beneficial. John Boyd's theories on the combat operations process highlight the importance of rapid and accurate decision making, a skill that can be directly applied to air show performances (Olsen, 2016). Models of a structured framework for problem-solving and decision making such as the 5P, the 3P, PAVE, CARE, TEAM, and DECIDE can be used (FAA, 2022).

As suggested by Orasanu-Engel and Mosier (2019), it will be beneficial to adopt training modules on the theoretical framework of naturalistic decision making or how people with domain expertise use their knowledge to make decisions, typically in time-pressured, dynamic, and safety-critical environments such as air show flying.

Models of aeronautical decision making characterized by situation assessment (SA) and choosing a course of action (CoA) must be well understood, and factors that make decisions difficult and contribute to error should be highlighted. Even though most air show flying tends to be individualized, effective crew decision making behaviors should be part of such training, including taskwork and teamwork built on shared mental models, team processes, and metacognitive skills needed for TEM.

Building on Experience for Hazardous Attitude Mitigation

The negative correlation between HA and experience implies that seasoned pilots could mentor newer pilots, imparting lessons learned. This mentorship model, as detailed in the works of Salas et al. (2006), could mitigate the development of HA among less experienced pilots. Experienced pilots can share their strategies for managing stress and maintaining focus, contributing to a safer flying environment.

Tailoring Safety Cultures

Considering that an RSC is crucial in shaping RP, it is essential for air show teams to tailor their safety culture to the unique challenges of display flying. As Guldenmund (2000) recommended, this could involve creating specific safety protocols and a reporting system that encourages continuous feedback and improvement. Tailored safety programs can address the specific risks associated with aerobatic displays and ensure that all team members are aligned in their approach to safety. Such a scaled approach aimed at building a proactive safety culture can be beneficial to the air show community (Karachalios, 2022).

Balancing Risk Tolerance

Since RT can influence pilots' decision making processes, especially when under the stress of performing intricate maneuvers, it is essential to strike a balance. Similar to elite sports practices to balance risk and performance, regular mental training could be adapted for pilots, as explored by Fletcher and Sarkar (2013). Techniques such as visualization, stress inoculation training, and resilience training can help pilots maintain optimal performance levels while managing risk effectively.

Implementing Comprehensive Training Regimes

Pilots would benefit from a comprehensive training regime that includes technical skills and cognitive-behavioral strategies to manage stress, focus, and decision making under the unique pressure of air show performance. Such approaches are detailed in Staal (2004), discussing stress management for aviators. Training programs should integrate physical, mental, and technical aspects to prepare pilots for the demands of air show flying, ensuring they can perform safely and effectively.

In summary, to enhance safety and performance, air show performers could benefit from an integrated approach that includes skills development, cognitive-behavioral strategies, and a strong safety culture. This integrated approach helps to ensure that as performers push the limits of aviation to entertain and inspire, they do so with the highest regard for their own safety, their team's, and the spectators'.

By synthesizing these insights into training and operational protocols, air show teams can enhance the safety and performance of their pilots. For example, regular workshops on MF and resilience, led by experts in the field, could be mandatory for pilots. Simulated training sessions designed to confront pilots with a variety of emergency scenarios can also help them build resilience and adaptability.

Practical Proposals for Enhancing Safety and Performance for Air Show Performers

To enhance safety and performance during air shows, pilots can implement various practical strategies. These proposals are grounded in research and best practices from aviation and other high-performance fields.

Pre-Flight Mental Rehearsal

Pilots should engage in cognitive practice by visualizing their routines from start to finish, focusing on precision and potential contingencies. This technique has been shown to improve performance under pressure by enhancing mental preparation and reducing anxiety (Cumming & Hall, 2002). Mental rehearsal helps pilots anticipate and plan for unexpected events, increasing their readiness and confidence.

In-Flight Breathing Techniques

Controlled breathing can help maintain calmness and focus. Techniques such as tactical breathing or box breathing, commonly used by military personnel, can manage stress and improve cognitive clarity (Grossman & Christensen, 2004). Implementing these techniques can help pilots stay composed during high-stress maneuvers, ensuring better control and decision making.

Checklists and Cue Cards

In-cockpit checklists and cue cards for complex maneuvers ensure that all steps are executed correctly and safely. Gawande (2009) highlights the effectiveness of checklists in high-stakes environments like surgery. Similarly, pilots can use checklists to ensure thorough pre-flight checks and in-flight procedures, reducing the likelihood of errors and omissions.

Crew Resource Management (CRM)

For multi-crew or formation teams, CRM principles can be practiced in-flight to improve communication, decision making, and team coordination, which are vital for synchronized aerial displays (Salas et al., 2006). Effective CRM involves clear communication, mutual support, and shared situational awareness, which is essential for maintaining formation integrity and safety.

Error Management Training

Training in real-time error detection and management can help pilots quickly rectify mistakes before they escalate. Helmreich and Foushee (1993) emphasize the importance of developing skills to recognize and correct errors promptly. Such training can enhance pilots' ability to maintain safety and performance even when unexpected issues arise.

Safety Calls

Implementing a system where any team member can call for a "knock-it-off" or "abort" if they perceive an unsafe condition is crucial. This practice, similar to those used in test pilot operations (Morley et al., 2021), ensures that safety takes precedence over performance. Empowering team members to halt maneuvers when necessary can prevent accidents and maintain overall team safety.

Fatigue Monitoring

Utilizing technology or self-assessment tools to monitor signs of fatigue pre-flight and during performances allows pilots to take necessary actions to ensure safety (Caldwell et al., 2009). Fatigue can significantly impair performance, so regular monitoring and proactive management are essential to maintain alertness and responsiveness.

Debriefing Sessions

Immediate post-flight team debriefs can be instrumental in identifying performance issues, safety concerns, and areas for improvement. Following practices from HROs (Dekker, 2002), these sessions encourage open communication and continuous learning. Debriefs provide a platform for constructive feedback and collective problem-solving.

Continuous Learning Loop

Establishing a routine where each flight's data, such as GPS tracks and cockpit recordings, are reviewed to learn and refine maneuvers promotes a culture of continuous improvement. By systematically analyzing performance data, pilots can identify trends, refine techniques, and enhance their overall skill set.

CORRELATION HEATMAPS FOR DEMOGRAPHICS

The study explored how demographic variables (air show flying experience, military or civilian flying experience, age, educational background, and marital status) relate to RSC, safety risk parameters, and MF within the international air show community. The research question guiding this analysis was, "*What are the differences in resilient safety culture, risk factors, mindfulness, and hazardous attitudes in air show performers based on demographic variables?*" probing an answer to the paradox that Barker (2003) suggested that exists in air show performers' demographic characteristics. Barker advocated that extensive flying experience does not necessarily prevent accidents, unlike in general aviation. Barker (2020a, b) further emphasized that flying experience alone is not a safeguard against accidents in air shows due to cognitive and physiological challenges at high G-forces and roll rates.

Similarly to the correlation heatmap for the risk factors, a heatmap visualized the comprehensive associations between the research variables and these demographic groups (Figure 3.3). This data was created by triangulating quantitative and qualitative

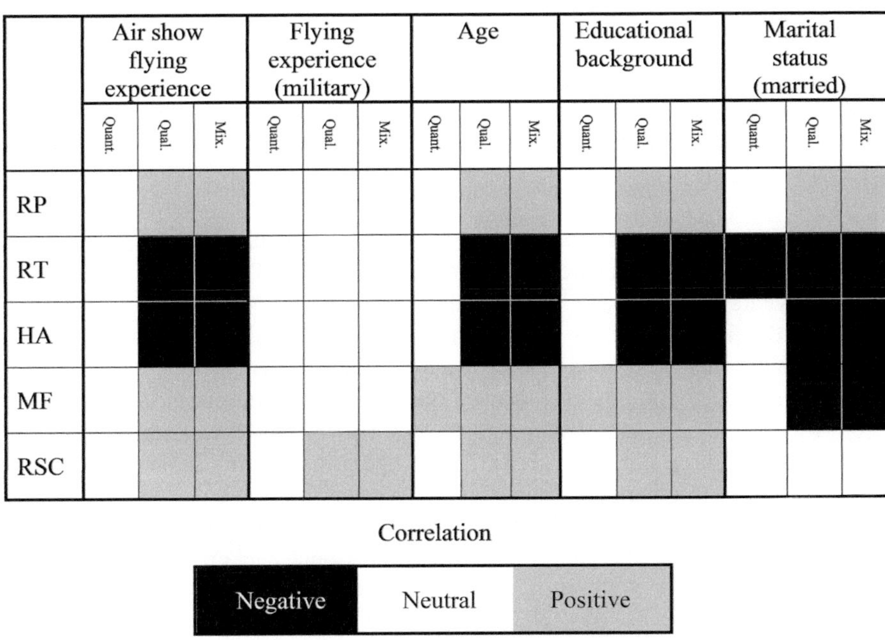

FIGURE 3.3 Correlation heatmap: Joint display of findings for research variables including demographic groups.

data into a mixed set. This mixed-methods approach generated robust and relevant inferences, providing a detailed and nuanced understanding of how demographic factors interact with critical safety and performance variables in the air show community (Younas et al., 2021). The 75 correlations between the research variables and demographic groups were displayed as a matrix of color tiles. Each tile represented a correlation coefficient derived from quantitative (Quant), qualitative (Qual), and final mixed (Mix) results following data triangulation and synthesis.

The provided heatmap is a correlation matrix displaying the relationships between various demographic and risk factors within the context of air show performers.

A breakdown of what the colors represent and a general analysis of each correlation is as follows:

- Black tiles represent a negative correlation between the variables. This suggests that as one variable increases, the other tends to decrease.
- Grey tiles represent a positive correlation, indicating that as one variable increases, the other also tends to increase.
- White tiles signify a neutral relationship, meaning no significant correlation exists between the variables.

Summary of the Findings Leading to the Correlation Heatmap for Demographics

Quantitative Analysis to Research Question 2

During the quantitative part of the study, to explore the research question, a one-way ANOVA was employed to analyze differences in mean perception scores for respondents on outcome variables based on demographic groupings. Histograms confirmed a normal distribution, and the Levene test for homogeneity was applied to ensure normal variances. If normal variance assumptions were not met, a robust ANOVA was used.

- Air Show Flying Experience: One-way ANOVA results indicated no statistically significant differences in mean scores for RP, RT, HA, MF, or RSC based on years of air show flying experience.
- Aerobatics Experience (Civilian or Military): No significant differences in mean scores were found based on aerobatics experience.
- Age Group: The analysis showed no significant differences in mean scores for RP, RT, HA, and RSC based on age. However, MF was significant, F (4, 103) = 3.48, p = .011, with the 55–64 age group having higher mean scores compared to other age groups.
- Educational Background: No significant differences in mean scores for RP, HA, and RSC were found based on educational background. However, RT was significant, F (3, 102) = 2.76, p = .046, and MF was significant, F (3, 105) = 3.36, p = .022. Posthoc analysis showed that respondents with a master's degree had the lowest RT scores, indicating lower RT compared

to other educational levels. High school graduates had higher MF scores, suggesting higher MF compared to those with higher education levels.
- Marital Status: No significant differences in mean scores for RP, HA, and RSC were found based on marital status. However, RT was significant, F (5, 102) = 2.32, p = .049, and MF was significant, F (5, 105) = 3.77, p = .004. Posthoc analyses (Tukey and Bonferroni) indicated that single respondents had significantly higher RT scores compared to other groups, suggesting higher RT among single individuals. Married respondents had lower MF scores, indicating lower MF levels compared to other groups.

In summary, significant differences were found in MF based on age, RT, and MF based on marital status, and both RT and MF based on educational background. No significant differences were identified based on air show flying experience or aerobatics experience, i.e., military or civilian.

Qualitative Analysis Related to Research Question 2

The qualitative analysis utilized data collected from air show observations, semi-structured interviews, focus groups, documentary analysis, and factual air show data. This approach provided a comprehensive understanding of how demographic variables (such as air show flying experience, military or civilian flying experience, age, educational background, and marital status) relate to RSC, safety risk parameters, and MF within the international air show community.

- Air Show Flying Experience: The qualitative data revealed a positive correlation between air show flying experience and all the studied variables: RP, RT, HA, MF, and RSC. Participants with extensive air show flying experience demonstrated heightened RP, suggesting they are more aware of potential hazards while display flying. This experience also correlated with greater RT, indicating confidence in managing these risks. Additionally, experienced pilots displayed fewer HA, higher MF, and a more substantial commitment to an RSC.
- Aerobatics Experience (Civilian or Military): There was no observed correlation between military or civilian flying experience and RP, RT, HA, or MF. However, a positive correlation with RSC was noted. This suggests that, regardless of whether the pilots had military or civilian backgrounds, their overall flying experience did not significantly impact their RP, RT, HA, or MF. Nevertheless, their experiences positively influenced their commitment to maintaining an RSC.
- Age Group: The data indicated a positive correlation between age and RP, MF, and RSC, with older participants showing higher RP, greater MF, and a more robust RSC. Conversely, age was negatively correlated with RT and HA, with older pilots displaying lower RT and fewer HA. This suggests that as pilots age, they become more cautious and mindful, contributing to a more robust safety culture.

- Educational Background: Educational background positively correlated with RP, MF, and RSC. Individuals with higher education levels demonstrated greater RP, MF, and commitment to an RSC. However, there was a negative correlation between educational background and RT and HA. Participants with higher education levels exhibited lower RT and fewer HA, indicating that education may foster a more cautious and safety-conscious mindset.
- Marital Status: Marital status, particularly being married, positively correlated with RP but negatively with RT, HA, and MF. Married participants showed higher RP but lower RT, fewer HA, and lower MF. There was no significant correlation between marital status and RSC. This suggests that married pilots might be more risk-averse and less mindful, potentially due to family responsibilities influencing their approach to flying.

Interpretation of Heatmap Analysis

In the context of air show performers, the heatmap analysis provides crucial insights that can inform pilot selection, training, and performance enhancement strategies.

Air Show Flying Experience

Based on the heatmap and the supporting literature, we can draw several important conclusions about the impact of air show flying experience on performers' attitudes toward risk, safety, and performance:

Air Show Flying Experience and Risk Perception

The positive correlation between air show flying experience and RP aligns with studies suggesting that accumulated experience in flying contributes to a pilot's heightened awareness of potential dangers. As pilots gain experience in air shows, they develop a more nuanced understanding of the risks involved, not just from a theoretical standpoint but also from real-world exposure to dynamic flying conditions. This comprehensive RP is crucial for the safety of both the performer and the spectators. Experienced performers, like those cited by Bob Hoover, understand that learning from the history of aviation accidents is essential to avoid repeating past mistakes (Barker, 2020a).

Experience and Mindfulness

The positive correlation between flying experience and MF suggests that experienced pilots are more likely to engage in deliberate, present-moment awareness of their actions, which is vital for maintaining high-performance levels under the pressure of an air show environment. This MF contributes to better decision making, especially during complex maneuvers requiring split-second judgments (Joseph & Reddy, 2013). Mindful pilots can better manage stress and remain focused, which is crucial for executing precise and safe performances.

Experience and Resilient Safety Culture

The association between experience and an RSC highlights that veteran air show pilots play an instrumental role in developing and sustaining a safety-first mindset.

Their extensive experience allows them to adapt and recover from challenges, thereby enhancing the overall safety culture within the air show community (International Council of Air Shows [ICAS], 2018; UK CAA, 2022). These experienced pilots often serve as role models and mentors, promoting best practices and encouraging a culture of continuous improvement and vigilance.

Experience, Risk Tolerance, and Hazardous Attitudes

Conversely, the negative correlation between experience, RT, and HA reveals a paradox within the air show industry. While experience usually equates to competency in general aviation, it does not automatically translate to safer performances in air shows. Highly experienced pilots may develop a sense of overconfidence, leading to an increased willingness to push safety boundaries, which may result in reduced RT and potentially HA. This phenomenon is documented in accident reports where even seasoned performers sometimes take unnecessary risks, highlighting the need for continuous education and mentorship to mitigate overconfidence and complacency (Barker, 2003, 2020a; AAIB, 2017; NTSB, 2012).

Implications for Air Show Safety

The findings emphasize the importance of ongoing training and mentorship for all performers, regardless of experience level. Continuous learning and adherence to best practices are crucial for maintaining safety standards. Implementing regular safety briefings, workshops, and mentorship programs can help bridge the experience gap for novice performers while reinforcing the importance of *caution* and *humility* for seasoned veterans.

Flying Background

The study highlights a crucial distinction in the safety cultures based on the flying background of military and civilian air show performers, potentially impacting the overall safety outcomes in air shows.

Correlation with Resilient Safety Culture

While the study indicates no significant direct relationship between military or civilian backgrounds and risk-related factors such as perception, tolerance, and MF, it reveals an indirect link to developing RSCs. Military organizations' structured approach to safety through SMSs and comprehensive risk management contributes to fewer accidents and incidents than civilian performers (Barker, 2020a). The quantitative data revealing more accidents involving civilian performers further emphasizes this point. The extensive training requirements and operational risk management frameworks within the military provide a disciplined, safety-centric environment that fosters resilience (FAA, 2020b).

Civilian Flying Background and Safety Management Systems

In contrast, civilian performers often operate as individuals or within less structured frameworks, viewing formalized risk assessments and SMS as bureaucratic and not directly beneficial to their practical needs. However, if civilian performers adopt SMS practices appropriate to their scale of operations, they may significantly enhance their safety without undue administrative burden. This adaptability is supported by the FAA's recommendations for scalable SMS, allowing for tailoring to different operational sizes and types (FAA, 2020b).

Operational Risk Assessment

Operational risk assessments, commonly used in the military, offer a proactive strategy for identifying and managing risks before air shows. The accountability and clear risk acceptance levels provided by such frameworks ensure a systematic approach to safety. Civilian performers might benefit from adapting these strategies to their operations, emphasizing risk detection and informed decision making.

Recommendations Based on Flying Background

To bridge the safety gap between military and civilian performers, the study suggests several recommendations:

- Selective SMS Implementation: Civilian performers should consider integrating elements of SMS that directly apply to their operations, such as risk assessments and safety assurance practices.
- Mentorship and Knowledge Sharing: Civilian performers could benefit from mentorship programs facilitating knowledge transfer from military to civilian air show sectors.
- Safety Education: Educating civilian performers about the tangible benefits of a structured approach to safety could alter perceptions of SMS and risk matrices.
- Community Collaboration: The air show community should encourage collaboration and sharing of best practices, potentially through workshops and seminars promoting a unified safety approach.

Age

The data's association between age and various risk-related factors provides insight into how demographic elements can influence attitudes and behaviors in high-risk professions such as air show performance.

Positive Correlations with Age

The positive correlations of age with RP, MF, and RSC are consistent with the notion that pilots generally develop a more mature outlook toward risk as they age. They gain a deeper understanding of the consequences of their actions not just in the cockpit but also in broader life contexts. Their accumulated experiences over the years can lead to a more mindful approach, where they are present and aware of their environment, actions, and the potential outcomes of decisions. This mature RP is crucial in high-stakes environments like air shows, where split-second decisions can have significant consequences (Gibson et al., 2013).

Additionally, older pilots may contribute more to an RSC within their organizations or the air show community at large. Their experience allows them to identify potential hazards and adapt to changing circumstances, fostering an environment that prioritizes safety and effectively responds to challenges (Hallahan et al., 2004).

Negative Correlations with Age

Conversely, the negative correlation between age and HA and RT aligns with previous findings suggesting a more conservative approach to risk as one age. Older pilots may have a greater awareness of their personal limitations and a heightened sense of mortality, leading them to take fewer unnecessary risks. They often prefer safer, more calculated flying displays and may be less inclined to engage in the aggressive or boundary-pushing maneuvers favored by younger pilots.

However, it is crucial to consider physiological factors associated with aging, such as reflexes, vision, and overall health status, which can impact a pilot's performance and risk assessment abilities. While these factors may not have been the primary focus of the study, they are critical considerations in the medical certification and overall evaluation of a pilot's fitness to perform at air shows (NTSB, 2012).

Implications for the Air Show Industry

The insights drawn from the correlation between age and risk-related factors emphasize the need for the following:

- Age-Appropriate Training: Tailoring training programs to accommodate older pilots' changing needs and capabilities, ensuring they remain sharp and competent within the bounds of safety.
- Knowledge Transfer: Leveraging the experience of older pilots to mentor younger ones, thus transferring valuable knowledge about risk management and safety practices.
- Health and Wellness Programs: Implementing health and wellness programs aimed at maintaining the physiological well-being of older pilots, thereby extending their ability to fly safely.

The findings highlight the importance of age as a determinant in the risk behavior of air show performers. The industry must recognize the value that mature pilots bring in terms of experience and risk management while also acknowledging and adapting to the physical realities of aging. Through thoughtful integration of these insights, the air show community can continue to enhance safety and performance standards for all pilots, regardless of age.

Educational Background

The correlations between educational background and various risk-related factors suggest that higher levels of education among air show performers are associated with more sophisticated perceptions and management of risk and more profound MF and commitment to safety culture.

Positive Correlations with Educational Background

The positive relationship between educational background and RP indicates that pilots with higher education levels may likely have a deeper understanding of the complexities of risk. They may possess a broader knowledge base, enabling them to evaluate potential hazards more comprehensively and adhere to a more systematic approach to safety (Chionis & Karanikas, 2018).

Similarly, educational attainment appears to enhance MF among air show performers, likely due to training that fosters critical thinking, awareness, and focused attention. These skills are essential in the high-concentration environment of air shows, where attentiveness to each moment can be the difference between a successful performance and a mishap.

A solid educational background also seems to contribute positively to an RSC. Pilots with advanced degrees might be better equipped to understand and implement safety protocols, contribute to safety education within the community, and advocate for accountability to aviation authorities and the air show audience.

Negative Correlations with Educational Background

Conversely, the negative correlation between educational background, HA, and RT is revealed. It suggests that those with higher educational achievements may inherently be more risk-averse, potentially due to a higher level of risk literacy and a greater understanding of the potential consequences of their actions. This trait can be crucial for ensuring safety during performances, as it may discourage engaging in overly risky maneuvers that could endanger the pilot and spectators.

Implications for Air Show Safety and Training

These findings have important implications for safety and training within the air show community:

- Tailored Education: There is a need for tailored educational programs at air show conventions to bridge the knowledge gap between pilots of varying educational backgrounds. Workshops, seminars, and webinars can help enhance their understanding of risk management and safety practices.
- Mentorship Programs: Pilots with lower levels of formal education could benefit significantly from mentorship by more experienced pilots with advanced degrees. These relationships can facilitate knowledge transfer, promoting a culture of safety and professionalism across the board.
- Continuous Learning: Encouraging continuous learning and professional development among air show performers can help maintain high safety standards. This could include offering further education and training opportunities in the latest safety protocols and risk assessment methodologies.

It was also found that educational background is a significant factor in shaping the risk behaviors and safety culture among air show performers. The industry should recognize the value of formal education in enhancing safety and work to ensure that all performers, regardless of their educational level, have access to the resources and support necessary to perform safely and effectively. By fostering an environment of continuous learning and mentorship, the air show community can elevate safety standards and ensure that performers of all backgrounds can succeed and contribute positively to the industry.

Marital Status

The analysis of the correlation between marital status and risk-related behaviors among air show performers suggests nuanced influences on safety and performance:

Positive Correlation with Risk Perception

We infer that married individuals tend to have a heightened perception of risk, which could be attributed to their responsibilities toward their families. This increased awareness may make them more cautious and conservative in their decision making process, preparing for and executing their air show performances. The positive association with RP aligns with studies by Hallahan et al. (2004) and Aumeboonsuke and Caplanova (2021), suggesting that individuals with familial obligations may exhibit greater risk sensitivity.

Negative Correlations with Hazardous Attitudes, Mindfulness, and Risk Tolerance

Interestingly, the negative correlation between marital status and HA suggests that being married is associated with a decrease in tendencies that could lead to unsafe practices. However, the negative correlation with MF may indicate that marital responsibilities could potentially distract a performer, impacting their focus and presence of mind. This effect on MF could, in turn, influence their RT, making them either more cautious due to their responsibilities or less attentive to risks due to distractions.

The lack of association between marital status and RSC indicates that marriage does not necessarily affect one's contribution to or participation in a safety-focused environment within the air show community.

Implications for Air Show Performers' Mental Preparedness

Given these findings, mental preparedness training becomes essential, especially for performers whose marital status may influence their approach to risk. Andersen et al. (2016) highlight the importance of mental training in preparing individuals to manage their personal circumstances effectively while maintaining professional focus and safety awareness.

Lessons Learned and Mitigation Methods

Sharing lessons learned from performers preoccupied with family concerns in safety workshops, such as those conducted by ICAS and European Airshow Council (EAC), could be beneficial. These discussions could lead to the development of methods to mitigate the impact of such personal concerns on performance. Creating support groups for single performers might also help them enhance their RP and lower their RT (Nosita et al., 2020).

Preoccupation with Family Issues

Although married pilots may enjoy a more stable life, family issues and concerns could impact their level of MF. This preoccupation may translate into increased RT as they may feel the pressure to perform despite these distractions. Including family-related risk factors in the ICAS risk assessment matrix (International Council of Air Shows, 2012) is evidence of the air show community's recognition of the importance of family considerations.

Marital status has complex and multifaceted impacts on air show performers' attitudes toward risk and safety. While married pilots may have higher risk awareness due to family obligations, they must also manage potential distractions effectively. As the air show community continues to evolve, recognizing and addressing the unique needs of performers based on their marital status will be essential for maintaining high safety standards and supporting performers' well-being.

Part-Time and Full-Time Employed Air Show Performers

The dynamics between full-time and part-time air show performers bring forth distinct challenges and operational pressures, particularly in the military context, where the majority of solo display pilots undertake air show responsibilities alongside their operational duties. This situation often increases workload, fatigue, and stress, requiring strategic management and support.

Table 3.4 highlights the key aspects of the military against civilian part-time air show performers.

This table delineates the fundamental challenges and operational dynamics faced by part-time air show performers in both military and civilian sectors. While both groups share common issues like fatigue management and the need for maintaining skill currency, the context in which these challenges manifest and are managed can differ significantly due to the nature of their primary employment and their available support structures.

Military Air Show Performers

Full-time display pilots, typically part of structured military aerobatic teams like the United States Air Force (USAF) Thunderbirds or the Royal Air Force (RAF) Red Arrows, are entirely dedicated to their air show roles. This full-time commitment provides comprehensive training, extensive practice, and adequate recovery time, ensuring peak performance and safety. The focused nature of their job minimizes the risks associated with operational fatigue and stress (Jensen, 1995; Wald, 2004).

In contrast, part-time display pilots in the military balance their regular duties with air show performances. This dual role can lead to additional stress and fatigue due to the extra workload. The complexity of transitioning between operational flying and the precision required for air show performances can physically and mentally strain the pilots (Caldwell, 2005; Caldwell & Caldwell, 2003).

TABLE 3.4
Key Aspects of Military Against Civilian Part-Time Air Show Performers

Aspect	Military Part-Time Performers	Civilian Part-Time Performers
Primary Role	Operational military duties alongside air show performances.	Commercial/business aviation duties alongside air show performances.
Fatigue Management	Must balance operational commitments with display preparation, risking fatigue.	Balancing commercial flying with aerobatic performances, leading to fatigue.
Risk Tolerance	High-risk military operations may influence risk assessment in air show flying.	Commercial flying risk management differs from high-risk aerobatic flying.
Currency and Recency	Maintaining proficiency in both operational and display flying.	Need to stay current in commercial flying and aerobatic skills.
Stress and Distractions	Operational pressures combined with the stress of display performances.	Juggling commercial flying responsibilities with the demands of air shows.
Support and Training	Military infrastructure for training and support but with operational priorities.	Depends on personal initiative for training, with some support from air show organizations.
Scheduling	Military scheduling with added air show commitments.	Commercial flight schedules with additional air show performances.

The dual role of part-time display pilots necessitates a careful balancing act between operational commitments and the demands of air show flying. This situation can potentially affect their performance in both arenas due to the increased risk of fatigue and stress-related issues (Harris, 2011; Muermans, 2002). The need for robust support mechanisms, such as adequate rest, mental health resources, and efficient scheduling, is critical to mitigate these risks (Caldwell, 2005; Caldwell & Caldwell, 2003).

Military organizations must implement strategies that prioritize the pilots' health and operational safety to support part-time display pilots. These strategies might include limiting the number of performances, ensuring adequate downtime between operational duties and air show commitments, and providing comprehensive wellness and mental health support services (Muermans, 2002; Wald, 2004).

Civilian Air Show Performers

Civilian air show performers, many of whom engage in air show flying on a part-time basis while maintaining full-time careers in commercial or business aviation, face unique challenges. These "weekend flyers" must manage the dual responsibilities of their primary aviation roles and their passion for air show performance, dealing with fatigue, RT, currency, recency issues, and the associated stresses and distractions.

Balancing Dual Roles

Part-time air show performers often juggle their commitments to regular employment, typically as airline or business jet pilots, with their engagements in air shows. This balancing act can lead to compounded fatigue due to irregular working hours and the intense physical and mental demands of aerobatic flying. Managing these dual roles requires exceptional time management and personal discipline to ensure that neither role negatively impacts the other.

Fatigue and Risk Tolerance

Fatigue is a significant risk factor that can diminish a pilot's performance and reaction times, potentially increasing RT. A fatigued pilot may underestimate the risks of complex aerobatic maneuvers or challenging weather conditions, compromising safety. This risk is particularly acute for part-time performers who might not have the same recovery time as full-time pilots focused solely on air shows.

Currency and Recency Issues

Maintaining currency and recency in their commercial flying duties and air show performances is a critical challenge for part-time civilian air show pilots. The distinct skill sets required for each role necessitate continuous practice and training. The sporadic nature of air show flying can exacerbate these challenges, with pilots needing to ensure they remain proficient in the specific demands of aerobatic performance. Regularly scheduled practice sessions and adherence to strict training regimens are essential to maintaining these dual competencies.

Stress and Distractions

The dual commitments also bring about stress and potential distractions, with pilots needing to mentally switch between the operational protocols of commercial aviation and the dynamic nature of air show flying. This mental shift can strain cognitive resources, impacting decision making and performance. Effective stress management techniques and mental conditioning are critical to navigating these transitions smoothly.

Mitigation Strategies

Addressing these challenges requires comprehensive strategies, including strict adherence to rest regulations, proactive management of schedules to ensure adequate downtime, and continuous training to maintain proficiency in commercial and aerobatic flying. Aviation authorities and air show organizations should provide support and guidance to help manage the unique pressures part-time civilian air show performers face. Specific strategies might include the following:

- Structured Scheduling: Implementing structured scheduling to balance work and air show commitments, ensuring adequate rest periods and minimizing fatigue.
- Regular Training: Continuous and rigorous training programs to keep skills sharp in both commercial aviation and aerobatic performance.

- Stress Management Programs: Providing access to stress management programs and psychological support to help pilots cope with the dual demands of their roles.
- Mentorship and Peer Support: Establishing mentorship and peer support systems to provide guidance and share best practices for balancing dual roles.

In summary, part-time civilian air show performers face significant challenges requiring diligent management of fatigue, risk, and skill maintenance. A supportive framework by aviation authorities and air show organizations, coupled with self-regulation by the pilots, is crucial to ensure safety and performance efficacy in the demanding realm of air show flying.

FURTHER KEY INSIGHTS: ENHANCING AIR SHOW SAFETY AND PERFORMANCE

The word cloud, Figure 3.4, generated from the study (Karachalios, 2022), derived from semi-structured and focus group interview sessions, highlights key themes and concepts central to the discourse on safety and performance in air show

FIGURE 3.4 Word cloud for air show safety, Karachalios (2022).

flying. Examining the prominence and frequency of these terms provides a deeper understanding of the factors that influence the practice and culture of air show performers, particularly concerning demography, risk factors, and resilient safety.

Central Themes

Time emerges as a critical element in various aspects of air show performance. Effective time management is essential for activities such as briefings, practice sessions, and training. These elements ensure that pilots can prepare thoroughly for their routines, addressing both the planning and execution phases of their performances. The accumulation of experience over time is highlighted, underscoring its importance in enabling pilots to handle complex maneuvers and make informed decisions under pressure.

The term *safety* is prominently featured, reflecting its paramount importance in air show operations. Safety encompasses personal safety for the pilots and public safety for the spectators. A culture of safety is integral, highlighting the need for a collective commitment to safe practices and continuous improvement in safety protocols. This culture ensures that safety is ingrained in every aspect of air show performance, from planning to execution.

Risk is another central theme in air show flying, with terms like "risk" and "attitude" indicating pilots need to manage and mitigate risks effectively. RP and RT are critical factors influencing decision making and performance, requiring pilots to remain constantly aware of and responsive to potential hazards. Assessing and managing risk is a critical component of maintaining safety during performances.

Training is crucial for developing and maintaining the skills necessary for safe and effective air show performances. The word cloud emphasizes the importance of ongoing training and practice to hone pilot skills continually. Training is not a one-time event but a continuous process that ensures pilots are always prepared for the demands of air show flying. Briefing sessions are essential to this training, providing pilots with a clear understanding of the plan and potential challenges they may face.

Supporting Concepts

The importance of adhering to established *rules* and maintaining control over the aircraft during performances is highlighted. These elements ensure consistency and safety in high-risk environments. Following rules and maintaining control are fundamental principles that underpin safe and successful air show performances.

The term *military* suggests that many air show performers draw on military aviation experience, contributing to a disciplined approach to flying and safety. This military background often brings a high level of training and discipline that is beneficial in air show flying. The presence of "*competition*" underscores the competitive nature of air shows, where pilots strive for excellence while balancing safety and performance demands. This competitive spirit drives pilots to continuously improve their skills and performance.

External factors such as *weather* and environment are crucial considerations for air show performers. Pilots must adapt to changing conditions and manage the risks associated with different weather patterns. Understanding and anticipating these factors are essential for planning and executing safe performances.

The terms *community* and *organizers* point to the collaborative nature of air shows. A strong sense of community and effective coordination among organizers, pilots, and support staff are vital for successful and safe performances. This collaboration ensures that all aspects of the air show are well-coordinated and that safety is a shared responsibility.

The word *experienced* signifies the critical role that experience enhances safety and performance in air show flying. Experienced pilots bring a wealth of knowledge and practical insights that are invaluable for managing risks and making informed decisions. Their deep understanding of air shows, energy management, and emergency procedures allows them to navigate complex situations effectively.

Experienced pilots also contribute significantly to an RSC. Their accumulated knowledge and skills are often shared with less experienced colleagues through mentorship and training programs. This knowledge transfer helps build a strong safety culture where best practices are continuously reinforced and improved upon.

DATA-DRIVEN STRATEGIES FOR ENHANCED RESILIENT SAFETY AND AIR SHOW PERFORMANCE EXCELLENCE

In air shows' dynamic and high-stakes environment, leveraging data-driven insights is pivotal for cultivating resilient safety protocols and achieving performance excellence. The intersection of various factors like air show flying experience, military and civilian background, age, educational level, and marital status plays a crucial role in shaping these strategies.

Air Show Flying Experience

Air show flying experience, positively correlated with RP, underscores the importance of accumulated flight hours in honing pilots' awareness of potential dangers (Cak et al., 2019). However, the inverse relationship with RT points to the necessity of continuous education, even for veteran performers, to prevent overconfidence and complacency (Barker, 2020a). Continuous learning programs are essential to maintain a high safety standard and adapt to evolving aerial dynamics.

Flying Background

The study reveals a more robust RSC among military pilots, which is attributed to their rigorous SMS training. Civilian performers can adopt these structured approaches to enhance safety practices (FAA, 2020b). Tailoring SMS training for civilian pilots to fit the air show context can bridge the gap between military precision and civilian flexibility in risk management.

Age-Related Dynamics

Signified by a positive correlation between age and RP, the maturation process suggests that older pilots possess refined risk assessment skills. Conversely, the negative correlation between HA and RT indicates a cautious approach to decision making (Gibson et al., 2013). Implementing age-specific safety workshops can provide age-appropriate training and mitigate physiological challenges associated with aging (NTSB, 2012).

Educational Background's Impact

Higher education enriches pilots' understanding of risk management and safety protocols (Chionis & Karanikas, 2018). Educational enhancement initiatives, such as workshops and seminars, can elevate pilots' safety knowledge and operational efficiency, irrespective of their educational attainment (Van Damme & Zahner, 2022).

Marital Status Considerations

Married pilots exhibit a heightened RP, potentially due to family responsibilities influencing their approach to risk (Aumeboonsuke & Caplanova, 2021). Family-centric support programs and focused mental preparedness training can help balance personal concerns with professional obligations, enhancing MF and reducing HA (Andersen et al., 2016).

Implementing Data-Driven Strategies

A comprehensive approach to enhancing air show safety and performance should integrate these insights, creating a robust framework for individual and collective growth. Strategies like individualized risk management plans, community engagement forums, and feedback mechanisms can foster a culture of safety and excellence. These initiatives should be dynamic, evolving with ongoing research and feedback from the air show community.

Data-Driven Strategies Enhanced by A-SMS

Integrating the concept of Airshow Safety Management Systems (A-SMS) proposed (Karachalios and Adjekum 2023) into the strategy framework for enhancing resilient safety and performance in air shows offers a targeted approach to managing the unique risks associated with aerobatic displays.

Given the variance in air show flying experience, implementing an A-SMS can provide a structured yet flexible safety framework. This system would account for the diverse range of experiences, from novices to veterans, ensuring that each performer's specific needs and risks are addressed effectively. The A-SMS can facilitate continuous learning and adaptation, which is vital for maintaining high safety standards across all experience levels.

A-SMS aligns with the previously discussed factors in the following manner:

Bridging Military and Civilian Practices

The A-SMS framework can serve as a bridge between the structured safety protocols typical of military backgrounds and the more varied approaches seen in civilian air show performances. By adopting best practices from both realms, the A-SMS can offer a comprehensive safety management solution that enhances risk management and operational safety for all performers.

Age-Adaptive Safety Protocols

The A-SMS can include age-adaptive safety protocols to cater to the physiological and cognitive changes associated with aging. This tailored approach ensures that older performers benefit from safety measures that consider potential age-related challenges, enhancing their performance, safety, and longevity in the air show industry.

Educational Integration in A-SMS

Educational background significantly impacts risk management and safety culture. The A-SMS can incorporate educational programs and training modules that elevate performers' understanding of SMS parameters such as policy and objectives, safety risk management, safety risk assurances, safety promotion, safety culture, continuous monitoring, and improvements regardless of their initial education level. This integration ensures a uniformly high level of safety awareness and competency among all performers (Karachalios & Adjekum, 2023).

Marital Status and Family-Centric Safety Planning

Recognizing the impact of marital status on performers' risk attitudes, the A-SMS can include family-centric safety planning. During individual risk profiling before a flight, flight safety risk tools must capture and assess the psycho-social effects of marital status. This approach acknowledges the additional responsibilities and potential distractions married performers face, offering support and resources to manage these effectively and maintain focus during performances.

Implementing A-SMS in the Air Show Industry

The A-SMS presents a robust framework for addressing the nuanced safety challenges in air show environments. By integrating A-SMS into the broader strategy for enhancing safety and performance excellence, the air show industry can achieve a standardized yet flexible approach to safety management tailored to the unique demands and risks of aerobatic performances.

Adopting voluntary, cost-effective, and phased A-SMS across the air show industry that is scaled to each operator's complexity and can track key safety performance metrics is critical. A-SMS can standardize safety practices and create a cohesive, community-wide approach to risk management. Regular workshops, seminars, and training sessions based on the A-SMS principles can facilitate knowledge sharing and collective adherence to best safety practices.

Using industry subject-matter experts to provide implementation facilitation and learning from other aviation service providers such as airlines, air traffic management, and the military, who may have a lot of veritable data on the successes and limitations of SMS, can be helpful.

It is needless to say that safety is a *shared responsibility* requiring continuous monitoring and improvements. The path to resilient safety and performance excellence in air shows is multifaceted, requiring a nuanced understanding of various demographic and experiential factors. By adopting data-driven strategies addressing these aspects, the air show industry can achieve higher safety standards and operational success, ensuring thrilling yet secure experiences for performers and spectators.

REFERENCES

Abraha, H. H., & Liyanage, J. P. (2014, August). Review of theories and accident causation models: Understanding of human-context dyad toward the use in modern complex systems. In *Proceedings of the 7th World Congress on Engineering Asset Management (WCEAM 2012)* (pp. 17–32). Cham: Springer International Publishing.

Adjekum, D. K., & Fernandez Tous, M. (2020). Assessing the relationship between organizational management factors and a resilient safety culture in a collegiate aviation program with safety management systems (SMS). *Safety Science, 131*, 104909. https://doi.org/10.1016/j.ssci.2020.104909

Adjekum, D. K., Keller, J., Walala, M., Young, J. P., Christensen, C., DeMik, R. J., & Northam, G. J. (2015). Cross-sectional assessment of safety culture perceptions and safety behavior in collegiate aviation programs in the United States. *International Journal of Aviation, Aeronautics, and Aerospace, 2*(3). https://doi.org/10.15394/IJAAA.2015.1074

Air Accidents Investigation Branch. (2017). *Aircraft accident report: AAR 1/2017, G-BXFI, 22 August 2015*. GOV.UK. www.gov.uk/aaib-reports/aircraft-accident-report-aar-1-2017-g-bxfi-22-august-2015

Akselsson, R., Ek, Å., Koornneef, F., Stewart, S., & Ward, M. (2009). Resilience safety culture. In *17th World Congress on Ergonomics*. IEA. Accessed 13/06/2024. https://portal.research.lu.se/en/publications/resilience-safety-culture

American Psychological Association. (2002, October). *The road to resilience. Monitor on Psychology, 33*(10). www.apa.org/monitor/oct02/pp

Andersen, J. P., Papazoglou, K., Gustafsberg, H., Collins, P., & Arnetz, B. (2016, March). Mental preparedness training. *FBI Law Enforcement Bulletin*. https://leb.fbi.gov/articles/featured-articles/mental-preparedness-training

Anes, V., Morgado, T., Abreu, A., Calado, J., & Reis, L. (2022). Updating the FMEA approach with mitigation assessment capabilities—A case study of aircraft maintenance repairs. *Applied Sciences, 12*(22), 11407.

Arnett, J. J., Offer, D., & Fine, M. A. (1997). Reckless driving in adolescence: 'State' and 'trait' factors. *Accident Analysis & Prevention, 29*(1), 57–63.

Aumeboonsuke, V., & Caplanova, A. (2021). The impact of marital status on risk tolerance. *Economic Research-Ekonomska Istraživanja, 34*(1), 2896–2912.

Banghart, M., Babski-Reeves, K., Bian, L., & Strawderman, L. (2018). Subjectivity in failure mode effects analysis (FMEA) severity classification within a reliability centered maintenance (RCM) context. *International Journal of Aviation, Aeronautics, and Aerospace, 5*(1), 2.

Barker, D. (2003). *Zero error margin: Airshow display flying analysed*. Freeworld Publications.

Barker, D. (2020a). *Anatomy of air show accidents* (1st ed., Vol. 1). International Council of Air Shows, International Air Show Safety Team.

Barker, D. (2020b). Air show accidents: A statistical and preventative perspective [video recording]. ICAS Convention. www.youtube.com/watch?v=29zT0edEAUQ&t=877s

Bell, M. (2020). *The resilient organization: How adaptive cultures thrive even when strategy fails*. Kogan Page Publishers.

Bhandari, S., & Hallowell, M. (2022). Influence of safety climate on risk tolerance and risk-taking behavior: A cross-cultural examination. *Safety Science, 146*, 105559. https://doi.org/10.1016/j.ssci.2021.105559

Breakwell, G. M. (2007). Individual and group differences in risk perception. In *The Psychology of Risk* (pp. 44–77). Cambridge University Press.

Brugnara, G., Pedrinolla, V., Venturelli, M., Cortesi, M., Galazzo, I. B., & Grassi, B. (2022). Human factors in aviation safety: A multidisciplinary perspective. *Safety Science, 146*, 105564.

Cairns, J., Cox, T., & O'Brien, M. (2008). Managing offshore safety: A survey of the workforce. *Safety Science, 46*(3), 304–320.

Cak, S., Say, B., & Misirlisoy, M. (2019). Effects of working memory, attention, and expertise on pilots' situation awareness. *Cognition, Technology & Work, 22*, 85–94. https://doi.org/10.1007/s10111-019-00551-w

Caldwell, J. L. (2005). Fatigue in aviation. *Travel Medicine and Infectious Disease, 3*(2), 85–96.

Caldwell, J. A., & Caldwell, J. L. (2003). Fatigue in military aviation: An overview of US military-approved pharmacological countermeasures. *Aviation, Space, and Environmental Medicine, 74*(7), C39–C51.

Caldwell, J. A., Caldwell, J. L., & Schmidt, R. M. (2009). Alertness management strategies for operational contexts. *Sleep Medicine Reviews, 13*(1), 27–36.

Carlson, C. S. (2012). *Effective FMEAs: Achieving safe, reliable, and economical products and processes using failure mode and effects analysis* (pp. 21–51). John Wiley & Sons, Inc. https://doi.org/10.1002/9781118312575

Causse, M., Chua, Z. K., & Rémy, F. (2019). Influences of age, mental workload, and flight experience on cognitive performance and prefrontal activity in private pilots: A fNIRS study. *Scientific Reports, 9*, 7688. https://doi.org/10.1038/s41598-019-44082-w

Chan, W. T., & Li, W. C. (2023). Development of effective human factors interventions for aviation safety management. *Frontiers in Public Health, 11*, 1144921. https://doi.org/10.3389/fpubh.2023.1144921

Chen, C.-F., & Chen, S.-C. (2014). Measuring the effects of safety management system practices, morality leadership and self-efficacy on pilots' safety behaviors: Safety motivation as a mediator. *Safety Science, 62*, 376–385. https://doi.org/10.1016/j.ssci.2013.09.013

Chionis, D., & Karanikas, N. (2018). The impact of education on aviation safety. *Safety, 4*(3), 39.

Chionis, D., Karanikas, N., Iordan, A.-R., & Svensson-Dianellou, A. (2022). Contribution of risk perception and communication in aviation safety events. *Transportation Research Record, 2676*(3), 405–416. https://doi.org/10.1177/03611981211051617

Christian, M. S., Bradley, J. C., Wallace, J. C., & Burke, M. J. (2009). Workplace safety: A meta-analysis of the roles of person and situation factors. *Journal of Applied Psychology, 94*(5), 1103–1127.

Conner, D. R. (1993). *Managing at the speed of change: How resilient managers succeed and prosper where others fail*. Villard Books.

Cookson, S. (2019). Intercultural communication on the flight deck: A review of studies in aviation. https://doi.org/10.1007/978-3-319-93885-1_8

Cooper, M. D. (2000). Towards a model of safety culture. *Safety Science, 36*(2), 111–136. http://dx.doi.org/10.1016/S0925-7535(00)00035-7

Civil Aviation Safety Authority. (2025). *Advisory circular AC 91-21 v2.3: Air displays* (File ref D25/74814). www.casa.gov.au/sites/default/files/2022-11/advisory-circular-91-21-air-displays.pdf
Cumming, J., & Hall, C. (2002). Deliberate imagery practice: The development of imagery skills in competitive athletes. *Journal of Sports Sciences, 20*(2), 137–145.
Dekker, S. (2002). *The field guide to human error investigations*. Ashgate Publishing.
Dekker, S. (2012). *Just culture: Balancing safety and accountability* (2nd ed.). CRC Press. https://doi.org/10.4324/9781315251271
Diehl, A. E. (2002). *Silent knights: Blowing the whistle on military accidents*. Brassey's Inc.
Dismukes, R. K., Berman, B. A., & Loukopoulos, L. D. (2007). *The limits of expertise: Rethinking pilot error and the causes of airline accidents*. Ashgate.
Doerr, A. J. (2020). When and how personality predicts workplace safety: Evaluating a moderated mediation model. *Journal of Safety Research, 75*, 275–283.
Drinkwater, J., & Molesworth, B. R. (2010). The effects of situational and personal characteristics on flight instructor risk perception and risk tolerance. *Aviation Psychology and Applied Human Factors, 1*(1), 37–43.
Dunning, D. (2011). The Dunning–Kruger effect: On being ignorant of one's own ignorance. In J. M. Olson & M. P. Zanna (Eds.), *Advances in experimental social psychology* (Vol. 44, pp. 247–296). Academic Press. https://doi.org/10.1016/B978-0-12-385522-0.00005-6
EASA. (2022). EASA publishes study of socio-economic factors in relation to aviation safety. Retrieved from www.easa.europa.eu
Ebrahim, Y., Molesworth, B. R. C., & Molloy, O. (2022). The link between individual predictors of risk and risky flight behavior: A systematic review. *The International Journal of Aerospace Psychology, 33*(1), 35–60. https://doi.org/10.1080/24721840.2022.2139700
Ebrahim, Y., Molesworth, B. R. C., & Rantz, W. (2021). Risk-taking propensity: A comparison between pilots and members of the general population. *The International Journal of Aerospace Psychology, 32*(2–3), 138–151.
Federal Aviation Administration (FAA). (n.d.). *Flight risk assessment tools*. General Aviation Joint Steering Committee Safety Enhancement Topic (SE 42). Retrieved from: www.faasafety.gov/files/events/WP/WP27/2018/WP2781452/SE_Topic_15-08[909].pdf
Federal Aviation Administration. (2020a, March 3). *AC 121-43: Mentoring training for pilots in command*. www.faa.gov/regulations_policies/advisory_circulars/index.cfm/go/document.information/documentID/1037177
Federal Aviation Administration (FAA). (2020b). *Safety management systems for aviation service providers*. Federal Aviation Administration.
Federal Aviation Administration. (2023). *Pilot's handbook of aeronautical knowledge* (FAA-H-8083-25C). www.faa.gov/regulations_policies/handbooks_manuals/aviation/faa-h-8083-25c.pdf
Fletcher, D., & Sarkar, M. (2013). Psychological resilience: A review and critique of definitions, concepts, and theory. *European Psychologist, 18*(1), 12–23.
Foster, R. A., & Adjekum, D. K. (2022). A qualitative review of the relationship between Safety Management Systems (SMS) and safety culture in multiple-collegiate aviation programs. *Collegiate Aviation Review International, 40*(1), 63–94. Retrieved from http://ojs.library.okstate.edu/osu/index.php/CARI/article/view/8511/7734
Foster, R. A., & Adjekum, D. K. (2023). Evaluating the effects of safety management systems (SMS) on safety culture factors in collegiate aviation operations: A structural equation modeling (SEM) approach. *International Journal of Aviation, Aeronautics, and Aerospace, 10*(2). https://doi.org/10.58940/2374-6793.1788
Fredrickson, B. L. (2001). The role of positive emotions in positive psychology: The broaden-and-build theory of positive emotions. *American Psychologist, 56*(3), 218–226.

Gautam, A., & Mathur, R. (2018). Influence of mindfulness on decision making and psychological flexibility among aircrew. *Journal of Psychosocial Research, 13*(1), 199–207. http://search.proquest.com/openview/373d0975cc4b9935d9b34fa759552ebe/1?pq-origsite=gscholar&cbl=506336

Gawande, A. (2009). *The checklist manifesto: How to get things right*. Metropolitan Books.

Gibson, R. J., Michayluk, D., & Van de Venter, G. (2013). Financial risk tolerance: An analysis of unexplored factors. *Financial Services Review, 22*(1), 23–50.

Grossman, D., & Christensen, L. W. (2004). *On combat: The psychology and physiology of deadly conflict in war and in peace*. PPCT Research Publications.

Grote, G. (2012). Safety management in different high-risk domains – All the same? *Safety Science, 50*(10), 1983–1992.

Guldenmund, F. W. (2000). The nature of safety culture: A review of theory and research. *Safety Science, 34*(1–3), 215–257.

Hallahan, T. A., Faff, R. W., & McKenzie, M. D. (2004). An empirical investigation of personal financial risk tolerance. *Financial Services Review, 13*(1), 57–78.

Harris, D. (2011). *Human performance on the flight deck*. Ashgate.

Heese, M., Kallus, W., & Kolodej, C. (2014, February). Assessing behaviour towards organizational resilience in aviation. In *5th REA Symposium managing trade offs* (pp. 67–74).

Helmreich, R. L., & Foushee, H. C. (1993). Why CRM? Empirical and theoretical bases of human factors training in aviation. In E. L. Wiener, B. G. Kanki, & R. L. Helmreich (Eds.), *Cockpit resource management* (pp. 4–53). Academic Press.

Helmreich, R. L., & Merritt, A. C. (1998). *Culture at work in aviation and medicine: National, organizational, and professional influences*. Ashgate.

Heyvaert, M., Hannes, K., Maes, B., & Onghena, P. (2011). Critical appraisal of mixed methods studies. *Journal of Mixed Methods Research, 5*(3), 207–227.

Hofstede, G. (1984). *Culture's consequences: International differences in work-related values*. SAGE Publications Ltd.

Hofstede, G. (2001). *Culture's consequences: Comparing values, behaviors, institutions and organizations across nations* (Vol. 41). SAGE Publications Ltd.

Hollnagel, E. (2010). *Safety-I and safety-II: The past and future of safety management*. Ashgate Publishing Ltd.

Hollnagel, E., Pariès, J., Woods, D. D., & Wreathall, J. (2011). *Resilience engineering in practice: A guidebook*. Ashgate Publishing, Ltd.

Hunter, D. R. (2002). Risk perception and risk tolerance in aircraft pilots (No. DOT/FAA/AM-02/17). United States Department of Transportation. Federal Aviation Administration. Office of Aviation. Civil Aerospace Medical Institute.

Hunter, D. R. (2006). Risk perception among general aviation pilots. *International Journal of Aviation Psychology, 16*(2), 135–144. https://doi.org/10.1207/s15327108ijap1602_1

ICAO. (2018). *Safety management manual* (Doc. 9859, 4th ed.). International Civil Aviation Organization. Accessed 12/06/2024.

Insua, D. R., Alfaro, C., Gómez, J., Hernández-Coronado, P., & Bernal, F. (2016). A framework for risk management decisions in aviation safety at state level. *Reliability Engineering and System Safety, 179*, 74–82. https://doi.org/10.1016/j.ress.2016.12.002

International Council of Air Shows. (2012, March 8). ICAS individual risk management worksheet. ICAS. https://airshows.aero/ViewDoc/2781

International Council of Air Shows. (2018). *Air show performers safety manual*. International Council of Air Shows.

Jarvis, S. (2009, December). Human factors—Performer safety debrief. ICAS Convention 2009, Las Vegas. https://publications.airshows.aero/AS1Q21/13/#zoom=z

Jensen, R. S. (1995). *Pilot judgment and crew resource management*. Avebury Aviation.

Jensen, R. S., & Benel, R. A. (1977). Judgment evaluation and instruction in civil pilot training. *Human Factors, 19*(1), 1–10.

Ji, M., You, X., Lan, J., & Yang, S. (2011). The impact of risk tolerance, risk perception and hazardous attitude on safety operation among airline pilots in China. *Safety Science, 49*(10), 1412–1420. https://doi.org/10.1016/j.ssci.2011.06.007

Joseph, J., & Reddy, V. (2013). Risk perception and risk-taking behavior among helicopter pilots in the Indian army. *Indian Journal of Aerospace Medicine, 57*(1), 1–10.

Kabat-Zinn, J. (1994). *Wherever you go, there you are: Mindfulness meditation in everyday life*. Hyperion.

Kaminska, A., Irwin, A., Ray, D., & Flin, R. (2021). Pilot is a pilot? Exploration of effects of professional culture in helicopter pilots. In N. L. Black, W. P. Neumann, & I. Noy (Eds.), *Proceedings of the 21st Congress of the International Ergonomics Association (IEA 2021)*. Lecture Notes in Networks and Systems, vol. 221. Springer. https://doi.org/10.1007/978-3-030-74608-7_83

Kao, C. W., Lin, P. S., & Lai, C. C. (2008). Safety culture factors, group differences, and risk perception in five petrochemical plants. *Process Safety and Environmental Protection, 86*(6), 389–399.

Karachalios, M. (2022). *An evaluation of the relationships between resilient safety culture, safety risk parameters, and mindfulness in the international air show community* [PhD dissertation, University of North Dakota]. https://commons.und.edu/theses/4267/

Karachalios, M., & Adjekum, D. K. (2023). *Air show performers: Safety, risk management, and psychological factors*. CRC Press. https://doi.org/10.1201/9781003431879

Kaspers, S., Karanikas, N., Piric, S., van Aalst, R., de Boer, R. J., & Roelen, A. (2017). Measuring safety in aviation: Empirical results about the relation between safety outcomes and safety management system processes, operational activities and demographic data. In *The Seventh International Conference on Performance, Safety and Robustness in Complex Systems and Applications* (pp. 9–16). IARIA.

Kelly, S., & Efthymiou, M. (2021). Human factors in air traffic control: The importance of risk perception and situational awareness. *Safety Science, 139*, 105260.

Kern, T. (1995). *Darker shades of blue: A case study of failed leadership*. https://docslib.org/doc/11768753/darker-shades-of-blue-a-case-study-of-failed-leadership

Key, K., Hu, P., Choi, I., & Schroeder, D. (2023, May). Safety culture assessment and continuous improvement in aviation: A literature review. *FAA Report DOT/FAA/AM-23/13*. Accessed 13/06/2024. www.faa.gov/sites/faa.gov/files/23-13-Safety%20Culture%20Assessment%20and%20Continuous%20Improvement%20in%20Aviation-%20A%20Literature%20Review.pdf

Koglbauer, I. (2016). A system-theoretic model of threat and error management in flight instruction. https://doi.org/10.15405/epsbs.2016.12.32

Kouzes, J., & Posner, B. (2017). *The leadership challenge* (6th ed.). John Wiley & Sons.

Langer, E. J. (1989). *Mindfulness*. Addison-Wesley Publishing Company.

Leveson, N. (2012). *Engineering a safer world: Applying systems thinking to safety*. MIT Press.

Li, G., Baker, S. P., Grabowski, J. G., Qiang, Y., McCarthy, M. L., & Rebok, G. W. (2003). Age, flight experience, and risk of crash involvement in a cohort of professional pilots. *American Journal of Epidemiology, 157*(10), 874–880. https://doi.org/10.1093/aje/kwg071

Li, Y., Chen, H., Xin, X., & Ji, M. (2020). The influence of mindfulness on mental state with regard to safety among civil pilots. *Journal of Air Transport Management, 84*, 101768. https://doi.org/10.1016/j.jairtraman.2020.101768

Liao, M.-Y. (2015). Safety culture in commercial aviation: Differences in perspective between Chinese and Western pilots. *Safety Science, 79*, 193–205. https://doi.org/10.1016/j.ssci.2015.05.011

Luciani, F., Veneziani, G., Ciacchella, C., Rocchi, G., Reho, M., Gennaro, A., & Lai, C. (2022). Safety at high altitude: The importance of emotional dysregulation on pilots' risk attitudes during flight. *Frontiers in Psychology, 13*. https://doi.org/10.3389/fpsyg.2022.1042283

Martin, W. L. (2019). Crew resource management and individual resilience. In B. G. Kanki, J. Anca, & T. R. Chidester (Eds.), *Crew resource management* (3rd ed., pp. 207–226). Academic Press. https://doi.org/10.1016/B978-0-12-812995-1.00007-5

Martinussen, M., & Hunter, D. R. (2018). *Aviation psychology and human factors* (2nd ed.). CRC Press.

Matyukhina, O., Ordenov, S., Poda, T., & Sukhova, N. (2021). The effect of socio-cultural factors on the international flight safety research findings. *E3S Web of Conferences, 258*, 07001. https://doi.org/10.1051/e3sconf/202125807001

Mjøs, K. (2004). Basic cultural elements affecting the team function on the flight deck. *The International Journal of Aviation Psychology, 14*(2), 151–169. https://doi.org/10.1207/s15327108ijap1402_3

Molesworth, B., & Chang, B. (2009). Predicting pilots' risk-taking behavior through an implicit association test. *Human Factors, 51*, 845–857. https://doi.org/10.1177/0018720809357756

Morley, D., Newsome, J., Moore, D., Comeau, P., Neal, B. A., Stoliker, P. C., Karwal, A., Buck, C., Kirli, O., Rice, C. B., Henderson, F. P. J., Stevenson, J. R., & Mortensen, J. A. (2021). Flight test safety and risk management. *The NATO Science and Technology Organization, 32*, 210. https://apps.dtic.mil/sti/trecms/pdf/AD1183573.pdf

Mowbray, D. (2011). *Building resilience: Seven steps to achieving positive mental health in the workplace*. Mental Health at Work.

Muermans, G. A. (2002). Human factors in aerobatic flight: Stress, fatigue, and spatial disorientation. *Aviation, Space, and Environmental Medicine, 73*(12), 1211–1218.

National Academies of Sciences, Engineering, and Medicine. (2009). *Building a resilient workforce: Opportunities for the Department of Homeland Security*. The National Academies Press.

Nosita, F., Pirzada, K., Lestari, T., & Cahyono, R. (2020). Impact of demographic factors on risk tolerance. *Journal of Security & Sustainability Issues, 9*(4), 1327–1336. https://doi.org/10.9770/jssi.2020.9.4(18).

NTSB. (2012). *Medical certification of pilots*. National Transportation Safety Board.

Nuñez, B., López, C., Velazquez, J., Mora, O. A., & Román, K. (2019). Hazardous attitudes in US Part 121 airline accidents. In *20th International Symposium on Aviation Psychology* (pp. 37–42). https://corescholar.libraries.wright.edu/isap_2019/7

O'Hare, D. (1990). Pilots' perceptions of risks and hazards in aviation. *Aviation, Space, and Environmental Medicine, 61*(7), 611–618.

Olsen, J. A. (2016). Boyd revisited: A great mind with a touch of madness. *Air Power History, 63*(4), 7–16. www.jstor.org/stable/26276810

Orasanu, J., Fischer, U., & Davison, J. (2002). Risk perception: A critical element of aviation safety. *IFAC Proceedings Volumes, 35*(1), 49–58. https://doi.org/10.3182/20020721-6-ES-1901.01639

Orasanu-Engel, J., & Mosier, K. L. (2019). Flight crew decision making. In B. G. Kanki, J. Anca, & T. R. Chidester (Eds.), *Crew resource management* (3rd ed., pp. 139–183). Academic Press. https://doi.org/10.1016/B978-0-12-812995-1.00005-1

Pamplona, D., & Jorge, C. (2020). Does a fighter pilot live in the danger zone? A risk assessment applied to military aviation. *Transportation Research, 1*. https://doi.org/10.1016/j.trip.2020.100114

Papadakis, M. (2008). An initial study to discover the human factors underlying air display accidents [Unpublished master's thesis]. Cranfield University.

Paul, P. S., & Maiti, J. (2007). The role of behavioral factors on safety management in underground mines. *Safety Science, 45*(4), 449–471.

Pidgeon, N. F., & O'Leary, M. (2017). Organizational safety culture: Implications for aviation practice. In G. M. Davies (Ed.), *Aviation psychology in practice* (pp. 21–43). Ashgate.

Reader, T. (2016). Measuring and improving safety culture in the aviation industry. *London School of Economics*. Retrieved from www.lse.ac.uk

Reason, J. (1997). *Managing the risks of organizational accidents*. Ashgate Publishing.

Renn, O., & Rohrmann, B. (2000). Cross-cultural risk perception: State and challenges. In O. Renn & B. Rohrmann (Eds.), *Cross-cultural risk perception*. Technology, Risk, and Society, vol 13. Springer. https://doi.org/10.1007/978-1-4757-4891-8_6

Rhodes, N., & Pivik, K. (2011). Age and gender differences in risky driving: The roles of positive affect and risk perception. *Accident Analysis & Prevention, 43*(3), 923–931.

Riviere, L. A., Merrill, J. C., Wilk, J. E., & Adler, A. B. (2011). Marital status & risk behaviors among US soldiers. Retrieved June 12, 2024. https://apps.dtic.mil/sti/tr/pdf/ADA582880.pdf

Rochlin, G. I. (1999). Safe operation as a social construct. *Ergonomics, 42*(11), 1549–1560.

Roussanov, N., & Savor, P. (2014). Marriage and managers' attitudes to risk. *Management Science, 60*(10), 2496–2508.

Saivasan, R., & Lokhande, M. (2022). Influence of risk propensity, behavioral biases and demographic factors on equity investors' risk perception. *Asian Journal of Economics and Banking, 6*(3), 373–403. https://doi.org/10.1108/AJEB-06-2021-0074

Salas, E., Wilson, K. A., Burke, C. S., & Wightman, D. C. (2006). Does crew resource management training work? An update, an extension, and some critical needs. *Human Factors, 48*(2), 392–412.

Scharf, M. T., & Cross, J. D. (2019). Analysis of low-time pilot attitudes in university aviation association member flight. *Collegiate Aviation Review International*. https://doi.org/10.22488/okstate.19.100216

Schopf, A. K., Stouten, J., & Schaufeli, W. B. (2021). The role of leadership in air traffic safety employees' safety behavior. *Safety Science, 135*, 105118. https://doi.org/10.1016/j.ssci.2020.105118

Schwarz, M., Kallus, K. W., & Gaisbachgrabner, K. (2016). Safety culture, resilient behavior, and stress in air traffic management. *Aviation Psychology and Applied Human Factors, 6*(1), 1–11. https://doi.org/10.1027/2192-0923/a000091

Serfontein, E., & Govender, K. K. (2021). The relationship between resilience and organizational control systems in the South African aviation industry. *Journal of Transport and Supply Chain Management, 15*. https://doi.org/10.4102/jtscm.v15i0.598

Shirali, G. A., Mohammadfam, I., & Ebrahimipour, V. (2016a). Safety culture and resilience engineering: A symbiotic relationship. *Safety Science, 86*, 202–210.

Shirali, G. A., Shekari, M., & Angali, K. A. (2016b). Quantitative assessment of resilience safety culture using principal components analysis and numerical taxonomy: A case study in a petrochemical plant. *Journal of Loss Prevention in the Process Industries, 40*, 277–284. https://doi.org/10.1016/j.jlp.2016.01.007

Sicard, B., Taillemite, J. P., Jouve, E., & Blin, O. (2003). Risk propensity in commercial and military pilots. *Aviation, Space, and Environmental Medicine, 74*(8), 879–881.

Siegrist, M., & Árvai, J. (2020). Risk perception: Reflections on 40 years of research. *Risk Analysis, 40*, 2191–2206. https://doi.org/10.1111/risa.13599

Signoret, J. P., & Leroy, A. (2021). Fault tree analysis (FTA). In *Reliability assessment of safety and production systems*. Springer Series in Reliability Engineering. Springer. https://doi.org/10.1007/978-3-030-64708-7_16

Şimşekoğlu, Ö., Nordfjærn, T., Hezaveh, A. M., Mamdoohi, A. R., & Rundmo, T. (2013). Risk perceptions, fatalism, and driver behaviors in Turkey and Iran. *Safety Science, 59*, 187–193.

Sjöberg, L., Moen, B. E., & Rundmo, T. (2004). *Explaining risk perception: An evaluation of the psychometric paradigm in risk perception research*. Rotunde.

SKYbrary. (2021). Socio-cultural effects on the flight deck. Retrieved from www.skybrary.aero

Staal, M. A. (2004). *Stress, cognition, and human performance: A literature review and conceptual framework*. NASA Ames Research Center.

Thompson, M. (1995, May 29). Way, way off in the wild blue yonder. *Time*. http://content.time.com/time/subscriber/article/0,33009,982358,00.html

Tränkle, U., Gelau, C., & Metker, T. (1990). Risk perception and age differences. *Accident Analysis & Prevention, 22*(2), 119–125.

Turner, M. J. (1995). Psychological factors associated with performance limits in aviation (ERA Report 95-14). Embry-Riddle Aeronautical University.

United Kingdom Civil Aviation Authority. (2025). *CAP 403: Flying displays and special events: Safety and administrative requirements and guidance* (Edition 22). www.caa.co.uk/our-work/publications/documents/content/cap-403/

U.S. Air Force. (1994). *Summary of AFR 110-14 USAF Accident Investigation Board Report* (Aircraft Accident Investigation Report 110–14). United States Air Force. www.baaa-acro.com/crash/crash-boeing-b-52h-stratofortress-fairchild-afb-4-killed

Van Damme, D., & Zahner, D. (Eds.). (2022). *Does higher education teach students to think critically?*. OECD Publishing. https://doi.org/10.1787/cc9fa6aa-en

Vredenburgh, A., & Cohen, H. (1995). Does culture affect risk perception? *Proceedings of the Human Factors and Ergonomics Society Annual Meeting, 39*, 1015–1019. https://doi.org/10.1177/154193129503901511

Wald, A. (2004). Managing pilot stress. In *Aviation Resource Management: Proceedings of the Fourth Australian Aviation Psychology Symposium* (Vol. 2, pp. 175–184). Ashgate Publishing Ltd.

Wang, L., Gao, S., Hong, R., & Jiang, Y. (2023). Effects of age and flight exposure on flight safety performance: Evidence from a large cross-sectional pilot sample. *Safety Science, 165*, 106199. https://doi.org/10.1016/j.ssci.2023.106199

Wang, L., & Zhang, J. (2020). The effect of psychological risk elements on pilot flight operational performance. *Human Factors and Ergonomics in Manufacturing & Service Industries, 30*(3), 3–13. https://doi.org/10.1002/hfm.20816

Weick, K. E., & Sutcliffe, K. M. (2007). *Managing the unexpected: Resilient performance in an age of uncertainty*. Jossey-Bass.

White, M. P., Eiser, J. R., & Harris, P. R. (2011). Risk perceptions of road users: A comparison of perceived accident likelihood and injury severity. *Risk Analysis, 31*(7), 1258–1268.

Wiegmann, D. A., & Shappell, S. A. (2003). *A human error approach to aviation accident analysis: The human factors analysis and classification system*. Ashgate.

Xu, Q., Wu, Y., Wang, M., Liu, B., Jiang, J., You, X., & Ji, M. (2022). The relationship between sense of calling and safety behavior among airline pilots: The role of harmonious safety passion and safety climate. *Safety Science, 150*, 105718.

You, X., Ji, M., & Han, H. (2013a). The effects of risk perception and flight experience on airline pilots' locus of control with regard to safety operation behaviors. *Accident Analysis & Prevention, 57*, 131–139. https://doi.org/10.1016/j.aap.2013.03.036

You, X., Ji, M., & Han, H. (2013b). The impact of risk perception and risk tolerance on safety behavior: A study of airline pilots in China. *Safety Science, 57*, 173–181.
Younas, A., Ali, P., & Fazal, A. (2021). The framework of triangulation in mixed methods research: A conceptual model. *International Journal of Social Research Methodology, 24*(5), 659–673.
Zhang, Y., Dong, C., Guo, W., Dai, J., & Zhao, Z. (2022). Systems theoretic accident model and process (STAMP): A literature review. *Safety Science, 152*, 105596.
Zhao, Y., Zhu, K., Zhang, J., et al. (2024). Exploring the measurement of psychological resilience in Chinese civil aviation pilots based on generalizability theory and item response theory. *Scientific Reports, 14*, 1856. https://doi.org/10.1038/s41598-024-52229-7

FURTHER READINGS

Hobbs, A., & Avers, K. B. (2011). *Managing fatigue in aviation.* Ashgate Publishing, Ltd.
Reinhart, R. O. (2008). *Basic flight physiology.* McGraw Hill Professional.
Rodrigues, C. C., & Cusick, S. K. (2012). *Commercial aviation safety.* McGraw-Hill Education.

4 The Odyssey of Pilot Selection

Manolis Karachalios

Contributor
Daniel Kwasi Adjekum

INTRODUCTION

As we conclude "Part I: Pilot Selection—The Foundation of Excellence," we arrive at this Chapter, a fitting culmination of our exploration into the meticulous process of selecting air show performers. Titled "The Odyssey of Pilot Selection," this Chapter captures the essence of our journey through the intricate and challenging pathways that aspiring pilots traverse to achieve excellence in air show performance.

The term "Odyssey" was chosen deliberately to evoke the epic nature of the pilot selection process. Much like the ancient Greek hero Odysseus, who embarked on a long and arduous journey filled with trials and triumphs, aspiring pilots undergo a similarly transformative experience. The selection process is not merely a series of evaluations; it begins a much larger journey, marking the start of an aspiring pilot's transformation into a distinguished air show performer.

This odyssey metaphor highlights several key aspects of the pilot selection process. First, there are the challenges and trials. Just as Odysseus faced numerous challenges on his journey, aspiring pilots can encounter rigorous tests of their skills, knowledge, and endurance. The selection process involves psychological profiling, stringent physical assessments, exhaustive skill evaluations, and continuous performance reviews. Each step is designed to push candidates to their limits, ensuring only the most capable and resilient are chosen.

Moreover, Odysseus's journey was one of personal growth and transformation, similar to that of an aspiring air show performer. Through the selection process, pilots develop their technical abilities, mental fortitude, decision making skills, and leadership qualities. This transformation is crucial for performing complex aerobatic maneuvers safely and spectacularly.

The odyssey of pilot selection also requires unwavering commitment and perseverance. Aspiring pilots must demonstrate a deep dedication to their craft, often sacrificing personal time and enduring intense training regimens. This relentless pursuit of excellence sets the extraordinary performers apart from the merely competent.

Furthermore, like Odysseus' voyage, the journey of an air show performer is ongoing. Even after selection, pilots continue to refine their skills, adapt to new

DOI: 10.1201/9781003504801-6

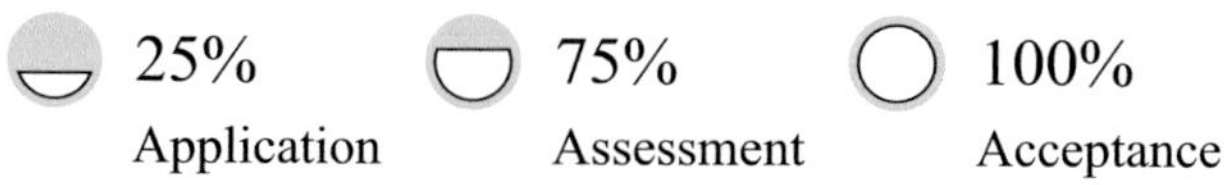

FIGURE 4.1 Path to becoming an air show performer.

technologies, and push the boundaries of aerial performance. The selection is merely the beginning of a lifelong odyssey of continuous improvement and innovation.

To better understand the progressive nature of the pilot selection process, the accompanying graph, Figure 4.1, illustrates the incremental journey toward becoming an air show performer. Initially, candidates undergo basic evaluations to ensure they meet minimum standards. This phase includes initial qualifications, basic physical fitness tests, preliminary interviews, and basic aptitude assessments, which comprise a quarter of the path to becoming an air show performer. Following this, the intermediate evaluation phase involves more comprehensive assessments, including detailed psychological evaluations, intermediate physical fitness tests, in-depth interviews, and initial practical tests to assess basic flying skills and decision making abilities.

The next stage, advanced assessment, intensifies the scrutiny of advanced skills and competencies. Candidates face challenging physical and mental tests and assessments of technical knowledge and practical flying skills. This stage often involves scenario-based evaluations to evaluate responses to complex and high-pressure situations.

Finally, after completing three-quarters of the effort to become an air show performer, the process culminates in the final selection stage, where the most demanding evaluations are conducted. Candidates undergo final practical flying tests, comprehensive psychological assessments, and evaluations of teamwork and leadership abilities, ensuring that only the most capable and well-rounded individuals are selected to become air show performers.

In this Chapter, we present three illuminating case studies. First, we explore the path of "Becoming the First F-16 Demo Pilot for the Hellenic Air Force," a personal narrative by the author. This case study offers an intimate glimpse into the challenges and triumphs of ascending to this prestigious role, showcasing the blend of skill, determination, and resilience required.

Next, we examine the Red Arrows in "The Transition from Seasoned Pilots to Precision Performers." This case study explores the transformation of experienced aviators into an elite aerobatic team renowned worldwide, emphasizing the rigorous training and unwavering discipline that define their journey.

Our third case study sheds light on Patty Wagstaff's journey to becoming a civilian air show performer, the often overlooked yet equally demanding path civilian pilots take. This narrative underscores the dedication, passion, and unique challenges those outside the military sphere face as they strive to captivate audiences with their aerial prowess.

In closing, we bring together the threads of these narratives, offering a comparative analysis of the distinct yet parallel journeys of military and civilian pilots. This synthesis represents the commonalities and divergences in their paths, highlighting the universal qualities that define excellence in air show performance.

COMMON PREREQUISITES FOR MILITARY DISPLAY PILOTS

Before analyzing the specific case studies, it is important to outline commonly applied, in the international military world, prerequisites for a military pilot to be considered for the role of a display pilot and the rigorous pathway they undertake to reach these levels.

Military Induction and Initial Training

The journey begins with the rigorous process of joining the military, where candidates undergo comprehensive evaluations, including physical, cognitive, and psychological assessments (Martinussen & Hunter, 2018). This phase ensures that only individuals with the highest adaptability, resilience, and leadership potential progress to flight training. The initial phase is crucial as it lays the foundation for a career in military aviation, filtering out those who may not withstand the demanding nature of this profession.

Advanced Flight Training and Specialization

Upon successfully entering the aviation branch, aspiring pilots face a demanding selection for flight training. Here, their aptitude for flying is assessed alongside their ability to endure military aviation's physical and mental rigors (Solen, 2021; Jenkins et al., 2022; Yang et al., 2022). Successful candidates are then exposed to advanced flight disciplines, including aerobatics, formation flying, and low-level navigation. Each discipline requires precision and exceptional situational awareness, which are essential for operational readiness and foundational for a future air show performance (Forces News, 2019).

Mastery of Flight Disciplines

As pilots progress through their training, they specialize in various flight disciplines, which are essential for operational readiness. Mastery in these areas is crucial for aspiring air show performers. Skills such as low-altitude step-down training and complex aerobatic maneuvers become second nature, forming the backbone of the pilot's expertise (International Council of Air Shows (ICAS), 2016, 2019). This training phase ensures that pilots can perform with the highest degree of accuracy and safety, which is crucial for the high-stakes environment of air shows.

Building Resilience and Mental Fortitude

Throughout their career, military pilots build resilience and mental strength, key attributes that enable them to navigate the high-risk, low-altitude environments of air shows (Karachalios, 2022). The challenges faced at each training stage contribute to their growth, equipping them with the robustness required for the air show arena. Resilience and mental fortitude are beneficial and essential for the intense demands and pressures of air show performances.

CASE STUDY 1: BECOMING THE FIRST SINGLE-SHIP FAST JET DEMONSTRATION PILOT FOR THE HELLENIC AIR FORCE, A PERSONAL NARRATIVE BY THE AUTHOR

Before I could examine the journeys of other demo pilots in this book, I had to immerse myself in the rigorous selection process and critically analyze my own odyssey within the Hellenic Air Force (HAF). This introspection is crucial, as it provides a foundation for understanding the challenges and triumphs others face in similar roles.

The HAF maintains rigorous standards for selecting its demonstration pilots, emphasizing a synthesis of skill, experience, and an innate passion for flight. My journey to becoming the first F-16 demo pilot for the HAF is a testament to these high standards. The HAF's leadership sought an individual who was proficient in flying the F-16 and possessed instructional prowess and extensive in-cockpit experience. Specifically, candidates must have logged over 1,000 flight hours in the F-16, underscoring a deep familiarity and nuanced control over the aircraft's capabilities.

This narrative examines the stringent selection process, highlighting the key attributes and experiences that shaped my successful candidacy. The path was marked by intensive training, continuous skill refinement, and a relentless commitment to excellence. By reflecting on my personal and professional journey, I gained insights essential for analyzing and appreciating the journeys of other demo pilots.

This case study provides an intimate glimpse into the demanding criteria set by the HAF, showcasing the dedication and resilience required to become the first F-16 demo pilot. It sets the stage for a deeper exploration of the processes and experiences that define excellence in air show performance.

THE SELECTION FRAMEWORK

The selection process deviated from conventional protocols; there were no formal interviews or public announcements, especially due to the short notice for the upcoming training by the USAF ACC F-16 Viper West Demo Team. The leadership, comprising the squadron and base commanders, relied on their intimate knowledge of their pilots' capabilities and professional demeanors to make their decisions. This approach reflects a deep-seated trust in the operational assessment mechanisms of the HAF and a belief in the commanders' judgment. The offer to become a demo pilot was presented straightforwardly, highlighting the recognition of the pilot's skill and dedication to aviation excellence.

Accepting the role of the demo pilot was a natural progression, driven by a lifelong dedication to aviation and a commitment to showcasing the prowess of the HAF. The selection was an acknowledgment of the pilot's flying skills and a testament to their reputation and standing within the Air Force community. The appointment came with the responsibility of representing the HAF globally, necessitating a commitment to perfection, discipline, and the continuous honing of aerobatic skills.

The selection process also included a backup pilot from the same squadron, ensuring the continuity and robustness of the demo team's operations. Following a similar vetting process, another pilot from the second squadron in Souda was selected. These selections underline the strategic approach of the HAF in creating a resilient and proficient demo team capable of executing complex aerial maneuvers with precision and flair.

Analytical Perspective

This narrative serves as a case study in military precision, strategic planning, and the cultivation of elite talent within the HAF. The direct selection method, including streamlined formal interviews, signifies a unique trust-based approach, where the proven record and the commanders' firsthand experiences with the pilots guide their choices.

The journey to becoming the HAF's first single-ship fast jet demonstration pilot encapsulates a career built on dedication, skill, and a deep-rooted passion for flight. It also reflects the broader ethos of the HAF, where excellence in aviation is nurtured, recognized, and celebrated, setting a benchmark for future generations of pilots aspiring to reach the zenith of aerobatic performance in the service of their country.

Training and Operational Foundation

The path to this esteemed position began with rigorous training within the HAF, where foundational skills were honed and operational competencies solidified. As an F-16 pilot, I had to accumulate extensive flight hours, surpassing the requisite 1,000-hour threshold, demonstrating technical proficiency, and accumulating a deep understanding of the aircraft's capabilities and limitations.

Mentorship and Leadership

Instructor qualification emerged as a crucial criterion, signaling flying expertise and the ability to lead, teach, and maintain the high standards expected of a demo pilot. This aspect of the selection underscores the HAF's emphasis on continuous learning and leadership within its ranks, ensuring that those chosen to represent the Air Force in such high-profile roles can contribute to their peers' development.

The Selection Ethos

The HAF's selection process, characterized by its directness and reliance on the intimate knowledge of the candidates' capabilities and conduct, reflects a broader organizational culture that values trust, familiarity, and proven competence. The expedited formal interviews in this process suggests confidence in the existing evaluation and monitoring systems that continuously assess the performance and potential of HAF pilots.

Building the Demo Team

The selection of a backup pilot and another demo pilot from a different squadron illustrates the comprehensive approach to building a robust, skilled, and cohesive demonstration team. This strategic selection ensures the continuity of excellence in the team's performances and fosters a supportive environment among the pilots.

Reflection and Future Implications

The process of becoming the first single-ship fast jet demo pilot for the HAF, therefore, is a personal achievement and a representation of the HAF's commitment to showcasing aerial excellence on a global stage. It sets a precedent for future selections, emphasizing the importance of operational experience, instructional capability, and a deep-rooted commitment to the ethos of the Air Force.

CASE STUDY 2: RED ARROWS, THE TRANSITION FROM SEASONED PILOTS TO PRECISION PERFORMANCE

The Red Arrows, the United Kingdom's Royal Air Force's esteemed display squadron, boasts an assembly of Hawk T1A jets piloted by the RAF's crème de la crème. The path to earning a coveted spot in this squadron is a testament to excellence, with a stringent selection procedure that filters out all but the finest.

For the Royal Air Force's elite Red Arrows, transitioning from highly experienced frontline pilots to precision aerobatic performers is an odyssey of rigor, discipline, and skill refinement. With the start of the training season, the newest members of this prestigious team began the arduous process of learning and perfecting complex maneuvers as a synchronized unit.

The "new signings"—pilots accustomed to the demands of operational flying in conflict zones like Iraq and Syria—must now adapt to the unique challenges of aerobatic flying. The team leader, a pilot with extensive operational experience, demonstrates the focused nature of their flying in the Red Arrows. While operational sorties could last for hours, the intense concentration required for a 20–25-minute display is of a different caliber altogether.

Every year, the squad sees significant changes, especially at the front of the formation. The spotlight is on the new team lead, which brings operational sharpness to the team. Behind him are Red Two and Red Three, collectively offering over a decade of flying experience. Their addition to the team heralds a blend of excitement and trepidation as they undertake the journey to embody the precision and discipline synonymous with the Red Arrows.

Moreover, the role of Red Ten, the squadron's supervising coach, is essential. His responsibilities are vast, ranging from supervision to public relations and from commentating for the audience to meticulously planning the display season.

The Prerequisites for Prospective Red Arrows Aviators

Becoming a member of this elite squadron is not merely a job but an honor reserved for those who meet an extensive list of prerequisites. Candidates are drawn exclusively from the ranks of serving Royal Air Force (RAF) pilots. The Red Arrows refresh their lineup every year, bringing in three new aviators to continue the squadron's legacy of precision flying.

To even be considered for this elite group, a candidate must (Royal Air Force, United Kingdom, 2024):

- Serve actively as an RAF pilot.
- Accumulated over 1,500 flight hours.
- Have at least one operational tour under their belt.
- Receive an above-average assessment in their current flying role.

This ensures a blend of youthful vigor and seasoned expertise, crucial for the demanding performances of the Red Arrows.

The Burden of Selection

The selection process for the Red Arrows is as demanding as it is prestigious. Out of all applicants, only a handful—nine, based on merit and expertise—are shortlisted for the grueling week-long selection gauntlet.

This gauntlet pushes candidates to their limits through:

- Rigorous flight assessments.
- In-depth interviews.
- Critical peer evaluations.

The successful trio who emerges from this arduous process are then granted the privilege to serve as the squadron's pilots, with designations ranging from "Red 2" to "Red 10". The roles of "Officer Commanding" and the craved position of "Red 1—Team Leader" are appointed through a specialized recruitment path. The Team Leader role is even more selective, requiring a minimum of a three-year tour in an operational capacity as a baseline.

All in all, the journey to join the ranks of the Red Arrows is not for the faint of heart. It is a path paved with exactitude, grueling trials, and the relentless pursuit of aerial perfection. For those who succeed, the skies become a canvas for their skill, a spectacle for the world to witness.

The Selection Week

The journey to becoming a part of the esteemed Red Arrows, the United Kingdom's premier aerobatic team, unfolds as a gripping saga of dreams, pressure, and camaraderie. For six hopeful candidates, the dream of joining this elite squadron culminates in an intense selection week in Cyprus, where the team conducts its training. These pilots, shortlisted from a pool of the RAF's finest, vie for just two available spots, each aiming to replace members concluding their three-year tenure.

For these candidates, the selection week is a high-stakes affair, with relentless scrutiny under the watchful eyes of current team members and evaluators. Despite their ample experience in fast jet operations and possessing over 1,500 flying hours each, the transition from conducting air strikes to performing public displays is a monumental shift. It's not just about flying; it's about executing with theatrical precision, blending raw piloting skills with the artistry of aerobatics.

One candidate described the experience (Forces News, 2013) as being under a constant spotlight, a sentiment echoed by others who find the balance between the tension of assessment and the enjoyment of flying with peers. This week-long trial is not just about individual performance but also about how well each candidate integrates with the existing team dynamics.

In-Depth Interviews

The interview process probes deeper than the candidates' flying competence; it explores their psychological readiness, leadership potential, and commitment to the team's ethos. During these in-depth discussions, candidates are questioned on their career motivations, understanding of aerobatic discipline, and their perspective on teamwork and leadership. The interviews are crucial in determining whether a pilot possesses the mental fortitude and the collaborative spirit essential for the synchronized artistry that the Red Arrows embody.

Rigorous Flight Assessments

In the flight assessments, candidates are subjected to a series of flying tests that evaluate not just their piloting skills but their ability to perform complex aerobatic maneuvers with precision and safety. These tests are designed to simulate the challenging conditions of air show performances, including tight formations, low-altitude flying, and split-second timing. Candidates must demonstrate impeccable control, situational awareness, and adaptability to changing scenarios, mirroring the exacting standards of the Red Arrows' displays. This is where the pressure intensifies, with each maneuver, each formation, and each moment critiqued for precision and flair.

Social Integration: Beyond the Cockpit

However, flying prowess alone doesn't secure a place in the squadron. The social aspect is equally critical. Through activities like bowling, candidates' personalities are assessed to ensure they can gel with the existing team culture and ethos.

This social vetting is essential, given the close-knit nature of the Red Arrows, where trust and camaraderie are paramount for both in-air safety and ground-level interactions.

Critical Peer Evaluations

Peer evaluations add another dimension to the selection process, where the candidates' ability to integrate and operate within a team is scrutinized. Current team members assess candidates on various attributes such as teamwork, communication skills, and their potential to contribute positively to the squad's dynamic. This peer review ensures that new recruits not only match the technical and mental criteria but also fit seamlessly into the close-knit fabric of the Red Arrows, maintaining the squadron's camaraderie and collective spirit.

More Than Just Display Pilots

The selection process shows a critical reality: the chosen candidates must be more than just skilled pilots. They must be ambassadors of the Red Arrows' legacy, carrying forward a tradition of excellence both in the skies and on the ground. The departing members, while nostalgic about their experiences, recognize the necessity of making way for fresh talent, ensuring the continued vibrancy and dynamism of the team.

Ultimately, the selection is about filling positions and continuing a legacy. The six candidates, each capable and accomplished, face a week that tests their skills, resilience, and ability to integrate into a fraternity known for its precision and camaraderie.

Becoming a Red Arrow is a delicate balance of technical skill, rapid adaptability, and social integration, with each element scrutinized to ensure that only the best and most compatible candidates ascend to this prestigious team. The selection process is a *holistic assessment*, ensuring that those who don the red suit are exemplary pilots and team players capable of upholding the legacy of the Red Arrows.

The Synchro Pair Selection Process

The Synchro Pair pilots, designated as Red 6 and Red 7, are an integral part of the Red Arrows aerobatic display team. Apart from flying in formation with the main team, they perform dynamic and opposing maneuvers to showcase the capabilities of their aircraft (Aircrew Interview, 2019). Their displays keep the audience engaged while the main formation prepares for the next maneuver (see Figure 4.2).

Therefore a critical aspect of the Red Arrows' pilot selection is the process of choosing the Synchro Pair pilots. Transitioning to the Synchro Pair, particularly moving from a position like Red 3 to the Synchro role, involves a meticulous and layered selection process. The selection for these roles is not solely based on flying competence but also on the ability to cultivate a profound trust and understanding between the two pilots. This trust is essential for executing the high-speed, close-proximity passes defining Synchro Pair's performances. Therefore, in the Red Arrows, Red 6, who leads the Synchro Pair, selects the new Red 7.

FIGURE 4.2 Red Arrows Synchro Pair at RIAT 2018. (Copyright: Tim Felce.)

The Journey from Red 3 to Synchro

Usually, the experience of moving from the position of Red 3 to becoming part of the Synchro Pair encapsulates the additional responsibilities and the heightened level of flying discipline required. In their second year with the team, a pilot could be selected as Red 7, automatically progressing to Synchro Leader in the third year. This progression also grants the Synchro Leader the prerogative to select their wingman.

Training and Trust

The selection is intricately tied to the Synchro Leader's assessment of who they believe can be trained effectively and can swiftly rectify any mistakes. This choice reflects the paramount importance of trust and mutual understanding in executing the close proximity aerobatics that define the Synchro Pair's displays. The training for the Synchro Pair is intense, involving three flights per day, pushing the pilots and aircraft to their limits to perfect their display routines.

Emotional and Professional Milestone

The announcement of the Synchro Pair selection is a significant event within the team, often revealed during a cocktail party following an air show, marking it as a momentous occasion for the team members, a pinnacle moment filled with joy and the realization of the additional flying discipline that awaited them.

CASE STUDY 3: PATTY WAGSTAFF, A CIVILIAN DISPLAY PILOT'S EVOLUTION

The journey to becoming a civilian air show performer, unlike their military counterparts, follows a distinct trajectory with different requirements and challenges. Taking the example of Patty Wagstaff (see Figure 4.3), a legendary air show pilot (International Council of Air Shows Foundation, 2007; National Aviation Hall of Fame, 2024), we can trace the path of progression and the meticulous preparation required to achieve air show excellence.

Early Beginnings and Initial Training

Patty Wagstaff's aviation journey began with a strong passion for flying, leading her to take basic flight lessons and earn her pilot's license. Civilian display pilots often start with a deep-seated interest in aviation, progressing from basic flight training to obtaining private and commercial pilot licenses. This initial phase is crucial, laying the foundation for advanced aerobatic skills (Wagstaff & Cooper, 1997).

Advanced Aerobatic Training

After mastering basic flying, Wagstaff dove into aerobatics, a discipline requiring precision, control, and a thorough understanding of aircraft limits. Civilian pilots like her often seek specialized aerobatic training schools to learn complex maneuvers and enhance their flying skills. This stage is vital for developing the technical competence necessary for air show performances (International Council of Air Shows (ICAS), 2016).

Gaining Qualifications and FAA Authorization

The Federal Aviation Administration (FAA) utilizes Categories of Aerobatic Competency (CATs) to authorize pilots for air show performances at varying altitudes, ensuring pilots demonstrate proficiency and safety in executing aerobatic maneuvers (International Council of Air Shows (ICAS), 2019). The system is structured to gradually advance pilots through levels as they gain experience and demonstrate their competency.

New pilots start at Level 4, where they must maintain a minimum altitude of 800 feet. As they gain experience and demonstrate their abilities, they can progress to lower altitude levels, ultimately reaching Level 1, which allows for unlimited altitudes.

Experienced pilots who have progressed through the categories and have demonstrated their proficiency in a safe manner are authorized to perform at lower altitudes (e.g., 250 feet or no limit in the minimum allowed altitude above the surface). Conversely, new pilots must start at a higher minimum altitude (800 feet) and demonstrate their proficiency in executing aerobatic maneuvers within these specified altitudes to then aim for lower-altitude authorizations.

FIGURE 4.3 Patty Wagstaff, Air Show Hall of Fame inductee 2006. (Copyright: Patty Wagstaff.)

Overcoming Challenges and Achieving Excellence

The odyssey for civilian air show performers like Wagstaff involves not only acquiring flying skills but also overcoming numerous challenges. For women and minorities in particular, the journey includes breaking through gender and racial biases within the industry. Early in her career, Patty was told she couldn't become an airline pilot because she was a girl, reflecting the gender biases in the industry (French, 2023). Despite this, she resolved to prove her capabilities and educate others on why she could succeed, maintaining a sense of humor along the way.

Logistically, performers face challenges with securing practice areas, managing extensive paperwork, and fulfilling contractual obligations promptly. Effective communication and collaboration with event organizers are essential to overcome these hurdles, as outlined by event organizers, emphasizing the importance of planning, safety compliance, and maintaining professional standards to achieve excellence in performances (Flight Global, 2018; International Council of Air Shows, 2023).

Financial challenges are significant in the world of air show pilots. The cost of maintaining, rebuilding and insuring their aircraft can easily consume the earnings from performances. To offset these expenses, performers often rely on sponsorships, which typically offer parts and discounts rather than cash. Additionally, many performers need to have high-paying day jobs to support their air show careers during the initial years. This financial strain requires careful budgeting and the ability to secure and maintain sponsor relationships (Scott, 2005).

The Role of FAA and Mentorship in Progression

Finding the right community and support network is also crucial, as pursuing a career in aviation cannot be done in isolation.

For civilian pilots, progressing through the FAA's CATs and achieving the pinnacle of air show performance is a testament to their dedication, skill, and passion for aviation. The mentorship they receive and the regulatory framework they navigate shape their journey, ensuring that they not only perform spectacularly but also adhere to the highest safety standards.

In summary, the journey of a civilian display pilot like Patty Wagstaff, from initial flight training to becoming an air show performer, encapsulates a unique odyssey filled with rigorous training, mentorship, and continuous aerobatic skill development. Overcoming the challenges and meeting the strict requirements of FAA authorizations exemplify the commitment and resilience needed to excel in the air show arena.

THE ODYSSEY OF AIR SHOW PERFORMERS: CIVILIAN AND MILITARY PATHWAYS

The journey to becoming an air show performer is characterized by a rigorous training regime, a strong passion for flying, and an unwavering commitment to excellence. This path is divided into civilian and military routes, each with distinct challenges, training methodologies, and goals.

Civilian air show performers often embark on their journey out of sheer love for aviation, leading them to private flight training (Szurovy & Goulian, 1996). Their path is self-directed and requires navigating through various levels of flight instruction, culminating in specialized aerobatic training. The progression in this realm is not only about mastering the aircraft but also about developing a unique presence in the air show circuit, where individual flair and creativity in performance are highly valued.

The experience and growth of civilian performers are significantly shaped by mentorship and practical exposure, typically gained through participation in aerobatic clubs and regional air shows. These platforms are crucial for performers to accumulate flight hours and refine their skills in aerobatic maneuvers.

Regulatory compliance, particularly with the FAA in the United States, is a critical aspect of a civilian air show performer's career. The FAA outlines specific aerobatic competency endorsements, necessitating pilots to demonstrate their prowess at higher altitudes before progressing to lower, more spectator-friendly levels (FAA, 2023).

For civilian performers, the journey involves substantial personal investment, as they are responsible for funding their training, aircraft acquisition, and maintenance (Wagstaff, 2013). Despite these financial hurdles, civilian pathways offer greater flexibility in routine design and performance style, allowing for a more personalized expression in their aerial displays.

In contrast, the military route to becoming an air show performer is marked by structured and comprehensive training within the armed forces. Military pilots undergo extensive training that includes aerobatics, formation flying, and low-level navigation, laying the groundwork for their future in air show performances (Martinussen & Hunter, 2018). The selection of elite aerobatic teams like the Red Arrows or Blue Angels is highly competitive based on operational experience and flight skills (Hunter & Hawker, 2009).

Military performers benefit from the resources and support of the armed forces, providing access to advanced aircraft and maintenance facilities. However, their creative autonomy is often limited by the collective objectives and traditions of the military and the aerobatic team.

The military pathway presents challenges that are more institutional and less financial, with the rewards including the honor of representing the military and showcasing aerial skills on a national or international stage (Martinussen & Hunter, 2018).

Table 4.1's comparison reveals the distinct yet equally demanding paths to becoming an air show performer, highlighting the unique challenges and rewards inherent in both civilian and military routes. While civilian performers navigate a journey marked by personal investment and creative expression, military performers follow a trajectory defined by disciplined training and operational excellence.

TABLE 4.1
Paths to Becoming Civilian and Military Air Show Performers

Aspect	Civilian Performers	Military Performers
Training and Pathway	Self-driven, diverse training routes.	Structured, uniform training within the military.
Experience Building	Competitions, regional shows, self-mentorship.	Operational missions, team-based training.
Regulatory Adherence	FAA regulations, progressive licensing.	Military standards and protocols.
Financial Investment	Significant personal investment.	Funded by the military.
Creative Freedom	High, individual routine design.	Limited, team-focused routines.
Resource Access	Dependent on personal resources.	Access to military aircraft and maintenance support.
Performance Style	Individualistic and varied.	Uniform and disciplined.
Career Progression	Based on personal skill development and networking.	Defined by military service and team selection.

REFERENCES

Aircrew Interview (Director). (2019, August 12). RAF Red Arrows, with Mike Ling- Part 1 [Video]. YouTube. www.youtube.com/watch?v=MFlhyO6WDuk

Federal Aviation Administration (FAA). (2023). *Order 8900.1, Volume 5, Chapter 9, Section 1: Issue/Renew/Reevaluate/Rescind a Statement of Aerobatic Competency (SAC).* Dynamic Regulatory System. https://drs.faa.gov/browse/excelExternalWindow/drsdocid106601797520230109222008.0001

Flight Global. (2018, March 14). Patty Wagstaff, member of US Aerobatic team and founder of Aerobatic School. Flight Global. www.flightglobal.com/interview-patty-wagstaff-member-of-us-aerobatic-team-and-founder-of-aerobatic-school/126869.article

Forces News (Director). (2013, May 17). Pilots compete for Red Arrows place [Video]. YouTube. www.youtube.com/watch?v=TD-LWv8V2WE

Forces News. (2019, April 9). Longest-serving Red Arrow leaves RAF after "incredible" career. www.forcesnews.com/news/longest-serving-red-arrow-leaves-raf-after-incredible-career

French, R. (2023, April 8). Patty Wagstaff reflects on her aviation career. *FOX 13 News*; FOX 13 Tampa Bay. www.fox13news.com/news/patty-wagstaff-reflects-on-path-that-led-to-her-to-becoming-one-of-the-nations-top-air-show-pilots

Hunter, J., & Hawker, J. (2009). *Red arrows: The royal air force aerobatic team in action.* Touchstone Books.

International Council of Air Shows Foundation. (2007, September 27). Patty Wagstaff, Air Show Hall of Fame. Retrieved from https://web.archive.org/web/20070927091607/www.icasfoundation.org/hall_fame/2006/wagstaff.htm

International Council of Air Shows (ICAS). (2016). *Air show performers safety manual* (p. 27). ICAS. www.airshows.aero/GetDoc/3444

International Council of Air Shows (ICAS). (2019, January 1). ACE Manual—Revision 9. International Council of Air Shows.

International Council of Air Shows (ICAS). (2023, November 27). What event organizers want performers to know. *Airshow Digest*. https://airshowdigest.aero/2023/11/27/what-event-organizers-want-performers-to-know/

Jenkins, P. R., Caballero, W. N., & Hill, R. R. (2022). Predicting success in United States Air Force pilot training using machine learning techniques. *Socio-Economic Planning Sciences, 79*, 101121.

Karachalios, M. (2022). *An evaluation of the relationships between resilient safety culture, safety risk parameters, and mindfulness in the international air show community* [PhD dissertation, University of North Dakota]. https://commons.und.edu/theses/4267/

Martinussen, M., & Hunter, D. R. (2018). *Aviation psychology and human factors*. CRC Press.

National Aviation Hall of Fame. (2024). Patricia "Patty" Wagstaff. Retrieved from https://nationalaviation.org/enshrinee/patricia-patty-wagstaff/

Royal Air Force. (2024). Red Arrows, Royal Air Force. Royal Air Force. www.raf.mod.uk/display-teams/red-arrows/

Scott, P. (2005, May). So you want to be an airshow pilot. *Smithsonian Magazine*. www.smithsonianmag.com/air-space-magazine/so-you-want-to-be-an-airshow-pilot-8329275/

Solen, D. (2021). Initial fighter pilot training in the PLA Air Force. China Aerospace Studies Institute. Air University.

Szurovy, G., & Goulian, M. (1996). *Advanced aerobatics* (1st ed.). McGraw-Hill Professional.

Wagstaff, P. (2013, March 5). The life of an air show pilot. *Plane & Pilot Magazine*. www.planeandpilotmag.com/article/the-life-of-an-air-show-pilot-2/

Wagstaff, P., & Cooper, A. L. (1997). *Fire and air: A life on the edge* (1st ed.). Chicago Review Press.

Yang, L., Zhang, Y., Zhang, Y., Liao, Y., Du, J., & Geng, X. (2022). Research on training method of information processing ability of military pilots. In *Man-Machine-Environment System Engineering: Proceedings of the 21st International Conference on MMESE* (pp. 753–759). Springer Singapore.

FURTHER READINGS

Culver, C. (n.d.). The best judges of airshow performers: Other performers. *Smithsonian Magazine*. Retrieved July 21, 2024, from www.smithsonianmag.com/air-space-magazine/best-judges-airshow-performers-other-performers-180955089/

Gibson, K. B. (2013). *Women aviators: 26 stories of pioneer flights, daring missions, and record-setting journeys*. Chicago Review Press. https://archive.org/details/womenaviators26s0000gibs/page/156/mode/2up

Part II

Training

Sharpening the Edge

Manolis Karachalios

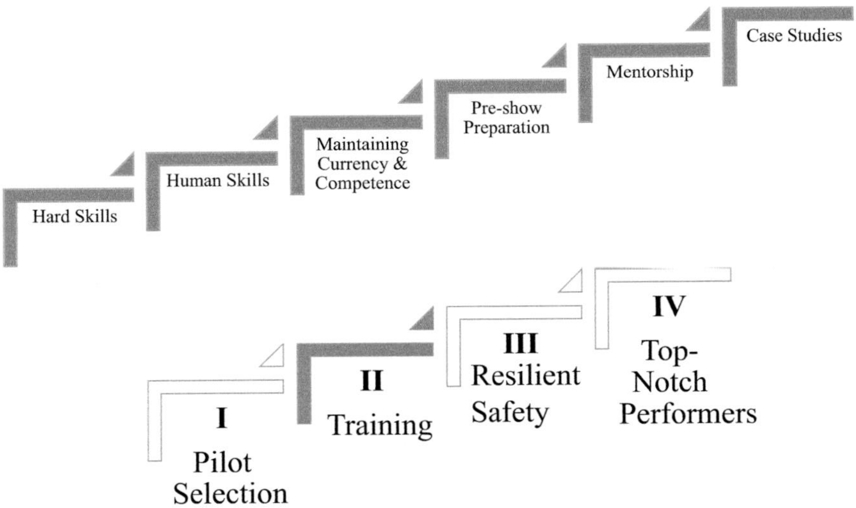

FIGURE PII.1 Steps to air show excellence: Training.

DOI: 10.1201/9781003504801-7

THE AIR SHOW WEEKEND

Understanding the requisite skills that need to be developed under training for an air show performer necessitates a thorough comprehension of its demanding time schedule, which most commonly takes place during a weekend, requiring them to operate in peak performance under high-pressure conditions. The air show weekend can be likened to a sports championship finals, where the focus shifts exclusively to execution, leaving no room for further training.

Detailed planning is required before the air show weekend, which usually begins on Thursday, the arrivals day. This initial phase emphasizes organizational and logistical competencies. Pilots must ensure that all equipment necessary, such as ribbon cut poles, smoke oil, and pyrotechnics, for the flying display is appropriately arranged and that preliminary preparations are completed. Proficiency in flight coordination and operations management is imperative during this stage to ensure a seamless start to the event.

Friday is dedicated to the flying display rehearsal and public engagements. Throughout the day, pilots participate in sponsor meetings, media interactions, and validation flights. These activities demand a synthesis of advanced aerobatic skills, effective communication, and sponsor relations. The training regimen leading up to this day includes advanced aerobatics, media training, and strategies for sponsor engagement. These preparatory sessions are crucial for fine-tuning maneuvers and building relationships with key stakeholders. The evening activities often involve social events and possibly flying activities for a specific audience, i.e., sponsors, highlighting the importance of networking and social engagement. Therefore, display pilots are necessary to demonstrate adeptness in public relations and enhance their social skills to foster rapport with peers and industry leaders.

The weekend reaches its zenith on Saturday and Sunday with open to the public shows. Morning activities involve compulsory safety briefings and final preparations, demanding strict adherence to protocols and a comprehensive understanding of the sequence of events that will take place. Training in regulatory compliance and safety briefing protocols is essential to ensure all safety measures and regulations are meticulously followed. During the day, pilots perform for the general public and engage in interviews and autograph sessions. These activities necessitate expertise in aerobatic flying and public engagement. Preparatory training could include public speaking and autograph session management, equipping pilots to deliver impeccable performances while effectively engaging with the audience, an ability crucial as it enhances the overall experience and reputation of the performer.

Sunday evenings are frequently reserved for attending air show social events and preparing for departure or the subsequent show. This period emphasizes the significance of networking and maintaining stamina, which is essential for sustaining energy levels and remaining connected within the aviation community. Training in social engagement and stress management is paramount during these intervals. Effective stress management techniques, such as mindfulness and relaxation exercises, help pilots maintain focus and composure throughout the weekend.

Interspersed between shows, pilots address logistics and administrative tasks. Activities such as ferrying aircraft and equipment to the next show, managing administrative responsibilities, and coordinating with event organizers require careful logistical planning and administrative expertise. Aircraft ferrying logistics and business administration are vital to ensure seamless transitions between events and maintain operational readiness. Also, the need for efficient logistical planning is essential to minimize downtime and ensure that all necessary resources are available for the next performance.

By examining the detailed itinerary of an air show weekend (see a summary in Table PII.1), one can appreciate the extensive array of skills an air show performer must cultivate. Training extends beyond mere aerobatic flying proficiency to involve comprehensive preparation for all facets of the event, including logistical planning, media interactions, public engagement, and stress management. This holistic approach ensures that when the weekend arrives, display pilots are thoroughly

TABLE PII.1
Example of Air Show Weekend Activities, Required Skills, and Related Training

Day/Time	Activity	Required Skills	Related Training/ Preparation
Thursday	Venue arrival	Organizational skills	Logistics and operations management training.
Friday	Rehearsal/ validation flights and media engagements	Precision flying, media communication	Advanced aerobatics training, media training.
Saturday and Sunday	Compulsory safety briefing	Protocol adherence, timeline comprehension	Safety briefing protocols, regulatory compliance.
	Open shows for the general public	Aerobatic performance, public engagement	Public speaking, autograph session management.
	Social events	Networking, stamina	Stress management, social engagement techniques.
In-between Shows	Ferry aircraft and administrative tasks	Logistical planning, administrative efficiency	Aircraft ferrying logistics, business administration.
	Maintenance scheduling	Technical knowledge	Maintenance planning and oversight, technical troubleshooting.
	Event coordination communication	Communication, negotiation	Professional communication, negotiation tactics.
Monday	Venue departure	Organizational skills	Logistics and operations management training.

equipped to deliver ideal performances, captivating audiences while upholding the highest standards of safety and professionalism. The rigorous and multifaceted nature of this training underlines the passion, commitment, and dedication required to excel as an air show performer, making the air show weekend proof of their outstanding skills and thorough preparation.

MINI-STEPS

The detailed schedule of an air show weekend provides a core context for understanding the extensive training regimen required for air show performers. Each aspect of the weekend, from the initial arrival to the final departure, necessitates a comprehensive set of skills. The journey from raw talent to enhanced artistry in air show performance is facilitated through a series of critical phases—a sequence of *mini-steps*—that form the basis of a performer's professional evolution.

Building on the robust foundation established in Part I: Pilot Selection, Part II: Training assumes paramount importance. This phase transforms the raw material of talent, identified through rigorous selection criteria, into the polished artistry of an air show performer. The concept of *mini-steps* is central to the display-flying training philosophy. Each *mini-step* represents a discrete, essential increment in the broader trajectory toward aerial mastery. Similar to the crafting of a complex timepiece, where each gear must be placed with precision and care, the development of an air show performer's skills comprises numerous such gears, each needing to mesh flawlessly with the others.

These *mini-steps* require performers to engage fully with each new piece of knowledge or skill to internalize the learning until it becomes second nature. This process is not to be rushed; it demands patience to pause and evaluate, ensuring that the lessons of each step are not merely learned but deeply ingrained. Each *mini-step* is designed to address a specific aspect of the skills required, ensuring that pilots are well-prepared for every challenge they might encounter during an air show weekend.

Resilience in training is also recognized as vital. Progression sometimes requires stepping back, reassessing, and re-engaging with the fundamentals. These moments, far from being setbacks, are essential recalibrations in any learning process, particularly in an area where the margin for error is minimal. Ultimately, the ability to adapt and refine skills continuously until retirement is crucial for maintaining the high standards expected of air show performers.

Part II will chart the sequence of *mini-steps* with an eye toward the combined picture they form. Through these measured, deliberate stages, pilots transform their initial potential into the polished performance of air show performer professionals. This is not an effortless journey; instead, it is a rigorous, disciplined march toward excellence—a march that, though comprising individual steps, results in a remarkable leap in capability and performance.

DISTINCTIVE NATURE OF AIR SHOW PERFORMANCE TRAINING

Compared to other forms of aviation training there is a distinctive nature of air show performance flight training. Unlike the relatively predictable flight paths and procedures, characteristic of commercial aviation, air show flying is dynamic and requires the execution of complex maneuvers at low altitudes with precision and elegance. Training must infuse performers with the ability to make split-second decisions, adapt to rapidly changing conditions, and maintain composure under the pressure of performance in front of an audience. Each *mini-step* in training amplifies these competencies, carefully preparing pilots for an environment where the margin for error is minimal and the demands for skill and concentration are at their zenith.

With no surprise, the aircraft used in air shows are often highly specialized for aerobatic performance, possessing capabilities that far exceed those of standard commercial planes. Training must, therefore, emphasize aircraft handling skills that leverage the unique aerodynamics of these machines. Pilots learn to navigate not just in three-dimensional space but within the added dimensions of performance artistry that such aircraft enable. Moreover, the low-level flying inherent in air shows presents its own set of challenges and risks, markedly different from the high-altitude cruising of commercial flights. Training must incorporate plans for dealing with reduced reaction times and increased potential for spatial disorientation, ensuring that pilots can execute their routines with both safety and spectacle in mind.

The *mini-steps* in training thus take on even greater significance within the context of air show performance. They must be carefully calibrated to address these unique factors, building a pilot's capability to perform not just technically demanding maneuvers but to do so in a manner that is both safe for the performer and enthralling for the audience. This is the delicate balance that air show performer training seeks to achieve—a balance where the pursuit of excellence meets the uncompromising demands of safety and showmanship.

Furthermore, in contrast to the conventional flight training for airline operations, which increasingly relies on automation, the regimen of air show performers diverges significantly. While commercial pilots are trained to become adept at managing and relying on sophisticated automated systems that govern modern airliners (Casner & Schooler, 2014; Chaparro et al., 2023; Chialastri, 2012; De Boer & Dekker, 2017; Hughes et al., 2009), air show pilots must hone a vastly diverse set of skills. Automation has undoubtedly enhanced the efficiency and safety of commercial travel (Billings, 1996; Parasuraman & Byrne, 2002), but it has also engendered a training environment where manual flying skills can take a back seat to systems management (Dehais et al., 2015; Dekker, 2002; National Transportation Safety Board (NTSB), 2014). Nevertheless, the introduction of compulsory upset prevention and recovery training (UPRT) from numerous aviation authorities and airlines has aimed to reverse a trend of manual flying skills atrophy for airline pilots (Advani et al., 2016; Degtyarev et al., 2021; Hanusch, 2017; Schroeder, 2016; Sumwalt, 2003). In the domain of air show aviation, however, the pilot's connection with the aircraft is immediate and unfiltered as low-altitude aerobatics demand an acute awareness and a hands-on approach that automation cannot replicate. Training, therefore, is intensely focused on developing and maintaining these manual flying skills—skills that must be executed with precision within a highly dynamic and sometimes unforgiving environment.

Each *mini-step* in the training process of an air show performer is thus carefully designed to build and refine these manual skills. It is a curriculum that respects the traditional stick-and-rudder expertise, yet also embraces modern advancements that aid in performance and safety. This blend ensures that display pilots achieve the finesse and adaptability required to captivate spectators, all while navigating an arena where the interplay between human skill and machine capability is on full display.

On the whole, there is an immense amount of information about aviation training that is beyond the scope of this publication. In the further readings section, insightful literature is available for review, while the interested reader is also referred to the regulations and publications contained in the recommended resources.

TRANSITIONING FROM A NON-AIR SHOW AVIATOR TO AN AIR SHOW PERFORMER

Transitioning from a non-air show aviator to an air show performer necessitates acquiring a specific set of skills and adaptations. Building upon the concept of *mini-steps* outlined earlier, this process involves methodically planned phases of training to ensure a seamless and effective transition. Table PII.2 outlines the essential skills and training areas that a pilot, traditionally not involved in air shows, would need to acquire or enhance to transition successfully into an air show performer. Such training program emphasizes a shift from automated flight operations and broader mission objectives to the precision and manual dexterity required for aerobatic performance.

TABLE PII.2
Comparative Training Focus of Air Show Performers versus Other Categories of Aviators

Category	Air Show Performers	Other Categories of Aviators (e.g., Commercial, Military)
Training Focus	Manual flying skills, advanced aerobatics, precision maneuvering.	Systems management, automation proficiency, standard operating procedures.
Flight Environment	Low altitude, highly dynamic, visually oriented.	Varied altitudes, from low-level tactical operations (military) to high-altitude cruising (commercial); environment dictated by mission or route.
Aircraft Type	Specialized aerobatic planes with high maneuverability.	Wide range from commercial airliners equipped with advanced automation to military aircraft designed for specific operational capabilities.
Skill Emphasis	Tactile manipulation of flight controls and throttle, energy management, innovation in performance, quick reaction time, timely decision making, stress management, social engagement.	Systems management, navigation, communication, and coordination with air traffic control, crew resource management.

TABLE PII.2 (Continued)
Comparative Training Focus of Air Show Performers versus Other Categories of Aviators

Category	Air Show Performers	Other Categories of Aviators (e.g., Commercial, Military)
Safety Protocols	Comprehensive safety protocols, situational awareness, emergency maneuvering, reliance on flying skills.	Comprehensive safety protocols, including automation for monitoring, emergency procedures, and reliance on air traffic management systems.
Performance Objectives	Entertainment, demonstration of flying skills and aircraft capabilities.	Safe, efficient transport of passengers/cargo (commercial); completion of mission objectives (military).
Dependency on Automation	Low; emphasis on manual control and pilot skill.	High; extensive use of automated systems for flight management, navigation, and control.
Operational Pressures	Performing complex maneuvers consistently and safely in front of an audience.	Managing long-duration flights, passenger comfort, and strict operational schedules (commercial); mission success under potentially adverse conditions (military).
Regulatory Environment	Specific regulations governing air shows and aerial events.	Comprehensive regulatory frameworks focused on public safety, airspace management, and environmental considerations; military operations subject to additional layers of operational control.
Professional Development	Continual refinement of aerobatic skills and performance techniques.	Ongoing training in systems management, regulatory changes, and advancements in aviation technology.

REFERENCES

Advani, S., Schroeder, J., & Burks, B. (2016). Global implementation of upset prevention & recovery training. https://doi.org/10.2514/6.2016-1430

Billings, C. E. (1996). Benefits and costs of aviation automation. In *Aviation automation*. CRC Press.

Casner, S. M., & Schooler, J. W. (2014). Thoughts in flight: Automation use and pilots' task-related and task-unrelated thought. *Human Factors, 56*(3), 433–442. https://doi.org/10.1177/0018720813501550

Chaparro, A., Miranda, A., & Grubb, J. (2023). Aviation displays: Design for automation and new display formats. In J. R. Keebler, E. H. Lazzara, K. A. Wilson, & E. L. Blickensderfer

(Eds.), *Human factors in aviation and aerospace* (3rd ed., pp. 341–371). Academic Press. https://doi.org/10.1016/B978-0-12-420139-2.00014-9

Chialastri, A. (2012). Automation in aviation. In F. Kongoli (Ed.), *Automation*. InTech. https://doi.org/10.5772/49949

De Boer, R., & Dekker, S. (2017). Models of automation surprise: Results of a field survey in aviation. *Safety, 3*(3), Article 3. https://doi.org/10.3390/safety3030020

Degtyarev, V., Mashoshin, O., & Degtyareva, A. (2021). Upset recovery training for civil aviation pilots. *Civil Aviation High Technologies*. https://doi.org/10.26467/2079-0619-2021-24-1-8-15

Dehais, F., Peysakhovich, V., Scannella, S., Fongue, J., & Gateau, T. (2015). "Automation surprise" in aviation: Real-time solutions. *Proceedings of the 33rd Annual ACM Conference on Human Factors in Computing Systems*, 2525–2534. https://doi.org/10.1145/2702123.2702521

Dekker, S. W. A. (2002). The re-invention of human error. https://citeseerx.ist.psu.edu/viewdoc/download?doi=10.1.1.522.4705&rep=rep1&type=pdf

Hanusch, M. (2017). *Manual flying skills: Airline procedures and their effect on pilot proficiency*. City University of London.

Hughes, J. S., Rice, S., Trafimow, D., & Clayton, K. (2009). The automated cockpit: A comparison of attitudes towards human and automated pilots. *Transportation Research Part F: Traffic Psychology and Behaviour, 12*(5), 428–439. https://doi.org/10.1016/j.trf.2009.08.004

National Transportation Safety Board (NTSB). (2014). Descent below visual glidepath and impact with seawall Asiana Airlines Flight 214 Boeing 777-200ER, HL7742 San Francisco, California July 6, 2013.

Parasuraman, R., & Byrne, E. A. (2002). Automation and human performance in aviation. In *Principles and practice of aviation psychology*. CRC Press.

Schroeder, J. (2016). Upset prevention and recovery training: A regulator update. https://doi.org/10.2514/6.2016-1429

Sumwalt, R. (2003). *Airplane upset recovery training: A line pilot's perspective*. Flight Safety Digest.

5 Hard Skills for Aerial Excellence

Manolis Karachalios

INTRODUCTION

> A hard skill is a technical or practical competency that can be objectively measured and is typically acquired through formal education, training programs, or on-the-job experience. These skills are essential for performing specific tasks and are often required for particular jobs or industries (Walden University, 2023; N.C. Cooperative Extension, 2023).

This Chapter explores the essential hard skills necessary for achieving aerial excellence as an air show performer with the objective of providing a comprehensive framework for developing the technical competencies necessary for air show performance.

This examination will cover the "*what*" (the specific skills needed), "*when*" (the phase of training during which these skills are developed), "*where*" (the training methods and environments used), and "*how*" (the process of skills development), supported by case studies from both the United Kingdom and the United States. Each of these areas will be explored in detail to provide a comprehensive understanding of the training process required to transform a pilot into an air show performer. The acquisition of these skills must follow a structured and incremental learning method, often referred to as a stepping stone approach (Cambridge English Dictionary, 2024; Ehrenfreund et al., 2012; Gulesci et al., 2021), ensuring that each phase of training builds upon the previous one in a systematic and coherent manner.

The "*what*" refers to the specific skills necessary for air show performance, including advanced aerobatic maneuvers, emergency procedures handling, and precision flying. These skills are identified based on the demands of air show routines and the need for exceptional levels of flying and safety. For instance, mastering complex aerobatic maneuvers requires not only technical proficiency but also an intuitive understanding of the aircraft's capabilities and limits.

The "*when*" addresses the timing and sequencing of training phases. Skills development is mapped out over a period, beginning with foundational knowledge and progressing through increasingly complex and demanding tasks. Early phases focus on theoretical knowledge and basic flight skills, while later stages introduce advanced techniques and performance-specific training. This phased approach ensures that pilots are not overwhelmed and can build confidence and competence gradually.

DOI: 10.1201/9781003504801-8

The "*where*" explores the training methods and environments used. Training typically takes place in a variety of settings, including classroom environments for theoretical knowledge, flight simulators for intermediate practice, and actual flight conditions for practical experience. Each environment is chosen to provide the most effective learning context for the skills being developed at that stage. For example, simulators offer a safe space to practice emergency procedures, while practical flying is essential for mastering in-flight maneuvers.

The "*how*" examines the process of skills development, emphasizing a step-by-step approach. This involves breaking down complex skills into manageable components and teaching these components incrementally. Each skill is practiced repeatedly until it becomes second nature, ensuring a solid foundation before moving on to more advanced tasks. This method also includes regular assessment and feedback, allowing pilots to identify areas for improvement and ensuring that each step is fully mastered before progressing.

WHAT: HARD SKILLS DESIRED

At first glance, one might consider that the skills required for air show performance are focused primarily on flying aerobatics. However, there are myriad other peripheral skills that support the precise and safe execution of a flying display. These skills are not only related to the specific type of aircraft performing—whether it be tail-wheel, vintage, or a fast jet—but also encompass a broader spectrum of competencies that ensure overall excellence and safety in performance.

Air show performances demand a blend of advanced technical skills and precise execution, all falling under the structured framework of the *5M* approach: huMan, Machine, Medium, Mission, and Management. Each element contributes to the overall safety, effectiveness, and spectacle of the performance.

HUMAN

Pilot Proficiency and Training

Pilots must have extensive experience and training in aerobatic maneuvers (Courtney, 2017; International Council of Air Shows (ICAS), 2016). This involves rigorous practice to master techniques such as loops, rolls, and spins, which are critical for performing complex aerobatic sequences. Additionally, formation flying requires precise coordination and communication skills to maintain tight formations and execute synchronized routines safely.

Aerobatic Skills

The cornerstone of air show flying is the ability to execute complex aerobatic maneuvers. This involves mastering a wide range of maneuvers, such as loops, rolls, spins, and hammerheads, each requiring precise control and acute spatial awareness. Pilots must be able to perform these maneuvers with precision, ensuring both safety and spectacle. The complexity of these maneuvers demands a deep understanding of the aerodynamic principles that govern flight dynamics at various speeds and attitudes (International Council of Air Shows, 2016).

Spin Training

Spin training is a critical component of aerobatic skill development (Federal Aviation Administration (FAA), 2016). Spins can occur unexpectedly, and recovery from a spin is a vital skill for any pilot performing aerobatics. This training teaches pilots to recognize the onset of a spin, understand the aerodynamic factors at play, and execute the correct recovery procedures efficiently. Most importantly, the mastery of spin recovery not only enhances safety but also builds confidence in handling advanced flight conditions.

Upset Prevention and Recovery Training (UPRT)

UPRT is essential for preparing pilots to handle unexpected and potentially hazardous flight conditions. This training focuses on the prevention and recovery from unusual flight attitudes and situations that could lead to loss of control. It encompasses a wide range of scenarios, from mild upsets to extreme maneuvers, and emphasizes the importance of maintaining composure and executing recovery techniques accurately (Advani & Field, 2011, 2011; Advani & Schroeder, 2016; Crider, 2008; Le Vie, 2016; Schroeder & Burke, 2016).

Aircraft-Specific Skills

Each type of aircraft—tailwheel, vintage, or fast jet—requires a distinct set of skills and knowledge (International Council of Air Shows, 2016). Tailwheel aircraft, for example, have different ground handling characteristics compared to tricycle gear aircraft, necessitating specialized training in takeoff, landing, and taxiing techniques. Vintage aircraft often have unique flight characteristics and systems that require pilots to adapt their flying techniques accordingly. For example, vintage fighter jets, such as the F-86, have limiting G restrictions for straight and rolling pullouts (Airplanes Online, n.d.; Werrell, 2013). Fast jets, on the other hand, demand proficiency in high-speed handling and advanced aerodynamics (Jones, 1958; Morris, 2010).

Formation Flying

Many air shows feature formation flying,[1] which requires precise coordination and disciplined communication between multiple aircraft (Formation and Safety Team (FAST, 2024). Pilots who participate in formation aerobatic teams must be skilled in maintaining specific positions relative to other aircraft, adjusting speeds, and performing synchronized maneuvers, a skill that needs a significant amount of flight training.

Physical Fitness and Conditioning

Mastering the aircraft is only part of the equation; pilots must also master their own physical conditioning (International Council of Air Shows, 2016). The physical demands of aerobatic flying, particularly the sustained exposure to high G-forces, require pilots to be in peak physical condition. Elevated levels of fitness and conditioning enhance G-tolerance, allowing pilots to maintain consciousness and perform effectively during high-G maneuvers. Regular cardiovascular training, strength training, and flexibility exercises are essential components of a pilot's fitness regimen. A well-conditioned body not only improves performance but also reduces the risk of G-induced loss of consciousness (Balldin, 1986; Crisman, R., & Burton, 1988; Bateman et al., 2006).

Machine

Mechanical and Technical Knowledge

A thorough understanding of the aircraft's mechanical systems, including engines, avionics, and structural components, is necessary for diagnosing and addressing any issues that may arise before or during a performance.

Aircraft Operation and Maintenance

In-depth knowledge of the aircraft's systems is essential. Pilots must perform detailed pre-flight checks and be capable of troubleshooting and addressing any mechanical issues. This ensures that the aircraft is in optimal condition for performance.

Customization and Calibration

Aircraft often require specific modifications and fine-tuning to enhance their performance capabilities. This includes adjustments to the avionics, engines, and flight control systems to suit the demands of aerobatic flying. For example, the installation of aerobatic sighting devices on aerobatic aircraft's wingtips that help pilots accurately perform vertical maneuvers by providing visual references, ensuring precise alignment with the horizon (Hirschman, 2022).

Medium

Environmental and Weather Analysis

Pilots must be adept at interpreting weather data and understanding its impact on flight performance (Barker, 2003). This skill allows them to adapt their routines to varying conditions, ensuring both safety and the quality of the performance.

Venue and Terrain Familiarization

Understanding the unique features of each air show venue, such as runway dimensions, surrounding terrain, and potential obstacles, is critical (International Council of Air Shows (ICAS), 2016; Barker, 2020a). This knowledge helps pilots plan their routines more effectively and anticipate any environmental challenges.

Mission

Show Planning and Execution

Detailed planning and rehearsal of aerobatic routines are crucial for ensuring precision and safety during air show performances (Barker, 2020b). This includes coordinating with other pilots and the ground crew to execute complex maneuvers seamlessly to deliver a visually stunning performance that captivates the audience.

Audience Engagement

Engaging the audience is a key aspect of air show performances. Pilots need to design routines that are not only technically impressive but also visually appealing; showmanship and elegance play a significant role in enhancing the audience's experience (Loudenslager, 2014).

Management

Operational Planning and Logistics

Effective management of schedules, logistics, and coordination with event organizers ensures the smooth execution of air show performances (Karachalios, 2022). This involves detailed planning to ensure that all necessary resources, including fuel, spare parts, and support personnel, are available.

Regulatory Compliance

Compliance with aviation regulations and safety standards is mandatory (International Council of Air Shows, 2019b; Karachalios, 2022; UK Civil Aviation Authority, 2024a). Pilots must adhere to airspace restrictions, maintain necessary certifications, and ensure that their aircraft meet all safety and maintenance requirements for both the safety of the performers and the audience.

4-D Training: Time as the Fourth Dimension in Air Show Performances

In air show performances, time management is a critical, yet often underappreciated, aspect that significantly elevates the level of difficulty. Similarly to physics, where the fourth dimension is time (Bork, 1964), display flying requires a *4-D* (four-dimensional) approach in training due to the precise synchronization of various activities during an air show, ensuring that each component of the performance happens at the exact scheduled moment, but also for formation flying time deconfliction is critical for both a safe (FAA, 2024) and entertaining performance (ICAS, 2016).

Air show performers adhere to a strict timeline that begins well before the actual performance. Pre-show briefings must occur at specific times to coordinate all air show participants, as well as all team members, and clarify the sequence of events. Pilots must initiate startup procedures promptly to align with their scheduled check-in times with the air boss, who coordinates the overall event.

The takeoff and flight routines are thoroughly timed to the second to maintain the flow of the show. During their performance, pilots execute maneuvers according to a pre-planned profile, often synchronized with music or other elements, requiring an in-depth understanding of the aircraft's capabilities and the ability to make real-time adjustments based on environmental factors such as wind speed and direction.

Precision in landing is equally crucial. Pilots must complete their routines and vacate the display area on schedule to ensure that subsequent acts can begin without delay. This level of time discipline ensures a seamless and continuous experience for the audience, maintaining the event's professional and entertaining atmosphere.

The importance of this time discipline is stressed by the need for pilots to develop strong time management skills. This involves not only technical flying skills but also the ability to coordinate and communicate effectively with the ground crew and other performers. The entire team must work together to ensure that every aspect of the performance aligns with the precise timing required.

For instance, the Pacific Airshow (2024) in Huntington Beach and Luke Days Airshow (2024) emphasize the importance of maintaining a tight schedule to ensure the smooth execution of performances. The Athens Flying Week Air Show (2025) also highlights the necessity of precise timing for a successful event, with a timetable that leaves no space for a gap between the flying events in order to keep the audience continuously engaged and entertained (see Figure 5.1).

Unique Aspect of Air Show Training

Time management in air shows is unique compared to other aviation activities due to the necessity of performing synchronized maneuvers within tight time frames. Pilots must execute their routines with exact precision, often coordinated down to the second, to ensure the show runs smoothly and safely, in synch with other formation members and music or supporting narration.

Overall, in aerial events, every aspect of the performance is choreographed, and pilots must be acutely aware of their timing. This includes the exact moment of takeoff, the duration of each maneuver, the time deconfliction with other aerial acts, and the precise timing for landing to avoid disrupting the sequence of events. The synchronization required in air show performances is akin to a finely tuned orchestra, where each element must come together perfectly to create a harmonious and impressive display.

Comparison with Other Aviation and High-Performance Activities

The precision and synchronization required in air show performances distinguish them from other aviation and performance activities, highlighting the unique demands placed on display pilots. Air show performances occur in dynamic and often unpredictable aerial environments, requiring pilots to adjust in real-time to varying conditions such as density altitude effect in aircraft performance, cloud ceiling, horizontal visibility, and both surface and upper winds (Barker, 2020a; Cox, n.d.; UK Civil Aviation Authority, 2018). This level of complexity demands not only exceptional flying skills but also meticulous training in time management to ensure safety and entertainment value.

Commercial Aviation

In commercial aviation, time management focuses on operational efficiency and maintaining schedules for takeoffs and landings (Ateş et al., 2018; Wickens, 2002). While precise timing is essential to keep flights on schedule, the level of synchronization required in air shows is not as critical. Delays or adjustments can be managed without the immediate safety and performance implications seen in air shows.

Athens Flying Week 2023 - Sunday 3 September 2023			
Time (Local=UTC+03 Hrs)	Duration (Minutes)	Air Show Performer	Flying Activity
11:00	0:05	Luca Bertossio (Italy)	Takeoff
11:05	0:03	Hellenic Skydivers (Greece)	Takeoff
11:08	0:10	Red Bull Paramotor - Dimitris Kolliakos (Greece)	Flying Display
11:18	0:10	Hellenic Skydivers (Greece)	Paradrop
11:28	0:12	Luca Bertossio (Italy)	Flying Display
11:40	0:15	WeFly! (Italy)	Flying Display
11:55	0:10	RC Kavala (Greece)	Flying Display
12:05	0:05	T-41 and P2002-JF (Hellenic AF)	Flyby
12:10	0:10	S-70B (Hellenic Navy)	Flying Display
12:20	0:25	Special Operation Forces (Hellenic NDGS)	Special Operation Forces Tactical Scenario
12:45	0:10	Super Puma (Hellenic AF)	Tactical Demonstration (CSAR)
12:55	0:15	Apache (Hellenic Army)	Flying Display
13:10	0:15	HH-139 (Italian AF)	Flying Display
13:25	0:16	Daedalus T-6 (Hellenic AF)	Flying Display
13:55	0:02	SKYVAN (BE)	Takeoff
13:57	0:03	F-4 Phantom (Hellenic AF)	Takeoff
14:00	0:12	Typhoon (Royal Saudi AF)	Flying Display
14:12	0:03	Rafale and Mirage 2000 (Hellenic AF)	Takeoff
14:15	0:12	F-4 Phantom (Hellenic AF)	Tactical Demonstration
14:27	0:03	AWACS, Rafale and F-16 Viper (Hellenic AF)	Flyby
14:30	0:07	Transport Aircraft and two Mirage 2000 (Hellenic AF)	Flyby
14:37	0:04	Rafale, F-16 Viper, two Mirage 2000, two F-4s (Hellenic AF)	Flyby
14:41	0:03	Rafale and two Mirage 2000 (Hellenic AF)	Land
14:44	0:02	SKYVAN (BE)	Land
14:46	0:12	Zeus F-16 (Hellenic AF)	Flying Display
14:58	0:45	**Patrulla Aguila (Spanish AF)**	Flying Display
15:43	0:12	F-16 Solo Display (Royal Danish AF)	Flying Display
15:55	0:35	**Saudi Hawks (Royal Saudi AF)**	Flying Display
16:30	0:16	F-35 Demo Team (United States AF)	Flying Display
16:46	0:15	Aegean Airlines and Olympic Air (Greece)	Flyby
17:01	0:12	Rafale Solo Display (French ASF)	Flying Display
17:13	0:12	Spitfire (Hellenic AF)	Flying Display
17:25	0:30	**Fursan Al Emarat (United Arab Emirates AFAD)**	Flying Display
Please Note: Airshow timetable is subject to change due to unforeseen weather conditions or operational requirements.			
(Version 5, Update: 2 August 2023)			

FIGURE 5.1 Example of air show timetable, Athens Flying Week (AFW) International Air Show 2023. (Copyright: Athens Flying Week International Air Show.)

Military Aviation

Military aviation involves rigorous time management, especially during coordinated missions and training exercises. However, the context differs as the primary goals are mission success and operational readiness rather than public entertainment. While precise timing is crucial, synchronization with other aircraft is often for strategic purposes rather than creating an audience spectacle (U.S. Navy, 2024; United States Air Force (USAF), 2022).

Performance Arts

In performance arts such as musical and dance, timing is also crucial. Performers must synchronize their actions with cues, music, and fellow performers to create a cohesive and engaging show (Orlandi et al., 2020; Shaffer, 1984). This requirement for precision is similar to air shows but occurs in a controlled environment rather than the dynamic and potentially hazardous setting of an air display.

Sports

In sports like gymnastics or synchronized swimming, athletes must perform routines with precise timing, often coordinated with music (Donti et al., 2014; Shaffer, 1984). This requires extensive practice and discipline, similar to air show performers. However, these activities are typically conducted in controlled environments where external variables, like weather, are not factors.

WHEN: PHASE OF TRAINING

The training phases for display pilots are crucial for ensuring proficiency, safety, and excellence in aerial performance. These phases include initial training and continuation training, both of which play distinct roles in a display pilot's career.

Initial Training

Initial training is the foundational phase where pilots acquire the essential skills and knowledge required for display flying. This training is comprehensive and covers both theoretical and practical aspects of display flying.

According to AFMAN 11-246 (USAF, 2022), initial training involves rigorous ground and flight training components. Ground training covers crucial areas such as aircraft systems, emergency procedures, and regulatory knowledge, including a thorough review of relevant regulations like FAA Order 8900.1 (Federal Aviation Administration (FAA), 2022) and AFMAN 11-246 itself, which outlines the standards for air show performances and safety requirements. For the United States Air Force (USAF) Air Combat Command (ACC) flight demonstrations, flight training is conducted under the supervision of experienced demo pilots, ensuring that new pilots gain practical experience with aircraft control limitations and performance characteristics specific to their demonstration profiles. New pilots are required to practice maneuvers at higher altitudes before gradually progressing to lower altitudes, ensuring a controlled and safe development of their aerobatic skills. This training also emphasizes recognizing and managing conditions that could necessitate maneuver aborts, which is critical for maintaining safety during complex aerobatic displays (USAF, 2022).

Initial training for a civilian air show performer is equally rigorous and comprehensive. It begins with mastering basic aerobatic maneuvers, progressing to advanced techniques under the guidance of experienced mentors. Pilots must obtain the necessary certifications and endorsements, focusing on safety and precision. They also need to learn about aircraft handling under various flight conditions, emergency procedures, and showmanship (ICAS, 2016; FAA, 2022).

Continuation Training

Continuation training is essential for maintaining and enhancing the skills acquired during initial training. This phase ensures that pilots remain proficient and up-to-date with the latest techniques and safety protocols.

Recurrent training, a key component of continuation training, includes annual recertification requirements. Pilots must complete a specified number of practice sorties, review safety procedures, and practice maneuver parameters. These sessions are meticulously documented in the pilot's grade book, ensuring continuous assessment of their performance and readiness (FAA, 2022; USAF, 2022).

Additionally, continuation training includes participation in advanced training courses such as the heritage flight training courses (HFTC). These courses involve training with both military and civilian aerobatic pilots, reinforcing skills in formation flying, communication protocols, and emergency response strategies. Such training helps pilots refine their individual flying skills and enhances their ability to perform aerobatics in formation, which are common in air shows (Formation and Safety Team (FAST), 2024; International Council of Air Shows, 2019a; USAF, 2022).

Formation and tail-chase training are particularly emphasized in continuation training. According to CAP 1724 (UK Civil Aviation Authority, 2024b), formation flying requires pilots to demonstrate exceptional situational awareness, precise energy management, and adherence to standard operating procedures during close formation maneuvers. This training ensures that pilots can safely fly formation during join-ups, position keeping, and breakaways, which are critical skills for performing complex aerobatic routines involving multiple aircraft (European General Aviation Safety Team, 2015).

Case Study: Annual Heritage Flight Training Course (HFTC), USAF

The Annual HFTC conducted by the USAF is a vital training program designed to ensure the proficiency and safety of both USAF demonstration aircrews and civilian pilots flying vintage aircraft in public air shows. Managed by ACC, the HFTC plays an essential role in the USAF Heritage Flight Program (USAFHFP), aiming to preserve the legacy of historical aircraft and foster a connection between modern USAF operations and their historical counterparts. The HFTC is specifically designed to train single-ship demonstration pilots and Air Force Heritage Flight Foundation (AFHFF) civilian warbird pilots to perform their Heritage Flight (HF) mission, as

FIGURE 5.2 The F-35 Lightning II in formation with a P-38 Lightning and two P-51 Mustangs during the 2016 Heritage Flight Conference. (Copyright: U.S. Air Force, photo by Staff Sgt. Staci Miller.)

seen in Figure 5.2, required for both initial certification and annual re-certification (USAF, 2022).

Initial training at the HFTC is comprehensive, covering both academic and practical flying components. ACC coordinates the availability of military and civilian aircraft to meet the rigorous training syllabus requirements. For initial certification, pilots must demonstrate significant flight experience, including a minimum of 500 hours in USAF fighter-type warbirds or a combination of 50 hours in these aircraft plus 2,000 hours in active duty USAF fighter-type aircraft. They must also possess a current Formation and Safety Training (FAST) formation flight certificate with a four-ship stamp or an equivalent FAA certification, a current Class II Medical Certificate, a Level II Statement of Aerobatic Competency (SAC) card or FAA equivalent low-level card, and a valid Commercial Pilot Certificate. Furthermore, pilots must have access to or own a World War II, Korean War, Vietnam War, or Desert Storm era US Army Air Corps or USAF pursuit/fighter/attack-type warbird (USAF, 2022).

The academic portion of the training includes detailed briefings on regulations, the USAF HF Briefing Guide, formation profiles, communications, join-up procedures, airspeed compatibility, sight pictures, formation changes, and flight break-up and landing procedures. These briefings ensure pilots are thoroughly prepared for the demands of HF operations, both theoretically and practically.

Continuation training is essential for maintaining certification and ensuring pilots remain proficient. Annual re-certification requires pilots to fly at least two training sorties during the annual HFTC, one of which must be a three or four-ship formation. Pilots unable to attend the annual HFTC must review ground training videos, receive a re-currency brief from ACC, and complete a training sortie before participating in public events to ensure pilots maintain their skills and stay current with procedural and regulatory updates.

Civilian pilots certified under the AFHFF program are authorized to fly recurring training flights in single-ship demo pilot aircraft. This integration allows them to become familiar with the flight characteristics of modern fighter aircraft involved in dissimilar formation flights, enhancing the overall performance during public air shows.

The HFTC emphasizes safety and community engagement. By ensuring all participating pilots can safely execute formation flights at public air shows, the training program supports recruitment efforts and fosters community relations. To achieve this, media personnel, both military and civilian, may participate in observation flights, provided the aircraft are not engaged in demonstration training or formation flights (USAF, 2022).

WHERE: TRAINING METHOD

The "*where*" aspect of training refers to the various methods and environments used to develop and maintain the proficiency of display pilots. Effective training methods are crucial to ensure that pilots can perform safely and effectively in the diverse and, often, challenging conditions encountered during air shows and other aerial events. These methods include a combination of ground-based theoretical instruction, simulator training, fitness training and conditioning, and advanced flight training, each playing a vital role in a display pilot's total training.

Ground Training: Setting the Foundations for Excellence in Air Show Performances

Ground training for air show performers is a multifaceted process integral to the cultivation of skilled, knowledgeable, and safety-conscious individuals. By focusing on human factors, personal qualities, practical skills, team dynamics, and risk management, this training lays the foundation for excellence in air show performances. It ensures that performers are not only equipped to thrill audiences with their aerial competence but also committed to upholding the highest standards of safety and professionalism in every display.

An essential role in the development and maintenance of proficiency for air show performers is played by various air show organizations, such as the International Council of Air Shows (ICAS), the European Airshow Council (EAC), the Fédération Aéronautique Internationale (FAI), and the British Air Display Association (BADA). These organizations are instrumental in delivering annual symposiums and workshops that focus on training and safety-related aspects for air show performers (Karachalios, 2022; Karachalios & Adjekum, 2023).

The ICAS conducts an annual convention that is one of the most significant gatherings for air show professionals globally. This convention includes comprehensive workshops and seminars addressing various facets of air show performance and safety. Topics covered range from advanced aerobatic techniques to the latest in safety protocols and risk management strategies. The ICAS convention provides a platform for performers to learn from leading experts, share best practices, and stay updated with regulatory changes and technological advancements (ICAS, n.d.).

Similarly, the EAC organizes an annual convention that brings together performers, organizers, and safety professionals from across Europe and beyond. The EAC convention features detailed sessions on safety management, pilot training programs, and the latest developments in air show regulations. By facilitating these discussions, the EAC helps ensure that air show performers are well-prepared to meet the highest standards of safety and performance (European Airshow Council (EAC), n.d.).

The FAI also plays a crucial role in promoting safety and excellence in air shows. Through its various committees and working groups, the FAI organizes workshops and training sessions that emphasize the importance of adherence to international safety standards and the implementation of best practices. These events provide valuable opportunities for performers to enhance their skills and knowledge in a collaborative environment (FAI, n.d.).

Then, the BADA focuses on the unique aspects of air shows in the United Kingdom. BADA's annual symposium is dedicated to discussing regulatory updates, safety initiatives, and training programs specific to the British air show community, involving both civilian and military delegates. By addressing the particular needs and challenges faced by UK performers, BADA ensures that its members are equipped with the latest information and tools to conduct safe and successful displays (BADA, n.d.).

Through these annual symposiums and workshops, these organizations foster a culture of continuous improvement and safety in the air show industry. These symposia put an emphasis on risk management and safety with performers being exposed to methods to identify, assess, and mitigate risks associated with flying displays, incorporating strategies to enhance safety without compromising the spectacle of their performance. These sessions utilize case studies and incident analyses to illustrate potential hazards and teach strategies for their prevention or management (Barker, 2020b; British Air Display Association (BADA), n.d.; European Airshow Council (EAC), 2020, 2023, 2024; International Council of Air Shows, n.d.; Karachalios, 2024). For example, analyzing past incidents where communication breakdowns led to near-misses in order to highlight the importance of clear and precise communication, reinforcing the importance of preparedness and proactive safety measures, and ensuring that pilots are equipped to handle unexpected situations.

Theoretical Training

In the context of air show performances, the convergence of skill, precision, and safety is not achieved spontaneously. It is the result of rigorous, well-structured ground training designed to refine the innate talents of performers and equip them with the knowledge and competencies necessary for excellence in this high-stakes field. Theoretical training provides the foundation upon which practical skills are built, emphasizing a comprehensive understanding of human factors, personal qualities, application of knowledge, team dynamics, and risk management to ensure peak performance and safety.

Regulatory Compliance and Safety Protocols Review

At the outset of their training, demonstration pilots continue into a comprehensive review of international aviation safety protocols and regulatory guidelines. This includes studying documents and directives from relevant aviation and air show organizations (EAC, 2020; FAA, 2022; UK Civil Aviation Authority (CAA), 2024a, b). Familiarity with these international standards is crucial for ensuring pilots can adapt to and comply with local regulations, irrespective of the country in which they are performing.

For example, in the United States, as a foundational aspect of ground training, demonstration pilots are required to review FAA Order 8900.1 (FAA, 2022), which covers critical regulatory guidelines and safety protocols, including sections on participant and aircraft eligibility, air show space requirements, minimum safety distances and altitudes, military performances, and the maneuvers packages approval process. Understanding these regulations is vital for ensuring compliance with FAA standards and fostering a productive relationship with the local Flight Standards District Office (FSDO). This training ensures pilots are well-versed in the regulatory framework governing air shows, facilitating a safer environment for performers and spectators alike.

Ground Safety Observer

The role of the ground safety observer is integral to the effective and efficient management of flying displays, serving as a crucial safety valve. This role, often filled by a team member, is essential for maintaining safety during air shows. Ground safety observers are responsible for monitoring the event from the ground, ensuring adherence to safety protocols, and communicating effectively with pilots. Their training includes understanding their responsibilities, mastering communication techniques, and learning the protocols for identifying and responding to safety concerns swiftly and efficiently (USAF, 2022).

Human Factors Training

Human factors training addresses the psychological and physiological aspects of flying, emphasizing the importance of mental fitness, decision making, and the impact of stress and fatigue on performance. Understanding the criticality of human factors is the cornerstone of ground training, as these factors play a pivotal role in the safety and success of flying displays. The UK Civil Aviation Authority (CAA) offers an online human factors course that encompasses a broad spectrum of psychological and physiological considerations significantly impacting display safety. This course exposes pilots to factors such as currency, sources of skill-based errors like negative transfer and distraction, sources of operational pressure like time and practical drift, and implementing measures to combat these issues (UK Civil Aviation Authority, 2024c).

G-awareness training is another critical component, focusing on the physiological effects of G-forces (Jouanin et al., 2005; Metzler, 2020; Whitley, 1997). Pilots learn techniques to mitigate G-induced loss of consciousness and the techniques for performing high-G maneuvers safely. For example, the Blue Angels' G-Tolerance Improvement Program (GTIP) focuses on enhancing pilots' ability to withstand high G-forces while flying without wearing anti-G pants.[2] This program combines anaerobic and aerobic exercises, emphasizing core and lower body strength. Additionally, it includes training with a high G-force centrifuge to simulate flight conditions, along with briefings on G-tolerance, the effects of high G-forces, and prevention of G-LOC or an almost G-induced loss of consciousness (A-LOC). The comprehensive approach aims to improve pilots' overall G-fitness, ensuring they can safely and effectively perform in high-G environment during air shows (Garcia, 2018; Aker, 2022).

Training on visual illusions is also integral, helping pilots recognize and mitigate the effects of visual illusions, particularly during low-altitude operations, such as false horizons, misjudged distances, and altitude misperceptions (Demir & Aydin, 2021; Muller, 2002). Understanding these illusions and learning to rely on both visual and instrument cues are crucial for maintaining orientation and preventing spatial disorientation (Song & Cui, 2017). Ground training scenarios that introduce these illusions enhance pilots' situational awareness and ability to maintain safety under various flight conditions.

Mindfulness training is increasingly recognized for its value in enhancing focus, reducing stress, and improving situational awareness among pilots (Darses et al., 2023; Karachalios, 2022; Karachalios & Adjekum, 2023; Meland et al., 2015). Techniques for maintaining mental clarity and emotional balance under pressure are taught, enabling pilots to remain calm and focused in high-stress situations and ultimately leading to optimum decision making and precise performance, which is essential for successful air show displays.

Aircraft Handling Characteristics

Ground training ensures that pilots are thoroughly familiar with the handling characteristics of their aircraft, including limitations, aerodynamic behaviors, and responses to control inputs under various conditions. This comprehensive understanding is critical for designing safe and spectacular display routines within the aircraft's performance envelope (ICAS, 2016).

The impact of low-altitude flight on aircraft performance constitutes a significant area of study in aviation, particularly for pilots involved in air shows and display routines. Flying at low altitudes introduces specific challenges and changes to aircraft performance due to variations in air density, temperature, and atmospheric pressure, which can affect aerodynamics, engine performance, and overall maneuverability (Barker, 2003; ICAS, 2016).

At lower altitudes, the increased air density results in greater aerodynamic forces on the aircraft. This means that an aircraft can generate more lift and drag at lower altitudes than it would at higher altitudes for a given airspeed. Pilots need to understand how this affects their control inputs and aircraft behavior during maneuvers (Flach & Warren, 1995; International Council of Air Shows, 2016; Karachalios & Adjekum, 2023).

Engine performance can also vary significantly with altitude. Most aircraft engines perform better at lower altitudes where the air is denser, providing more oxygen for combustion. This can result in increased thrust output, but pilots must be mindful of the potential for engine overheating and must manage their throttle settings carefully (Federal Aviation Administration (FAA), 2023, Chapter 4).

Maneuverability is directly influenced by the aircraft's aerodynamic properties, which change with altitude. At lower altitudes, the increased lift and drag can enhance maneuverability, allowing for tighter turns and more responsive handling. However, this can also lead to increased fuel consumption and a higher risk of overstressing the aircraft if maneuvers are not planned and executed with care (Barker, 2003; International Council of Air Shows, 2016).

Understanding fuel and center of gravity (C.G.) requirements is a key aspect of ground training. Pilots learn to calculate fuel consumption based on planned maneuvers, account for emergency reserves, and consider the impact of external factors, such as weather conditions, on fuel efficiency. Additionally, they are trained in the effects of C.G. considerations on aircraft performance, ensuring that the aircraft remains balanced and controllable during all phases of flight (FAA, 2023; ICAS, 2016).

Additionally, training on flight control limitations and performance characteristics is tailored to each demo profile. New demo pilots undergo rigorous sessions to understand how their aircraft behaves under various conditions and configurations, especially during extreme control inputs required to enter specific advanced aerobatic maneuvers. This knowledge is crucial for safe but also elegant performance execution.

Abort procedures for each maneuver and the total display are also thoroughly covered during ground training (USAF, 2022; ICAS, 2019b). Pilots learn common conditions that may necessitate aborting a maneuver or the entire flight, such as mechanical failures, adverse weather conditions, loss of situational awareness, or significant deviations from expected performance parameters. Additionally, specific steps for safely aborting maneuvers are covered, including technical execution, communication protocols, exiting the display airspace safely, and ultimately landing safely. It is emphasized to display pilots the need to respond quickly and decisively to such conditions, always prioritizing safety.

Low Altitude Training Academics for USAF ACC Demonstration Teams

Low Altitude Training (LOWAT) Academics are crucial in the ground training of the USAF's demonstration pilots, focusing on the essential aspects of operating in low-level display flying environments (USAF, 2022). This training involves mastering the LOWAT crosscheck, a technique for maintaining situational awareness of the aircraft's energy, i.e., airspeed and height while adhering to display lines during low-altitude display flying. Pilots learn to identify and mitigate hazards, which are more prevalent and dangerous at lower altitudes, such as bird strikes.

The training thoroughly covers handling emergency and abort procedures, emphasizing quick and effective responses to unexpected situations while ensuring clearance from the ground and obstacles. Additionally, pilots are educated on the effects of low altitude on aircraft performance, understanding the differences in how their fast jets fly, especially in terms of pitch rate,[3] G-onset rate,[4] airspeed[5] and G sustaining capability,[6] engine spool-up time,[7] and airspeed acceleration rate[8] at low altitudes compared to higher ones (Glenn Research Center, n.d.). Addressing visual illusions, particularly the challenges of low depth perception over water, the training teaches pilots how to recognize and counteract these deceptive perceptions, ensuring they maintain focus and execute maneuvers safely.

The Synergistic Role of Flight Simulators in Air Show Performer Training

In the world of air show performances, where pilots become artists and the sky their canvas, the integration of flight simulators into the training regimen has been nothing short of revolutionary. While it's crucial to understand that simulators augment rather than replace actual flying experiences, their contribution to enhancing pilot proficiency and safety is undeniable.

Augmenting Real-World Experience

Flight simulators are heralded not as a replacement but as a vital complement to actual flight training. They offer a risk-free platform for pilots to develop automatic responses to routine operations and critical incidents alike (Schofield, 2012). Simulators allow for repeated practice of specific maneuvers, building muscle memory that is invaluable during the high-pressure scenarios of an air show. This repetitive training fosters automaticity (Endsley, 1999), ensuring that pilots can perform complex routines with precision and grace under any conditions.

The Versatile Training Ground

The capability of flight simulators to mimic various operational scenarios is unparalleled. Pilots can virtually experience diverse airport environments, familiarizing themselves with different layouts and operational nuances before ever setting foot on the actual tarmac (Geyer & Biggs, 2018). This pre-acquaintance can significantly reduce pilot stress and enhance performance during shows. Furthermore, simulators can introduce pilots to a wide array of weather conditions, enabling them to adapt their flying techniques to handle relevant atmospheric challenges effectively (Geyer & Biggs, 2018). Most importantly, emergency procedure training within the simulator environment allows pilots to hone their problem-solving skills and emergency responses without the actual risks (Araiza-Alba et al., 2021).

Recognizing Limitations

Despite the significant advantages offered by flight simulators, it is essential to acknowledge their limitations. Simulators cannot fully replicate the physiological effects of flying, such as G-forces and the gyroscopic impact on the pilot's body, which are crucial elements of the real flying experience (Schofield, 2012). Additionally, while flight models within simulators strive for accuracy, there is a boundary to the realism they can achieve, especially concerning the nuanced aerodynamics of different aircraft.

A notable constraint is that simulators are often specific to particular aircraft models, making them excellent for practicing routines for single-ship demonstrations but less effective for formation flying training. The intricate choreography of formation flying, requiring precise spatial awareness and timing in relation to other aircraft, is a challenge to replicate in the simulated environment (Geyer & Biggs, 2018).

Nevertheless, Dahlstrom et al. (2002) examined mid-fidelity simulations and their effectiveness, suggesting that fidelity alone does not guarantee the transferability of skills to real-world settings and emphasized the need to evaluate simulator training based on its actual outcomes rather than just its resemblance to reality. For air show performers, this highlights the importance of focusing on how well the training enhances their ability to perform complex maneuvers safely and accurately in actual air shows rather than relying solely on the simulator's realistic appearance. This means that even with a flight simulator on a desktop computer, display pilots—especially those without access to high-fidelity sims, like those flying Extra 300s—can gain specific benefits such as practicing performance sequence, rehearsing maneuvers, and acclimating with the air show venue environment, ultimately enhancing their readiness and precision for their actual performance.

Toward a More Integrated Future

The future of flight simulators in air show performer training looks promising, with technological advancements bridging the gap between simulation and actual flight experiences. Innovations in virtual reality (VR) and augmented reality (AR) are expected to enhance the realism of simulated environments (Wallace et al., 2020; Xie et al., 2021), offering more immersive training experiences that can better replicate the physical sensations of flight and formation flying dynamics.

Harris's work (2011) offers critical insights into the nuanced benefits and challenges of simulator training. It emphasizes the importance of high-fidelity simulation environments that replicate the complexities of the flight deck, aiming to bridge the gap between simulated scenarios and real-world demands.

Evidence

Flight simulators have been fundamental in commercial and military aviation training for decades, advancing to become crucial in both aircraft design and operation. Their success has paved the way for applications in other fields, reinforcing the value of specific development and application rules learned from industrial usage (Sikder et al., 2013).

In the medical field, VR simulations for procedural training, such as bronchoscopy, have proven that focused instruction and practice enable novices to quickly acquire basic skills, performing at par with experienced practitioners. This suggests the potential for cross-disciplinary applications of VR training systems (Colt et al., 2001).

The aviation industry's embracement of simulation for training reflects a broader adoption of VR and simulation technologies across various sectors, from medical procedures to complex engineering tasks, highlighting the simulator's role in enhancing skill acquisition, reducing error rates, and improving safety protocols (Jentsch & Curtis, 2011).

Potential and Practical Pros and Cons

Analyzing the advantages and disadvantages of using simulators in air show performers' training programs reveals a balanced perspective on their effectiveness, as summarized in Table 5.1.

Fitness Training and Conditioning

To construct a case study on the importance of fitness training and conditioning for air show performers, similar to high-performance athletes, let's first lay out the key areas and insights from the available literature and then integrate the EASA guidelines as a practical application in this context.

Air show performers, much like high-performance athletes, require a prominent level of personal fitness to ensure optimal performance and safety during their routines. The European Union Aviation Safety Agency (EASA) emphasizes the need for display pilots to maintain a high degree of personal fitness, highlighting its importance not just for the execution of complex aerial maneuvers but also for overall well-being and mental focus (EASA, 2015).

TABLE 5.1
Advantages and Disadvantages of Using Simulators in Air Show Performers Training

Pros		Cons	
Aspect	**Description**	**Aspect**	**Description**
Safety	Simulators provide a risk-free environment to practice and perfect potentially dangerous maneuvers.	Cost	High-fidelity simulators are expensive to develop and maintain.
Efficiency	They allow for repeated practice without the logistical constraints of real-world settings.	Realism Gap	Despite advancements, a gap remains between simulated environments and the unpredictable nature of real-world conditions.
Skill Transfer	Skills learned in simulators have been shown to transfer to real-world tasks effectively.		

Fitness Training and Conditioning: Insights from Literature

- Aerobic Conditioning for Performance: Aerobic fitness is crucial for maintaining and generating energy levels during high-intensity efforts and ensuring effective recovery between performances (Polman et al., 2004; Stone & Kilding, 2009; Garcia, 2018). Regular aerobic conditioning significantly improves both aerobic function and performance in team sport athletes, which can be paralleled with the requirements of air show performers for sustained energy and focus during flights.
- Strength and Conditioning Program Benefits: The integration of strength and conditioning programs, even when conducted alongside regular training routines, has been shown to enhance physical fitness characteristics significantly (Stone et al., 2020). Such programs can improve upper and lower body strength, aerobic fitness, and potentially, by extension, the precision and control needed for aerial displays.
- Physical Fitness and Stress Management: Higher physical fitness levels can buffer stress symptoms and improve overall well-being, which is critical for air show performers who operate in high-stress environments. Fitness-related attenuations in trait anxiety suggest that a well-rounded physical conditioning program could enhance performance under stress (Taylor et al., 2008).

Practical Application: EASA Guidelines Integration

EASA's guidelines (2015) highlighting the criticality of physical and mental fitness for display pilots, suggesting a comprehensive approach to fitness training and conditioning that includes both aerobic and strength components tailored to the unique demands of flying displays. The guidelines stress the need for pilots to be free of personal worries and physically fit, indicating that a well-structured conditioning program could significantly contribute to achieving these standards.

Fitness Plan for Air Show Performers

Creating a fitness plan for air show performers based on the latest research involves considering exercises that improve physical fitness, particularly cardiorespiratory fitness, strength, endurance, and flexibility (Despres & Ewing, 1998; Lin et al., 2015; Bjelica et al., 2020; Myers et al., 2021). A detailed fitness plan based on insights from academic papers follows, focusing on improving the overall fitness of air show performers, which is essential for their demanding and physically intensive performances.

Weekly Fitness Plan

The goal of the fitness plan is to enhance physical preparedness to meet the demanding requirements of air show performances, focusing on cardiorespiratory fitness, strength, endurance, agility, and recovery.

The duration of the plan is 8–12 weeks, with periodic evaluations to adjust intensity and exercises based on progression.

WEEKS 1–4: FOUNDATION PHASE

- Cardiorespiratory Fitness: High-intensity interval training (HIIT) using air bikes twice a week. Sessions should include work intervals of 15–45 seconds to improve aerobic endurance and VO2 peak significantly (Schlegel et al., 2022)
- Strength Training: Circuit strength-training program three days per week, focusing on improving muscular strength necessary for handling the physical demands of display flying. This approach has shown significant improvements in physical fitness scores among Air Force personnel (Westcott et al., 2007)
- Flexibility and Recovery: Daily stretching routines and at least one yoga session per week to improve flexibility and aid in recovery (Bhardwaj, 2024).

WEEKS 5 TO 8: INTENSIFICATION PHASE

- Endurance Training: Incorporate longer endurance sessions on non-HIIT days to build up an aerobic base, focusing on activities like running, swimming, or cycling.
- Strength and Power: Increase the intensity of strength training sessions by incorporating plyometrics[9] and powerlifting[10] to enhance explosive power and muscular endurance.
- Mental and Cognitive Enhancement: Aerobic fitness has been associated with increased cognitive function, which is crucial for the mental demands of air show performances. Moreover, regular aerobic exercise can improve attentional control functioning (Colcombe et al., 2004; Perez et al., 2014).

WEEKS 9 TO 12: PEAK PHASE

- Specificity Training: Tailor fitness sessions more closely to mimic the physical demands experienced during air shows, including G-force resistance exercises and agility training (Tesch et al., 1983; Crisman & Burton, 1988; Lennemann et al., 2013).
- Recovery Management: Emphasize active recovery with muscle stretching, adequate sleep, and proper nutritional support to enhance fitness performance. In addition, an important aspect of fitness centers used by air show performers that can enhance comfort and efficiency is increasing air movement and controlling indoor temperature during exercise (Zhai et al., 2015).

WEEKLY TRAINING RECOMMENDATIONS

- Aerobic and Strength Training Sessions: On days without performances, incorporate aerobic and strength training sessions to improve overall physical condition. Resistance training is essential for building the muscular strength and endurance necessary for the rigors of air show performances.

- Personalized Monitoring: Continuously monitor the effectiveness of the fitness regimen and adjust based on performance feedback and individual needs. Personalized training regimens are more effective at meeting the unique demands of each performer.

Daily Fitness Plan during the Air Show Weekend

To optimize the performance and recovery of display pilots during an air show, a detailed daily fitness plan is equally crucial to the weekly fitness plan discussed in the previous paragraphs. This plan should address the unique demands placed on performers before, during, and after air shows, focusing on enhancing their strength, supporting their stamina, and accelerating their recovery processes.

BEFORE THE AIR SHOW

- Morning Cardiovascular Routine: Initiate the day with a session of HIIT, utilizing exercises like air bike sprints for 20 minutes. This approach is validated for improving both aerobic endurance and VO2 peak, which is crucial for the high-energy demands of air show performances (Schlegel et al., 2022
- Flexibility and Mobility Session: Follow HIIT with a dedicated 15-minute session aimed at improving flexibility and mobility. Engaging in activities that improve functional fitness can significantly benefit performance and reduce injury risk, as demonstrated in adults but applicable to the physically demanding routines of air show performers (Dobek et al., 2007)

DURING THE AIR SHOW

- Dynamic Warm-ups: Implement a 10-minute dynamic warm-up routine before the performance to prime the muscles and cardiovascular system for the demanding physical activity ahead. Such practices are supported by evidence indicating improved performance in physical activities following dynamic stretching and warm-ups (Baquet et al., 2001).
- Nutrition and Hydration: Emphasize proper hydration and nutrition throughout the show to maintain energy levels and concentration. Consuming snacks rich in proteins and carbohydrates can help sustain performance and energy (Astbury et al., 2014; Nielsen et al., 2020).

AFTER THE AIR SHOW/ POST-FLIGHT

- Cool Down and Recovery: Post-performance, a thorough cool-down, including light aerobic activity and stretching, can aid in recovery, facilitating the removal of lactate from the muscles and reducing soreness. Engaging in low-intensity recovery activities such as Pilates can further enhance flexibility and core strength, which are crucial for the demands placed on air show performers' bodies (Kloubec, 2010).

- Recovery Nutrition: Focus on a balanced meal post-performance to aid in recovery. A diet rich in proteins, carbohydrates, and fats is essential for repairing muscles and replenishing energy reserves (Heaton et al., 2017; Bonilla et al., 2021).

Fitness and Conditioning Requirements of Air Show Performers and Fighter Pilots

Conditioning for air show performers, similarly to fighter pilots, encompasses a broad spectrum of physical and mental preparedness strategies designed to enhance performance, endurance, and safety under the high demands of their roles. Both groups require a high degree of aerobic and anaerobic fitness, strength, endurance, and mental resilience to effectively handle the rigors of their respective aviation activities. Here are key insights drawn from the literature:

- Physical Fitness and Flight Performance: Higher physical fitness levels significantly enhance student pilots' flight operation abilities, potentially reducing training time and failure rates for aspiring fighter or civil aviation pilots and suggesting that air show performers might also benefit from enhanced aerobic capacity and endurance to sustain lengthy routines without fatigue (Choi, 2003).
- Special Physical Fitness Requirements: Naval carrier fighter plane pilots require rapid reaction capabilities, aviation endurance, resistance to vertigo, and quick environmental adaptation, indicating the need for tailored fitness programs to address the unique challenges of high-performance and high-stress flight conditions (U.S. Navy Aeromedical Reference and Waiver Guide, 2021.
- G-Tolerance Training: Essential for both air show performers and fighter pilots, G-tolerance training includes physical strength training and exercises like centrifuge training to acclimate the body to high-G environments. Resistance training enhances G-duration tolerance, directly impacting performance in display flying and aerial combat maneuvers. Regular physical activity, primarily aerobic, is crucial for operating in high +Gs environments, with improvements in muscular fitness, particularly axial strength, reducing the risk of spinal disorders (Crisman & Burton, 1988; Newman et al., 1999).
- Selection and Fitness: Israeli Air Force pilot candidates accepted into flight school exhibited higher aerobic capacity, anaerobic power, and lower body fat compared to those not accepted, emphasizing the role of physical fitness in the selection process, which is likely similarly important for air show performers (Hoffman et al., 1999).

In summary, while air show performers and fighter pilots operate under different flight conditions, their core fitness and conditioning requirements share significant similarities. Both groups need elevated levels of aerobic and anaerobic fitness, muscular strength and endurance, and specific conditioning to enhance G-tolerance and reduce injury risk. Tailored fitness programs that address these needs can improve performance, safety, and operational effectiveness for both air show performers and fighter pilots.

Case Study: Rafale Solo Display and Patrouille de France, French Air and Space Force

Incorporating a detailed fitness regimen into the routine of air show performers is a practice that is not only recommended by academic research but is also being implemented by elite air forces around the world. Notably, prestigious teams like the French Air and Space Force's Rafale Solo Display and the Patrouille de France Demonstration Teams emphasize the importance of physical fitness in enhancing performance and ensuring the safety of their pilots (Ministry of Defence, France, 2024a, 2024b). These teams have recognized that the intense physical demands placed on air show performers require a comprehensive approach to physical training and recovery. As a result, they have incorporated physiotherapists and physical trainers into their teams to provide specialized training regimens and recovery programs.

This commitment to fitness and well-being ensures that performers are at their physical peak, which is crucial for executing the precise and demanding maneuvers required during air shows. The involvement of professional physiotherapists and physical trainers helps in tailoring fitness programs to meet the specific needs of each performer, focusing on enhancing their cardiorespiratory fitness, muscular strength, flexibility, and overall endurance.

The practices of these elite Teams reinforcing the value of a holistic approach to fitness in the field of air show performance. By prioritizing physical training and recovery, they set a benchmark for air show performers worldwide, demonstrating that peak physical condition is integral to achieving excellence in aerial displays.

Flight Training: Advanced Flying Techniques and Maneuvers

Advanced flying techniques for air show performers require a combination of precision, creativity, and rigorous training to execute aerobatic maneuvers safely, elegantly, and spectacularly. According to Smithsonian Magazine (Scott, 2005), performers must make thrilling displays appear dangerous while maintaining utmost safety. This involves mastering maneuvers that, while appearing perilous to the audience, are controlled and well-rehearsed to minimize risk. Patty Wagstaff, one of the top air show pilots, emphasizes the importance of discipline and precision in aerobatics, noting that her routines demand exacting control to prevent accidents (French, 2023). Wagstaff's career also highlights the necessity for pilots to be adaptable and continuously improve their skills, whether through competing in aerobatic championships or transitioning to roles like aerial firefighting, where precision and quick thinking are equally critical.

Executing advanced demonstration maneuvers presents a unique set of challenges that require both the pilot and the ground crew to operate in perfect harmony, not limited to the following:

- Weather Conditions: Variability in weather conditions can significantly impact the execution of maneuvers. Pilots must be adept at adapting their performance to ensure safety and visibility (ICAS, 2016; Karachalios, 2022).
- Aircraft Limitations: Understanding the physical and mechanical limitations of the aircraft is crucial. Pilots must execute maneuvers within these constraints to prevent over-stressing the aircraft.
- Spectator Safety: Ensuring the safety of spectators is paramount. Pilots must execute maneuvers at safe distances, adhering to strict flight paths and altitude restrictions.

Common Aerobatic Maneuvers

There is a plethora of aerobatic maneuvers developed over the years, with the main limiting factor nowadays being the display pilot's imagination. However, a common pool of maneuvers is utilized during aerobatic training and is accepted by the World Air Sports Federation, the Fédération Aéronautique Internationale (FAI, n.d.), and adopted by the International Aerobatic Club in the United States. These maneuvers include aileron roll, barrel roll, slow roll, snap roll, point roll, avalanche, chandelle, competition turn, loop, eight-sided loop, square loop, English bunt, Cuban-eight, half Cuban eight, reverse half Cuban eight, hammerhead, humpty-bump, Immelmann, inside-outside eight, lazy-eight, rolling turn, spin, split-S, tailslide, and wing over (International Aerobatic Club (IAC), 2024a).

Each of these maneuvers has its unique characteristics and training benefits. For instance, aileron rolls and slow rolls help pilots develop precise control over the aircraft's roll axis, while loops and Cuban eights are excellent for practicing pitch control and energy management. The hammerhead maneuver, where the aircraft climbs vertically before pivoting 180 degrees on its yaw axis, is particularly challenging and tests a pilot's ability to manage power and rudder coordination. Meanwhile, maneuvers like the Immelmann and split-S are fundamental for understanding how to change flight direction efficiently while minimizing altitude loss.

Training in these maneuvers is crucial for air show performers as it builds the foundational skills required for more advanced aerobatics and complex routines. The mastery of basic aerobatic maneuvers not only enhances a pilot's technical proficiency but also their confidence in handling the aircraft under various flight conditions. This solid foundation allows for the safe and creative execution of more intricate and spectacular aerobatic displays that captivate audiences.

For those interested in further exploring the details and techniques of these maneuvers, additional resources and detailed manuals are available in the further readings section, as well as through the Fédération Aéronautique Internationale and the International Aerobatic Club, which provide comprehensive guidelines and standards for aerobatic training and performance.

Aresti System: The Aerocryptographic Language

The Aresti System, also known by its creator, Colonel José Luis Aresti, as the *Aerocryptographic Language* (Aresti System, 2024), revolutionized the standardization of aerobatic sequence notation and communication in competitive aerobatics.

Before the Aresti System, there was no universal standard for describing these sequences, leading pilots to use their own shorthand notations or verbal descriptions. This lack of standardization created significant challenges, particularly as the sport expanded globally and language differences posed additional barriers (IAC, 2016; Aresti System, 2024; IAC, 2024b)

Initial attempts to create notational systems for aerobatic maneuvers began in the 1920s, but it wasn't until 1961 that a comprehensive system was developed. Aresti published the Sistema Aerocryptographica Aresti, a catalog that meticulously detailed aerobatic maneuvers. Initially adopted in Spain, the system gained international recognition when the Aerobatics Commission (CIVA) of the Fédération Aéronautique Internationale (FAI) elected to use it at the 1964 World Aerobatic Championships in Bilbao, Spain. Since then, Aresti's symbolic notations have become the global standard for aerobatic sequence diagrams.

The Aresti System provides a clear and universal language for aerobatic figures, enabling pilots worldwide to understand and execute complex maneuvers with precision. It uses specific symbols to denote each maneuver, including the type, direction, and any additional elements like rolls or loops. This standardization has significantly enhanced the clarity and safety of aerobatic competitions.

For those involved in aerobatic flying—whether as participants, judges, or enthusiasts—the Aresti Aerobatic Catalogue is an indispensable resource. The catalog is available in versions tailored to various types of aerobatics, including power, glider, and remote-controlled (RC) model aerobatics, ensuring that all categories of the sport benefit from this standardized system. Despite the different versions, the organization and notation remain consistent, providing a reliable reference for the aerobatic community.

Understanding the Aresti System is crucial for anyone involved in aerobatic competition. It forms the foundation for designing, executing, and judging aerobatic sequences, ensuring that all involved speak the same "*aerobatic language*" and uphold the highest standards of performance and safety.

Application for Air Show Performers

Air show performers, much like competitive aerobatic pilots, rely heavily on the Aresti System to design and communicate their flight routines. The precise notation provided by the Aresti System allows air show pilots to conceptualize, choreograph, and execute advanced aerobatic maneuvers with a high degree of precision and safety.

For air show performers, mastering the Aresti System means being able to clearly convey their planned sequences to other pilots, air bosses, and event organizers, ensuring that every aspect of the performance is understood and coordinated. This is particularly important in multi-ship formations or when performing in conjunction with other aerial acts, as the use of a standardized notation system helps mitigate risks associated with misunderstandings or miscommunications during aerial events.

Furthermore, the ability to read and interpret Aresti diagrams allows display pilots to quickly adapt to changes or unexpected situations, such as immediately adjusting their routines to accommodate weather conditions. For example, in case of a change in wind direction, the pilot might alter the display profile's direction to have the main axis of their performance face the headwind, thus better controlling their position inside the aerobatic box[11].

Example of Display Sequence Depiction

The Aresti catalog contains a variety of symbols representing different flight maneuvers, each with specific instructions on how to execute them (Aresti System, 2024).

Key Points of the Aresti System

1. Basic Symbols:
   - Lines: Solid lines indicate positive or zero angle of attack (AoA), while dashed lines indicate negative AoA.
   - Angles: Represent horizontal, vertical, and 45 degree lines in various orientations.
   - Loops: Full loops are continuous curves; segments of loops are shown as arcs.
   - Start and End Symbols: A small, filled circle marks the start of a maneuver, and a short perpendicular line indicates the end.
2. Rotational Elements:
   - Aileron Rolls: Represented by curved arrows, indicating continuous rotation (slow rolls) or hesitation rolls (2-point, 4-point, 8-point).
   - Snap Rolls: Indicated by isosceles triangles, with filled triangles for negative snaps and unfilled triangles for positive snaps.
   - Spins: Denoted by right triangles, with filled triangles for inverted spins and unfilled triangles for upright spins.
3. Drawing Conventions:
   - Looping lines less than 180 degree are drawn as hard angles.
   - Entry and exit lines on the Y-axis are depicted as isometric lines (30 degree angle to horizontal).
   - Multiple rotational elements are limited to two per line segment, and elements of the same type must rotate in opposite directions if combined.

For example, the aerobatic flight profile shown in Figure 5.3 shows all the above elements. Using the Aresti System (2024), we can see that the sequence starts at the filled start circle with the number 1, indicating the first segment of the profile. Then, it transitions to a 90-degree upline (solid), indicating a positive AoA, which continues with a one-fourth roll to the left, followed by a three-fourth snap roll. After completing the snap roll, a positive loop is flown that ends at 45 degrees downline with two rolls to the left. The first segment is completed with a recovery to level flight in the opposite direction from where the sequence started.

A similar analysis can be used for the subsequent eight segments of this aerobatic flight profile using the Aresti System.

Understanding and accurately interpreting the Aresti *aerocryptographic system* is crucial for several reasons. It enables pilots to methodically prepare for

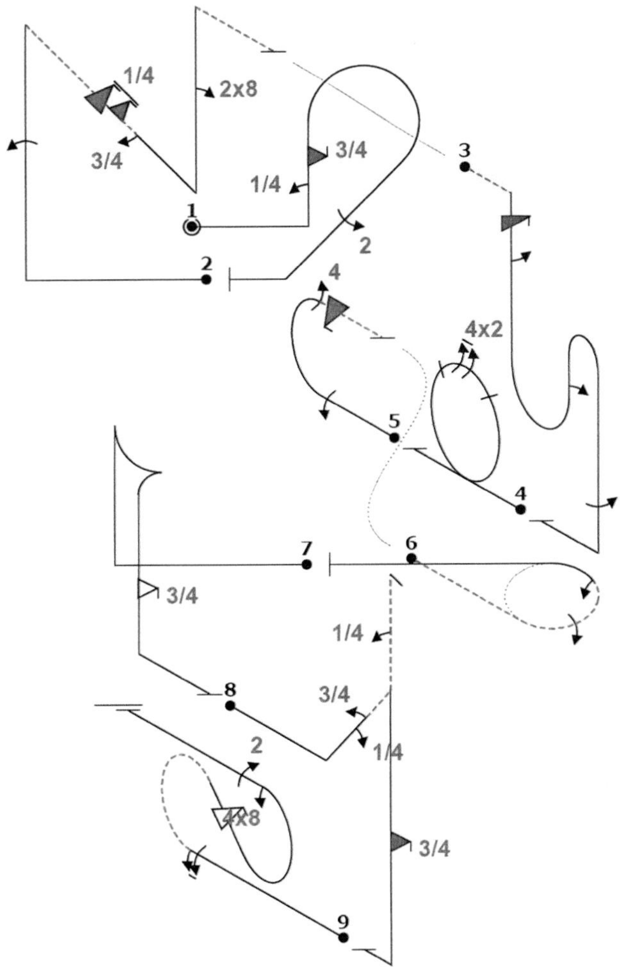

FIGURE 5.3 Example of aerobatic flight profile outlined with the Aresti System. (Copyright: International Aerobatic Club (IAC).)

complex routines by breaking down each maneuver into understandable components, allowing them to visualize and conceptualize the entire sequence before taking off. Additionally, this structured framework aids in studying and mastering each maneuver, while standardized symbols help pilots memorize sequences more efficiently, ensuring smooth and precise execution during performances. Furthermore, the Aresti System acts as a universal language, enabling effective communication between display pilots and air bosses, regardless of their native languages.

Furthermore, the display profile can be printed out and is often placed at eye level in the cockpit, providing pilots with an immediate visual reference if needed, either before or during the display flight. This common practice helps display pilots stay oriented and ensures that they can quickly recall the sequence of maneuvers, contributing to the overall safety and precision of their performance.

Designing Impactful Display Layouts: Considering Margins for Error

Air show performers, when designing a display profile and planning individual maneuvers, adhere to a set of generalized rules of thumb that significantly enhance safety, performance precision, and audience engagement, regardless of aircraft type (Barker, 2003; ICAS, 2016). These guidelines focus on establishing margins for altitude, airspeed, throttle setting, G-forces (Gs), reaction time, and human performance limitations, which are crucial for the safe execution of air show performances.

Altitude Margin

A minimum performance altitude is established for each maneuver well above the regulatory or organizational requirements. This margin serves as a safety buffer, allowing pilots room to recover in case of an error or unexpected aircraft behavior. For example, if the regulatory minimum is 500 feet, planning the maneuver to start at a minimum of 600 feet provides additional safety.

Airspeed Margin

Then, optimal entry and recovery airspeeds are decided to offer a balance between performance visibility and safety. They plan for a margin that allows the aircraft to be above stall speed and below maximum speed limits for the airframe during all phases of the maneuver. For instance, an additional 20%–30% above stall speed for entry is a commonly accepted starting point.

Throttle Setting Margin/Available Thrust/Torque

While specific throttle settings will vary greatly among different aircraft, it's essential to design maneuvers with flexibility in throttle use to accommodate adjustments for wind conditions, temperature, and aircraft weight. Therefore, display pilots always plan for a power reserve that can be applied quickly to exit a maneuver safely if needed.

G-Forces (Gs) Margin

Maximum G-forces expected during the maneuver are defined and display pilot's plan to operate within the aircraft's and their own G-force tolerances. Consequently, they design maneuvers that keep G-forces within a range that is comfortably below the aircraft's structural limits and the pilot's G-tolerance to mitigate risks of a G-LOC or airframe damage.

Time Safety Margin (TSM)

A concept that has been introduced in the flight-testing field has also been extensively utilized in the display flying discipline. The Time Safety Margin (TSM) is a critical safety metric developed to ensure risk reduction and aid in risk assessment for flight test maneuvers involving dives (Gray, 2016). It quantifies the time an aircraft can remain on its worst-case vector (a combination of lowest altitude and highest descent rate) before recovery becomes insufficient to prevent ground impact. This metric was created to address the limitations of traditional dive planning methods, which often relied on intuition and simple rules that did not account for the complexities of modern aircraft performance and recovery dynamics.

TSM Categories and Requirements

The TSM categories and their requirements are defined as listed in Table 5.2 (Gray, 2016).

Application in Display Profile Design

When designing display maneuvers, TSM is used to ensure that each maneuver has sufficient recovery time built into its profile. By calculating TSM, pilots can establish safety margins that consider various factors such as altitude, airspeed, dive angle, bank angle, roll rate, recovery G-forces (Nz), and aircraft-specific limitations. A commonly used rule of thumb in test flying for dive recoveries is the respected "*AGL/100 rule*," which considers that regardless of airspeed, the maximum allowable dive angle is determined by dividing the current AGL altitude by 100. For example, based on this rule, at 5,000 feet AGL, the max dive angle would be 50 degrees.

For example, when designing a maneuver for a fast jet aircraft performing a flying display that includes a 90-degree dive at 400 knots true airspeed (KTAS) and planning to recover at the minimum altitude of 500 feet above ground level (AGL), several factors must be considered. The aircraft will execute a 90-degree vertical dive, with the starting altitude for the dive determined based on the TSM calculation. The speed at the start of the dive will be 400 KTAS, and the planned recovery will end at 500 feet AGL (minimum altitude), with the recovery pull-out conducted at 5 Gs, as depicted in Table 5.3, the aircraft must begin the dive from at least 7,673 feet AGL, then do not delay the pull out of the dive before 4,972 feet (approximately 2,700 feet lost in 4 seconds with 400 KTAS), to ensure recovery at the minimum altitude with a 4-second TSM (see Figure 5.4). This ensures the maneuver includes a safety buffer, allowing for any unplanned delays, deviations, or distractions. It needs to be noted that beyond detailed planning, additional training and redundant recovery measures

TABLE 5.2
TSM categories

Category	TSM Range	Description
Routine	TSM ≥ 8 Seconds	Minimal risk, standard procedures suffice.
Focused	8 > TSM ≥ 4 Seconds	Additional training and planning are required.
Aided	4 > TSM ≥ 2.5 Seconds	Requires redundant recovery commands and simulation rehearsals.
Redundantly Aided	2.5 > TSM ≥ 1.5 Seconds	High risk, necessitates extensive training and backup systems.
Automatic Recovery	1.5 > TSM ≥ 0 Seconds	Extremely high risk, mandates automatic recovery systems.

TABLE 5.3
Example of TSM Calculation

TAS (knots)	Recovery Time (seconds)	TSM (seconds)	Total Time for Recovery (seconds)	Total Altitude Loss (feet)	Required Starting Altitude (feet AGL)
100	1.65	4	5.65	953.52	1,453.52
200	3.31	4	7.31	2,467.70	2,967.70
300	4.97	4	8.97	4,540.92	5,040.92
400	6.62	4	10.62	7,172.18	7,672.18
500	8.28	4	12.28	10,361.49	10,861.49

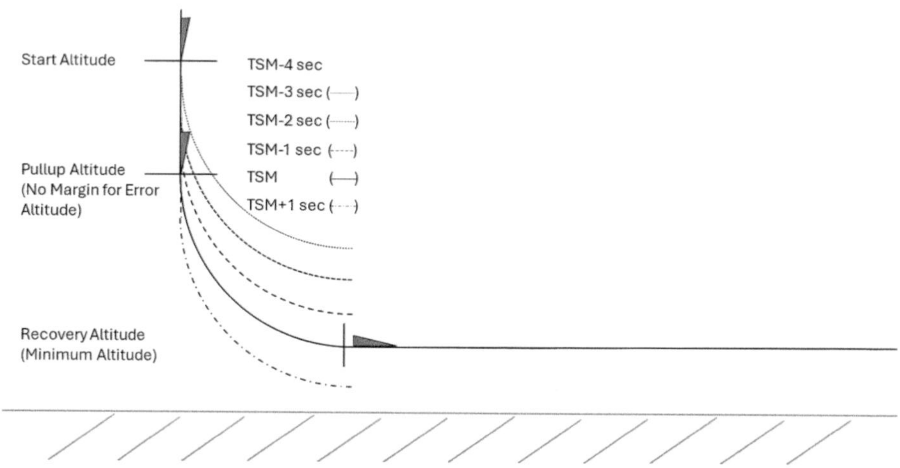

FIGURE 5.4 Example of dive recovery with TSM.

must be implemented to mitigate the increased risk associated with the maneuver's TSM category.

General Rules for Designing Displays and Maneuvers

The following rules are applied by air show performers to design display profiles and maneuvers that are spectacular, engaging, and uphold the highest safety standards, ensuring air shows remain a valued and safe form of entertainment, as well as an inspiration for aviation enthusiasts worldwide:

- Plan for Variability: Air show performers design each maneuver with adaptability in mind, allowing for adjustments based on real-time conditions such as weather changes or airspace restrictions.
- Safety Always: Safety is always prioritized over the show. While visually appealing maneuvers are essential, the safety of the pilot, spectators, and aircraft is never compromised.

- Pilot Proficiency: Display pilots tailor the maneuvers to the skill level of the pilot(s). Ensuring maneuvers are within the pilot's capabilities results in safer and more entertaining performances.
- Audience Perspective: Performers consider the visual impact of maneuvers from the audience's viewpoint. Therefore, they design maneuvers that are safe, comply with all safety margins, and are visually gripping from the ground.
- Brief and Debrief: Thorough pre-flight briefings and post-flight debriefings are integral to the design process. These sessions assess safety margins and allow for adjustments based on pilot feedback and observed performance.
- Regulatory Compliance: All maneuvers included in the display profile comply with local and international aviation regulations, including altitude restrictions, airspace use, and safety distances from the crowd area.

Case Study: Display Profile Options for the USAF ACC F-16 Demonstration Team

In fast jets, advanced flying techniques are incorporated during training sessions beyond the common aerobatic maneuvers. The model of flight training analyzed as a case study is that of the USAF ACC F-16 demonstration team, which requires pilots to exhibit not only technical proficiency but also creativity and situational awareness (USAF, 2022). The cognitive and physical demands on the pilot during these maneuvers are immense, as they must maintain precise control over the aircraft while executing maneuvers at varying speeds and altitudes.

The F-16 demonstration maneuver profiles, as detailed in Table 4.1 of the US Air Force Manual 11-246 Volume 1, dated March 18, 2022 (USAF, 2022), provide a structured sequence of aerobatic and precision flying tasks designed to showcase the agility, speed, and technical capabilities of the F-16 Fighting Falcon. These profiles are divided into two main categories: the High Show and the Low Show. The selection between these shows typically depends on the weather conditions, specifically the ceiling height (the height of the lowest layer of clouds above the ground that is either broken or overcast), which can impact the visibility and safety of the maneuvers.

High Show Profile

The High Show (see Figure 5.5) is performed under ideal weather conditions, allowing for the full range of high-altitude maneuvers:

1. Maximum Performance Takeoff and Climb to Cuban 8: A powerful takeoff with the use of afterburner followed by a steep climb into a Cuban 8 maneuver, demonstrating the F-16's power and agility.
2. High-Speed Flat Pass: A low-altitude, high-speed pass in front of the audience, demonstrating the aircraft's speed capabilities.
3. Triple Aileron Roll: Rapid consecutive rolls performed to display the aircraft's roll rate and the pilot's control.
4. High-G Turn: A turn executed under high G-forces, emphasizing the F-16's maneuverability and the structural and physical limits of both aircraft and pilot.

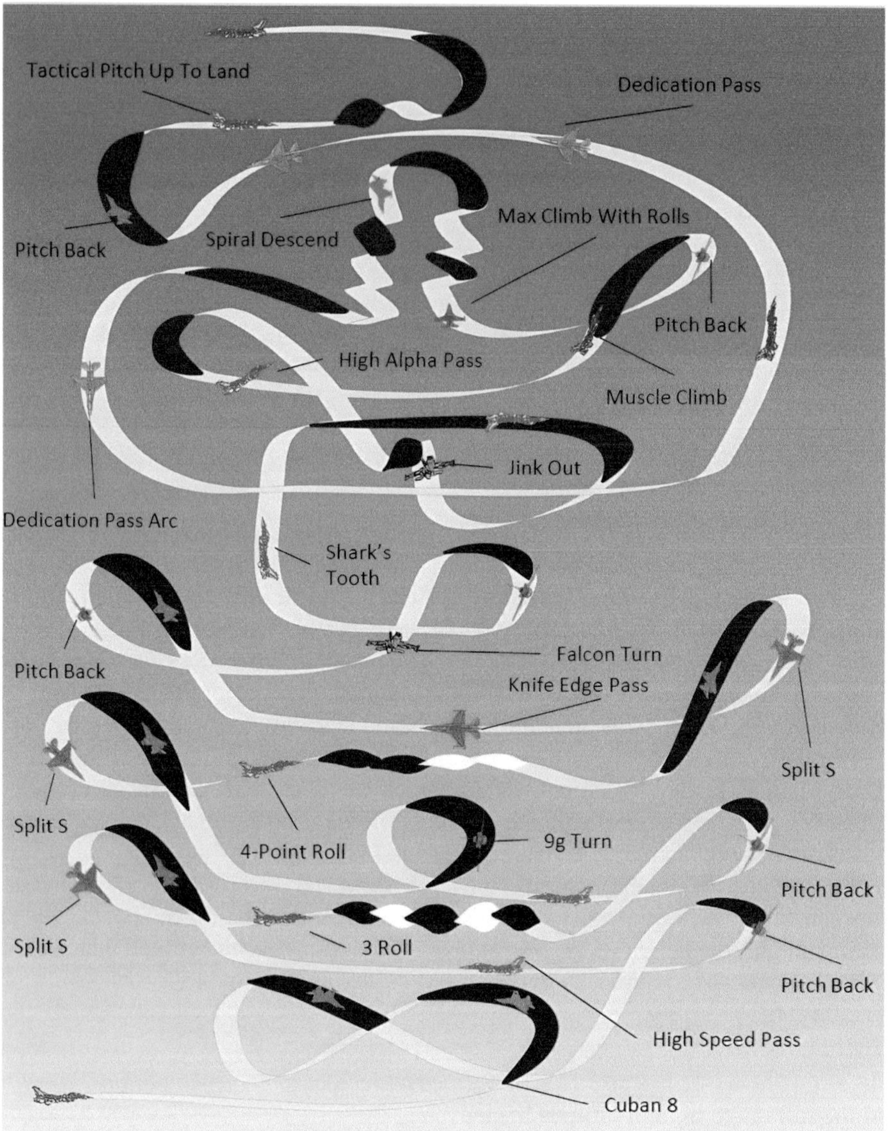

FIGURE 5.5 High show profile for F-16 demonstration flight. (Copyright: Hellenic Air Force.)

5. Four-Point Roll: The aircraft performs a roll, stopping momentarily every 90 degrees at each of the four points, highlighting precision flying.
6. Knife Edge Pass: Another low-altitude pass, allowing spectators to see the aircraft's profile.
7. Falcon Turn: A signature maneuver specific to the F-16, highlighting its turning capabilities.

8. Shark's Tooth: A maneuver that resembles the profile of a shark's tooth, combining vertical climbs and descents with sharp turns.
9. High Alpha Pass: Flying at a high AoA, demonstrating the F-16's stability at low speeds.
10. Muscle Climb: A powerful climb with the use of afterburner displaying the aircraft's thrust-to-weight ratio.
11. Maximum Performance Climb with Rolls: Combines a steep climb with consecutive rolls, emphasizing power and agility.
12. Spiral Descent: A controlled spiral descent demonstrating the pilot's skill in managing energy and direction.
13. Dedication Pass: A tribute pass, often in memory or honor of military service members.
14. Tactical Pitch-Up to Landing: A maneuver that transitions from a low pass to a sharp climb, followed by a setup for landing.

Low Show Profile

The Low Show is adapted for conditions where the ceiling is lower, impacting the safe execution of high-altitude maneuvers:

1. Takeoff to Level 8 (or Cuban 8): Depending on ceiling height, a modified takeoff leads into a level 8 or a Cuban 8 maneuver.
2. High-Speed Flat Pass through Knife Edge Pass: Similar to the High Show, emphasizing low-altitude skills and the F-16's performance capabilities.
3. Dedication Pass: Maintained from the High Show, paying tribute within the constraints of the weather.
4. Tactical Pitch-Up to Landing: Concluding the Low Show with a demonstration of tactical landing capabilities.

The choice between the High Show and Low Show ensures that, regardless of weather conditions, the audience can experience the capabilities of the F-16 and the skill of the demo pilot. These profiles not only entertain but also serve as a powerful recruitment and public relations tool for the US Air Force, highlighting the precision, agility, and power at the disposal of its aviators.

HOW: PRECISION AND AEROBATIC SKILLS DEVELOPMENT

This book does not provide detailed techniques on advanced aerobatics; rather, it introduces fundamental concepts and the "how" of precision and aerobatic skills development, with further details available in the further readings section. Understanding these basics is crucial as the training and skill development needs for air show performers vary significantly across distinct categories of pilots and aircraft. These needs are shaped by the specific demands of the aircraft being flown, the nature of the display (solo or formation), and the objectives of the performance. Below is an analysis of these needs across six categories of air show performers:

- Fast Jet Solo Display Pilots: These pilots must excel in handling high-speed jets independently, performing precision aerobatics, and managing high G-forces. Training often includes advanced simulators that mimic the aircraft's response to various inputs and scenarios, emphasizing solo performance dynamics. For instance, Gabriel and Burrows (1968) discussed improving pilots' time-sharing performance between intra- and extra-cockpit data sources, which is central for solo display pilots managing complex aircraft systems in the dynamic low-level environments where air shows take place.
- Fast Jet Demonstration Team Pilots: Pilots in aerobatic teams like the Blue Angels or Thunderbirds require synchronized flying skills, where precision and timing are paramount. Training includes practicing formation flying before moving to actual display training flights while focusing on maintaining strict formations at high speeds, which necessitates a deep understanding of both individual aircraft capabilities and team maneuvering dynamics.
- High-Performance Aerobatic Airplane Pilots: Flying aircraft, like the Zivko Edge 540, for aerobatics demands acute awareness of the aircraft's limits and capabilities. Pilots must undergo intensive training in executing advanced aerobatics with precision, often taking place in specialized aerobatic training facilities where pilots can safely push the aircraft to its limits.
- High-Performance Aerobatic Airplane Pilots in Formation Demo Teams: Combining the complexities of high-performance aerobatics with the precision of formation flying, these pilots train extensively in both individual aerobatic skills and team maneuvers. Teams such as The Flying Bulls (2024) showcase this dual expertise, where pilots perform both solo aerobatics and intricate formation flying.
- Helicopter Display Pilots: Unlike fixed-wing aircraft, helicopters offer unique flight characteristics that display pilots must master, including hovering, vertical takeoff and landing, and backward flight. Training focuses on exploiting these unique capabilities in displays, often using advanced flight simulators that can replicate the helicopter's flight dynamics accurately.
- Test Pilots Flying for Demonstration and Sales: Test pilots need comprehensive knowledge across various aircraft types, focusing on demonstrating an aircraft's capabilities to potential customers. This includes familiarization with the aircraft's performance envelope, the ability to perform under different conditions, and effective communication skills to highlight the aircraft's features during demonstrations (Flynn, 2021).

The Global Landscape of Aerobatic Training: A Comparative Analysis

The air show industry, a world where the confluence of skill, precision, and thrill captivates audiences worldwide, operates on the foundational principle of rigorous and systematic training. The structured progression of aerobatic training is assessed, drawing upon regulatory frameworks from the USAF's (2022) AFMAN 11-246, the

United Kingdom through CAP 1724 (UK Civil Aviation Authority, 2024b), and introducing the categorization of aerobatic competencies as outlined by the ACE Manual Revision 9 in the United States (International Council of Air Shows, 2019b).

Graduated Approach: Stages of Learning

The essence of these structured pathways lies in their phased, step-by-step approach, allowing pilots to gradually build upon their existing skills while continuously pushing their boundaries within controlled environments. Such a methodical progression is crucial for several reasons:

- Foundation Building: Initially, pilots are introduced to the fundamental principles of aerobatics, including the basics of aircraft control, the physics of flight, and the initial maneuvers that form the cornerstone of aerobatic flying.
- Incremental Skill Acquisition: As pilots advance through the training stages, they are gradually exposed to more complex maneuvers and scenarios, each designed to challenge their abilities and encourage mastery and confidence. This incremental approach helps pilots to assimilate new skills more effectively, minimizing the risk of overwhelming them with complexities beyond their current capability.
- Safety and Proficiency: By progressing through predefined stages, pilots have the opportunity to focus intensely on each aspect of their training, from basic maneuvers at higher altitudes to more intricate performances closer to the ground. This not only enhances their proficiency but also reinforces the paramount importance of safety.

The Role of Altitude in Training

Both the USAF's (2022) altitude step-down process and the CAP 1724 framework utilize altitude as a critical variable in training, setting specific minimums for various stages of aerobatic maneuvers. This careful regulation of altitude serves multiple purposes:

- Risk Mitigation: Higher altitudes provide a safety buffer, giving pilots more time and space to correct potential errors. As pilots become more skilled, this buffer is gradually reduced, simulating the conditions of competitive aerobatics and air shows while still prioritizing safety.
- Confidence Building: Starting at higher altitudes allows pilots to gain confidence in their abilities and their aircraft's capabilities without the immediate pressure of low-altitude performance. As their confidence grows, they can safely transition to more challenging maneuvers at lower altitudes.
- Skill Refinement: The reduction in altitude is not merely a progression in risk but also a refinement of skill. Executing advanced aerobatic maneuvers with precision at lower altitudes demands a higher degree of control and finesse, qualities that are essential for any top-tier aerobatic pilot.

USAF: Structured Training and Skill Progression

Aerobatic training encompasses a detailed, stepwise approach that progressively hones the pilot's skills. In the United States, the USAF (2022) delineates a clear altitude step-down process in the AFMAN 11-246, marking stages of skill development from foundational maneuvers at safer altitudes to advanced techniques closer to the ground.

Stage 1 and Stage 2: From Foundation to Mastery

The altitude step-down process is segmented into Stage 1 and Stage 2, starting with pilots performing maneuvers no lower than 500 feet AGL and progressing to more complex aerobatics at reduced altitudes. This transition is supported by off-station practices at higher altitudes and under stringent observation, emphasizing the gradual and controlled exposure to varied flying conditions and environments.

The UK Perspective: Categorization of Aerobatic Competence

Mirroring the structured approach seen in the United States, the United Kingdom's Civil Aviation Authority (UK CAA, 2024b) introduces a nuanced categorization of aerobatic skills through CAP 1724, offering a comprehensive framework for aerobatic display authorization, which spans several levels of expertise, from standard to unlimited aerobatics, and extends to both fixed-wing and rotary-wing aircraft.

Fixed Wing Aerobatics: A Spectrum of Skill Levels

- Standard Level (aS) Aerobatics: Marks the entry point for aerobatic display authorization, focusing on fundamental aerobatic maneuvers.
- Intermediate Level (aI) Aerobatics: A step up from the standard level, introducing more complex maneuvers and combinations.
- Advanced Level (aA) Aerobatics: For pilots demonstrating high proficiency in a wide range of demanding aerobatic skills.
- Unlimited Level (aU) Aerobatics: The pinnacle of aerobatic skill, with no restrictions on the complexity or type of maneuvers performed.

Rotary Wing Aerobatics: Evaluating Skills

For rotary-wing aircraft, the skill levels are categorized into standard (aRS) and advanced (aRA), each with its own set of evaluation criteria focusing on the pilot's ability to execute aerobatic maneuvers safely and accurately.

Beyond Solo Aerobatics: Formation and Tail Chase Authorizations

CAP 1724 extends the categorization to include formation flying and tail chase scenarios, introducing specific skill levels and evaluation criteria for these complex aerial disciplines. Formation flying is differentiated by the number of aircraft and the role of the pilot (leading or part of the formation), while tail chase authorizations assess the pilot's skill in both leading and performing aerobatics in a follow-the-leader style sequence.

United States's ACE Manual: Categorization and Qualification

The ACE Manual Revision 9 (ICAS, 2019a) further enriches the landscape of aerobatic training in the United States, introducing a dual framework of categorization by aircraft type and qualification by minimum altitude AGL for aerobatic maneuvers.

Aircraft Categories and Aerobatic Levels

- Aircraft Categories: The manual classifies aircraft into five categories based on their performance capabilities, ranging from sport aerobatics (Category A) to helicopter aerobatics (Category E), each category accommodating a diverse array of aerobatic interests and specializations.
- Aerobatic Levels: The qualification levels, defined by the minimum altitude AGL authorized for initiating and completing maneuvers, set clear benchmarks for skill progression. These levels range from Level 4 (800 feet) to Level 1 (unrestricted), offering pilots a pathway to gradually decrease their minimum performance altitude as their proficiency increases.

Initial Qualifications: Establishing Competency

The ACE Manual mandates a comprehensive evaluation process for initial qualifications, emphasizing the importance of both ground and flight assessments in accordance with its appendices. Key stipulations for initial applicants include:

- Level 4 SAC: New applicants without prior FAA or TC aerobatic competency are assigned a Level 4 qualification, setting an 800-foot minimum altitude for aerobatic maneuvers.
- Formation Aerobatics and Special Endorsements: Applicants seeking qualifications in formation aerobatics, dogfighting, or wing walking are required to undergo an evaluation at their respective solo aerobatic endorsement level, ensuring a thorough vetting process for specialized aerobatic disciplines.

Comparative Analysis

The three regulatory documents cover the scope of their areas of operations in display flying. Below is a comparative analysis of the scope and application, training and progression, qualification and certification, and safety and oversight of all three frameworks.

- Scope and Application: AFMAN 11-246 is specific to the USAF, focusing on military aviation and the training of demo pilots within a structured military environment. In contrast, the ACE Manual is broader, encompassing civilian aerobatic pilots within the United States and outlining qualifications for various levels of aerobatic competency. CAP 1724 applies to civilian aerobatic display pilots in the United Kingdom, providing guidelines for both fixed-wing and rotary-wing aircraft.

- Training and Progression: AFMAN 11-246 details a step-down altitude process for training, ensuring pilots incrementally develop their skills starting from higher, safer altitudes. The ACE Manual categorizes aerobatic competency by aircraft performance capabilities and pilot proficiency levels, emphasizing a progression through minimum altitude levels for aerobatic maneuvers. CAP 1724 categorizes skills based on the complexity of maneuvers and includes specific criteria for formation flying and tail chase scenarios.
- Qualification and Certification: Both AFMAN 11-246 and the ACE Manual include initial qualification processes involving ground and flight evaluations, but AFMAN 11-246 focuses on military demonstration pilots, whereas the ACE Manual addresses civilian aerobatic competency. CAP 1724 outlines a display authorization process with different skill levels for aerobatic and formation flying, tailored to the civilian air show industry in the United Kingdom.
- Safety and Oversight: All three frameworks prioritize safety, with AFMAN 11-246 and CAP 1724 detailing supervision requirements for practice and demonstrations. The ACE Manual emphasizes the importance of evaluating pilots at their designated aerobatic level before allowing progression to ensure competency at lower altitudes.

Case Study: International Aerobatics Club (IAC) – Achieving Precision and Aerobatic Skill Development

The history and contributions of the International Aerobatics Club (IAC) to aerobatic training in the United States are profound. Formed in 1970 under the leadership of Bob Heuer, the IAC aimed to formalize and promote grassroots aerobatics through a structured competition system. The organization has played a pivotal role in advancing aerobatic standards and safety, earning a reputation as a leader in the field of sport aerobatics. Over the decades, the IAC has organized numerous competitions and training events, significantly contributing to the development of aerobatic skills and the promotion of a robust safety culture in the United States (International Aerobatic Club (IAC), n.d.).

Nowadays, the IAC provides a structured and comprehensive pathway for mastering aerobatics, which is crucial for display pilots. Under the guidance and tuition of experienced instructors, pilots at the IAC progress through various levels of aerobatic training, honing their skills to achieve the highest standards of precision and performance. Its training programs, practical application of knowledge, continuous evaluation, and supportive community make it a case study for both current and future air show performers to hone their skills and achieve the highest standards of aerial performance.

The training at IAC is meticulously organized into five progressive levels: basic, sportsman, intermediate, advanced, and unlimited. Training begins at the basic level, where pilots are introduced to fundamental aerobatic maneuvers and essential safety protocols. This initial phase is designed to build a solid foundation, focusing on the development of core skills and an understanding of the physics and dynamics of aerobatic flight.

As pilots advance to the sportsman and intermediate levels, they are exposed to more complex maneuvers, enhancing their control and precision. These levels focus on refining techniques, increasing maneuver complexity, and integrating performance elements that are critical for air shows. This stage of training not only builds technical proficiency but also instills a deep sense of discipline and attention to detail, essential for achieving precision in aerobatics.

For those aiming to reach the highest levels of proficiency, the advanced and unlimited levels involve the most intricate and demanding maneuvers. This advanced training is more intensive and costly, requiring specialized aircraft and highly skilled instructors. The focus at these levels is on mastering advanced aerobatic sequences and maneuvers that require exceptional skill and precision. Pilots learn to handle aircraft capable of high-performance demands and to execute complex routines flawlessly, which are crucial for top-tier aerobatic performances.

A unique aspect of the IAC's approach is the practical application of theoretical knowledge. Training sessions are conducted in designated aerobatic zones or dedicated airfields, providing a controlled environment for practicing maneuvers. Additionally, regular evaluations and feedback from instructors are integral to the IAC's training methodology. Continuous assessment ensures that pilots refine their skills progressively, addressing any weaknesses and building on their strengths, which is vital for achieving and maintaining high levels of precision in aerobatic flying.

In addition to the structured training levels, the IAC also fosters a supportive community and networking opportunities. Pilots engage with fellow aviators, participate in club events, and attend workshops and seminars organized by the IAC, facilitating the exchange of best practices, staying updated on the latest safety protocols, regulatory changes, and advancements in aerobatic techniques.

For air show performers, the IAC provides an exemplary model of how structured training environments and methods contribute to the development and maintenance of pilot proficiency. By following this structured path and seeking additional instruction from experienced performers, air show performers can achieve excellence in aerobatics, ensuring both safety and performance in their flying endeavors.

NOTES

1 Formation is considered a group of aircraft flying in a coordinated pattern, while maintaining lateral, vertical, and/or time deconfliction (FAST, 2024).
2 Anti-G pants comprises of inflatable bladders around the legs and abdomen that apply pressure to prevent blood pooling during high G-forces, thus maintaining blood flow to the brain and preventing G-induced loss of consciousness (G-LOC) (Green, 2006).
3 Pitch rate is the rate at which an aircraft's nose moves up or down, changing its pitch angle, and is typically measured in degrees per second.
4 G-onset rate is the rate at which gravitational forces (G-forces) increase or decrease on an aircraft and its occupants, typically measured in Gs per second, i.e., 4 Gs in 2 seconds.

5 Airspeed sustaining capability refers to an aircraft's ability to maintain a consistent airspeed under various flight conditions, including during maneuvers and in different atmospheric environments.
6 G-sustaining capability refers to an aircraft's ability to maintain a constant level of G-force during sustained maneuvers, ensuring performance and control without excessive loss of speed or altitude.
7 Engine spool-up time is the duration it takes for an aircraft's engine to accelerate from idle to full thrust, affecting the responsiveness and performance during rapid changes in power demand.
8 Airspeed acceleration rate is the rate at which an aircraft can increase its speed, typically measured in knots per second, indicating the aircraft's ability to rapidly gain velocity.
9 Plyometrics, also known as jump training, involves explosive exercises designed to increase speed, power, and endurance, such as jumps and bounding movements that utilize rapid muscle contractions (Robinson, 2022).
10 Powerlifting is a strength sport that focuses on three main lifts: the squat, bench press, and deadlift, with the goal of lifting the maximum possible weight in each (Austin & Mann, 2021).
11 Aerobatic box is a designated airspace of 1,000 meters on each side where aerobatic competitions and displays take place (IAC, 2024a).

REFERENCES

Advani, S., & Field, J. (2011, August 8). Upset prevention and recovery training in flight simulators. *Guidance, Navigation, and Control and Co-Located Conferences. AIAA Modeling and Simulation Technologies Conference*, Portland, Oregon. https://doi.org/10.2514/6.2011-6698

Advani, S., & Schroeder, J. (2016). Global implementation of upset prevention & recovery training. *AIAA SciTech Forum*. https://doi.org/10.2514/6.2016-1430

Airplanes Online. (n.d.). F-86 Sabre jet fighter. Retrieved July 27, 2024, from www.airplanes-online.com/f86-sabre.htm

Aker, J. A. (2022, December 16). How Blue Angels and Thunderbirds keep flying 300 days a year. *DVIDS*. www.dvidshub.net/news/435455/blue-angels-and-thunderbirds-keep-flying-300-days-year

Araiza-Alba, P., Keane, T., Chen, W. S., & Kaufman, J. (2021). Immersive virtual reality as a tool to learn problem-solving skills. *Computers & Education, 164*, 104121. https://doi.org/10.1016/j.compedu.2020.104121

Aresti System. (2024). *Aresti system*. www.arestisystem.com/

Astbury, N., Taylor, M., French, S., & Macdonald, I. (2014). Snacks containing whey protein and polydextrose induce a sustained reduction in daily energy intake over 2 wk under free-living conditions. *The American Journal of Clinical Nutrition, 99*(5), 1131–1140. https://doi.org/10.3945/ajcn.113.075978

Ateş, S. S., Kafali, H., & ÇeliKtaş, M. (2018). Analysis of internal/external factors affecting time management and the reasons of delay in aviation. *Journal of Aviation, 2*(2), 105–118. https://doi.org/10.30518/jav.480981

Athens Flying Week International Air Show (AFW). (2025). *Athens Flying Week International Air Show*. www.athensflyingweek.gr/

Austin, D., & Mann, B. (2021). *Powerlifting: The complete guide to technique, training, and competition*. Human Kinetics.

Balldin, U. I. (1986). Factors influencing G-tolerance. *Clinical Physiology, 6*(3), 209–219. https://doi.org/10.1111/j.1475-097X.1986.tb00620.x

Baquet, G., Berthoin, S., Gerbeaux, M., & Praagh, E. (2001). High-intensity aerobic training during a 10 week one-hour physical education cycle: Effects on physical fitness of adolescents aged 11 to 16. *International Journal of Sports Medicine, 22*(4), 295–300. https://doi.org/10.1055/S-2001-14343

Barker, D. (2003). *Zero error margin: Airshow display flying analysed.* Freeworld Publications.

Barker, D. (2020a). *Anatomy of airshow accidents* (1st ed., Vol. 1). International Council of Air Shows, International Air Show Safety Team.

Barker, D. (Director). (2020b, December). Air show accidents: A statistical and preventative perspective [video]. ICAS Convention. www.youtube.com/watch?v=29zT0edEAUQ&t=877s

Bateman, W. A., Jacobs, I., & Buick, F. (2006). Physical conditioning to enhance +Gz tolerance: Issues and current understanding. *Aviation, Space, and Environmental Medicine, 77*(6), 573–580.

Bhardwaj, N. K. (2024). Yoga: Recovery tool for every athlete. *International Journal of Sports, Yoga and Physical Activity, 3*(1), 1–6.

Bjelica, B., Milanović, L., Aksović, N., Zelenović, M., & Božić, D. (2020). Effects of physical activity to cardiorespiratory changes. *Turkish Journal of Kinesiology, 6*(4), 164–174.

Bonilla, D. A., Pérez-Idárraga, A., Odriozola-Martínez, A., & Kreider, R. B. (2021). The 4R's framework of nutritional strategies for post-exercise recovery: A review with emphasis on new generation of carbohydrates. *International Journal of Environmental Research and Public Health, 18*(1), 103.

Bork, A. (1964). The fourth dimension in nineteenth-century physics. *Isis, 55*, 326–338. https://doi.org/10.1086/349865

British Air Display Association. (n.d.). *British Air Display Association.* www.bada-uk.com

Cambridge English Dictionary. (2024). *Stepping stone.* Cambridge University Press. https://dictionary.cambridge.org/dictionary/english/stepping-stone

Choi, S. (2003). The effect of student pilot's physical fitness on flight performance abilities. *Transactions of the Japan Society for Aeronautical and Space Sciences, 45*, 267–268. https://doi.org/10.2322/TJSASS.45.267

Colcombe, S., Kramer, A., Erickson, K., Scalf, P., McAuley, E., Cohen, N., Webb, A., Jerome, G., Marquez, D., & Elavsky, S. (2004). Cardiovascular fitness, cortical plasticity, and aging. *Proceedings of the National Academy of Sciences of the United States of America, 101*(9), 3316–3321. https://doi.org/10.1073/PNAS.0400266101

Colt, H., Crawford, S., & Galbraith, O. (2001). Virtual reality bronchoscopy simulation: A revolution in procedural training. *Chest, 120*(4), 1333–1339. https://doi.org/10.1378/CHEST.120.4.1333

Courtney, C. (2017, December 21). Aviation careers: Becoming an airshow pilot. *Disciples of Flight.* https://disciplesofflight.com/professional-airshow-pilot/

Cox, M. (n.d.). *Human factors in flying displays – Practical guidance for display pilots & flying display directors.* UK Civil Aviation Authority.

Crider, D. (2008). Need for upset recovery training. *AIAA Modeling and Simulation Technologies Conference and Exhibit.* American Institute of Aeronautics and Astronautics. https://doi.org/10.2514/6.2008-6864

Crisman, R., & Burton, R. (1988). *Physical fitness to enhance aircrew G tolerance.* https://doi.org/10.21236/ada204689

Dahlstrom, N., Dekker, S., van Winsen, R., & Nyce, J. (2002). Fidelity and validity of simulator training. In *Simulation in aviation training.* Routledge.

Darses, F., Bernier, M., Berthelot, V., Fornette, M.-P., Launay, Y., Dozias, B., Chastres, V., & Fournier, J. (2023). Effects of mindfulness training on decision making in critical and high-demand situations: A pilot study in combat aviation. *Safety Science, 166*, 106204. https://doi.org/10.1016/j.ssci.2023.106204

Demir, A., & Aydin, E. (2021). Vestibular illusions and alterations in aerospace environment. *Turkish Archives of Otorhinolaryngology, 59*, 139–149. https://doi.org/10.4274/tao.2021.2021-3-3

Despres, J., & Ewing, C. (1998). American College of Sports Medicine position stand: The recommended quantity and quality of exercise for developing and maintaining cardiorespiratory and muscular fitness, and flexibility in healthy adults. *Medicine and Science in Sports and Exercise, 30*(6), 975–991.

Dobek, J., White, K., & Gunter, K. (2007). The effect of a novel ADL-based training program on performance of activities of daily living and physical fitness. *Journal of Aging and Physical Activity, 15*(1), 13–25. https://doi.org/10.1123/JAPA.15.1.13

Donti, O., Donti, A., & Theodorakou, K. (2014). A review on the changes of the evaluation system affecting artistic gymnasts' basic preparation: The aspect of choreography preparation. *Science of Gymnastics, 6*, 63–72.

EASA. (2015). *EGAST Leaflet GA 11 Flying Displays for Pilots.*

Ehrenfreund, P., McKay, C., Rummel, J. D., Foing, B. H., Neal, C. R., Masson-Zwaan, T., Ansdell, M., Peter, N., Zarnecki, J., Mackwell, S., Perino, M. A., Billings, L., Mankins, J., & Race, M. (2012). Toward a global space exploration program: A stepping stone approach. *Advances in Space Research, 49*(1), 2–48. https://doi.org/10.1016/j.asr.2011.09.014

Endsley, M. R. (1999). Situation awareness in aviation systems. In J. A. Wise, V. D. Hopkin, & D. J. Garland (Eds.), *Handbook of aviation human factors* (pp. 257–276). Lawrence Erlbaum Associates

European Airshow Council. (n.d.). *European Airshow Council.* www.europeanairshow.org/

European Airshow Council (EAC). (2020). *Airshow manual (5.1).* European Airshow Council.

European Airshow Council (EAC). (2023). *EAC: Flying display directors safety seminar.* www.europeanairshow.org/copy-of-fast-jet-safety-workshop

European Airshow Council (EAC). (2024). *European Airshow Council (EAC) – Convention.* www.europeanairshow.org/convention

European General Aviation Safety Team. (2015). *Safety at flying displays and events: A guide for pilots.* European Union Aviation Safety Agency.

Federal Aviation Administration (FAA). (2016, January 6). *AC 61-67C – Stall and Spin Awareness Training with Change 1 & 2.* Federal Aviation Administration (FAA). www.faa.gov/regulations_policies/advisory_circulars/index.cfm/go/document.information/documentID/1028760

Federal Aviation Administration (FAA). (2022). *Order 8900.1A – Flight Standards Information Management System.* Federal Aviation Administration (FAA). www.faa.gov/regulations_policies/orders_notices/index.cfm/go/document.information/documentID/1041430

Federal Aviation Administration. (2023). *Pilot's handbook of aeronautical knowledge* (FAA-H-8083-25C). www.faa.gov/regulations_policies/handbooks_manuals/aviation/faa-h-8083-25c.pdf

Federal Aviation Administration (FAA). (2024). *Mass Aircraft Demonstrations at Aviation Events (Safety Alert for Operators (SAFO) SAFO24005).* Federal Aviation Administration (FAA). www.faa.gov/other_visit/aviation_industry/airline_operators/airline_safety/safo/all_safos/SAFO24005.pdf

Fédération Aéronautique Internationale. (n.d.). *Fédération Aéronautique Internationale*. www.fai.org

Flach, J. M., & Warren, R. (1995). Low-altitude flight. In *Local applications of the ecological approach to human-machine systems*. CRC Press.

Flynn, B. (2021, September 1). F-35 airshow, seeing is believing. Billie Flynn – My Blog. https://billieflynn.com/f-35-airshow-seeing-is-believing/

Formation and Safety Team (FAST). (2024). Formation and Safety Team (FAST). www.flyfast.org/

French, R. (2023, April 8). Patty Wagstaff reflects on her aviation career. *FOX 13 News*; FOX 13 *Tampa Bay*. www.fox13news.com/news/patty-wagstaff-reflects-on-path-that-led-to-her-to-becoming-one-of-the-nations-top-air-show-pilots

Gabriel, R., & Burrows, A. (1968). Improving time-sharing performance of pilots through training. *Human Factors: The Journal of Human Factors and Ergonomics Society, 10*, 33–40. https://doi.org/10.1177/001872086801000106

Garcia, S. (2018, August 26). The Blue Angel's G-tolerance improvement program: Measuring G-fitness. *Grind to Fly*. www.grindtofly.com/post/the-blue-angel-s-g-tolerance-improvement-program-measuring-g-fitness

Geyer, D. J., & Biggs, A. T. (2018). The persistent issue of simulator sickness in naval aviation training. *Aerospace Medicine and Human Performance, 89*(4), 396–405.

Glenn Research Center. (n.d.). *Guide to Aerodynamics*. Glenn Research Center, NASA. Retrieved July 27, 2024, from www1.grc.nasa.gov/beginners-guide-to-aeronautics/learn-about-aerodynamics/

Gray, W. R. (2016). Time safety margin: Theory and practice. *412th Test Wing Edwards Air Force Base, California Air Force Materiel Command United States Air Force*.

Green, N. D. (2006). Protection against long duration acceleration. In *Ernsting's aviation medicine* (pp. 159–167). Hodder Arnold.

Gulesci, S., Jindani, S., La Ferrara, E., Smerdon, D., Sulaiman, M., & Young, H. (2021). A stepping stone approach to understanding harmful norms (SSRN Scholarly Paper 3784002). https://papers.ssrn.com/abstract=3784002

Harris, D. (2011). *Human performance in the flight deck*. CRC Press.

Heaton, L. E., Davis, J. K., Rawson, E. S., Nuccio, R. P., Witard, O. C., Stein, K. W., …, & Baker, L. B. (2017). Selected in-season nutritional strategies to enhance recovery for team sport athletes: A practical overview. *Sports Medicine, 47*, 2201–2218.

Hirschman, D. (2022, January 5). Sighting devices. *AOPA*. www.aopa.org/news-and-media/all-news/2022/may/flight-training-magazine/how-it-works-sighting-devices

Hoffman, J., Kahana, A., Chapnik, L., Shamiss, A., & Davidson, B. (1999). The relationship of physical fitness on pilot candidate selection in the Israel Air Force. *Aviation, Space, and Environmental Medicine, 70*(2), 131–134.

International Aerobatic Club (IAC) (Director). (2016, January 23). The aerobatic textbook: Aresti symbols [video]. YouTube. www.youtube.com/watch?v=Kh6ggZDnGJw

International Aerobatic Club (IAC). (2024a). Aerobatic figures. International Aerobatic Club (IAC). www.iac.org/aerobatic-figures

International Aerobatic Club (IAC). (2024b). Intro to Aresti. International Aerobatic Club (IAC). www.iac.org/intro-aresti

International Council of Air Shows. (n.d.). *International Council of Air Shows*. https://airshows.aero/

International Council of Air Shows (ICAS). (2016). *Air show performers safety manual* (p. 27). ICAS. www.airshows.aero/GetDoc/3444

International Council of Air Shows (ICAS). (2019a). *ICAS air show manual* (7th ed.). International Council of Air Shows. https://airshowdigest.aero/wp-content/uploads/2019/01/ICAS-Air-Show-Manual-7th-Edition.pdf

International Council of Air Shows (ICAS). (2019b, January 1). ACE Manual—Revision 9. International Council of Air Shows.

Jentsch, F., & Curtis, M. (Eds.). (2011). *Simulation in aviation training*. Ashgate Publishing. https://doi.org/10.4324/9781315243092

Jones, W. (1958). Problems of high-speed flight. *Nature, 182*, 1582–1583. https://doi.org/10.1038/1821582b0

Jouanin, J.-C., Dussault, C., Tran, D., & Guézennec, C.-Y. (2005). Aerobatic flight effects on baroreflex sensitivity and sympathovagal balance in experienced pilots. *Aviation, Space, and Environmental Medicine, 76*(12), 1151–1155.

Karachalios, M. (2022). *An evaluation of the relationships between resilient safety culture, safety risk parameters, and mindfulness in the international air show community* [PhD dissertation, University of North Dakota]. https://commons.und.edu/theses/4267/

Karachalios, M. (2024). *Des Barker report 2023, airshow incident and accident review* (p. 12). European Airshow Council.

Karachalios, M., & Adjekum, D. K. (2023). *Air show performers: Safety, risk management, and psychological factors*. CRC Press. https://doi.org/10.1201/9781003431879

Kloubec, J. (2010). Pilates for improvement of muscle endurance, flexibility, balance, and posture. *Journal of Strength and Conditioning Research, 24*, 661–667. https://doi.org/10.1519/JSC.0b013e3181c277a6

Le Vie, L. R. (2016). Survey of quantitative research metrics to assess pilot performance in upset recovery (NASA/TM-2016-219191). https://ntrs.nasa.gov/citations/20160006310

Lennemann, L., Sidrow, K., Johnson, E., Harrison, C., Vojta, C., & Walker, T. (2013). The influence of agility training on physiological and cognitive performance. *Journal of Strength and Conditioning Research, 27*, 3300–3309. https://doi.org/10.1519/JSC.0b013e31828ddf06

Lin, X., Zhang, X., Guo, J., Roberts, C., McKenzie, S., Wu, W., Liu, S., & Song, Y. (2015). Effects of exercise training on cardiorespiratory fitness and biomarkers of cardiometabolic health: A systematic review and meta-analysis of randomized controlled trials. *Journal of the American Heart Association: Cardiovascular and Cerebrovascular Disease, 4*. https://doi.org/10.1161/JAHA.115.002014

Loudenslager, L. (Director). (2014, August 24). Leo Loudenslager [video]. YouTube. www.youtube.com/watch?v=92M2iCD_rQw

Luke Air Force Base. (2024). Luke Days 2024 Airshow Need-to-know info. Retrieved from www.luke.af.mil/

Meland, A., Fonne, V., Wagstaff, A., & Pensgaard, A. M. (2015). Mindfulness-based mental training in a high-performance combat aviation population: A one-year intervention study and two-year follow-up. *International Journal of Aviation Psychology, 25*(1), 48–61. https://doi.org/10.1080/10508414.2015.995572

Metzler, M. M. (2020). G-LOC due to the push-pull effect in a fatal F-16 mishap. *Aerospace Medicine and Human Performance, 91*(1), 51–55. https://doi.org/10.3357/AMHP.5461.2020

Ministry of Defence, France. (2024a, June 18). The Patrouille de France. www.defense.gouv.fr/air/tous-nos-dossiers/90-ans-larmee-lair-lespace-versailles/ambassadeurs-larmee-lair-lespace/patrouille-france

Ministry of Defence, France. (2024b, June 28). Aerobatics and Rafale solo display: The same mission. www.defense.gouv.fr/air/actualites/voltige-aerienne-rafale-solo-display-mission

Morris, P. (2010). The instability of high speed jets. *International Journal of Aeroacoustics, 9*, 1–50. https://doi.org/10.1260/1475-472X.9.1-2.1

Muller, T. U. (2002). G-induced vestibular dysfunction ('the wobblies') among aerobatic pilots: A case report and review. *Ear, Nose, & Throat Journal, 81*(4), 269–272.

Myers, J., Kokkinos, P., Arena, R., & LaMonte, M. J. (2021). The impact of moving more, physical activity, and cardiorespiratory fitness: Why we should strive to measure and improve fitness. *Progress in Cardiovascular Diseases, 64*, 77–82.

N.C. Cooperative Extension. (2023). *Youth: Hard and soft skills*. Retrieved from https://ces.ncsu.edu

Newman, D. G., White, S., & Callister, R. (1999). Patterns of physical conditioning in Royal Australian Air Force F/A-18 pilots and the implications for +Gz tolerance. *Aviation, Space, and Environmental Medicine, 70*(8), 739–744.

Nielsen, L., Lambert, M., & Jeppesen, P. (2020). The effect of ingesting carbohydrate and proteins on athletic performance: A systematic review and meta-analysis of randomized controlled trials. *Nutrients, 12*. https://doi.org/10.3390/nu12051483

Orlandi, A., Cross, E. S., & Orgs, G. (2020). Timing is everything: Dance aesthetics depend on the complexity of movement kinematics. *Cognition, 205*, 104446. https://doi.org/10.1016/j.cognition.2020.104446

Pacific Airshow. (2024). *2024 performers*. https://pacificairshowusa.com/2024-performers

Pérez, L., Padilla, C., Parmentier, F., & Andrés, P. (2014). The effects of chronic exercise on attentional networks. *PLOS One, 9*. https://doi.org/10.1371/journal.pone.0101478

Polman, R., Walsh, D., Bloomfield, J., & Nesti, M. (2004). Effective conditioning of female soccer players. *Journal of Sports Sciences, 22*, 191–203. https://doi.org/10.1080/02640410310001641458

Robinson, K. M. (2022, August 28). Plyometrics: What it is and how to do it. *WebMD*. www.webmd.com/fitness-exercise/a-z/what-is-plyometrics

Schlegel, P., Křehký, A., & Hiblbauer, J. (2022). Physical fitness improvement after 8 weeks of high-intensity interval training with air bike. *Sport Mont*. https://doi.org/10.26773/smj.221012

Schofield, D. (2012). Using simulators for training. In https://doi.org/10.4018/978-1-60960-762-3.CH015

Schroeder, J., & Burke, R. (2016). Upset prevention and recovery training—A regulator update. *AIAA SciTech Forum*. https://doi.org/10.2514/6.2016-1429

Scott, P. (2005, May). So you want to be an airshow pilot. *Smithsonian Magazine*. www.smithsonianmag.com/air-space-magazine/so-you-want-to-be-an-airshow-pilot-8329275/

Shaffer, H. (1984). Timing in musical performance. *Annals of the New York Academy of Sciences, 423*(1), 420–428. https://doi.org/10.1111/j.1749-6632.1984.tb23450.x

Sikder, S., Tuwairqi, K., Al-Kahtani, E., Myers, W., & Banerjee, P. (2013). Surgical simulators in cataract surgery training. *British Journal of Ophthalmology, 98*, 154–158. https://doi.org/10.1136/bjophthalmol-2013-303700

Song, X., & Cui, L. (2017). Mental training about aviation sports influences on spatial orientation of flying cadets. https://doi.org/10.2991/emcs-17.2017.266

Stone, B., Alvar, B., Orr, R., Lockie, R., Johnson, Q., Goatcher, J., & Dawes, J. (2020). Impact of an 11-week strength and conditioning program on firefighter trainee fitness. *Sustainability*. https://doi.org/10.3390/su12166541

Stone, N., & Kilding, A. (2009). Aerobic conditioning for team sport athletes. *Sports Medicine, 39*, 615–642. https://doi.org/10.2165/00007256-200939080-00002

Taylor, M., Markham, A., Reis, J., Padilla, G., Potterat, E., Drummond, S., & Mujica-Parodi, L. (2008). Physical fitness influences stress reactions to extreme military training. *Military Medicine, 173*(8), 738–742. https://doi.org/10.7205/MILMED.173.8.738

Tesch, P., Hjort, H., & Balldin, U. (1983). Effects of strength training on G tolerance. *Aviation, Space, and Environmental Medicine, 54*(8), 691–695.

The Flying Bulls. (2024). The Flying Bulls. www.flyingbulls.at/en/about-us
UK Civil Aviation Authority (CAA). (2018). *CAP 1694: Human factors in air displays: Transfer of behaviours and error path study*. UK Civil Aviation Authority. www.caa.co.uk/our-work/publications/documents/content/cap1694/
UK Civil Aviation Authority (CAA). (2024a). *CAP 403: Flying displays and special events: Safety and administrative requirements and guidance* (21st ed.). UK Civil Aviation Authority. www.caa.co.uk/our-work/publications/documents/content/cap-403/
UK Civil Aviation Authority (CAA). (2024b). *CAP 1724: Flying display pilot authorisation and evaluation: Requirements and guidance*. www.caa.co.uk/our-work/publications/documents/content/cap1724/
UK Civil Aviation Authority. (2024c, April). *Human factors in air displays* [video]. YouTube. www.youtube.com/playlist?list=PLig80S1opAEIavKrE0iFMvRWqV28964-p
U.S. Air Force (USAF). (2022). *Air Force Manual 11-246: Air Force fighter type aircraft demonstrations (A-10, F-16, F-22, F-35)* (Vol. 1). USAF. https://static.e-publishing.af.mil/production/1/af_a3/publication/afman11-246v1/afman11-246.pdf
U.S. Navy. (2024). *U.S. Navy Blue Angels*. www.blueangels.navy.mil/
U.S. Navy Aeromedical Reference and Waiver Guide. (2021). *Aviation physical standards*. www.med.navy.mil/Portals/62/Documents/NMFSC/NMOTC/NAMI/ARWG/Waiver%20Guide/01_Physical_Standards_210128.pdf
Walden University. (2023). What is the difference between hard skills and soft skills? Retrieved from https://lifelonglearning.waldenu.edu
Wallace, J., Hu, Z., & Carroll, D. (2020). Augmented reality for immersive and tactile flight simulation. *IEEE Aerospace and Electronic Systems Magazine, 35*, 6–14. https://doi.org/10.1109/MAES.2020.3002000
Werrell, K. P. (2013). *Sabres over Mig Alley: The F-86 and the battle for air superiority in Korea*. Naval Institute Press.
Westcott, W., Skaggs, J., Gibson, J., Annesi, J., Reynolds, R., & O'Dell, J. (2007). Comparison of two exercise protocols on fitness score improvement in poorly conditioned Air Force personnel. *Perceptual and Motor Skills, 104*, 629–636. https://doi.org/10.2466/pms.104.2.629-636
Whitley, P. E. (1997). Pilot performance of the anti-G straining maneuver: Respiratory demands and breathing system effects. *Aviation, Space, and Environmental Medicine, 68*(4), 312–316.
Wickens, C. D. (2002). Situation awareness and workload in aviation. *Current Directions in Psychological Science, 11*(4), 128–133. https://doi.org/10.1111/1467-8721.00184
Xie, B., Liu, H., Alghofaili, R., Zhang, Y., Jiang, Y., Lobo, F. D., & Yu, L. F. (2021). A review on virtual reality skill training applications. *Frontiers in Virtual Reality, 2*, 645153.
Zhai, Y., Elsworth, C., Arens, E., Zhang, H., Zhang, Y., & Zhao, L. (2015). Using air movement for comfort during moderate exercise. *Building and Environment, 94*, 344–352. https://doi.org/10.1016/J.BUILDENV.2015.08.026

6 Human Skills
The Cohesive Glue

Manolis Karachalios

INTRODUCTION

While technical proficiency in piloting is essential, the integration of human skills such as a passion for air shows, professional discipline, keenness to safety behavior, humility, resilience, personal responsibility, communication, teamwork, and emotional intelligence (Mayer et al., 1999, 2001) is equally pivotal. These skills enhance the spectator experience and ensure the seamless execution of air shows, contributing significantly to both safety and the overall success of the event. These skills are the glue that binds technical proficiency with emotional intelligence, enabling pilots to perform not only with accuracy but also with composure.

These competencies were chosen to be called "*human skills*," as suggested by Simon Sinek (2021), rather than "*soft skills*" to emphasize their universal importance for human interaction and development. The term "*soft*" might imply that these skills are less critical or less challenging to master than "*hard*" technical skills. However, in practice, human skills are often more difficult to develop and have a profound impact on a person's effectiveness and ability to lead. These skills encompass the ability to communicate effectively, collaborate with others, adapt to changing circumstances, and lead with empathy—capabilities that are essential not only in air shows but in all aspects of life.

An extensive analysis of the qualifications necessary for air show performers has already been provided in Chapter 1; however, in this Chapter we will investigate the human skills that are teachable during an air show performer's career, such as communication, leadership, and teamwork. The human skills that are related to personality traits might be difficult, even impossible, to adjust during the pre-display season period; these traits need to be already integrated into one's personality, and the selection process should be based on these, as discussed in Chapter 1.

To set a context for an effective application of necessary human skills, let's suppose the following fictional scenario of an air show performer like "Pete Peak," who successfully managed a performance amidst an unexpected thunderstorm.

> Pete was preparing for his solo aerobatic routine at the annual local air show. The weather forecast had predicted clear skies, but as the afternoon progressed, dark clouds began to gather on the horizon. Minutes before his scheduled

DOI: 10.1201/9781003504801-9

takeoff, a sudden thunderstorm rolled in, creating a potentially dangerous situation for the pilots and spectators.

Pete quickly assessed the rapidly changing weather conditions. Recognizing the threat posed by the approaching storm, he knew immediate action was needed to ensure everyone's safety. With years of experience in high-pressure situations during air shows, Pete relied on his leadership and communication skills to manage the unfolding crisis.

He swiftly contacted the air boss and air traffic control, requesting an update on the storm's development and expressing his concerns. Pete's clear and authoritative communication helped the air show team understand the severity of the situation. He recommended postponing the performance and implementing the air show's backup plan.

While waiting for the final decision from the air boss, Pete gathered the ground crew and informed them of the weather. He maintained a calm demeanor, which helped to alleviate the rising tension among the team.

As the storm intensified, visibility decreased, and the winds picked up. He monitored the weather radar, receiving real-time updates and keeping his ground crew updated.

The ground crew, inspired by Pete's composed leadership, worked efficiently to secure the equipment. His ability to delegate tasks and trust his team's expertise ensured that everyone knew what to do and remained focused.

Also, the air boss made the call to delay the performance until the storm passed. Pete's proactive measures had bought valuable time, preventing any accidents or injuries. Once the storm subsided, he led a thorough safety check of the aircraft and equipment, ensuring that everything was in top condition before resuming the show.

When the skies cleared, Pete was ready to perform. The audience, who had witnessed the seamless handling of the unexpected storm, cheered as Pete took off into the air. The successful performance that followed was a testament not only to his piloting skills but also to the critical human skills of leadership, communication, and teamwork that Pete had demonstrated.

From the above scenario, we can see that communication, teamwork, and effective decision making are important skills for air show performers. Training that could include CRM principles as well as decision making augmentation parts could be of significant importance (Flin et al., 2002; Grogan et al., 2004). Moreover, mindfulness is a critical tool for maintaining focus and managing stress that every air show performer needs to cultivate in order to manage both their aircraft and their stresses effectively (Karachalios & Adjekum, 2023). In the current Chapter we introduce a mindfulness training (MT) program that fits the needs of the intense flying environment so that air show performers can augment their flying skills.

Since air show performers interact with people either in the air or on the ground, media handling is an indispensable part of their training. Specific social media messages and appropriate communication with the public are necessary to pass their message on to their audiences and make themselves acceptable to the public. This is a skill that might be irrelevant to other aviators at a professional level, but for air show performers, it is essential since they need to market themselves appropriately to allow them an increased number of followers on their social media accounts so that they can augment either the recruitment efforts of their organization, in case of military display pilots, or their reach to markets for their sponsors, and as such to enhance their revenues and keep themselves in business, by covering their expenses while keeping a decent amount of revenue to be able to maintain their air show operations.

COMMUNICATION

In the sphere of air show performances, communication transcends the mere exchange of information; it is the lifeline that ensures safety, precision, and the seamless execution of complex maneuvers.

Communication in air show performances is as crucial as the aerobatic skills displayed. It not only ensures the safety of the maneuvers but also enhances the overall spectacle by enabling precise and coordinated execution. The integration of communication protocols and rigorous training programs is essential to address the inherent challenges and ensure that every performance is conducted with the utmost precision and safety.

The Imperative Role of Effective Communication

Effective communication is a fundamental component of safety and performance in air shows, particularly when considering the distinct roles and responsibilities of air show performers who fly solo (single-ship) against those who are part of a formation, either a demonstration team or a group of different aircraft that coordinate for an aerial event such as a flypast. Each group faces unique communication challenges that require tailored strategies to ensure flawless execution and audience engagement.

Single-Ship

Single-ship air show performers, often flying advanced aerobatic maneuvers, rely heavily on clear and concise communication with ground safety observers and flying display directors. For these pilots, the clarity of communication can directly impact their ability to perform safely and effectively. In addition, the role of communication equipment, such as high-quality headsets and noise-canceling technologies, cannot be overstated. Jang, Molesworth, and Burgess (2014) discuss how these technologies reduce the cognitive load on pilots by filtering out background noise, allowing them to focus on crucial communications and instructions, which is especially critical when flying at low altitude in a sterile from distractions environment.

Multi-Ship Formation

For air show performers who are part of a multi-ship formation, the complexity of communication multiplies as it involves not only ground-based communications but also intricate intra-team interactions. Effective communication among team members is crucial for maintaining formation integrity and ensuring the safety of complex, coordinated maneuvers. Estival and Molesworth (2020) and Molesworth and Estival (2015) highlight the potential for miscommunication due to factors like pilot workload, which can have heightened implications in a team environment where multiple aircraft operate in close quarters.

One critical aspect of communication among aircraft that perform in formation is the use of standardized communication protocols. These protocols help reduce ambiguity and ensure that all team members have a clear and consistent understanding of the tasks at hand. Kanki, Folk, and Irwin (1991) explore how predictable communication patterns can enhance team coordination and performance by ensuring that each pilot clearly understands their role and the roles of their teammates during every phase of the performance.

Comparison in Training and Preparedness

Both single-ship and multi-ship air show performers undergo rigorous communication training tailored to their specific performance contexts. This training includes mastering aviation-specific language, understanding international communication standards, and practicing emergency communication procedures. The goal is to ensure that every pilot, regardless of their role, can communicate effectively under normal and adverse conditions, thereby maintaining the highest levels of safety and performance (International Council of Air Shows [ICAS], 2019; UK Civil Aviation Authority [UK CAA], 2024).

Yet, communication doesn't finish during the training session; rather, it is an ongoing process essential for display pilots to maintain proficiency and allow them to adapt to evolving communication standards based on their performances. This training includes regular contingency scenarios that allow pilots to practice communication under simulated abnormal conditions, reflecting the dynamic and sometimes unpredictable nature of air shows (US Air Force [USAF], 2022). Such training helps teach a habit of precision and clarity in communication, which is crucial for both single-ship and multi-ship operations.

Moreover, both single-ship and multi-ship display pilots benefit from standardized communication protocols, which minimize the risk of misunderstandings and ensure a common operational picture. The establishment of these protocols involves predefined actionable radio calls, i.e., communication scripts and strict adherence to aviation phraseology, which are essential during both routine performances and unforeseen emergencies. Research by Molesworth and Estival (2015) suggests the need for pilots to adhere to standardized phraseology to mitigate risks associated with linguistic barriers (Shamsi et al., 2020; van Rosse et al., 2016; Zhang et al., 2022) and high workload scenarios.

Understanding the psychological underpinnings of effective communication also plays a critical role, especially in multi-ship formations where interpersonal dynamics significantly influence team performance. Training programs designed to enhance situational awareness and team cohesion help display pilots better manage stress and maintain concentration during complex maneuvers (Wickens, 2002; Salas et al., 2008; Fornette et al., 2012). Kanki, Folk, and Irwin (1991) discuss how predictability and regularity in communication contribute to effective team dynamics, which are vital for maintaining safety and performance in air shows.

Expanded Discussion on Communication Skills

Effective interpersonal communication is vital for air show performers, not only for the flawless execution of aerial maneuvers but also for building strong relationships with their team, organizers, other display pilots, the air boss, and, importantly, the audience. These communication skills enable performers to coordinate seamlessly, handle unforeseen events, and enhance the overall experience for everyone involved.

Air show performers must excel in collaborative communication skills to ensure safety and cohesion during performances. Looze et al. (2015) emphasize the significance of team communication skills in operations involving complex machinery like aircraft, highlighting that prosodic accommodation and turn-taking organization are effective metrics for evaluating these skills. These mechanisms are crucial for performers to maintain clear and effective communication under the dynamic and often noisy conditions of an air show.

Communication skills also extend to interactions with event organizers and peers, where credibility and compassion play pivotal roles. Apker et al. (2006) discuss how professional nurses communicate professionalism through collaboration, credibility, compassion, and coordination. These skills are directly applicable to air show performers, who must navigate the logistics of event planning, address safety concerns, and collaborate with a diverse team. Demonstrating these communicative skill sets helps in building trust and professionalism, which is essential for long-term relationships and successful event execution.

Moreover, interacting with the air boss and regulatory officials requires a prominent level of precision in communication and the ability to follow strict protocols. Effective listening and clear, concise speaking are critical, as these interactions often involve receiving and complying with complex directives under time constraints. The ability to communicate effectively with the air boss ensures that performances are conducted within the agreed parameters, enhancing safety and regulatory compliance.

The interaction with the audience, though less direct, is no less critical. Performers engage with the audience through both planned presentations and impromptu interactions. Here, interpersonal communication skills help performers convey their passion for aviation, educate the public about air show dynamics, and share safety measures. Effective communication fosters a greater connection with the audience, enhancing their experience and building a loyal fan base.

Studies show that team-building and communication training significantly improve interpersonal interactions within teams. Hapsari (2023) found that team-building training effectively improved interpersonal communication among team members, leading to better collaboration and performance. Such training could be invaluable for air show teams, enhancing their ability to perform cohesively and safely.

Moreover, Bambacas and Patrickson (2008) highlight specific interpersonal communication skills that enhance organizational commitment. These include the clarity and frequency of messages and the ability to lead collaboratively. For air show performers, these skills can ensure that every team member is aligned with the mission and committed to the group's success.

Engaging the Media: Upholding the Image of the Aviation Community

Within the realm of air show performances, engaging with the media requires a disciplined approach akin to the precision of flight maneuvers themselves. Air show performers apply a framework for media engagement to communicate effectively and uphold the image of the aviation community.

Pre-engagement preparation is likened to crafting a mission plan. Performers delineate their communication goals, understand the target audience, and construct key messages grounded in a comprehensive grip of relevant topics. Integral to this stage is relentless practice, which equips performers with the finesse to interact during various media scenarios.

In the message development phase, performers distill their communication to essential points designed to resonate with and be remembered by the audience. The construction of these messages demands brevity, memorability, and relevance (Alaviyya & Alisoy, 2003; Henley et al., 2008; Gligoric et al., 2019). Additionally, tailoring messages to fit different media channels—television, newspapers, or radio—is paramount, with repetition employed for emphasis and retention.

The art of message delivery incorporates techniques such as "bridging"—the skill to transition the conversation toward key messages—and "flagging"—a method to underscore critical points (Tumpey et al., 2019; Katz & Byrne, 2019). Delivery must radiate positivity, enthusiasm, and confidence, reflecting the values and professionalism of the air show community.

When conducting an interview, air show performers need to adhere to facts, stay within their scope of expertise, and prioritize truthfulness. In instances of uncertainty, acknowledging a lack of knowledge is preferable to speculation (Harvey, 2011). The narrative also cautions against the dismissive "no comment," advising instead to correct misconceptions and protect the integrity of information.

Furthermore, air show performers consider the professional nature of reporters and the permanence of public records. With the potential to be the sole military representative seen by the community, their interaction with the media carries significant weight. Therefore, the importance of utilizing public affairs as a resource is critical—reinforcing the notion that effective media engagement is a proactive, not a reactive, process.

Lastly, air show performers keep in mind the ethical considerations and the necessity of liaising with legal and public affairs advisors when dealing with contractors, fundraisers, or situations that might imply endorsements (Hoppe, 2018; MacQuillin & Sargeant, 2019).

Media Training for Air Show Performers

As an extension of the needed communication skills, air show performers use social media to reach their audience (Berriochoa, 2014). As such, they frequently receive media training, especially in military organizations, preparing them to interact effectively with various media outlets and handle interviews professionally (USAF, 2022). This training equips performers with the skills to manage their public image and communicate their experiences and insights about the air shows effectively.

This training also focuses on the effective management of media distractions, which is critical for air show performers to maintain mental focus and ensure safety. By implementing strategic practices such as social media blackouts,[1] targeted training on distraction management, and careful scheduling of media engagements, performers can minimize the impact of these distractions (Hayes et al., 2020, 2021, 2023).

Key Aspects of Media Training for Air Show Performers

Media training programs are designed to prepare individuals for a range of media interactions, including print, radio, online, and TV interviews. These programs are crucial for air show performers, who often face media exposure and need to present a positive image of their profession and the events they participate in. The training focuses on teaching performers how to handle questions, especially those that might be challenging or unexpected (Illman, 2006).

Such training helps performers feel more prepared and confident, which translates into better communication and a more positive portrayal in the media. This is particularly important when performers need to discuss complex maneuvers or respond to questions about the risks associated with air shows.

Air show performers sometimes face situations that require careful communication, such as accidents or incidents during shows. Media training includes preparing them to handle sensitive topics and to communicate effectively under pressure, ensuring that information is conveyed accurately and responsibly (Olajiga et al., 2024).

One of the central components of media training is helping performers identify and clarify the key messages they wish to convey. This includes articulating the excitement and appeal of air shows, the importance of safety measures, and the skill involved in performing (Wendling et al., 2013). Effective media training aids performers in consistently delivering these messages during their interactions with the press.

Managing Media Distractions for Air Show Performers

Media engagement is essential for promoting air shows and enhancing the public profiles of performers. However, interactions with the media can also present significant challenges, particularly distractions that may affect performers before or during their displays. Effective management of these distractions is crucial to maintain focus and ensure safety during performances (Karachalios, 2022; Karachalios & Adjekum, 2023).

Interviews and media interactions scheduled close to performance times can lead to decreased concentration (Hayes et al., 2021). The need to switch focus between engaging with the media and preparing for flights can dilute the mental focus required for the safe performance of complex maneuvers (Weiss et al., 2008).

Handling probing media questions, especially those related to the risks or past incidents of air shows, can increase stress levels and consume mental resources essential for performance, potentially affecting a performer's readiness and mental state (Young & Dulewicz, 2005).

Air show performers learn to control and avoid media distractions by implementing media blackouts before performances that ensure that performers have uninterrupted time to prepare and focus. Another measure for effective distraction management applied by air show performers is to adhere to the "*30-minute bubble*[2]" rule (Barker, 2020) or the "*sacred 60-minute*[3]" policy (Hollowell, 2012), a method that emphasizes the importance of creating a focused environment to optimize performance and ensure safety during high-stress and high-precision activities like air shows. This approach is supported by research indicating that pre-performance routines and reduced external interference significantly enhance focus and performance outcomes (Beilock, 2010; Barker, 2003, 2020; Hayes et al., 2021, 2023; Karachalios, 2022).

Media training for air show performers can include specific techniques to manage distractions. Training in mindfulness and stress-reduction techniques can help performers quickly refocus after media interactions, minimizing stress and mental clutter (Kabat-Zinn, 1990).

Then, assigning specific team members to handle the bulk of media interactions reduces the cognitive load on performers, especially those close to show times. This strategy allows performers to maintain focus on their routines without the added pressure of media engagements (Hanton et al., 2004).

Also, having a set of pre-prepared responses can help performers handle media interactions more efficiently. This reduces the mental effort required during interviews and helps maintain focus on upcoming performances (Weinberg & Gould, 2023).

Where possible, scheduling media engagements after performances can alleviate the pressure on performers, allowing them to concentrate fully on their routines without the distraction of pending media interactions. This timing allows performers to engage with the media without compromising the safety and quality of the air shows (Hayes et al., 2020).

Case Study: Pre-Show Distraction

However, there are limits to public engagement, especially before performing a display. During his first international air show at Radom Airshow 2011, in Poland, the author, even if he had received and acknowledged all the appropriate media training, felt so absorbed by the festive atmosphere and the sense of responsibility to represent his Air Force and Nation that he needed to be pulled away from the public by his wife and colleagues. This excitement led him to overlook the importance of returning to his pre-flight "*bubble*" on time. Engaging with the public can wait until after a successful flight to ensure safety and focus.

LEADERSHIP SKILLS FOR AIR SHOW PERFORMERS

Leadership within the air show community encompasses a diverse set of skills aimed at not only managing personal performance but also influencing a wide range of stakeholders, including team members, air show organizers, regulators, and the public.

Developing leadership skills in air show performers, whether for formation leaders or solo display pilots, revolves around core principles such as safety, precision, and respect (Grote, 2016; Bastola, 2020; Waltz, 2022). Leaders in these roles must exemplify a proactive approach to safety, conducting rigorous pre-flight checks and debriefings to learn from any mistakes or oversights. Experienced leaders implement safety protocols and introduce innovative methods to enhance team safety, ensuring every aspect of the performance prioritizes the well-being of the team and themselves (Skybrary, 2023; DEKRA, 2023).

Precision in air shows involves not only perfect execution of maneuvers but also synchronicity with teammates and timing that aligns with planned choreography. For solo display pilots, precision means executing maneuvers with impeccable timing and accuracy. Leaders and solo performers develop these skills through regular, structured practice sessions and clear communication during both the planning and performance stages, that ensures flawless execution, while creating a seamless and breathtaking display for the audience (Karachalios, 2022).

Respect within the team is cultivated by leaders who demonstrate fairness, transparency, and open communication (Abrams et al., 2003). Ensuring every team member's voice is heard and valued strengthens team bonds and enhances collective performance. Solo display pilots also exhibit respect by coordinating with ground crews and supporting staff, ensuring smooth operations. Effective leaders and solo pilots are approachable and empathetic, adept at resolving conflicts and boosting team morale, fostering an environment where every member feels appreciated and motivated (DEKRA, 2023).

Air show leaders and solo display pilots must possess exceptional flying skills and the ability to make decisive decisions without putting anyone at risk (Barker, 2003; Karachalios, 2022). They must avoid unnecessary risks while ensuring the well-being and happiness of everyone involved in the stressful flying and working environment. Additionally, these leaders need outstanding organizational and planning skills, encompassing project management and overseeing financial, time, and administrative aspects of their teams or personal schedules.

Effective leadership also requires the ability to inspire not only their team but also the public (Conger, 1991). For military teams, leaders bear the responsibility of identifying and mentoring the right person to succeed them as team leaders. Solo pilots must inspire through their individual excellence and professional conduct. Ultimately, air show leaders and solo pilots must excel as display pilots and exceptional managers, setting an example that others can follow and trust (Skybrary, 2023; Karachalios, 2022).

As leaders develop, they gain deeper cognitive structures, enabling them to navigate the complex interpersonal dynamics of their teams more effectively. This involves understanding individual strengths and weaknesses, adapting communication styles to meet diverse needs, and fostering an environment where trust and collaboration are paramount. Emotional intelligence is crucial for effective leadership in air show teams and for solo pilots, allowing them to manage their emotions and respond to others' emotions. Leaders with high emotional intelligence maintain team harmony and motivation, especially in high-pressure situations common in air shows (Goleman, 1999; George, 2000; Skybrary, 2023).

Managing stress is another critical aspect of leadership in air show teams and for solo performers. The high-stakes environment demands that leaders remain calm and composed under pressure. Effective team leaders and solo display pilots employ various stress management techniques, such as mindfulness and mental rehearsal, to maintain their focus and augment their decision making abilities during performances (Rumbold et al., 2012; Karachalios, 2022; Karachalios & Adjekum, 2023). They also encourage their team members to adopt similar practices, creating a resilient team capable of handling intense demands (DEKRA, 2023).

Personal responsibility and leadership skills are enduring attributes crucial for air show performers. According to a study conducted by the author in 2022, there is a consensus among interviewees regarding the necessity of these traits. One interviewee noted, "*Unfortunately, it comes down to the individual pilots to execute on the day. The air show fraternity can only do so much, but it is an incredibly high-risk environment*," highlighting the critical role individual accountability plays in ensuring the safety and success of performances.

A leader in an air show team or a solo display pilot not only directs flight maneuvers but also sets the tone for safety and precision. For example, before a show, the lead pilot conducts a detailed briefing, emphasizing the sequence of maneuvers and highlighting any potential safety concerns based on the day's weather and equipment checks. This practice prepares the team, builds trust, and reinforces the leader's commitment to each member's safety. Effective leaders like Jacques Bothelin, an acclaimed aerobatic team leader, often share their experiences and insights, fostering a culture of continuous improvement and learning within the team (Cunha & Louro, 2000; Akdere & Egan, 2020).

Leadership within the Team

Air show performers must lead by example, fostering a culture of safety, precision, and mutual respect among their peers (Figure 6.1). Lord and Hall (2005) suggest that leadership skill development is a progression from novice to expert, where each level demands different knowledge and skills. For air show performers, transitioning to leadership roles means mastering not only the technical aspects of their performances but also the interpersonal dynamics that ensure team cohesion and performance.

The journey of an air show performer from a team member to a leader is marked by significant growth and the acquisition of a range of competencies that extend far beyond flying skills. Leadership within this context requires an exemplary commitment to the highest standards of safety, an unyielding dedication to precision in performance, and the nurturing of a culture that values mutual respect and team unity.

As an example of leadership progression in air show performers of display teams, we can consider the following case. Initially, a novice air show performer focuses on honing flying skills, learning basic maneuvers, and understanding the

FIGURE 6.1 Brazilian Air Force's Esquadrilha da Fumaça (Smoke Squadron) flight demonstration team is preparing for an air show in Pirassununga, Brazil, 2017. (Copyright: Lucas Rizzi/Shutterstock.com.)

dynamics of air show routines. As they gain more experience, the transition toward intermediate levels involves deeper engagement with more complex aerobatics and increased responsibilities such as mentoring newer team members or participating in the planning of routines. Once reaching expert status, usually after the first or second display season, a performer often takes on leadership roles, such as opposing solos lead pilot or display coordinator, where they are responsible not only for their own performance but also for overseeing the safety and effectiveness of the entire team's display. They must ensure that all team members are well-prepared, that routines are executed flawlessly, and that every safety measure is meticulously followed.

Engaging with Organizers and Regulators

Effective leadership in the air show community extends well beyond the cockpit and aerobatic performances. It involves strategic interactions with organizers and meticulous compliance with regulatory bodies, which are essential for the smooth execution of air shows.

Effective leaders understand the importance of building robust relationships with air show organizers and regulatory bodies. Mumford et al. (2000) emphasize that leadership involves problem-solving, social judgment, and knowledge, which are crucial when negotiating terms of performance, ensuring compliance with safety regulations, and scheduling events.

Air show leaders work closely with event organizers to ensure that every aspect of the show aligns with both the performers' capabilities and the event's objectives. A performer with strong leadership skills will proactively engage with organizers to align on the show's themes, participant expectations, and logistical needs. For example, they might work closely with organizers to choreograph their performance to fit within the broader event schedule, ensuring their acts complement other segments, which can enhance the overall spectator experience.

Regulatory compliance is another critical area where the leadership skills of air show performers are put to the test. Display pilots must ensure that all aspects of the show comply with local, national, and international aviation regulations. This requires a thorough understanding of these regulations and the ability to apply this knowledge effectively to maintain high safety standards. They often take the initiative to stay updated on regulatory changes and lead their teams in adapting to these changes swiftly and efficiently. For instance, when introducing a new aerobatic maneuver, a leader must work closely with regulators to demonstrate that the maneuver meets safety standards and to secure the necessary approvals. Another practical example of effective leadership in this context could be the coordination between demo team leaders and regulatory bodies during international air shows, where regulations might differ significantly from the performers' home country. Team leaders must understand these differences and ensure that their teams make the necessary adjustments to their display routines.

Public Engagement: A Key Leadership Role for Air Show Performers

In the air show community, the role of performers extends beyond executing breathtaking aerobatics. It involves significant interaction with the public, where performers must not only entertain but also engage, educate, and inspire (see Figure 6.2). Leaders in this field use public engagement as a strategic tool to foster a deeper connection with the audience, influence public perception of aviation, and encourage the next generation of aviators.

Public engagement for air show performers is vital for several reasons: it inspires future generations, educates on safety and the science behind aviation, and represents the aviation community. Engaging with the public also fosters interest in STEM (Science, Technology, Engineering, and Mathematics) fields, showcasing the technical skills and knowledge required for aviation (Ayrshire Daily News, 2024; Rich Goodwin, 2024; European Airshow Council [EAC], 2024). Through demonstrations and interactions, performers can highlight the importance of STEM education, especially young people, ignite a passion for aviation, and perhaps inspire future careers in aerospace (DEKRA, 2023). As visible representatives of the broader aviation community, performers help shape public perceptions about the professionalism and integrity of pilots and engineers (Karachalios, 2022).

Kouzes and Posner, in "The Leadership Challenge," discuss how effective leaders inspire a shared vision and encourage the heart, which is essential for engaging with diverse audiences (Kouzes & Posner, 2017). These principles are directly applicable to air show performers, who must not only perform but also connect emotionally with

FIGURE 6.2 Patrouille de Suisse pilots sign autographs for the public after their display at the "Connected" event in Kloten, Switzerland, 2023. (Copyright: Angelique Clic/Shutterstock.com.)

their audience, sharing their enthusiasm and love for aviation in a way that is accessible and motivating.

Expanding on engagement strategies, demo team leaders can organize interactive sessions or educational seminars at air shows. These platforms allow performers to engage in deeper conversations with attendees, demonstrate aspects of their aircraft, and discuss the principles of flight and aerodynamics. Many performers extend their reach through social media, where they can engage with fans, share behind-the-scenes content, and promote upcoming shows. This ongoing engagement helps build a community of followers who feel a personal connection to the performers and a personal stake in their success. Engaging with local communities through school visits or participation in local events can enhance public relations and provide role models for youth (Royal Air Force [RAF], 2018; Wear News, 2024). These activities also reinforce the performers' roles as ambassadors of the aviation industry.

Innovative public engagement examples can be drawn from initiatives like those discussed by Sayer et al. (2014), where ecology was brought to music festivals. Performers could leverage similar creative engagement strategies, such as combining air shows with other cultural or educational events to reach wider and more diverse audiences. Additionally, the "FUTURES on Air" project highlights the potential of using community radio for public engagement, suggesting that performers could participate in or host radio shows to discuss various aspects of aviation and air show performances (Watson, 2021).

The integration of public engagement into the roles of air show performers enriches the spectator experience and plays a critical role in the sustainability of the air show industry. By adopting and adapting the strategies outlined by Kouzes and Posner (2017), performers can effectively inspire, educate, and connect with the public, ensuring their role as leaders extends far beyond the aerial maneuvers they perform. This approach not only elevates their personal profiles but also serves to foster a vibrant and engaged aviation community.

TEAMWORK AS AN ESSENTIAL HUMAN SKILL FOR AIR SHOW PERFORMERS

Teamwork for air show performers transcends simple collaboration. It is a critical human skill that not only ensures the flawless execution of sophisticated display profiles but also upholds safety, optimizes performance, and nurtures a supportive environment among team members. Insights from various high-risk and high-reliability professions, such as healthcare and air traffic controllers, provide a rich understanding of how teamwork functions in contexts that mirror the operational and safety demands of air shows (Leonard et al., 2004; Malakis et al., 2010).

Multi-Dimensional Importance of Teamwork

The importance of teamwork in air shows is multifaceted. Firstly, it ensures the flawless execution of aerial routines. Complex flight maneuvers require precision and synchronicity, with each member's role being interdependent. This necessitates absolute trust and flawless communication among team members. The coordination observed in well-executed air shows is a direct result of rigorous practice and the team's ability to work cohesively under pressure.

Secondly, teamwork is essential for upholding safety. In air shows, the slightest miscommunication can lead to critical situations. Effective teamwork involves continuous communication, routine safety checks, and mutual oversight, where each team member remains vigilant for potential safety issues and collaborates to mitigate risks.

Additionally, teamwork optimizes performance by allowing each team member to contribute their best work. This includes planning and executing complex flight patterns, timing maneuvers perfectly with music or other performance elements, and ensuring that each segment of the show complements the others. Beyond the technical aspects, teamwork significantly enhances overall performance.

Moreover, nurturing a supportive environment is crucial in high-stress settings like air shows. Camaraderie and mutual support among team members enhance morale, reduce stress, and boost overall performance. A supportive team environment provides a safety net for performers, knowing their team is there to support them through every step of their performance.

Expanded Perspective on Teamwork

Insights into effective teamwork extend beyond aviation and are echoed in other high-risk professions requiring precision, safety, and excellent coordination. For example, in healthcare, surgical teams rely on teamwork for patient safety and successful operations. Studies have shown that better teamwork in operating rooms can significantly reduce errors and improve patient outcomes (Malakis et al., 2010; Rouse & Rouse, 2004).

Similarly, emergency services such as firefighters and emergency response teams depend on strong teamwork for quick and efficient responses to emergencies. The ability to work well together under extreme pressure can often mean the difference between life and death. In military operations, impeccable teamwork is essential for mission success. Military exercises and operations require strategic planning, trust, and communication, mirroring the teamwork necessary in air shows (Stevens & Campion, 1994; Hoegl & Gemuenden, 2001).

Communication and coordination in teams are discussed by Rouse and Rouse (2004), who emphasize that effective communication and precise coordination are fundamental to successful performances. These elements are directly applicable to air show teams, where precise timing and flawless execution are paramount.

The knowledge, skills, and abilities for teamwork are highlighted in the Teamwork Test developed by Stevens and Campion (1994), which supports the essential components of teamwork, such as mutual support and shared objectives. These components are vital for air show teams, where each member must synchronize their actions with others to ensure safety and spectacle.

The "Big Five" of teamwork, identified by Burke (2005), includes adaptability and team orientation. These traits are crucial for air show teams as they navigate the complexities of their performances, adapting to changing conditions and maintaining a cohesive group dynamic.

Finally, the impact of teamwork on innovative projects is demonstrated by Hoegl and Gemuenden (2001), who suggest that high-quality teamwork, marked by balanced contributions and mutual support, significantly enhances the success of innovative projects. This concept is applicable to air shows, where innovation in routine design and performance execution can distinguish one team from another.

Core Aspects of Teamwork in Air Shows

Teamwork is essential for the precise execution of choreographed maneuvers, a defining feature of air shows. The Blue Angels, the U.S. Navy flight demonstration squadron, exemplify this, with each member maintaining exact distances and perfect alignment during formations. Such precision requires each pilot to not only perform their role flawlessly but also communicate and adjust continuously based on their teammates' actions. Therefore, effective communication and coordination are paramount, allowing the team to execute complex maneuvers with split-second accuracy (Salas et al., 2008).

Furthermore, in an environment where the margin for error is minimal, the safety of both performers and spectators is paramount. Effective teamwork ensures rigorous adherence to safety protocols. For example, before each show, teams engage in comprehensive briefings to review the day's conditions, potential risks, and each member's responsibilities. During the show, pilots rely on the ground crew for real-time information and alerts, creating a network of support aimed at preventing accidents. Consequently, this systematic approach to safety is critical in mitigating risks and ensuring a secure environment for all participants (Schmutz, Meier, & Manser, 2019).

Teamwork also enhances performance by allowing each member to leverage their strengths, resulting in a performance that exceeds the sum of its parts. An aerobatic team might include pilots with different specialties—some excelling in solo aerobatics while others specialize in formation flying. Together, they design a program that highlights each pilot's strengths, maximizing the overall impact of the performance, ensuring that with this collaborative approach, each aspect of the show is optimized, leading to a more compelling and cohesive display (Salas et al., 2008).

Additionally, the high-pressure environment of air shows can be intensely stressful. A supportive team atmosphere helps mitigate this stress, allowing performers to focus on their roles without additional anxiety. Experienced pilots mentor newer team members and offer advice and encouragement, which not only improves performance but also strengthens team bonds. Such a supportive dynamic fosters a sense of security and camaraderie, which is essential for maintaining superior performance under pressure (Espevik et al., 2021).

Practical Examples from Air Shows

In air shows, the timing of takeoffs and landings is meticulously planned and coordinated among all team members to ensure both visual appeal and safety. This coordination requires pilots to communicate effectively, adjusting their timing based on others' actions and instructions from the control tower. During unexpected situations, such as engine failure, the immediate and coordinated response of the entire team is crucial. Pilots, ground crew, and emergency responders must work seamlessly together to manage the situation, illustrating the vital role of teamwork in crisis management.

Moreover, post-performance debriefs are essential for continuous improvement. Teams engage in sessions discussing what went well and identifying areas for improvement, reinforcing each member's understanding of their experiences and challenges during the performance, a process of reflection and feedback that is critical for refining their techniques and ultimately enhancing future performances.

Teamwork in the Pre-Display Season and Planning Phase of Air Shows

Teamwork in air shows begins long before the moments of flight, during meticulous preparation and planning stages. This phase is critical as it sets the groundwork for the success and safety of the performance. Effective teamwork during this period helps distribute the workload, reduce stress, and ensure all team members are well-prepared, promoting a peer-review culture[4] that fosters excellence both on the ground and in the air (Andrade et al., 2020; Burgess et al., 2021; Burns et al., 2019; Hinrichs et al., 2022).

During the pre-display season, tasks range from routine maintenance of aircraft to the design of the display profile. Effective teamwork helps distribute these tasks evenly, preventing burnout and ensuring no single individual is overwhelmed. This distribution optimizes performance and enhances safety by reducing the risk of errors due to fatigue (Schmutz, Meier, & Manser, 2019). As soon as the new display season commences, preparations for an air show is a high-pressure task that can lead to significant stress. A supportive team environment helps manage these pressures by allowing team members to share concerns and solutions, providing emotional support, and ensuring everyone feels part of a cohesive unit (Weir, 2018).

Moreover, during the planning stages, having a peer-review system where team members can critique and provide feedback on each other's responsibilities is invaluable. Especially in high-stakes environments, such as air show performances, promoting a just culture where team members feel safe to report issues and learn from mistakes without fear of undue blame is crucial. This process ensures all aspects of the show are thoroughly vetted and optimized, leading to higher quality and safer performances (Espevik et al., 2021), while with this culture, trust, and openness within the team are enhanced, leading to more effective problem-solving, innovation, and efficiency (Salas et al., 2008).

Prior to the show, tasks such as safety checks, mechanical repairs, and practice flights need to be scheduled. Effective teamwork ensures these tasks are assigned according to each member's skills and workload capacity, optimizing efficiency and efficacy. Therefore, decisions about routine adjustments based on weather conditions or technical issues are best made collaboratively, ensuring all perspectives are considered and leading to more robust choices. In addition, conducting regular technical reviews and safety audits during the planning phase involves teamwork to ensure all technical standards are met and safety protocols are followed. According to Weick and Sutcliffe (2007), such reviews are critical in high-reliability organizations for maintaining safety and performance.

Team-based simulations and rehearsals are crucial for preparing for contingencies. Research by Müller et al. (2009) stresses the importance of simulation training in enhancing team performance under stress, particularly in fields like clinical and aviation, where team coordination is essential.

Lastly, since the air show community is international and many performers are invited to showcase their skills in other countries, it is imperative that teams need to participate in training programs focused on communication and cultural issues they may encounter with other performers and organizers. Such an approach can significantly improve teamwork by aligning understanding and expectations across diverse groups (Alsabri et al., 2020).

AIR SHOW-FOCUSED CONCEPTS

In the air show industry, even though all aviation principles are as is, due to the specificity and the small periodicity of such an event, compared to the daily operations of an airline company, a "*twist*" in three important concepts is provided in the following pages of the Chapter. Similarly to the air show safety management system (A-SMS), introduced in the first book of the series (Karachalios & Adjekum, 2023), the introduction of the Air Show Resource Management (ARM), the Air Show Decision Making (AsDM), and the Air Show Mindfulness Training (AMT), based on the Crew Resource Management (CRM), Aeronautical Decision Making (ADM), and the MT, respectively, scales down the size of vast concepts and focuses in the needs of the air show performer community. The goal is to make such concepts easily applicable and ultimately adopted in the daily air show operations, regardless of whether there is a civilian solo display pilot or a military multi-ship demonstration team.

Air Show Resource Management (ARM)

Air Show Resource Management (ARM) is a specialized adaptation of the well-established CRM principles, tailored specifically to meet the unique demands and challenges faced by air show performers. ARM encompasses a comprehensive set of practices and protocols designed to enhance safety, performance, and teamwork among all participants involved in air shows, from solo display pilots to multi-aircraft demonstration teams. Key components of ARM include effective communication, meticulous team coordination, informed decision making, acute situational awareness, stress management, and efficient time management. These elements work synergistically to ensure that air shows are not only spectacular and engaging but also conducted in a manner that prioritizes the safety of both performers and spectators.

Crew Resource Management (CRM) in Air Show Performers

> The Federal Aviation Administration (FAA) describes CRM as the effective use of all available resources—information, equipment, and people—to achieve safe and efficient flight operations (FAA, 2001).

Before we analyze the ARM, the concept of CRM needs to be understood as a critical training component for air show performers, encompassing both solo display pilots and demonstration team pilots. CRM training emphasizes essential non-technical skills such as communication, teamwork, decision making, and situational awareness, which are vital for enhancing safety and efficiency in high-stakes environments. This framework has been adapted for air show contexts to form the ARM.

CRM Training Components for Air Show Performers

CRM training highlights the importance of effective communication and teamwork. These skills are particularly vital for demonstration teams that must perform coordinated maneuvers with precision. For solo display pilots, CRM enhances interaction with support crews and air traffic controllers, ensuring clear communication and mutual understanding. Research by Fisher et al. (2000) indicates that CRM training significantly increases crew awareness and promotes team concepts, suggesting similar benefits for air show performers.

Maintaining situational awareness is another crucial aspect of CRM training. This helps pilots stay aware of their surroundings and the state of their aircraft, which is essential during air shows that involve complex flight patterns and require elevated levels of concentration. Flin et al. (2002) highlight that CRM's focus on situational awareness can significantly improve operational performance and safety in high-reliability environments.

Furthermore, effective decision making is a key focus of CRM training, enabling pilots and their teams to make quick and accurate decisions under pressure. This ability is crucial during unplanned situations or deviations during performances. Helmreich and Wilhelm (1991) demonstrate that CRM training improves decision making and crew coordination, supporting its application in complex team environments like air shows.

CRM training also provides strategies for managing and mitigating errors, which is essential for air show performers operating in environments where the margin for error is minimal. Baker et al. (1993) discuss the benefits of CRM training using simulation-based scenarios to practice error management and enhance CRM skills.

Implementing CRM training for air show performers involves regular, structured training sessions that include simulations and role-playing scenarios. These sessions should be tailored to the specific dynamics of air show performances, considering the unique challenges faced by both solo and team air show performers.

Key Components of the ARM

Effective communication is fundamental in ARM, ensuring that all team members, including pilots, ground crews, and support staff, are consistently informed about flight plans, safety protocols, and any changes in conditions or schedules. This clear communication helps prevent misunderstandings and errors during the high-pressure situations inherent in air shows. ARM focuses on optimizing the coordination among various team members, including precise execution of aerial maneuvers, synchronized timing between pilots, and efficient cooperation between ground crews, such as aircraft engineers, safety observers, air bosses, and flying teams. Such coordination is crucial for maintaining the flow and safety of the air show.

Additionally, ARM emphasizes the importance of sound decision making processes involving the assessment of risks, management of unforeseen changes, and making quick, informed decisions that affect the safety and success of the air show. Decision making within ARM is supported by comprehensive situational awareness, including understanding the current conditions of the aircraft, weather, and airspace.

Maintaining elevated levels of situational awareness is equally critical in ARM. Performers and crews must be acutely aware of their environment, including other aircraft, weather conditions, audience locations, and emergency access routes, essential for ensuring the safety and success of the performance. By understanding the position and movement of other aircraft, performers can avoid mid-air collisions and coordinate complex maneuvers effectively. Monitoring weather conditions helps in anticipating and mitigating risks associated with sudden changes that could affect flight dynamics. Being aware of audience locations ensures that maneuvers are performed at safe distances, protecting spectators from potential hazards.

After each performance, ARM includes detailed debriefings where teams review what went well and what could be improved. These sessions focus on constructive feedback and lessons learned, which are crucial for continuous improvement and maintaining high-performance standards. Given the high-risk nature of air shows, managing stress is an integral part of ARM. Training in stress management techniques helps performers and crew maintain their focus and decision making abilities under pressure (Chiesa & Serretti, 2009).

Implementing ARM involves developing tailored training programs that focus on essential areas such as communication, coordination, decision making, and stress management, specifically designed to meet the unique demands of air show environments. Consequently, this proactive approach to managing both human and technical resources establishes a benchmark for excellence in the air show industry, ensuring safety and efficiency at all levels. Moreover, by addressing the specific needs of performers and crews, these programs foster a culture of continuous improvement and professional development, thereby enhancing overall performance and reliability in high-stakes scenarios.

Time as a Critical Resource

Air show resource management (ARM) adapts the principles of CRM to the unique requirements of air shows, with a particular emphasis on one critical resource that differentiates ARM from CRM: *time*. In the fast-paced and tightly scheduled environment of air shows, time management is crucial. The ability to perform precise maneuvers within a specific timeframe coordinate with multiple elements of the show simultaneously and adhere to strict schedules impacts not just the safety but also the overall success and fluidity of the performance.

Each segment of an air show is meticulously timed, often down to the second. This precision ensures that the sequences flow smoothly from one to the next without unexpected delays or overlaps that could confuse the audience or disrupt other scheduled activities. For instance, a solo aerobatic performance might be scheduled between team formations, requiring exact start and end times to facilitate seamless transitions. ARM requires exceptional synchronization between aircraft flying in formation, ground crews, and air bosses. This synchronization helps manage the *domino effect*,[5] where a delay in one act can impact all subsequent performances. Effective ARM training focuses on developing strategies that enhance team coordination within these tight time limits.

Given the time-sensitive nature of air shows, ARM emphasizes the need for quick and effective decision making. Pilots and crews must be able to make rapid assessments and decisions in response to real-time developments such as sudden weather changes or mechanical issues. The speed at which these decisions are made can be critical to maintaining the schedule and safety of the show.

Additionally, efficient and timely communication is pivotal in ARM. Information about changes in flight conditions, timings, or display routines needs to be communicated swiftly and clearly across all levels of the team. This ensures that all members, from pilots to logistical support, are consistently aligned with the current status and any adjustments in the schedule.

ARM also involves dedicated times for debriefing and feedback after each performance. These debriefs are crucial for identifying any issues that occurred and implementing lessons learned in subsequent performances, often on the same day.

To effectively manage time as a critical resource, air show performers and their teams can undergo specialized ARM training in the same principles as the CRM training with an emphasis on time and stress management. Time management training could focus on prioritizing tasks under pressure and quickly adapting to changes without compromising safety or performance (Häfner & Stock, 2010). Stress management techniques help performers maintain composure and clarity when making quick decisions (McClernon et al., 2011; Karachalios, 2022).

Ultimately, for air show performers, time is not just a resource but a critical asset that must be expertly managed in such a time-sensitive operational environment to ensure the success of the event and the safety of the performers and the public.

Examples of Time Management in ARM

During an air show, effective time management, as an aspect of ARM, is crucial to ensure the seamless execution of performances and the safety of both performers and spectators.

In timed aerobatic sequences, each aerial activity must begin and end precisely on time to allow for sufficient separation between air show performers inside the showbox while maintaining the overall show schedule. This meticulous timing ensures that each aerial activity flows smoothly into the next, preventing any overlaps or gaps that could disrupt the audience's experience or compromise safety. Pilots, safety observers, and air bosses work in harmony, coordinating their efforts to achieve the exact timing required for each maneuver.

During emergencies, such as a sudden mechanical failure or an unexpected bird strike, the quick, coordinated response of safety observers, air bosses, and pilots is essential for ensuring safety and minimizing disruption. Effective ARM training prepares teams to make rapid assessments and decisions, enabling them to handle such situations efficiently. This swift response capability is crucial in maintaining the safety of the performers and the audience while ensuring the show can continue with minimal interruption.

Back-to-back performances demand exact timing to maintain the continuous flow of the show. For instance, a solo jet performance may be scheduled to do a flypast immediately after a vintage propeller aircraft display. The timing between these aerial activities is critical due to the aircraft's differing speeds and airspace requirements. ARM ensures that as soon as one performance ends, the next is ready to begin without delay while maintaining sufficient time and altitude deconfliction, with the ultimate goal of optimizing the audience's experience while maintaining a smooth progression of events.

In addition, weather-related schedule adjustments require rapid reassessment and communication to re-coordinate planned activities. ARM involves continuously monitoring weather conditions and having contingency plans in place to adjust the schedule as needed. For example, if a high-altitude act needs to be delayed due to approaching weather, this decision must be communicated instantly to all performers and support staff to re-coordinate the activities accordingly.

Furthermore, quick debriefing sessions after performances allow for immediate feedback and improvements. These sessions are essential for identifying any issues that occurred and implementing lessons learned in subsequent performances, often on the same day, reinforcing continuous improvement and maintaining high-performance standards throughout the event.

Air Show Decision Making (AsDM)

In the dynamic low-level environment of display flying, air show performers must persevere under pressure caused by a plethora of factors, requiring a specialized approach to decision making. To cater to these unique challenges, the AsDM concept has been developed specifically for air show performers. This model enhances traditional ADM frameworks to accommodate the specific needs and pressures of air show environments, emphasizing rapid response, precision, and team synchronization.

Before we further explain the AsDM concept, we need to remind the ADM, which is defined by the Federal Aviation Administration (2023, Chapter 2) as "*a systematic approach to the mental process used by pilots to consistently determine the best course of action in response to a given set of circumstances.*" For air show performers who operate in an environment where precision and safety are paramount, mastering ADM is crucial. This skill set is essential for ensuring that each performance is both spectacular and safe, as it encompasses a range of critical mental processes from situational awareness to risk management.

At the core of AsDM for air show performers is the need for *exceptional* situational awareness. Performers must maintain a constant awareness of their environment, the status of their aircraft, the weather, and other external factors such as crowd distribution in the air show box. This heightened awareness is vital, especially when weather conditions can change abruptly, requiring immediate adjustments to the performance plan to ensure safety.

Risk assessment and management also play pivotal roles in AsDM. Air show performers regularly evaluate the risks associated with each maneuver they perform. This continuous assessment helps in making informed decisions that balance the inherent risks with the potential rewards of executing daring maneuvers. Effective risk management involves not only recognizing potential hazards but also having contingency plans that can be executed immediately if necessary (Karachalios & Adjekum, 2023).

Managing stress effectively is another important aspect of AsDM. The high-pressure environment of air shows can affect performers' judgment and reaction times. Techniques such as mental rehearsals, controlled breathing, and physical exercises are employed to manage stress. These techniques help performers maintain mental clarity and focus, which are essential for making sound decisions quickly and efficiently during a show (Karachalios, 2022).

Furthermore, communication and coordination within the team are indispensable for successful air show performances. Clear and concise communication ensures that all team members, including pilots, ground crews, and controllers, are synchronized and can respond effectively to any changes or emergencies. This is particularly important for teams performing complex, coordinated maneuvers, where timing and precision are critical for safety (Barker, 2003; ICAS, 2019).

Lastly, the use of structured decision making models such as the FAA's (2023) *DECIDE* model—*Detect, Estimate, Choose, Identify, Do*, and *Evaluate*—provides a methodical approach to handling unexpected situations during air shows. This model guides performers in detecting problems, estimating their impact, choosing the best course of action, identifying how to implement these actions, doing what is necessary, and finally evaluating the outcome. This structured approach helps maintain consistency and reliability in decisions, enhancing both safety and quality of performance.

Key Components of the AsDM

AsDM is designed to support performers in making quick and correct decisions that ensure safety and excellence in their performance. Unlike broader ADM models that are applicable across several types of aviation, AsDM is tailor-made for the specific scenarios encountered during air shows, such as closely timed maneuvers, audience considerations, and the unique dynamics of performing in a highly visible public forum.

The key components of the AsDM are listed below:

- Pre-Flight Preparation
  This initial phase involves comprehensive planning and preparation before the air show. Performers review the show's timetable and sequence, discuss weather conditions, and go over safety protocols crucial for setting the foundation for effective decision making.
- Real-Time Situational Awareness
  At the heart of the AsDM model is an enhanced focus on situational awareness. Performers must maintain a heightened awareness of their environment, including the positions of other aircraft, the behavior of the crowd, and any potential hazards.
- Dynamic Risk Assessment
  Given the ever-changing conditions of air shows, dynamic risk assessment is a continuous process. Performers assess and reassess risks in real-time, considering factors such as aircraft performance, weather changes, and their physical and mental status.
- Communication and Coordination
  Effective communication and coordination among all team members is critical. The AsDM model emphasizes the importance of clear, concise, and

standardized communication to ensure that all team members can make coordinated decisions quickly.

- Decision Execution
  This component focuses on the implementation of decisions made during the air show. It involves not only an individual performer but also extends to formation flying, with the coordinated efforts of all members to ensure that decisions are executed smoothly and effectively. In the execution process, the role of the air boss and the safety observer is equally important to ensure the smooth implementation of the decision making process.
- Post-Performance Review
  After each performance, a thorough debriefing is conducted to discuss what went well and what could be improved, a phase essential for continuous learning and improvement, which helps refine the air show performer's decision making process for future shows.

LOOP Model for Air Show Decision Making

The introduction of the *LOOP* model is not merely about adding another acronym to the extensive list of aviation terminologies; rather, its purpose is far more practical and impactful (see Figure 6.3). The *LOOP* model serves as a memorable mnemonic that directly activates ADM process among air show performers.

The core idea behind the *LOOP* model is to provide a straightforward and intuitive framework that performers can easily recall and apply during the high-pressure, fast-paced scenarios typical of air shows. Each element of the model—*Lookout, Orient, Optimize, and Perform*—is designed to streamline the decision making process, ensuring that safety and performance are maintained at the highest levels. Additionally, the choice of the acronym *LOOP* is inspired by the loop maneuver, a fundamental aerobatic figure that all aerobatic pilots practice during their first-ever

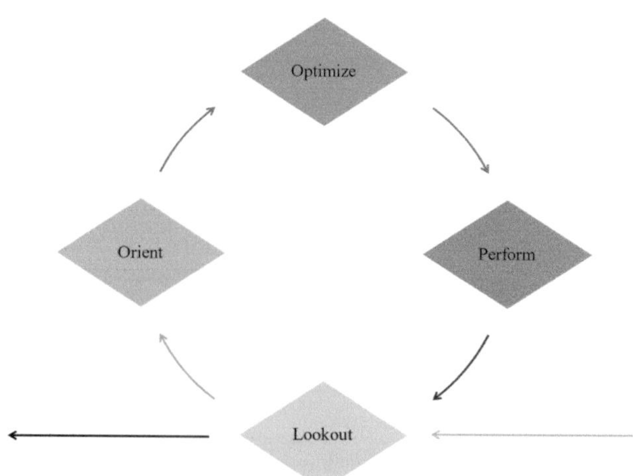

FIGURE 6.3 The LOOP model for Air Show Decision Making.

aerobatic flight, further enhancing the model's relevance and memorability for air show performers.

The simplicity and relevance of the *LOOP* model resonate particularly well with the practical nature of flying. As usually acknowledged in the fighter pilot community, when all else fails or when options seem limited, "*Go for a Loop!*" This piece of advice underlines the essence of the *LOOP* model: when faced with uncertainty or a hazard, revert to the basics of the model to navigate through challenges effectively.

The *LOOP* model is designed as a continuous decision making cycle, essential for air show performers to ensure ongoing safety and optimal performance. Once a performer completes one iteration of the cycle—by looking out for potential threats, orienting themselves to the situation, optimizing their response, and performing the chosen action—they must immediately initiate another *LOOP* cycle. This ongoing process allows for constant reassessment and adjustment in response to any new changes or challenges that arise during the display. For example, after adapting their maneuvers to account for a sudden shift in wind direction, the performer needs to re-enter the Lookout phase to scan for any further environmental changes or hazards. By perpetually cycling through the *LOOP* steps, performers maintain high situational awareness and remain alert to handle the dynamic conditions of the low-level air show environment, ensuring both their safety and the safety of the audience.

L—Lookout

Continually observe and monitor the surrounding environment, including other aircraft, weather conditions, and crowd placement.

O—Orient

Quickly orient to new or developing situations by assessing how changes impact the current flight path and routine. Orientation involves understanding the relationship between the aircraft, the environment, and the planned performance.

If unexpected weather conditions arise, the performer must orient themselves to these changes by adjusting their routine to maintain safety and precision in their maneuvers.

O—Optimize

Optimize and evaluate available options for action in response to observed changes or threats. This step is crucial for developing a contingency plan.

For example, if a display pilot identifies, before executing a half Cuban eight maneuver, that the wind has shifted toward the crowd, they need to optimize their flying profile to maintain a safe distance from the spectators without violating the display lines.

P—Perform

Execute the chosen action with precision and confidence. This step completes this iteration of the decision making process and starts another *LOOP* cycle for any subsequent decisions needed during the display.

For instance, in the previous example, after deciding on the safest course of action to start the half Cuban eight maneuver with increased separation from the crowd due to the changed wind direction, the performer executes the maneuver with precision. This might involve altering the applied G-forces while maintaining communication with the safety observer and other team members if they are part of an aerobatic team.

Air Show Mindfulness Training (AMT)

An additional aviation concept that is modified to meet the needs of the air show performers' community is Mindfulness Training (MT), which could be shaped for the needs of air show performers in the form of AMT. Mindfulness and its dedicated training offer significant benefits for air show performers, whose roles demand extreme precision, focus, and mental resilience. By merging foundational mindfulness principles with specific strategies tailored to the unique challenges of air shows, this section highlights how mindfulness can enhance performance, ensure safety, and promote well-being among high-performance professionals (Karachalios, 2022; Karachalios & Adjekum, 2023).

Mindfulness is defined as "*the practice of paying attention purposefully, in the present moment, and nonjudgmentally*" (Kabat-Zinn, 1994). For air show performers, mastering mindfulness is crucial not only for enhancing performance but also for ensuring safety, as it helps maintain situational awareness and quick, accurate responsiveness during complex aerial maneuvers (Karachalios, 2022; Karachalios & Adjekum, 2023).

Research in similar high-performance settings, such as elite athletes, suggests that MT can significantly reduce errors, enhance stress management, and improve recovery from mistakes (Myall et al., 2022). If implemented, these practices could analogously transform team dynamics and the performance of display pilots.

The Role of AMT in Enhancing Safety and Performance

Extensive research supports the idea that MT significantly improves attention regulation, stress resilience, and emotional control (Holzel et al., 2011; Karachalios, 2022). These benefits are particularly pertinent for air show performers, for whom sharp mental focus and emotional steadiness can mean the difference between a successful performance and a catastrophic error.

Mindfulness Techniques for AMT

Mindfulness techniques adapted to fit the needs of air show performers and be formally set for developing an AMT program are the following:

Pre-Flight and In-Flight Mindfulness Practices

- Pre-flight Briefings: Integrating mindfulness exercises into pre-flight briefings can help ensure that all team members begin their performance with an aligned focus (Kabat-Zinn, 1994).

- In-Flight Mini-Mindfulness: Employing brief, focused breathing or mental anchoring during quieter moments can help pilots maintain calm and focus throughout the show (Karachalios & Adjekum, 2023).

Post-Flight Practices

Engaging in reflective practices after flights, such as mindfulness-based debriefing sessions, allows performers to process experiences constructively, fostering resilience (Kabat-Zinn, 1994).

Daily Mindfulness Integration

Daily practices such as meditation, yoga, or body scans can significantly enhance pilots' overall mental readiness and health, promoting sustained performance readiness (Kabat-Zinn, 1994).

Special Considerations for High-G Maneuvers

High-G maneuvers require pilots to manage intense physical stress. Mindfulness techniques such as controlled breathing and mental anchoring are crucial for managing these physiological challenges and maintaining mental clarity under pressure (Meland et al., 2015).

Building a Mindful Team Culture

Collective mindfulness is essential for maintaining effective team dynamics. Regular AMT sessions can help synchronize team members' focus and awareness, enhancing cohesion and supportive interactions among the team (Kabat-Zinn, 1994).

NOTES

1 A social media blackout for distraction control refers to the intentional discontinuation of social media access to reduce distractions and improve focus on tasks. This can be a personal decision to disable notifications, reduce screen time, or temporarily deactivate accounts, aimed at mitigating the negative impacts of social media on productivity and mental well-being (American Psychiatric Association, 2022).

2 This rule stipulates that 30 minutes before a performance, air show performers must enter a "bubble" of focused preparation. During this time, they avoid any distractions, such as conversations, phone calls, or other activities that could divert their attention from the task at hand. This period is dedicated to mental rehearsal, checking equipment, and ensuring everything is in order, thereby reducing the risk of mistakes during the performance.

3 Similar to the 30-minute bubble, this policy extends the distraction-free preparation period to 60 minutes before a performance. The idea is to create a "sacred" time frame where the performer isolates themselves from any potential interruptions. This longer period allows for a more thorough preparation, giving the performer ample time to mentally and physically prepare for the show, thus enhancing performance and safety.

4 Peer review culture in a team context refers to the practice where team members critically evaluate each other's work to maintain quality, provide constructive feedback, and support

collective improvement and accountability; with no blame intentions but rather a focus on growth.

5 The domino effect refers to a chain reaction where a single event sets off a sequence of related events, each one causing the next in succession, much like a line of falling dominoes (van Leeuwen, 2010).

REFERENCES

Abrams, L. C., Cross, R., Lesser, E., & Levin, D. Z. (2003). Nurturing interpersonal trust in knowledge-sharing networks. *Academy of Management Perspectives, 17*(4), 64–77.

Akdere, M., & Egan, T. (2020). Transformational leadership and human resource development: Linking employee learning, job satisfaction, and organizational performance. *Human Resource Development Quarterly, 31*(4), 393–421.

Alaviyya, N., & Alisoy, H. (2023). Anaphora in text: Echoing words, amplifying messages. *Experimental Physics, 106*, 106–113. https://doi.org/10.5281/zenodo.10369937.

Alsabri, M., Boudi, Z., Lauque, D., Dias, R., Whelan, J., Östlundh, L., Alinier, G., Onyeji, C., Michel, P., Liu, S., Camargo, C., Lindner, T., Slagman, A., Bates, D., Tazarourte, K., Singer, S., Toussi, A., Grossman, S., & Bellou, A. (2020). Impact of teamwork and communication training interventions on safety culture and patient safety in emergency departments: A systematic review. *Journal of Patient Safety, 18*, e351–e361. https://doi.org/10.1097/PTS.0000000000000782

American Psychiatric Association. (2022, February 11). 6 tips to help take control of your social media use and improve well-being. American Psychiatric Association. www.psychiatry.org:443/news-room/apa-blogs/tips-to-take-control-of-your-social-media-use

Andrade, M. S., Kakegawa, Y., & Johnson, J. (2020). The use of peer review in student teams: An impetus to improved performance? *Journal of Higher Education Theory & Practice, 20*(14), 32–48.

Apker, J., Propp, K., Ford, W., & Hofmeister, N. (2006). Collaboration, credibility, compassion, and coordination: Professional nurse communication skill sets in health care team interactions. *Journal of Professional Nursing, 22*(3), 180–189. https://doi.org/10.1016/J.PROFNURS.2006.03.002

Ayrshire Daily News. (2024, July 25). Jet Pitts aerobatic aircraft named as STEM display for 2024 airshow. *Ayrshire Daily News*. www.ayrshiredailynews.co.uk/news/jet-pitts-aerobatic-aircraft-named-as-stem-display-for-2024-airshow

Baker, D., Prince, C., Shrestha, L., Oser, R., & Salas, E. (1993). Aviation computer games for crew resource management training. *The International Journal of Aviation Psychology, 3*, 143–156. https://doi.org/10.1207/S15327108IJAP0302_4

Bambacas, M., & Patrickson, M. (2008). Interpersonal communication skills that enhance organisational commitment. *Journal of Communication Management, 12*, 51–72. https://doi.org/10.1108/13632540810854235

Barker, D. (2003). *Zero error margin: Airshow display flying analysed*. Freeworld Publications.

Barker, D. (2020). *Anatomy of airshow accidents* (1st ed., Vol. 1). International Council of Air Shows, International Air Show Safety Team.

Bastola, D. (2020). The relationship between leadership styles and aviation safety: A study of aviation industry. *Journal of Air Transport Studies*. https://doi.org/10.38008/jats.v11i1.155

Beilock, S. (2010). *Choke: What the secrets of the brain reveal about getting it right when you have to*. https://doi.org/10.5860/choice.48-7212

Berriochoa, M. (2014, August 1). Navigating the maze: Integrating social media into your air show business operations. *Air Show Digest*. https://airshowdigest.aero/2014/08/01/navigating-the-maze-integrating-social-media-into-your-air-show-business-operations/

Schmutz, J. B., Meier, L. L., & Manser, T. (2019). How effective is teamwork really? The relationship between teamwork and performance in healthcare teams: A systematic review and meta-analysis. *BMJ Open, 9*(9), e028280. https://doi.org/10.1136/bmjopen-2018-028280

Burgess, A., Roberts, C., Lane, A., Haq, I., Clark, T., Kalman, E., Pappalardo, N., & Bleasel, J. (2021). Peer review in team-based learning: Influencing feedback literacy. *BMC Medical Education, 21*. https://doi.org/10.1186/s12909-021-02821-6

Burke, S. (2005). Is there a "Big Five" in teamwork? *Small Group Research, 36*(4), 555–599.

Burns, J., Miller, T., Weiss, J., Erdfarb, A., Silber, D., & Goldberg-Stein, S. (2019). Just culture: Practical implementation for radiologist peer review. *Journal of the American College of Radiology, 16*(3), 384–388. https://doi.org/10.1016/j.jacr.2018.10.021

Chiesa, A., & Serretti, A. (2009). Mindfulness-based stress reduction for stress management in healthy people: A review and meta-analysis. *Journal of Alternative and Complementary Medicine, 15*(5), 593–600. https://doi.org/10.1089/acm.2008.0495

Conger, J. (1991). Inspiring others: The language of leadership. *Academy of Management Perspectives, 5*, 31–45. https://doi.org/10.5465/AME.1991.4274713

Cunha, P., & Louro, M. (2000). Building teams that learn. *Academy of Management Perspectives, 14*, 152–153. https://doi.org/10.5465/AME.2000.2909848

DEKRA. (2023). *Top 10 effective safety leadership skills*. DEKRA. www.dekra.com/en/top-10-effective-safety-leadership-skills/

Espevik, R., Johnsen, B., Saus, E., Sanden, S., & Olsen, O. (2021). Teamwork on patrol: Investigating teamwork processes and underlying coordinating mechanisms in a police training program. *Frontiers in Psychology, 12*. https://doi.org/10.3389/fpsyg.2021.702347

Estival, D., & Molesworth, B. (2020). Errors in air-ground pilot communication: An experimental study. *The ESPecialist, 41*(3). https://doi.org/10.23925/2318-7115.2020v41i3a5

European Airshow Council (EAC). (2024). European Airshow Council (EAC)-convention. European Airshow Council. www.europeanairshow.org/convention

Federal Aviation Administration. (2001, February 8). *Crew resource management (CRM) training* (Advisory Circular No. 120-51D). U.S. Department of Transportation. www.faa.gov/documentLibrary/media/Advisory_Circular/AC_120-51D.pdf

Federal Aviation Administration. (2023). *Pilot's handbook of aeronautical knowledge* (FAA-H-8083-25C). www.faa.gov/regulations_policies/handbooks_manuals/aviation/faa-h-8083-25c.pdf

Fisher, J., Phillips, E., & Mather, J. (2000). Does crew resource management training work? *Air Medical Journal, 19*(4), 137–139. https://doi.org/10.1016/S1067-991X(00)90006-3

Flin, R., O'Connor, P., & Mearns, K. (2002). Crew resource management: Improving teamwork in high reliability industries. *Team Performance Management, 8*, 68–78. https://doi.org/10.1108/13527590210433366

Fornette, M., Bardel, M., Lefrançois, C., Fradin, J., Massioui, F., & Amalberti, R. (2012). Cognitive-adaptation training for improving performance and stress management of air force pilots. *The International Journal of Aviation Psychology, 22*, 203–223. https://doi.org/10.1080/10508414.2012.689208

George, J. (2000). Emotions and leadership: The role of emotional intelligence. *Human Relations, 53*, 1027–1055. https://doi.org/10.1177/0018726700538001

Gligoric, K., Anderson, A., & West, R. (2019). Causal effects of brevity on style and success in social media. *Proceedings of the ACM on Human-Computer Interaction, 3*, 1–23. https://doi.org/10.1145/3359147

Goleman, D. (1999). What makes a leader? *Harvard Business Review, 76*(6), 93–102. https://doi.org/10.4135/9781446213704.n9

Grogan, E., Stiles, R., France, D., Speroff, T., Morris, J., Nixon, B., Gaffney, F., Seddon, R., & Pinson, C. (2004). The impact of aviation-based teamwork training on the attitudes of health-care professionals. *Journal of the American College of Surgeons, 199*(6), 843–848. https://doi.org/10.1016/J.JAMCOLLSURG.2004.08.021

Grote, G. (2016). Leading high-risk teams in aviation. *Team Performance Management, 8*, 22–35. https://doi.org/10.1108/S1479-357120160000008006

Häfner, A., & Stock, A. (2010). Time management training and perceived control of time at work. *The Journal of Psychology, 144*, 429–447. https://doi.org/10.1080/00223980.2010.496647

Hanton, S., Mellalieu, S., & Hall, R. (2004). Self-confidence and anxiety interpretation: A qualitative investigation. *Psychology of Sport and Exercise, 5*, 477–495. https://doi.org/10.1016/S1469-0292(03)00040-2

Hapsari, M. (2023). Team building training to improve interpersonal communication among operators at PT Komatsu Indonesia. *Academic Journal of Psychology and Counseling*. https://doi.org/10.22515/ajpc.v4i1.7371

Harvey, W. (2011). Strategies for conducting elite interviews. *Qualitative Research, 11*, 431–441. https://doi.org/10.1177/1468794111404329

Hayes, M., Filo, K., Geurin, A., & Riot, C. (2020). An exploration of the distractions inherent to social media use among athletes. *Sport Management Review, 23*, 852–868. https://doi.org/10.1016/j.smr.2019.12.006

Hayes, M., Filo, K., Riot, C., & Geurin, A. (2021). Using communication boundaries to minimize athlete social media distractions during events. *Event Management*. https://doi.org/10.3727/152599521X16106577952l5

Hayes, M., Filo, K., Riot, C., & Geurin, A. (2023). Addressing athlete social media distractions during events: Examining effective and ineffective approaches. *Event Management, 27*(5), 763–780.

Helmreich, R., & Wilhelm, J. (1991). Outcomes of crew resource management training. *The International Journal of Aviation Psychology, 1*(4), 287–300. https://doi.org/10.1207/S15327108IJAP0104_3

Henley, N., Donovan, R., & Francas, M. (2008). Developing and implementing communication messages. In *Communication for Behavior Change* (pp. 433–447). Springer.

Hinrichs, L. A., Judd, D. L., Hernandez, M., & Rapport, M. J. (2022). Peer review of teaching to promote a culture of excellence: A scoping review. *Journal of Physical Therapy Education, 36*(4), 293–302.

Hoegl, M., & Gemuenden, H. (2001). Teamwork quality and the success of innovative projects. *Organization Science, 12*(4), 435–449. https://doi.org/10.1287/ORSC.12.4.435.10635

Hollowell, D. (2012, August 1). Putting accident/incident analysis to work for our air show family. *Air Show Digest*. International Council of Air Shows.

Hölzel, B., Carmody, J., Vangel, M., Congleton, C., Yerramsetti, S., Gard, T., & Lazar, S. (2011). Mindfulness practice leads to increases in regional brain gray matter density. *Psychiatry Research: Neuroimaging, 191*, 36–43. https://doi.org/10.1016/j.pscychresns.2010.08.006

Hoppe, E. A. (Ed.). (2018). *Ethical issues in aviation*. Routledge.

Illman, J. (2006). Training for interviews with the media. *The Psychiatrist, 30*, 272–274. https://doi.org/10.1192/PB.30.7.272

International Council of Air Shows (ICAS). (2019, January 1). *ACE manual—Revision 9*. International Council of Air Shows.

Jang, R., Molesworth, B., Burgess, M., & Estival, D. (2014). Improving communication in general aviation through the use of noise cancelling headphones. *Safety Science, 62*, 499–504. https://doi.org/10.1016/J.SSCI.2013.09.012

Kabat-Zinn, J. (1990). *Full catastrophe living: Using the wisdom of your body and mind to face stress, pain, and illness*. Delacorte.

Kabat-Zinn, J. (1994). *Wherever you go, there you are: Mindfulness meditation in everyday life*. Hyperion.

Kanki, B., Folk, V., & Irwin, C. (1991). Communication variations and aircrew performance. *The International Journal of Aviation Psychology, 1*, 149–162. https://doi.org/10.1207/S15327108IJAP0102_5

Karachalios, M. (2022). *An evaluation of the relationships between resilient safety culture, safety risk parameters, and mindfulness in the international air show community* [PhD dissertation, University of North Dakota]. https://commons.und.edu/theses/4267/

Karachalios, M., & Adjekum, D. K. (2023). *Air show performers: Safety, risk management, and psychological factors*. CRC Press. https://doi.org/10.1201/9781003431879

Katz, S., & Byrne, S. (2019). Cognitive bridging: Using strategic communication to connect abstract goals with the means to achieve them. *Health Communication, 34*, 484–499. https://doi.org/10.1080/10410236.2018.1428848

Kouzes, J., & Posner, B. (2017). *The leadership challenge* (6th ed.). John Wiley & Sons.

Leonard, M., Graham, S., & Bonacum, D. (2004). The human factor: The critical importance of effective teamwork and communication in providing safe care. *Quality and Safety in Health Care, 13*, i85–i90. https://doi.org/10.1136/qshc.2004.010033

Looze, C., Vaughan, B., Kelly, F., & Kay, A. (2015). Providing objective metrics of team communication skills via interpersonal coordination mechanisms. *Interspeech, 2015*, 1226–1230. https://doi.org/10.21437/Interspeech.2015-309

Lord, R., & Hall, R. (2005). Identity, deep structure and the development of leadership skill. *Leadership Quarterly, 16*, 591–615. https://doi.org/10.1016/J.LEAQUA.2005.06.003

MacQuillin, I., & Sargeant, A. (2019). Fundraising ethics: A rights-balancing approach. *Journal of Business Ethics, 1–12*. https://doi.org/10.1007/S10551-018-3872-8

Malakis, S., Kontogiannis, T., & Kirwan, B. (2010). Managing emergencies and abnormal situations in air traffic control (part II): Teamwork strategies. *Applied Ergonomics, 41*(4), 628–635. https://doi.org/10.1016/j.apergo.2009.12.018

Mayer, J. D., Caruso, D. R., & Salovey, P. (1999). Emotional intelligence meets traditional standards for an intelligence. *Intelligence, 27*(4), 267–298. https://doi.org/10.1016/S0160-2896(99)00016-1

Mayer, J. D., Salovey, P., Caruso, D. R., & Sitarenios, G. (2001). Emotional intelligence as a standard intelligence. *Emotion, 1*(3), 232–242. https://doi.org/10.1037/1528-3542.1.3.232

McClernon, C., McCauley, M., O'Connor, P., & Warm, J. (2011). Stress training improves performance during a stressful flight. *Human Factors: The Journal of Human Factors and Ergonomics Society, 53*, 207–218. https://doi.org/10.1177/0018720811405317

Meland, A., Fonne, V., Wagstaff, A., & Pensgaard, A. M. (2015). Mindfulness-based mental training in a high-performance combat aviation population: A one-year intervention study and two-year follow-up. *International Journal of Aviation Psychology, 25*(1), 48–61. https://doi.org/10.1080/10508414.2015.995572

Molesworth, B., & Estival, D. (2015). Miscommunication in general aviation: The influence of external factors on communication errors. *Safety Science, 73*, 73–79. https://doi.org/10.1016/J.SSCI.2014.11.004

Müller, M., Hänsel, M., Fichtner, A., Hardt, F., Weber, S., Kirschbaum, C., Rüder, S., Walcher, F., Koch, T., & Eich, C. (2009). Excellence in performance and stress reduction during two different full-scale simulator training courses: A pilot study. *Resuscitation, 80*(8), 919–924. https://doi.org/10.1016/j.resuscitation.2009.04.027

Mumford, M., Zaccaro, S., Connelly, M., & Marks, M. (2000). Leadership skills: Conclusions and future directions. *Leadership Quarterly, 11*, 155–170. https://doi.org/10.1016/S1048-9843(99)00047-8

Myall, K., Montero-Marín, J., Gorczynski, P., Kajee, N., Sheriff, R., Bernard, R., Harriss, E., & Kuyken, W. (2022). Effect of mindfulness-based programmes on elite athlete mental health: A systematic review and meta-analysis. *British Journal of Sports Medicine, 57*, 99–108. https://doi.org/10.1136/bjsports-2022-105596

Olajiga, O. K., Olu-lawal, K. A., Usman, F. O., & Ninduwezuor-Ehiobu, N. (2024). Conceptual framework for effective communication strategies in high-risk industries: Insights from the energy sector. *World Journal of Advanced Engineering Technology and Sciences, 11*(2), 080–090.

Rich Goodwin. (2024). Rich Goodwin airshow. Rich Goodwin Airshow. www.richgoodwinairshows.com

Rouse, W., & Rouse, R. (2004). Teamwork in the performing arts. *Proceedings of the IEEE, 92*, 606–615. https://doi.org/10.1109/JPROC.2004.825880

Royal Air Force (RAF). (2018). Red Arrows visit children at two hospitals ahead of Christmas. Royal Air Force. www.raf.mod.uk/display-teams/red-arrows/news/red-arrows-visit-children-at-two-hospitals-ahead-of-christmas/

Rumbold, J., Fletcher, D., & Daniels, K. (2012). A systematic review of stress management interventions with sport performers. *Sport, Exercise, and Performance Psychology, 1*, 173–193. https://doi.org/10.1037/A0026628

Salas, E., DiazGranados, D., Klein, C., Burke, C., Stagl, K., Goodwin, G., & Halpin, S. (2008). Does team training improve team performance? A meta-analysis. *Human Factors: The Journal of Human Factors and Ergonomics Society, 50*, 903–933. https://doi.org/10.1518/001872008X375009

Sayer, E., Featherstone, H., & Gosling, W. (2014). Sex & bugs & rock 'n roll—Getting creative about public engagement. *Trends in Ecology & Evolution, 29*(2), 65–67. https://doi.org/10.1016/j.tree.2013.12.008

Shamsi, H., Almutairi, A., Mashrafi, S., & Kalbani, T. (2020). Implications of language barriers for healthcare: A systematic review. *Oman Medical Journal, 35*, e122–e122. https://doi.org/10.5001/omj.2020.40

Sinek, S. (Director). (2021, February 26). There's no such thing as "soft skills" [Video]. YouTube. www.youtube.com/watch?v=o9uzJ0LgvT0

Skybrary. (2023). *Leadership (OGHFA BN)*. Flight Safety Foundation; EUROCONTROL. www.skybrary.aero/index.php/Leadership_(OGHFA_BN)

Stevens, M., & Campion, M. (1994). The knowledge, skill, and ability requirements for teamwork: Implications for human resource management. *Journal of Management, 20*, 503–530. https://doi.org/10.1177/014920639402000210

Tumpey, A. J., Daigle, D., & Nowak, G. (2019). Communicating during an outbreak or public health investigation. In S. A. Rasmussen & R. A. Goodman (Eds.), *The CDC field epidemiology manual* (pp. 243–259). Oxford University Press.

UK Civil Aviation Authority (UK CAA). (2024a). *CAP 1724: Flying display pilot authorisation and evaluation: Requirements and guidance*. www.caa.co.uk/our-work/publications/documents/content/cap1724/

US Air Force (USAF). (2022). *Air Force Manual 11-246: Air Force fighter type aircraft demonstrations (A-10, F-16, F-22, F-35)* (Vol. 1). USAF. https://static.e-publishing.af.mil/production/1/af_a3/publication/afman11-246v1/afman11-246.pdf

van Leeuwen, J. M. J. (2010). The domino effect. *American Journal of Physics, 78*(7), 721–727.

van Rosse, F., de Bruijne, M., Suurmond, J., Essink-Bot, M.-L., & Wagner, C. (2016). Language barriers and patient safety risks in hospital care: A mixed methods study. *International Journal of Nursing Studies, 54*, 45–53. https://doi.org/10.1016/j.ijnurstu.2015.03.012

Waltz, R. J. (2022). *An evaluation of the relationships between safety management systems (SMS), transformational safety leadership, self-efficacy, and safety performance metrics in a 14 Code of Federal Regulation (CFR) Part 121 airline: A mediation analysis* (Doctoral dissertation, The University of North Dakota).

Watson, R. (2021). Futures on air: Community radio and public engagement. *Interactions: Studies in Communication & Culture*. https://doi.org/10.1386/iscc_00040_1

Wear News. (2024). *FSUSD students inspired by the United States Navy Blue Angels presentation*. Fairfield-Suisun Unified School District. www.fsusd.org/article/1515730

Weick, K., & Sutcliffe, K. (2007). *Managing the unexpected: Resilient performance in an age of uncertainty* (2nd ed.). https://doi.org/10.5860/choice.45-3293

Weinberg, R. S., & Gould, D. (2023). *Foundations of sport and exercise psychology*. Human Kinetics.

Weir, K. (2018, September). What makes teams work? *Monitor on Psychology, 49*(8), 46–55. www.apa.org/monitor/2018/09/cover-teams

Weiss, S., Reber, A., & Owen, D. (2008). The locus of focus: The effect of switching from a preferred to a non-preferred focus of attention. *Journal of Sports Sciences, 26*, 1049–1057. https://doi.org/10.1080/02640410802098874

Wendling, C., Radisch, J., & Jacobzone, S. (2013). *The use of social media in risk and crisis communication. OECD Working Papers on Public Governance*. https://doi.org/10.1787/5k3v01fskp9s-en

Wickens, C. (2002). Situation awareness and workload in aviation. *Current Directions in Psychological Science, 11*(4), 128–133. https://doi.org/10.1111/1467-8721.00184

Young, M., & Dulewicz, V. (2005). A model of command, leadership and management competency in the British Royal Navy. *Leadership & Organization Development Journal, 26*, 228–241. https://doi.org/10.1108/01437730510591770

Zhang, D., Jiang, Z., Xie, Y., Wu, W., Zhao, Y., Huang, A., Li, T., & Ba-Thein, W. (2022). Linguistic barriers and healthcare in China: Chaoshan vs. Mandarin. *BMC Health Services Research, 22*(1), 376. https://doi.org/10.1186/s12913-022-07744-6

7 Maintaining Currency and Proficiency

Manolis Karachalios

INTRODUCTION

Repetitio est mater studiorum.
Repetition is the mother of learning.
Latin Proverb

As the ancient Latin proverb wisely states, *Repetitio est mater studiorum*—repetition is the mother of learning (Repetitio est mater studiorum, n.d.). This principle is particularly vital for air show performers, ensuring both the safety and spectacle of aerial displays. Therefore, they need to stay both current and proficient to fly their display profile precisely, comfortably, and elegantly. Currency refers to the necessity for pilots to regularly engage in flying to keep their skills sharp and reactions instinctual. Proficiency goes a step further, encompassing a pilot's ability to perform flight maneuvers with precision, confidence, and safety, particularly under the unique air show environment.

Air show performers must remain at the peak of their capabilities due to the dynamic nature of their performances, which involve complex maneuvers close to the ground and near other aircraft when conducting formation flying. It is not enough to simply log sufficient flight hours; continuous skill enhancement tailored to their performance routines is crucial.

Currency
Typically quantified by regulatory and organizational standards, this requires pilots to complete a specific number of flight hours within a certain timeframe. For air show performers, currency also involves regular rehearsals of display routines to maintain each maneuver's precision and safety since skill deteriorates and needs to be refreshed. Ultimately, currency is about whether a pilot has flown an airplane recently proportionate to their understanding of that airplane.

DOI: 10.1201/9781003504801-10

Proficiency
This involves a deeper level of mastery over flying skills. Proficiency is measured by the ability to not only fly but also perform with an exceptional skill that ensures safety, impresses audiences, and meets strict air show standards.

The 2018 report "Human Factors in Air Displays: Transfer of Behaviours and Error Path Study" suggests currency as a crucial factor influencing error likelihood in display flying, according to display pilots and expert stakeholders (UK Civil Aviation Authority, 2018). Also, in the 2022 study (Karachalios, 2022), many interviewees regarded the regulatory minimums on currency as merely a starting point, suggesting that prudent individuals should exceed these benchmarks, even if they face practical and financial challenges, such as accessing aircraft and covering costs.

OLYMPIANS ON WINGS

Air show performers, often regarded as the *Olympians on Wings*, can only display their skills to audiences when they are at their sustainably-peak physical and mental performance. Similar to sprinters and marathon runners, who must maintain exceptional physical fitness to excel, air show performers must uphold rigorous physical conditioning to withstand the intense G-forces encountered during low-level aerobatics. Furthermore, akin to archers who require precise focus and control, these aviators must possess sharp cognitive abilities to make split-second decisions while maneuvering at high speeds.

Research highlights the importance of mental and physical preparedness in achieving peak performance, as seen in Olympic athletes (Johns Hopkins University, 2018). It cannot be overstressed that, like Olympian athletes, air show performers maintain optimal physical condition and adapt their training to individual needs to mitigate the risks associated with high G-forces. Tailored training regimens are crucial for ensuring that every performance is conducted safely and remains visually impressive.

SKILL DECAY

Skill decay represents a significant risk in professions requiring elevated levels of technical expertise and operational precision, such as aviation, medicine, or any field involving complex procedural knowledge (Arthur et al., 1998; Landman et al., 2022). In such contexts, the concepts of currency, recency, and proficiency are vital to ensure that individuals maintain their ability to perform their duties safely and effectively.

Risk of Skill Decay

Skill decay refers to the gradual loss of learned skills or knowledge when they are not used frequently. It occurs because neural pathways, reinforced through practice, begin to weaken over time if they are not regularly engaged. This can lead to slower response times, increased errors, and a need for retraining. In high-stakes environments, such as piloting aircraft at air shows, skill decay can have severe consequences, including reduced safety and performance inefficiencies (Arthur et al., 1998; Childs & Spears, 1986; Haslbeck & Hoermann, 2016; Klostermann et al., 2022).

The Need for Currency

Currency in a professional context refers to the consistent engagement in specific tasks or procedures to maintain a standard level of competency. This practice is essential to ensure that professionals do not just possess theoretical knowledge but also retain the practical and procedural memory required to execute their tasks effectively under various conditions. For pilots, maintaining currency involves adhering to a prescribed minimum number of flight hours or operational activities within a certain timeframe, which helps keep their skills sharp and ensures that they are always prepared to handle the complexities and demands of flying.

The concept of currency highlights the importance of ongoing practice and engagement in one's profession. According to Wilson (1993), the value of currency lies in its ability to keep professionals adept and responsive in their roles, ensuring that their expertise remains relevant and effective over time. This is particularly crucial in high-stakes professions like aviation, where the consequences of skill degradation can be significant.

Recent studies, such as those by Olaganathan and Amihan (2021), have examined the impact of external factors, such as the COVID-19 pandemic, on pilot proficiency. Their research indicates that prolonged periods of inactivity or reduced flight operations can adversely affect pilots' skills, posing potential risks, emphasizing the critical need for regular practice and adherence to currency requirements to mitigate such risks.

Furthermore, Jensen (1995) emphasizes the role of pilot judgment and Crew Resource Management in maintaining currency, while Harris (2011) also points out that maintaining currency ensures that pilots remain competent and confident in their abilities.

The Importance of Recency

Recency, a concept closely related to currency, emphasizes the significance of the time elapsed since an individual last performed a specific task. While currency focuses on regular engagement in tasks to maintain competency, recency highlights the value of recent practice in ensuring optimal performance. This distinction is crucial because recent experiences are more strongly retained in memory and can be more readily recalled during task execution.

For example, a pilot who has recently performed takeoffs and landings within the past month is likely to exhibit higher proficiency and confidence in these operations compared to a pilot who has not flown for several months. This phenomenon occurs because recent practice reinforces procedural memory and enhances the ability to perform tasks smoothly and efficiently. Conversely, a lack of recent practice can lead to "*rustiness*"—a state where initial inefficiency and errors are more likely due to a lapse in muscle memory and cognitive sharpness.

Ruthruff, Remington, and Johnston (2001) discuss how switching between cognitive tasks involves both top-down and bottom-up processes, with recent experiences playing a significant role in how well individuals adapt to new or resumed activities. Their study suggests that recent engagement in tasks helps to streamline cognitive processes, making it easier to switch and adapt to different tasks with minimal error.

Additionally, research by Scofield et al. (2020) demonstrates that repetition and recency in task performance enhance recognition memory. Their findings indicate that recent repetitions of tasks lead to stronger behavioral and electrophysiological responses, underscoring the importance of recency in maintaining task proficiency.

Ebbatson et al. (2010) explored the relationship between manual handling performance and recent flying experience in air transport pilots. Their study found that pilots with recent flight experience exhibited better manual handling skills, reinforcing the idea that recency is critical for maintaining high levels of performance in practical tasks.

Moreover, Sikora, Hari, and Hanusch (2020) highlight the role of human factors and Crew Resource Management techniques in mitigating risks associated with loss of control in flight (LOC-I) accidents. Their work stresses the importance of recency in maintaining crew coordination and effective decision making during critical flight operations.

Proficiency: Beyond Currency and Recency

While currency and recency are crucial for maintaining basic operational abilities, proficiency encompasses a higher standard of performance. Proficiency means not only performing tasks correctly but doing so efficiently, confidently, and under varying circumstances. It demands more than just meeting the minimum requirements for currency; it involves ongoing training, practice, and evaluation to ensure a high level of skill and adaptability.

Proficiency is achieved through a combination of advanced simulations, continual learning, and rigorous performance evaluations. This comprehensive approach ensures that professionals are not just competent but excel in their roles. They can handle complex situations, adapt to new challenges, and integrate new knowledge and techniques into their practice effectively.

For instance, the work by Saastamoinen and Maunula (2021) emphasizes the value of increasing fighter pilots' proficiency through specialized training programs. Their research highlights how advanced simulations and ongoing education play a critical role in ensuring that personnel can perform their duties with high efficiency and effectiveness.

Similarly, Keller et al. (2020) discuss the justification and development of competencies in collegiate aviation flight programs. They stress the need for a structured approach to training that goes beyond basic currency and recency, focusing on developing a broad set of skills that enable pilots to handle diverse and challenging situations.

Wright (1973) explored the retention of flying skills and refresher training requirements, noting that proficiency flying—a regular, structured practice designed to maintain high skill levels—significantly impacts a pilot's ability to perform under pressure, underscoring the importance of regular, advanced training sessions to keep skills sharp and ready for any situation.

The Aircraft Owners and Pilots Association (AOPA) differentiates between currency and proficiency, emphasizing that while currency ensures pilots meet regulatory requirements, proficiency involves maintaining a higher level of skill through continual practice and learning. According to AOPA (2016), proficiency is about being prepared for the unexpected, mastering the aircraft, and understanding the environment in which one operates.

Proficiency, therefore, involves a commitment to ongoing education and practice beyond the minimum standards. It means engaging in advanced training, such as flight simulators that replicate various scenarios, attending workshops and courses to learn new techniques, and undergoing regular performance assessments to identify and address areas for improvement. This commitment ensures that professionals remain not only competent but also capable of excelling in their roles and adapting to new and challenging circumstances.

EXAMPLES OF CURRENCY AND PROFICIENCY MAINTENANCE

Maintaining currency and proficiency among pilots, particularly those who participate in air shows, involves a diverse array of activities designed to ensure they remain at the peak of their abilities. These activities include regular practice, advanced simulations, peer reviews, competency checks, and continuous education.

Regular training sessions allow pilots to practice display profiles, refine their skills, and adapt their maneuvers based on varying weather conditions and aircraft performance. This consistent practice helps maintain their currency by ensuring they meet the minimum flight hours and operations required to stay sharp. Additionally, display pilots utilize flight simulators to rehearse their routines in a controlled environment, which is particularly useful for practicing responses to emergency scenarios without the risk associated with real-world flights.

Engaging in peer review and feedback is a critical aspect of maintaining proficiency. By having fellow pilots assess each other's performances, pilots can gain valuable insights into their flying techniques. This collaborative approach helps identify areas for improvement and enhances the quality of their displays. Constructive feedback from peers ensures that pilots continually evolve and perfect their skills. Regular competency checks conducted by qualified instructors are also essential for maintaining high standards. These evaluations often include both ground and flight tests, ensuring that pilots can execute their routines safely and effectively. Competency checks verify that pilots not only meet the basic requirements of currency but also demonstrate a high level of proficiency in their flying skills.

Attending workshops and clinics led by veteran air show performers provides pilots with opportunities to learn new tips on managing performance stressors, perfecting complex maneuvers, and handling specific flying conditions like low-altitude maneuvers and aerobatics. These educational sessions are tailored to address various aspects of flying that are critical to successful and safe performances in air shows (International Council of Air Shows, n.d.; British Air Display Association, n.d.; European Airshow Council, n.d.). The psychological aspects of performance, such as managing anxiety and mental preparation, are crucial for flawless execution in front of large audiences (Karachalios & Adjekum, 2023). Studies by Driskell et al. (1994) indicate that mental practice can significantly enhance performance, emphasizing the importance of visualization and mental rehearsal. Similarly, Murray (2007) highlights the role of mental preparation in flying, suggesting that pilots who engage in mental practices are better equipped to handle the pressures of performance.

Managing performance anxiety is another crucial aspect of proficiency. Research by Roland (1994) on how professional performers manage performance anxiety provides valuable insights for display pilots. Techniques such as deep breathing, visualization, and positive self-talk can help air show performers maintain composure and confidence during their performances and execute them with precision and elegance.

CURRENCY REQUIREMENTS

Table 7.1 outlines the currency requirements for various aviation authorities and specific pilot scenarios.

The table consolidates various currency and proficiency requirements for air show and display pilots as specified by different aviation regulatory bodies, including the UK CAA, USA FAA, and USAF Air Combat Command (ACC). Additionally, it is listed in Table 7.1, the commonly accepted currency requirements for fast jet solo display pilots. Each set of guidelines addresses distinct aspects of pilot readiness, emphasizing the importance of continuous practice and evaluation to maintain safety and performance standards in aviation displays. Below is a detailed analysis.

UK CAA

The currency requirements detailed in the table for UK CAA (2025) display pilots highlight a rigorous framework designed to ensure that pilots participating in air shows and display events are not only compliant with regulatory standards but also proficient in their flying skills.

Key Elements of the Currency Requirements

- Structured Frequency and Recency
  The stipulation that pilots complete multiple routines within these intervals ensures ongoing engagement with the technical and performance aspects of their aircraft and display routines. The 90-day requirement builds a consistent practice schedule, while the 30-day requirement ensures that skills are sharp and recently validated just before a display.

TABLE 7.1
Currency Requirements for Various Aviation Authorities

Authority/ Organization	Requirement	Details
UK CAA[1] (UK CAA, 2025)	Initial DA holders	Must attend a Display Symposium within the first 12 months of the point of issue of the DA.
	General Display Aircraft (800 hp or greater, 2730 kilogram or greater, jet-powered, or turbo-prop) and Aerobatic Display Aircraft (standard skill level)	Three complete display routines flown or practiced within 90 days of display. One complete display routine flown or practiced in the DA category within 30 days of display.
	Aerobatic Display Aircraft (intermediate, advanced, unlimited)	Three complete display routines flown or practiced within 90 days of display. One complete display routine flown or practiced at appropriate aerobatic skill level in DA category within 30 days of display.
USA FAA (FAA, 2022)	Currency	A pilot is considered current if they have flown either a practice performance or an air show performance in the same aircraft, level, and endorsements within the previous 45 days.
	Proficiency	A pilot is considered proficient if they hold a current and valid SAC card for the aircraft category, level, and endorsements being flown.
USAF ACC (USAF, 2022)	Demo and HF pilot proficiency/currency	• Demo practices should include low profiles that flow less frequently. • A minimum of one low profile every 45 days, and one HF performance every 45 days for ACC demo pilots and every 90 days for AFHFF pilots. • HF pilots' currency is evaluated at least 14 days prior to the scheduled HF performance. • Non-current pilots must regain currency via HF practice procedures.
	Demo pilots	• To maintain currency, each pilot must fly a minimum of one demo every 15 calendar days. • If this period is exceeded, the next demo practice is limited to no lower than 1,000 feet AGL target and 900 feet AGL minimum on all maneuvers. A supervisor must be present.
	Abort currency	Practice pilot abort procedures during practice demos at least once every 60 days.

TABLE 7.1 (Continued)
Currency Requirements for Various Aviation Authorities

Authority/ Organization	Requirement	Details
Fast Jet solo display pilots (Commonly Accepted Practice)	Regular season	• Prior to an air show, the display must be flown live twice within one week at home base. • Currency flights include one per week during the season and at least one sortie every 15 days at 500 feet.
	15–30 days	• If the break between air shows is more than three weeks, a simulator session and a display rehearsal at home base are required. • One sortie at 1,000 feet and one at 500 feet, or one dual sortie with another demo pilot.
	Greater than 30 days	One sortie between 5,000 feet and 1,000 feet, and one sortie at 500 feet.

- Differentiation by Aircraft Type and Power
  The table applies the same frequency of practice to all aircraft but distinguishes by noting that high-performance aircraft (800 hp or greater, 2730 kilogram or greater, jet-powered, or turbo-prop) also fall under these stringent requirements. This categorization acknowledges that more powerful and potentially more complex aircraft require rigorous and frequent handling to maintain safety and performance standards.
- Aerobatic Skill Levels
  The uniform requirements across various skill levels (from standard to unlimited) reinforce the principle that the inherent risks associated with aerobatic maneuvers necessitate continuous practice, irrespective of the pilot's experience level. Advanced skills and unlimited categories, which likely involve more complex and risky maneuvers, require pilots to maintain their proficiency through consistent practice, ensuring safety and technical precision.
- Experience Level
  The requirement for new Display Authorization (DA) holders to attend a Display Symposium (UK CAA, 2025) within the first 24 months signifies the importance of formal education and awareness about air show safety and operational standards. This helps ensure that newly certified pilots are well-versed in safety protocols and performance expectations.

USA FAA

- Currency
  The FAA mandates that pilots must have flown a practice or actual performance in the specified aircraft, certified level, and with the correct endorsements

within the last 45 days to be considered current (International Council of Air Shows, 2019). This frequency ensures pilots remain familiar with the aircraft and display routines, minimizing the risk of errors due to unfamiliarity or lack of recent experience.

- Proficiency
  Holding a current and valid SAC card, as per FAA (2023) regulations, ensures that pilots not only meet the minimum requirements but also maintain ongoing proficiency in their designated aircraft category and level. This card acts as a certification of the pilot's capability to operate safely within specific aviation contexts.

USAF ACC

- Demo and HF Pilot Proficiency/Currency
  The USAF (2022) has established this set of requirements, which is particularly stringent, focusing on regular practice of less frequently performed maneuvers (low profiles) and mandatory performance frequencies for both demo and Heritage Flight (HF) routines. The periodic evaluation and mandatory practices for currency recovery highlight a proactive approach to ensuring pilot readiness and the safe execution of maneuvers.
- Abort Currency
  The requirement to practice abort procedures every 60 days, with initiated calls during practice demos, underscores the emphasis on emergency preparedness, ensuring pilots are ready to handle unexpected situations effectively.

Fast Jet Solo Display Pilots (Commonly Accepted Practice)

- Seasonal and Interval Requirements
  These requirements demonstrate a structured approach to maintaining pilot readiness relative to their performance schedule. The specific stipulations for sorties at different altitudes and frequencies based on the time elapsed since the last air show ensure that pilots are continuously practicing and evaluating their skills under varying conditions.

General Observations

The regulations designed to govern pilot proficiency are meticulously structured to ensure not only frequent but also highly targeted practice sessions and performance evaluations. This regimen is crucial, as it keeps pilot skills honed and ensures that pilots are continuously prepared for the fast-paced and often high-risk environment of air shows. By requiring pilots to engage in routine practice, these guidelines help maintain a consistent level of excellence and operational readiness that is necessary for the unique challenges of flying display.

These frequent practice sessions do more than just reinforce basic flying skills; they are also crucial for enhancing pilot adaptability. By mandating that pilots practice under a variety of conditions and at different intervals, the regulations aim to prepare pilots for any situation they might encounter during a display. This could range from sudden adverse weather conditions to mechanical issues. Consequently, the ability to adapt quickly and effectively is vital, ensuring pilots can manage these challenges without compromising safety.

Furthermore, the regulations establish a robust framework for safety and compliance monitoring. This includes specific oversight roles such as a wing commander designee, who oversees demonstrations for the USAF ACC, ensuring that each performance adheres to the highest safety standards. Additionally, the requirement for pilots to hold current certifications, such as the Statement of Aerobatic Competency (SAC) card required by the FAA, serves a dual purpose. First, it verifies that the pilot has met all the necessary qualifications and is proficient in their craft. Second, it acts as a mechanism for continuous performance monitoring, ensuring that pilots not only meet but maintain the standards required for safe and successful air show participation.

Overall, these comprehensive regulations form an integrated approach to pilot training and performance evaluation. They emphasize not just the mechanical aspects of flying but also the mental and situational preparedness that is essential for air show performers. By focusing on frequent practice, adaptability, and strict compliance with safety standards, these guidelines ensure that display pilots are not only technically proficient but also equipped to handle the complexities of performing in air shows, thus safeguarding the success and safety of these high-profile events.

CONTINUOUS LEARNING AND SKILL REFRESHMENT

Research detailed in FAA AC 61-98D indicates that loss of control (LOC) in flight is frequently a result of inadequate pilot proficiency (FAA, 2018). Unexpected situations that surpass a pilot's current capabilities can emerge without warning. Recognizing this, the UK CAA (2025) asserts that the prescribed currency requirements should be seen as the bare minimum for display currency. Pilots are particularly encouraged to undertake extensive practice during the off-season or in pre-season preparations to uphold a high standard of safety (Baines Simmons, 2024). Additionally, if a display routine is new or if a pilot's experience only meets the minimal currency requirements, the UK CAA recommends stricter practice and display minima.

Similarly, the FAA (2018) strongly advises that general aviation pilots not only meet but also strive to surpass the regulatory minimum requirements for currency. This proactive approach includes participating in additional flight training sessions and advanced maneuver practices that go beyond routine requirements, thereby significantly enhancing pilot safety and performance capabilities.

PROFICIENCY REQUIREMENTS: A BALANCED CHASE FOR EXCELLENCE

In the realm of air shows, currency represents the bare minimum requirement for participating in a flying display, primarily to ensure safety. However, what air show performers truly aspire to achieve is proficiency. Proficiency enables pilots not only to operate the aircraft but also to make critical, timely decisions in response to environmental changes, mechanical issues, or personal factors. This level of skill ensures that performers have the mental capacity not just to react but to excel under pressure.

It is crucial for pilots to maintain focus and discipline, consistently performing at the peak of their capabilities—not just during training sessions or theoretically on paper, but through the mental resilience required to sustain this superior performance. They must remain within a "*buffer zone*" between safety and excellence, avoiding complacency until the very moment the engine shuts down and they step out of the cockpit. The accolades and admiration from spectators can create an *Icarian illusion*[2] of invulnerability, which can lead to catastrophic outcomes if not managed properly.

This ongoing high standard of mental and physical performance is crucial not only during the visible part of the display but also from engine startup to shut down. An air show encompasses much more than the 10–12 minutes of aerial maneuvers; it includes every moment the aircraft is operating. This extended period demands sustained elevated levels of human performance to ensure a safe and successful display.

INSIDIOUS THREATS TO AIR SHOW PERFORMERS

The air show community is well-acquainted with the common risks and errors associated with human factors, and the related literature is often filled with final accident reports attributing accidents to a vaguely stated "*pilot*" or "*human error*." These incidents frequently involve experienced display pilots, leaving many questions unanswered (Barker, 2020; Karachalios, 2022). However, there are still insidious threats for air show performers that have not yet received the appropriate emphasis. In the following section, we strive to shed light on these critical factors, which need to be considered during both training and actual flying displays.

Mental Workload and Air Show Decision Making Degradation During Aerobatics

Performing aerobatics at air shows places an extraordinary amount of mental and physical strain on pilots (Karachalios & Adjekum, 2023). The combination of high G-forces, rapid rolls, visual illusions, and physiological reactions not only challenges a pilot's flying skills but also significantly impacts their cognitive functions, imposes a significant mental workload on pilots, and affects their decision making abilities. This high cognitive load can lead to decision making degradation, which can impact performance and safety. Understanding the factors that contribute to this degradation is crucial for improving pilot training and enhancing safety measures.

Dahlstrom et al. (2011) explored this phenomenon by recording psychophysiological data during aerobatic training. Their study, published in The International Journal of Aviation Psychology, highlights how the cognitive and physical demands of aerobatics affect pilots' performances. The researchers measured various indicators of mental workload, including heart rate variability, skin conductance, and subjective workload assessments, to understand how these factors influence decision making.

The findings from Dahlstrom et al. (2011) suggest that as mental workload increases, pilots may experience a decline in their ability to make rapid and accurate decisions. This is particularly concerning in aerobatics, where even minor errors can lead to significant safety risks. The study indicates that during high-stress maneuvers, pilots' cognitive resources are heavily taxed, which can result in slower reaction times, reduced situational awareness, and a higher likelihood of errors.

Effects of Aerobatic Flight Conditions on Mental Workload

The mental workload in aerobatic flying is compounded by the need to perform complex maneuvers with precision while navigating variable environmental conditions and maintaining strict safety protocols. This workload is analogous to multitasking under extreme stress, where cognitive resources are stretched to their limits. Researchers such as Wickens (2002) have identified multiple dimensions of cognitive load, including the processing of spatial and temporal information, which are highly relevant to aerobatics. Adler and Benbunan-Fich (2012) highlighted the detrimental effects of multitasking on performance, noting that simultaneous tasks can lead to decreased efficiency and increased errors, a concern for aerobatic pilots who must manage multiple inputs and actions seamlessly. Koch et al. (2018) further examined the cognitive structure and flexibility required in multitasking, emphasizing the plasticity needed to switch between tasks without loss of performance (Adler & Benbunan-Fich, 2012; Koch et al., 2018).

High G-forces significantly increase the mental workload of pilots by requiring greater attention and effort to maintain control and orientation during flight. These forces can lead to decreased cerebral perfusion, which impairs cognitive functions and decision making abilities. Consequently, pilots may experience visual impairments such as grayout, where vision fades due to decreased blood flow to the eyes, G-induced Loss of Consciousness (G-LOC), when the brain's blood supply becomes critically low, and Almost Loss of Consciousness (A-LOC), which involves a significant but not complete reduction in awareness (Shender et al., 2003; Kumar et al., 2020; Karachalios & Adjekum, 2023). These physiological challenges necessitate that pilots maintain high levels of physical fitness and undergo rigorous training to recognize and mitigate the effects of high G-forces. As Campbell and Bagshaw (2002) discuss, understanding human performance limitations in aviation is crucial for developing effective training and safety protocols.

Aerobatic maneuvers, especially rapid rolls, often induce spatial disorientation, making it difficult for pilots to discern their position relative to the ground. This potential disorientation increases their reliance on visual lookout to maintain orientation, thereby elevating the mental workload (Gibb et al., 2011). Additionally, high G stresses may also lead to a loss of peripheral vision, further reducing their ability to visually assess the flying environment, forcing pilots to depend on pre-planned routines for their flying displays (Green, 2018).

The body's physiological response to stress, characterized by increased heart and breathing rates, can further complicate these challenges. The rapid breathing associated with high stress might not supply sufficient oxygen to crucial areas like the brain and eyes (Guézennec et al., 2001), exacerbating cognitive deficits and impairing visual capabilities during flight (Samardzić et al., 2017).

In conditions of extreme stress and impaired cognitive functioning, pilots may revert to automatic behaviors—deeply ingrained reactions that might not be appropriate for the current circumstances. This "*fight instead of fly*[3]" phenomenon can lead pilots to make decisions that prioritize immediate physical responses over strategic flying decisions, potentially compromising both safety and performance (Ercan & Gunduz, 2020).

Factors Contributing to Decision Making Degradation

Several factors contribute to decision making degradation during air show performances, affecting display pilots' ability to make swift and accurate decisions under high-pressure conditions. These factors include the following:

- Complexity of Maneuvers: The complexity of aerobatic maneuvers requires precise control and timing. Each maneuver requires pilots to process a vast amount of sensory information quickly and accurately (Orasanu et al., 2001; Barker, 2020).
- Physical Stress: The physical stress of handling high G-forces can impair cognitive function, reducing a pilot's ability to make quick decisions (Ercan & Gunduz, 2020).
- Environmental Variability: Changes in weather and wind conditions can alter the dynamics of flight, requiring constant adjustment and re-evaluation of planned maneuvers (Gultepe, 2023).

Impact on Decision Making

The impact of a high mental workload on decision making can be detrimental, leading to reduced situational awareness, slower reaction times, and increased potential for errors.

- Reduced Situation Awareness
  As the mental workload increases, situational awareness decreases, leading pilots to miss critical cues about their environment or aircraft performance (Endsley, 1995). This decline in situational awareness can be particularly hazardous during display flying at the low-level air show environment, where the ability to accurately perceive and interpret information is crucial for maintaining safety. Wickens (2002) highlights that increased workload can overwhelm a pilot's cognitive resources, resulting in diminished attention and slower reaction times. Svensson et al. (1997) further illustrate that in combat aircraft, information complexity and high mental workload can severely impact performance as pilots struggle to process multiple streams of data simultaneously.
- Increased Reaction Times
  Increased cognitive load can significantly slow a pilot's reaction times, which is critical when split-second decisions are needed (Kahneman, 1973). This

delay in response can have severe implications during heavy-workload and fast-paced flight situations during display flying where immediate actions are necessary to ensure safety and performance. Morris and Leung (2006) studied pilot mental workload and found that high cognitive demands can impair performance, highlighting the need for effective workload management strategies. Additionally, Zhu et al. (2023) utilized portable electroencephalography[4] (EEG) devices to identify pilots' mental workload under different flight phases, demonstrating that higher cognitive load phases are associated with slower reaction times. Causse, Peysakhovich, and Fabre (2016) also observed that high working memory load impairs cognitive functions, such as language processing during simulated piloting tasks, further indicating that elevated cognitive demands can hinder quick and accurate decision making.

- Potential for Error
  The likelihood of errors increases as mental workload exceeds a pilot's capacity to process information effectively (Sarter & Woods, 1995). When cognitive demands surpass a pilot's available resources, critical tasks may be overlooked or improperly executed, leading to potentially dangerous situations. Hidalgo-Muñoz et al. (2018) found that during realistic flight simulations, increased cognitive workload and time-on-task effects were associated with heightened emotional states and cardiovascular stress, which in turn elevated the risk of errors. Furthermore, Lobo et al. (2016) demonstrated that cognitive workload classification using eye-tracking and EEG data could effectively predict error-prone conditions, emphasizing the importance of monitoring and managing cognitive load to minimize errors.

Mitigating Decision Making Degradation

To address the challenges of decision making degradation, particularly under high-stress conditions, several strategies can be implemented to maintain cognitive function and performance. These strategies include the following:

- Enhanced Training
  Enhanced training, particularly simulation-based programs that focus on cognitive skills, is essential for helping pilots manage high-workload situations more effectively (Salas & Burke, 2002). Research by Mohanavelu et al. (2020) demonstrates the value of cognitive workload analysis in flight simulator environments for fighter aircraft pilots, highlighting the benefits of such training in preparing for real-world scenarios. Moreover, Mark et al. (2022) explore neuroadaptive training via functional near-infrared spectroscopy (fNIRS) in-flight simulators, stressing the advancements in adaptive training techniques that respond to pilots' cognitive states. Additionally, Koglbauer (2015) emphasizes the importance of training for the prediction and management of complex and dynamic flight situations, further supporting the need for comprehensive and adaptive simulation-based training programs in aviation (Mohanavelu et al., 2020; Mark et al., 2022; Koglbauer, 2015).

- Cognitive Aids
  Cognitive aids, such as checklists, Aresti cards, and advanced avionics systems, like GPS navigation and head-up displays, play a crucial role in reducing pilots' mental workload by transferring some cognitive demands to technological support (Wickens, 2002). These aids help pilots by systematically organizing information and tasks, allowing for better focus on critical flight operations. For instance, Risko and Gilbert (2016) discuss cognitive offloading, where tools and technologies assist in managing and processing information, thereby enhancing performance and reducing cognitive strain. Similarly, Hales and Pronovost (2006) highlight the effectiveness of checklists in error management and performance improvement, suggesting their importance in high-stakes environments such as aviation. These cognitive aids not only enhance safety but also improve overall efficiency by ensuring that essential procedures are followed systematically and consistently (Risko & Gilbert, 2016; Hales & Pronovost, 2006).
- Stress Management Programs
  Stress management programs are essential for enhancing physical and mental resilience, thereby improving cognitive function under stress (Ercan & Gunduz, 2020). Such programs often include techniques such as mindfulness, physical fitness routines, and cognitive-adaptation training specifically designed for high-stress professions like aviation. For instance, Fornette et al. (2012) explored cognitive-adaptation training for Air Force pilots, finding significant improvements in both performance and stress management. Similarly, Zhang and Zhang (2022) introduced psychological immunity training for pilots aimed at bolstering their mental resilience against the psychological demands of flying. Additionally, Hormeño-Holgado and Clemente-Suárez (2019) studied the psychophysiological responses of professional pilots to various combat jet maneuvers, emphasizing the importance of tailored stress management strategies in maintaining optimal cognitive and physical performance (Fornette et al., 2012; Zhang et al., 2022; Hormeño-Holgado & Clemente-Suárez, 2019).

Helme(n)t Resilience Model for Air Show Performers

To enhance the mental resilience and performance of air show pilots, particularly when faced with the challenges of high G-forces and the complexities of aerobatic maneuvers, a comprehensive "*Helme(n)t Resilience Model*" can be conceptualized (see Figure 7.1), combining the concepts of a *helmet* for protection ("*Hel*") and *mental* resilience ("*me(n)t*"). This model metaphorically equips pilots with an integrated system of protection and support, similar to a pilot's helmet, which not only guards but also enhances their capabilities through various components: the shell, the visor, the chin strap, and a microphone.

Components of the Helme(n)t Resilience Model

The Shell: Training and Practice

The shell represents the core of the pilot's mental armor, signifying robust and continuous training and practice. This includes routine simulation of aerobatic maneuvers

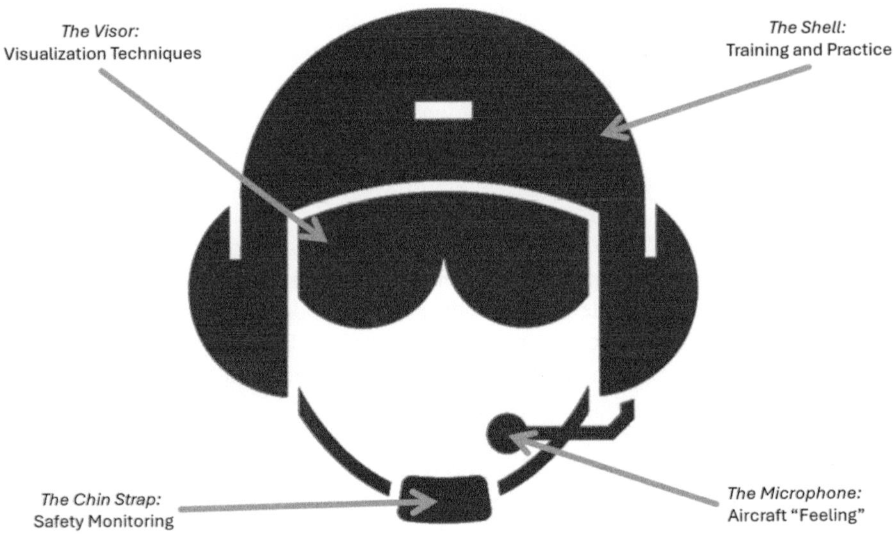

FIGURE 7.1 The *Helme(n)t Resilience Model.*

and exposure to high G-forces in controlled environments to build physical tolerance and muscle memory. Training should also incorporate cognitive-behavioral techniques to enhance mental resilience, enabling pilots to maintain focus and decision making capabilities under stress.

The Visor: Visualization Techniques

The visor represents the utilization of visualization techniques that are crucial for enhancing the pilot's mental and physical preparedness. This component includes practices such as mindfulness, guided imagery, and stress management strategies. Pilots employ these visualization techniques during pre-flight routines to mentally rehearse maneuvers, anticipate potential scenarios, and maintain peak performance states. Such practices help solidify mental clarity and foster a state of relaxation, akin to how a visor provides unobstructed vision and protection from external disturbances.

The Chin Strap: Safety Monitoring

The chin strap represents the safety mechanisms in place, provided by a safety observer or an FDD—a "*human in the loop*" who continuously monitors the pilot's status and the external environment. This role is crucial for maintaining safety as it ensures real-time communication and intervention if the pilot shows signs of stress or if performance deviates from safety parameters. The observer acts as a safeguard, ensuring that all safety protocols are followed and providing a lifeline to the pilot in cases of emergency.

The Microphone: Aircraft "Feeling"

The microphone stands for the continuous "dialogue" between the aircraft and the pilot, allowing for an ongoing assessment of the aircraft's energy status and engine performance. This "communication" is critical, as it enables the pilot to receive instant feedback on the mechanical and aerodynamic aspects of the flight. In addition, monitoring systems should provide clear and concise information so the pilot can make informed decisions quickly, enhancing their ability to respond to the aircraft's needs and external conditions effectively.

Example Scenario

Consider a pilot named "Tony Top" preparing for an air show performance. By applying the *Helme(n)t Resilience Model*, Tony undergoes rigorous simulation training to build physical tolerance and muscle memory (the shell component). This is supplemented with cognitive-behavioral techniques to manage stress. Before the flight, Tony uses visualization techniques to mentally rehearse the maneuvers, building confidence and mental clarity (the visor component).

During the performance, a safety observer monitors Tony's status and provides real-time feedback, ensuring all safety protocols are adhered to (the chin strap component). Additionally, the aircraft's monitoring systems continuously feed data to Tony, allowing for quick adjustments to any changes in performance or conditions (the microphone component).

This comprehensive approach ensures that Tony is mentally resilient, physically prepared, and able to maintain high performance and safety throughout the air show.

G-Force Impact on Air Show Performers' Cognitive Functions

Research indicates that high G-forces have detrimental effects on cognitive functions. A 2011 study by Banks et al. found that high G-forces impair various cognitive functions, including sensing, perception, reaction time, and decision making. Another study by Green in 2016 highlighted that prolonged +Gz exposure leads to decreased brain blood pressure while increasing it in the lower body, further impairing cognitive functions.

In 2005, the UK Royal Air Force conducted a study led by Slungaard et al., which concluded that the incidence of G-induced Loss of Consciousness (G-LOC) and A-LOC among aircrew could be significantly reduced. This reduction was achieved through interventions such as formalized G-theory classroom lectures, the proper execution of the Anti-G Straining Maneuver (AGSM), and undergoing centrifuge training.

Ercan and Gunduz (2020) examined a cross-section of aerobatic pilots and found that resistance to G-forces positively correlated with pilots' readiness and the countermeasures they took. They determined that a decrease in resistance leads to greater G-induced cognitive impairment. They also noted that changes in blood pressure impact cerebral perfusion and disrupt cortical functions. Interestingly, the study suggested that older and more experienced pilots experience less cognitive impairment due to better adaptation and coping mechanisms developed over the years. Height was found to have a minimal effect on G-tolerance. Cardiovascular changes were identified as the primary cause of cognitive dysfunctions under +Gz exposure.

Additional studies support these findings. Morrison et al. (1994) discussed the cumulative effects of +Gz on cognitive performance, emphasizing the long-term impact of repeated G-force exposure. Croft et al. (2021) explored how acceleration-induced reductions in retinal and cerebral oxygenation affect human performance, further highlighting the physiological impacts of G-forces.

Recent research by Radstake et al. (2023) on F-16 fighter jet pilots demonstrated neuroplasticity as a response to continuous high G-force exposure, suggesting potential long-term adaptive changes in the brain. Furthermore, Rickards and Newman (2005) surveyed operational fighter pilots and found significant visual and cognitive disturbances caused by G-forces, revealing the widespread impact of high G environments on pilot performance.

Overall, the body of research consistently shows that high G-forces can significantly impair cognitive functions, but proper training, experience, and physiological adaptations can mitigate some of these effects.

Case Study: Shoreham Airshow Crash and the Relevance of Cognitive Impairment Due to G-Forces

On August 22, 2015, a tragic accident occurred at the Shoreham Airshow in West Sussex, England, involving a Hawker Hunter T7 jet, resulting in 11 fatalities and several injuries on the ground. The pilot, who survived, was performing a loop maneuver that ended in the aircraft crashing onto the A27 highway. Investigations suggested that cognitive dissonance[5] resulted in errors in judgment about the aircraft's altitude and speed during the maneuver which were significant contributory factors to the crash (Air Accidents Investigation Branch [AAIB], 2017).

Relevance of Cognitive Impairment Due to G-Forces

The Shoreham Airshow crash provides an ideal case study for examining the impact of G-forces on cognitive functions during aerobatic maneuvers.

- G-Force Exposure during the Loop Maneuver

Loop maneuvers expose pilots to significant G-forces, which can lead to cognitive impairments such as G-induced Loss of Consciousness (G-LOC) or slowed reaction times (Green, 2018), which could impact a pilot's ability to manage critical flight parameters.

If the pilot experienced cognitive difficulties due to G-forces during the Shoreham Airshow, this might have affected his ability to accurately judge the aircraft's altitude and speed, both of which were noted as insufficient by investigators (AAIB, 2017).

- Pilot's Decision Making Capabilities

High G-forces are known to impair the prefrontal cortex, which is responsible for decision making and risk assessment (Samardzić et al., 2017). This impairment could contribute to poor decision making, such as failing to abort a dangerous maneuver in a timely manner.

The pilot's decisions during the loop maneuver at Shoreham suggest a potential lapse in situational awareness and risk perception, possibly exacerbated by the physiological effects of G-forces.

- Aircraft's Altitude and Speed

Effective execution of a loop maneuver demands precise control over altitude and speed. The Shoreham crash investigation reported that the aircraft commenced the maneuver at an altitude of about 200 feet with a speed of 185 knots—parameters that were too low and slow for safe completion (AAIB, 2017).

The Shoreham Airshow crash serves as a poignant example of the considerable influence of G-forces on pilot cognitive function, especially during intricate aerobatic maneuvers. The implementation of strategies centered on comprehensive training, vigilant monitoring, and thorough assessment is essential to augmenting safety and performance, thereby mitigating the risk of analogous incidents in the future. This case emphasizes the necessity for an integrative approach to pilot preparedness, one that thoroughly addresses the physiological demands associated with aerobatic aviation.

Key Insights from Research and Their Relevance to Air Show Performers

Age and Experience Impact on Cognitive and Physiological Responses

Research indicates that older and more experienced pilots exhibit greater resilience in their physiological responses to G-forces, evidenced by less variation in mean arterial pressure following training. More specifically, Ercan and Gunduz (2020) found that resistance to G-forces positively correlated with pilots' readiness and countermeasures, indicating that experience helps mitigate some negative effects of G-forces on cognitive functions. However, these pilots may experience diminished cognitive accuracy under G-stress compared to their younger counterparts.

Furthermore, Kumar (2023) studied the effects of age and flying experience on heart rate responses during high-G exposure, finding that experienced pilots showed more stable heart rate responses. In addition, Rangel et al. (2023) explored neuro-cardiovascular responses to sympathetic stimulation in fighter pilots, further supporting the notion that experience and age influence physiological resilience under G-stress.

Height and G-Tolerance

The study by Ercan and Gunduz (2020) reveals a positive correlation between height and G-tolerance among pilots, although the effects are minor. Furthermore, Balldin (1986) discussed various factors influencing G-tolerance, emphasizing that individual physiological characteristics play a crucial role. He noted that taller individuals might have different blood distribution patterns under G-forces, affecting their tolerance levels. Kasture et al. (2020) further explored this by examining the correlation of age, height, and gender with +Gz tolerance among healthy Indian participants, confirming that height is a contributing factor, though its impact is relatively minor compared to other variables.

Understanding these nuances allows for the development of more effective training regimens tailored to individual needs. By incorporating height and other physiological factors into training and display routine planning, air show performers can achieve improved G-tolerance and consequently maintain higher cognitive function during aerobatic maneuvers.

Physiological Basis of G-Force Impact

CARDIOVASCULAR CHANGES

High G-forces can cause significant cardiovascular changes, notably increasing blood pressure and altering blood flow patterns. These changes can severely reduce the amount of oxygen reaching the brain, a condition known as reduced cerebral perfusion (Kumar et al., 2020). Cerebral perfusion is crucial because the brain requires a constant and adequate supply of oxygen to maintain optimal function. When this supply is compromised, brain function deteriorates, affecting cognitive abilities such as decision making, reaction time, and overall mental clarity.

During aerobatic maneuvers, the high G-forces can lead to G-induced Loss of Consciousness (G-LOC) or A-LOC. A/G-LOC occurs when cerebral perfusion is significantly decreased to the point where the brain cannot sustain consciousness. However, before reaching this critical threshold, pilots often experience diminished cognitive abilities due to reduced blood flow (Kumar et al., 2020; Ercan & Gunduz, 2020). This intermediate stage is characterized by symptoms such as tunnel vision, lightheadedness, and confusion, which can impair a pilot's ability to perform complex tasks and respond to emergency situations.

Ossard et al. (1994) studied the response of human cerebral blood flow to +Gz accelerations and found that significant reductions in cerebral blood flow occur under these conditions. This reduction directly impacts cognitive function, as the brain's ability to process information and make decisions is compromised by the lack of adequate oxygen and nutrient supply.

Additionally, Convertino (1998) discussed physiological adaptations to high sustained +Gz acceleration, emphasizing the body's efforts to maintain adequate blood flow and oxygen delivery to critical organs, including the brain. These adaptations, while beneficial, are often insufficient to completely prevent cognitive impairments during high-G maneuvers.

Furthermore, Fong and Fan (1997) provided an overview of the physiological effects of sustained high +Gz forces, highlighting that the cardiovascular system's response to these forces includes significant stress on the heart and blood vessels. Such stress can lead to both acute and chronic changes in cardiovascular function, which further complicates the maintenance of cerebral perfusion and overall cognitive function during high-G exposure. Newman (2015) also examined the physiological effects of high G flight and discussed various countermeasures. Effective strategies such as the AGSM (Whitley, 1997), physical conditioning, and specialized training programs are essential to enhance display pilots' tolerance to G-forces and protect against the adverse effects on the cardiovascular system and brain function, leading to cognitive impairments and the risk of G-LOC.

COGNITIVE DYSFUNCTIONS

The prefrontal cortex, responsible for higher cognitive functions such as judgment and rational thinking, is particularly vulnerable to oxygen deprivation. When G-forces reduce blood flow to the brain, the functionality of this region deteriorates, leading to impaired decision making (Samardzić et al., 2017). This reduction in cerebral perfusion can manifest in symptoms such as tunnel vision, confusion, and disorientation, further impairing a pilot's ability to make rational decisions—a critical requirement during air shows (Green, 2018).

As cerebral perfusion decreases, pilots may also experience short-term memory loss and delayed reaction times, which further reduce their ability to respond effectively to rapidly changing flight conditions (Green, 2018). These cognitive impairments pose significant risks during high-stress maneuvers, where quick thinking and precise actions are essential.

Moreover, high G-forces trigger the body's stress response, leading to the release of adrenaline and other stress hormones. While this response can be advantageous in immediate danger scenarios by enhancing physical readiness, it is counterproductive in situations requiring careful decision making. The surge of stress hormones can disrupt rational thought processes, exacerbating the cognitive impairments caused by reduced cerebral perfusion (Samardzić et al., 2017).

Methods to Enhance G-Tolerance and Cognitive Function during Flying Displays

Regular medical monitoring and cardiovascular conditioning should be integral components of pre-show preparations to ensure that performers' bodies are optimally prepared for the physiological stresses of aerobatic performances. Research indicates that post-training improvements in cognitive performance underscore the efficacy of specific G-tolerance training in mitigating cognitive declines (Bateman et al., 2006). Such training, involving controlled exposure to G-forces through centrifuge-based programs, can help pilots adapt physiologically by improving blood flow management during aerobatic displays.

Cardiovascular conditioning enhances overall cardiovascular health, allowing pilots to better manage the physiological demands of high G-forces. Regular medical monitoring ensures that any underlying health issues are identified and managed before performances, reducing the risk of adverse events during flight. Centrifuge-based training programs are specifically designed to simulate the high G-forces experienced during aerobatic maneuvers. These programs help pilots acclimate to the stresses of G-forces, improving their ability to maintain blood flow and oxygen delivery to the brain. This adaptation is crucial for maintaining cognitive function during high-stress situations.

Preconditioning strategies, as discussed by Kilduff et al. (2013), can also enhance physical performance on the day of competition. Techniques such as specific warm-up exercises and controlled exposure to mild stressors can prepare the body to handle the extreme conditions of aerobatic displays more effectively. Cognitive resilience training can prepare pilots to maintain mental clarity under the physical stress of high G-forces, as suggested by Fornette et al. (2012). This type of training focuses on mental exercises and stress management techniques that enhance a pilot's ability to think clearly and make sound decisions even when under significant physical strain.

Furthermore, aircraft systems such as G-suits are designed to maintain blood flow to the brain by exerting pressure on the body. Enhancements in G-suit technology, as suggested by Ercan and Gunduz (2020), could further support cognitive functions by better supporting cerebral perfusion. Advanced G-suits can dynamically adjust pressure to optimize blood flow, thereby reducing the risk of G-LOC and cognitive impairments. Advanced systems capable of monitoring vital signs and providing real-time feedback on physiological status can be highly beneficial. Implementing these systems allows display pilots to make immediate adjustments or interventions during their performances based on real-time data, thereby enhancing cognitive function and ultimately improving the safety of their displays (Ercan & Gunduz, 2020; Roza, 2024; Wylie, 2024).

Papadelis et al. (2008) discussed the development of an intelligent wearable platform for real-time pilot health telemonitoring. Such platforms can provide continuous health data, alerting ground crews and pilots to any physiological abnormalities that need immediate attention. This technology not only enhances safety but also contributes to overall performance by ensuring pilots remain within safe physiological limits during high-stress maneuvers.

Implications for Display Pilot Selection and Training Based on G-Force Impact

The findings from studies offer significant implications for selecting and training air show performers, particularly concerning their exposure to high G-forces during performances. Below is an examination of how these insights can be integrated into selection criteria and training programs to enhance safety and performance.

Pilot Selection

All efforts to enhance display pilots' G-tolerance should be initiated at the early stages of their selection. Regardless of the traditional norms often favored tall and aesthetically appealing pilots, especially from military backgrounds, research indicates that pilots of shorter stature may have a physiological advantage in high-G environments due to a shorter cardiovascular system, which can better maintain blood flow to the brain during high-G maneuvers (Kasture et al., 2020). Modern display pilot selection criteria should prioritize G-tolerance over physical appearance. Candidates should undergo rigorous G-tolerance testing and be selected or approved based on their ability to withstand high G-forces effectively, irrespective of height or appearance (Gillingham, 1987).

In the selection of display pilots, it is crucial to consider a balanced mix of age and experience, as older pilots may handle physiological stress better, while younger pilots tend to maintain higher cognitive performance under G-stress. A diverse team can leverage the strengths of both groups. Comprehensive cardiovascular and neurological assessments should be included in the selection process to determine a pilot's fitness and resilience to G-forces, ensuring all pilots meet a high standard of G-tolerance. Additionally, incorporating physiological profiling, such as height and cardiovascular health, into selection criteria can help tailor training programs to individual needs, thereby optimizing performance under G-forces.

Pilot Training

Pilot training should incorporate a possible mandatory centrifuge training for all pilots to enhance their G-tolerance, simulating the specific G-forces encountered during air shows to improve physiological and cognitive resilience. Ongoing training programs should be established, with regular updates based on the latest research and technological advances in G-force mitigation. Specific exercises designed to improve cognitive function under stress, such as decision making role-playing and scenarios that simulate multitasking during aerobatic maneuvers at higher altitudes, should be included. A cardiovascular fitness program tailored to improving heart and lung function is critical for maintaining cerebral perfusion under high G-forces, with regular monitoring and adjustments to address emerging health concerns or performance issues.

Additionally, fostering a team environment where pilots can share experiences and strategies for dealing with G-forces is essential. Veteran pilots can mentor newer pilots, providing insights and techniques that have proven effective, while a feedback loop within training sessions should be created to allow for adjustments based on pilot experiences and suggestions, ensuring the training program remains dynamic and responsive to pilots' actual needs.

A practical technique for self-monitoring performance and mental capacity levels, which could be integrated into display flying, training, rehearsing, and recurrency programs, is to *verbalize out loud* all maneuver gates and even the flight parameters for the maneuvers to be flown. This method allows display pilots to preemptively assess their mental capacity and workload levels, enabling them to take appropriate actions such as raising display altitudes, maintaining a wider lateral separation from the display line or the flight leader if in formation, or replacing a maneuver with a level pass or a vertical maneuver to a level turn. Incorporating this *verbalization* technique into regular training sessions helps pilots monitor their G-tolerance relative to their cognitive functioning, providing a real-time assessment of their mental state under stress. Furthermore, keeping the intercom microphone on at all times during in-flight video recording can facilitate post-flight debriefings by identifying any lapses in verbalization that could have resulted from heavy workloads, fatigue, or distractions, as well as by monitoring their Anti-G Straining Maneuver through their breathing rate (Karn, 2022; Pooja et al., 2022; Whitley, 1997).

Display Profile Design

The design of the display profile should be meticulously crafted to accommodate both flying deviations and potential cognitive errors, ensuring that safety and performance are maintained at all times. A standard display profile, incorporating tested and approved maneuvers, can significantly minimize the mental workload on pilots, allowing them to focus on precise execution and situational awareness.

Furthermore, standardizing maneuvers reduces cognitive load by enabling pilots to rely on muscle memory and pre-learned responses. For example, a well-designed display profile might include a sequence of aerobatic maneuvers such as loops, rolls, and lazy-eight turns that have been thoroughly tested for their safety margins and predictability. These maneuvers should be arranged in a logical and repetitive pattern, reducing the cognitive load on pilots by creating a familiar and predictable environment. This can help pilots anticipate and react more effectively to any unexpected situations.

Moreover, incorporating buffer zones within the display profile can provide pilots with the necessary space to recover from minor deviations without compromising safety. For instance, if a pilot experiences a momentary loss of cognitive clarity due to high G-forces, having additional altitude or lateral separation from the crowd can allow for safe recovery and continuation of the display.

Another critical aspect is the inclusion of cognitive aids (McLaughlin & Byrne, 2020; Banks et al., 2020), such as pre-briefed visual markers or timing cues that pilots can use to maintain orientation and timing throughout the performance. These aids can act as cognitive anchors, helping pilots stay on track and reducing the likelihood of cognitive overload.

Finally, continuous verbalization of maneuver gates and flight parameters, as discussed earlier, can be integrated into the display profile. This practice not only helps in real-time self-monitoring of G-tolerance and cognitive functioning but also facilitates comprehensive post-flight assessment.

Complacency

Complacency in aviation is a subtle yet profound danger that can creep into a pilot's routine, potentially leading to catastrophic outcomes. Often born out of routine and confidence, complacency diminishes a pilot's vigilance, clouding judgment and eroding the ingrained safety behaviors that are crucial for flying (Dekker, 2003; Hyten & Ludwig, 2017; Mahlfeld et al., 2011; Moray & Inagaki, 2000; Neff, 2022).

At its core, complacency occurs when a pilot feels overly comfortable or unconcerned about the potential dangers of flying, typically due to repeated experiences without incident. This false sense of security can lead pilots to cut corners, ignore standard procedures, or overlook critical warnings. For instance, a display pilot might decide to skip certain pre-flight checks after hundreds of successful air show performances, mistakenly believing that previous successes guarantee future safety.

Understanding Complacency in Air Shows

Complacency in display flying stems from a variety of sources, primarily the confidence that develops from repeated successful performances and encouraging feedback from the crowd (Karachalios, 2022). This overconfidence may lead pilots to bypass vital safety checks or underestimate the complexity of maneuvers, believing their extensive experience alone will ensure safety. For example, a veteran display pilot might forego a detailed review of weather conditions at a stage display site or aircraft maintenance logs, relying on past experiences to guide them through the performance.

Factors Leading to Complacency among Display Pilots

Several factors contribute to complacency among display pilots. Familiarity with performance routines can dull a pilot's perception of risk, leading to reduced attentiveness to variations in aircraft behavior or external conditions (Barker, 2003). Confidence from previous successes can create a false sense of security, causing pilots to push their aircraft beyond safe limits to thrill an audience (Barker, 2020). Routine operation environments, such as regular performances at the same venues, can make emergency procedures feel unnecessary, causing crucial response skills to atrophy. Additionally, in the highly independent environment of air shows, especially for solo display pilots, essential feedback that highlights creeping inefficient habits or complacency may be lacking.

Consequences of Complacency in Air Show Flying

The consequences of complacency in air show flying can be severe. Overlooking pre-flight checks or ignoring subtle signs of mechanical issues can lead to in-flight emergencies. A complacent mindset can delay a pilot's reaction to unexpected events, which is critical in the high-stakes environment of an air show. Each act of negligence, no matter how small, significantly increases the risk of incidents during performances, where there is little margin for error.

Combating Complacency in Aerial Displays

To combat complacency, air show performers need targeted strategies that maintain their edge and ensure operational safety. Regular simulator training to rehearse emergency scenarios and rare failures helps keep pilots sharp and ready for any eventuality (Karachalios & Adjekum, 2023). Routine peer reviews and performance audits by experienced pilots or coaches can identify and correct complacency before it impacts a show, breaking any potential ego-inflated mindset that could lead to unnecessary risks. Also, occasionally introducing new maneuvers can help maintain a pilot's focus and engagement.

What is also critical to be encouraged is continuous self-reflection and mindfulness, with pilots regularly assessing their own attitudes toward risk and safety (Karachalios, 2022; Karachalios & Adjekum, 2023). The use of onboard cameras in the cockpit and automated assessment of aerobatic maneuvers (David et al., 2022) can be essential for this purpose, allowing for thorough post-flight analysis and fostering a culture of mindfulness about performance and decision making processes (Hyten & Ludwig, 2017).

Crowd-Pleasing Mentality

The *crowd-pleasing* mentality in air show performers (Fusco, 2018) is an insidious threat that subtly undermines their decision making and safety awareness. This issue extends beyond simple performance errors, touching deeply into the philosophy of display flying training and the very essence of pilot judgment during high-pressure public events. Ricardo Traven (Barker, 2020), a highly experienced demonstration pilot with a background in both fighter and test piloting, offers profound insights into how conventional training methodologies for air shows can alter a pilot's ingrained habits of safe and professional flight conduct.

Traven argues that traditional air show training, where pilots repeatedly practice the same routine at decreasing altitudes until the final show altitude is achieved, fundamentally changes the way pilots think and make decisions under pressure. This type of training emphasizes repetition to the point of perfection, much like training a pet to perform a trick without understanding the broader implications or potential dangers of the action. The analogy Traven uses is particularly telling, a trained dog will chase a ball across a busy street if commanded to fetch, regardless of the risk. Similarly, he suggests that air show pilots, through their repetitive training, may develop a "*fetch*" mentality—focused solely on executing the routine to please the crowd, potentially at the expense of their own safety.

This "*go-fetch*" syndrome, as Traven describes it, means that during an air show, a pilot might undertake maneuvers that they would normally consider too risky or unsafe under other circumstances. He shares a personal anecdote to illustrate this point: *he once proceeded with a show despite knowing the weather conditions were below acceptable limits, driven by the strong desire to please the crowd—a decision he later recognized as compromised judgment.*

The critical issue here is that the intense focus on pleasing the audience can overshadow a pilot's professional judgment. This shift is particularly dangerous in air show environments where the margin for error is slim, and the consequences of a lapse in judgment can be fatal. Traven's observations suggest that even the most experienced and professional pilots are not immune to this shift in mindset, and it has indeed been cited in several accident investigations where pilots' actions were otherwise inexplicable.

Consequently, the "*crowd-pleasing*" mentality observed in air show performers highlights a critical need to revamp training methods, shifting from purely repetitive mechanical execution to a more comprehensive training model that prioritizes situational awareness, risk assessment, and decision making. This adjustment is essential to prepare pilots not only for flawless performances but also for making crucial safety decisions under pressure.

Given that the *crowd-pleasing* mentality, also identified as *social facilitation bias*, was discussed extensively in the first book of this series (Karachalios & Adjekum, 2023), it is crucial to further explore its implications in the context of air show pilot training. This exploration is vital for enhancing the understanding of how pilots can be influenced by audience presence, leading to potential risks, if not responsibly managed.

Social facilitation bias—the tendency for individuals to perform differently when observed by others (Bond & Titus, 1983)—can significantly influence pilot behavior during air shows. For pilots, the presence of an enthusiastic crowd can enhance performance on routine or well-rehearsed tasks but can also heighten the risk of errors in more complex maneuvers due to increased pressure to impress (Karachalios, 2022).

Training Implications

To mitigate the risks associated with social facilitation bias, pilot training programs should adopt a multifaceted approach that balances the thrill of performance with the paramountcy of safety. Training should incorporate psychological resilience techniques to help pilots manage the stress and excitement of performing in public. Simulated stress-inducing scenarios can condition pilots to maintain composure under the watchful eyes of an audience. Additionally, utilizing cognitive-behavioral strategies to help pilots recognize and control the influence of the audience is crucial. By training pilots to actively acknowledge and manage this bias during flights, they are encouraged to continuously reevaluate risks and decision making processes. Frequent reviews and feedback, highlighted in the first book of this series (Karachalios & Adjekum, 2023) as essential for continuous improvement, can counteract the effects of social facilitation bias. Lastly, emphasizing decision making and safety during the debriefing reinforces the importance of prioritizing these over *crowd-pleasing* actions.

Expanded Training with Decision Making Gates

Training programs should be designed to include *decision making gates*, which are predefined criteria that guide display pilots to make clear decisions, such as whether to continue with a planned maneuver or to abort based on current conditions. These *decision making gates* ensure that every action taken by a pilot during an air show is both deliberate and safe. Enhanced scenario-based simulations can integrate specific decision points, allowing pilots to practice responding to variable and unexpected situations. For example, simulations could introduce sudden weather changes, engine malfunctions, or airspace-related unexpected events. Each scenario would have built-in decision gates like cloud ceiling limits or wind speed thresholds, which require a *go/no-go decision*.[6]

Explicit training on decision making while being observed can help pilots make safer choices. By simulating audience presence during training and providing real-time feedback, pilots can grow accustomed to the demands of performance and safety. During these training sessions, pilots must assess the situation in real-time and make decisions based on safety protocols. This could mean choosing to abort a flight if the wind speeds exceed safe limits or adjusting a flight path in response to unforeseen mechanical warnings.

Beyond Crowd-Pleasing Mentality: Entertainer Mindset

The difference between a *crowd-pleaser mentality* and an *entertainer mindset* is pivotal for the high-risk environments of air shows. Both aim to captivate audiences, yet they differ in their approach to risk, decision making, and prioritization of safety, as illustrated in Table 7.2.

TABLE 7.2
Comparison of Crowd-Pleaser Mentality and an Entertainer Mindset

Category	Crowd-Pleaser Mentality	Entertainer Mindset
Focus	Primarily, it is about exceeding audience expectations, often at any cost.	On engaging the audience within the bounds of safety and professionalism.
Risk Management	Often takes unnecessary risks to impress the audience.	Manages risks carefully, ensuring all safety measures are met.
Decision Making	Influenced by immediate audience reactions, may prioritize spectacle over safety.	Decisions are well-rounded, considering audience engagement, personal capability, and situational awareness.
Performance Approach	May perform beyond safe operational parameters to deliver the expected thrill.	Prepares extensively, ensuring performances are exciting yet safe.
Long-term Impact	It can lead to short-term audience approval but at higher risks, potentially hazardous.	Builds a reputation for reliability and safety, leading to sustained success and trust.
Safety Prioritization	Lower, as the thrill and approval are often placed above safety.	High, with a strong commitment to maintaining safety standards.

The *crowd-pleaser mentality*, often driven by *social facilitation bias*, focuses primarily on meeting or exceeding audience expectations, sometimes at any cost. This mentality is marked by a powerful desire to impress and gain approval, which can lead to taking *unnecessary* risks or irresponsibly pushing boundaries.

In contrast, the *entertainer mindset* is about delivering a performance that is awing the crowds but within the confines of safety and professionalism. Performers with this mindset aim to please the audience without compromising safety or professional standards. They *balance* the need to entertain with a strong commitment to excellence, safety, and ethical practices.

While both approaches seek to captivate and satisfy spectators, the *entertainer mindset* emphasizes *sustainability* and *safety*. The *crowd-pleaser mentality* may lead to *short-term* gains in audience approval but carries a higher risk. In contrast, the *entertainer mindset* builds a reputation for trustworthiness and safety, leading to *longer-term* success.

"Perfection-itis" Syndrome in Air Show Pilots: A Double-Edged Sword

The drive for perfection is a common trait among pilots in the air show community. This striving for excellence is often celebrated as a hallmark of professionalism and skill. However, when the pursuit of a flawless performance becomes *obsessive*, it transforms into what can be termed, similarly to an illness, "*perfection-itis*[7]" syndrome. This condition represents a significant psychological challenge for display pilots, with both practical and safety implications.

Nature of Perfection-itis

Perfection-itis in air show pilots is characterized by an excessive fixation on achieving an absolutely flawless routine. While the pursuit of perfection might initially seem like a positive attribute, it becomes problematic when display pilots are unwilling to tolerate even minor deviations that fall within acceptable safety margins. This mindset can compel pilots to make unnecessary and sometimes hazardous corrections during their performance.

Implications for Performance

The primary concern with *perfection-itis* is the increased risk it poses. Pilots stricken with this *syndrome* might react to minor errors or deviations by making dramatic corrections that are disproportionate to the issue. For instance, a slight misalignment in formation or a minor timing issue could prompt a pilot to adjust their speed or trajectory more aggressively than needed, potentially leading to unstable flight conditions.

In their quest for perfection, air show performers may prioritize the aesthetic or technical aspects of a maneuver over safety considerations. This can lead them to perform closer to the ground, execute tighter turns, or push their aircraft beyond recommended operational limits.

Hardy (1992) discusses the impact of psychological stress on performance and injury in sports, noting how stress can compromise both physical and cognitive abilities, thereby increasing the risk of errors and accidents. Similarly, Allen et al. (2013) found that emotions correlated with perceived mental effort and concentration disruption in adult sports performers, suggesting that the mental effort to maintain perfection can interfere with performance.

Habay et al. (2021) conducted a systematic review on mental fatigue and sport-specific psychomotor performance, concluding that mental fatigue impairs performance, which is crucial for pilots who require peak cognitive and physical functioning. Yoshie et al. (2009) explored stress responses in competitive settings, highlighting how stress affects motor and autonomic functions, which are critical for precise maneuvers in aviation.

Furthermore, Kenny et al. (2004) examined music performance anxiety and occupational stress among opera chorus artists, finding a relationship between perfectionism, anxiety, and performance. This research is relevant to display pilots, as similar mechanisms of stress and perfectionism can affect their performance.

Additionally, Mandrick et al. (2016) investigated neural and psychophysiological correlates of human performance under stress and high mental workload, indicating that stress impacts neural functioning and performance. Sandi (2013) also reviewed the effects of stress on cognition, emphasizing how stress can impair cognitive processes crucial for decision making and performance.

Marcora et al. (2009) demonstrated that mental fatigue impairs physical performance in humans, underscoring the importance of managing mental workload to maintain high levels of performance in demanding environments such as air shows.

These findings collectively highlight the risks associated with *perfection-itis* and the importance of balancing performance aspirations with safety considerations since the mental burden of striving for an unattainable ideal can lead to increased stress levels during performances. Such heightened stress can ultimately impair cognitive function, leading to decision fatigue, slower reaction times, and a decreased ability to manage unexpected events effectively.

Strategies for Management

Addressing *perfection-itis* in display pilots requires several strategies to adjust their approach to performance and safety. Setting realistic and attainable performance goals can help reduce the pressure to achieve perfection, especially considering the demanding low-level air show environment. Training should focus on methods to enhance pilots' resilience during variations in display flying due to aircraft performance deviations, weather changes, and potential unplanned external factors that can distract pilots.

A balanced approach that prioritizes safety over flawlessness is essential. Therefore, it is crucial to plan and prepare a display profile that includes *energy gates*,[8] which incorporate airspeed, altitude, and G-force *buffers*, as opposed to *energy points*,[9] which specify exact speeds, altitudes, and G-forces (see Figure 7.2).

Extremely valuable can be the incorporation of constructive debriefing sessions after performances, which can provide display pilots with a balanced assessment of their performance and highlight both strong and areas for improvement. It is crucial that feedback emphasizes the importance of maintaining and accepting *energy* or *position margins* in case of formation flying and explains the risks associated with unnecessary corrections.

In their study, Driskell et al. (1994) suggested that mental practice significantly enhances performance across various tasks, a finding highly applicable to display flying as a tool to mitigate *perfection-itis* by allowing pilots to accept and manage imperfections, fostering a balanced approach that prioritizes safety. Integrating

Energy Gate	**Energy Point**
• Margin for Altitude Deviation • Margin for Airspeed Deviation • Margin for Gs Deviation • Margin for Power Deviation	• No Margin for Altitude Deviation • No Margin for Airspeed Deviation • No Margin for Gs Deviation • No Margin for Power Deviation
Energy Gate provides safety margins for critical parameters.	*Energy Point allows no deviation, indicating a critical state.*

FIGURE 7.2 Energy gates versus energy points during display flying.

mental practice with constructive feedback during debriefings can further enhance continuous improvement and performance quality.

While the pursuit of perfection is admirable and inspiring in air show pilots, it is important to recognize when this pursuit becomes detrimental. Unmanaged perfectionism, referred to as *perfection-itis*, can compromise both safety and performance quality; therefore, it is crucial to recognize and manage this tendency for optimal display flying.

Human Performance Limitations

Understanding human performance limitations is vital for air show performers. Their routines are highly demanding and stressful, placing significant strain on their bodies both physically and mentally. This section of the book looks into a range of factors that can affect performance, including blood sugar levels, oxygen saturation (SpO2), thermal stress, hormonal balances, and other biochemical influences.

Blood Sugar Levels

The International Council of Air Shows (ICAS) *Air Show Performers Safety Manual* (2016) offers a thorough guide on how blood sugar levels affect a display pilot's performance. Managing blood sugar is essential for withstanding the G-forces encountered during high-G maneuvers, especially when facing low blood sugar situations or hypoglycemia,[10] which can reduce a pilot's G-tolerance, increasing the risk of blackouts or loss of consciousness during flight. This is particularly important for display pilots who are exposed to sustained extreme G-forces, both positive and negative.

The ICAS manual emphasizes that keeping blood sugar levels stable is crucial for physical performance and cognitive function. Low blood sugar can cause dizziness, weakness, and impaired thinking (Malouf & Brust, 1985), all of which can jeopardize acute decision making, performance, and ultimately flight safety. Additionally, the manual advises against using sugary soft drinks to quickly boost blood sugar levels since these drinks can raise blood sugar rapidly and also result in lower alertness and increased fatigue, also known as a "*sugar crash*" (Mantantzis et al., 2019).

Instead, it is recommended that display pilots eat regular, balanced meals during the air show weekend to keep blood sugar levels steady. A diet that includes complex carbohydrates, proteins, and healthy fats provides a consistent energy supply, crucial for regulating blood sugar and ensuring sustained energy without the drawbacks associated with sugary foods and drinks. For example, whole grains, non-fatty meats, nuts, and vegetables can help maintain stable blood sugar levels and offer sustained energy. Research by Kerksick et al. (2018) highlights the importance of balanced nutrition for athletic performance, noting that regular intake of nutrient-rich foods helps maintain blood sugar levels and supports overall health. Similarly, Clark and Mach (2016) found that athletes who consume balanced meals with sufficient protein and complex carbohydrates have better endurance and performance stability.

Additionally, fluctuations in blood sugar can impair cognitive performance, affecting attention, decision making, and reaction time, which are crucial for pilots who need to stay highly aware and make quick decisions (Lieberman, 2003). Eating regular, balanced meals helps prevent the cognitive impairments associated with low blood sugar, ensuring that display pilots can stay alert and focused during their performances.

Furthermore, studies have demonstrated the effects of blood sugar on pilot performance. Johnson et al. (2007) found that low blood glucose negatively impacts pilot performance, underscoring the need for proper nutritional strategies. Research by Kirwan et al. (2001) and Moitzi and König (2023) on the dietary habits of pilots supports the recommendation of moderate and high glycemic index[11] meals to optimize metabolism and exercise performance. Additionally, Gannon and Nuttall (2004) highlighted the benefits of high-protein, low-carbohydrate diets in maintaining blood glucose control, which can be particularly advantageous for managing energy levels and cognitive function during demanding air show profiles.

The Impact of Low Levels of Oxygen Saturation (SpO2) in Display Pilots' Performance

SpO2 measures the percentage of hemoglobin in the blood saturated with oxygen (Hafen & Sharma, 2018). Maintaining adequate SpO2 levels is critical for optimal brain function, with recommended levels for pilots generally above 92% to ensure safe performance (Humphreys et al., 2005). Levels below this threshold can impair cognitive functions, and when SpO2 levels drop to around 85% or lower, pilots are at significant risk. Immediate corrective action, such as using in-flight supplemental oxygen, is recommended to avoid serious impairments or health risks (Post et al., 2002; Kourtidou-Padadelis et al., 2008; Edvardsen et al., 2012; Johannigman et al., 2015). Hence, low SpO2 levels, referred to as hypoxemia (Dempsey & Wagner, 1999; Rodríguez-Roisin & Roca, 2005; Sarkar et al., 2017), can significantly impact the cognitive performance of air show performers.

Given that display flying primarily occurs at low altitudes, where atmospheric oxygen levels are typically sufficient, SpO2 levels become a pragmatic and immediate concern for air show performers compared to hypoxia,[12] which is a significant risk at high altitudes (Shaw et al., 2021). Consequently, maintaining optimal SpO2 is crucial for ensuring peak mental and physical performance (Besch et al., 1994; Samuel & Franklin, 2008).

Memory functions are also adversely affected by low oxygen levels. A study by Post et al. (2002) indicates that hypoxemia can impair both short-term and long-term memory. Pilots experiencing reduced SpO2 may struggle to recall important procedures or sequence details during aerobatic routines, leading to potential errors and highlighting the risks for air show performers who depend on sharp memory for their display routines.

Regular monitoring of SpO2 levels is essential for assessing a pilot's readiness and overall health, particularly when they are required to perform multiple display flights, such as during training sessions. Keeping track of SpO2 levels helps ensure that pilots are physically prepared for the demands of their performances, thereby reducing the risk of unexpected cognitive and physical impairments during flight (Dempsey & Wagner, 1999). Santos et al. (2022) suggest that wearable pulse oximeters can detect hypoxemia with accuracy degraded by motion. Therefore, display pilots can continuously monitor their blood SpO2 with the use of such devices, ensuring it stays within the recommended range, typically above 92%.

If SpO2 levels begin to drop, pilots must take proactive measures to restore adequate oxygenation. For instance, pilots can use pure oxygen or temporarily reduce physical exertion to restore adequate oxygen levels (Kourtidou-Papadeli et al., 2008). In scenarios where pilots are required to perform display flights or rehearsals after cross-country flights, maintaining optimal SpO2 levels becomes even more critical. The physical exertion and stress associated with long flights can deplete energy reserves and lower SpO2 levels. Regular breaks between sorties, during which pilots can rest and rehydrate, are essential for maintaining adequate oxygenation and overall physiological function. Proper hydration is vital as it supports optimal blood volume and circulation, which in turn aids oxygen delivery to tissues (Lieberman, 2003).

Thermal Stress Effects

Thermal stress[13] significantly impacts human performance (Driskell & Salas, 1996), especially for display pilots who often operate in extreme environmental conditions. This stress can arise from both high and low temperatures, affecting physiological and cognitive functions crucial for safe and effective performance (Enander & Hygge, 1990).

Several studies have explored the effects of thermal stress on human performance. For example, Hancock and Vasmatzidis (2003) discuss the relationship between temperature and human performance, highlighting the detrimental effects of both heat and cold stress on cognitive and physical functions. Similarly, Parsons (1999) provides a comprehensive review of human thermal environments and their impact on performance, emphasizing the importance of managing thermal stress in occupational settings. Additionally, Pilcher, Nadler, and Busch (2002) examined how heat stress affects cognitive performance, finding significant decrements in memory and attention tasks under high-temperature conditions.

When pilots are exposed to hot temperatures, their core body temperature can rise, leading to hyperthermia.[14] Hyperthermia affects cardiovascular function by reducing blood flow to the brain and muscles, which impairs physical performance and increases fatigue (Hancock & Vasmatzidis, 2003). Cognitive functions such as attention, memory, and decision making are also adversely affected by hyperthermia, which is particularly concerning for display pilots who require elevated levels of concentration and quick decision making abilities (Parsons, 1999). For example, during the summer of 2022 at the Farnborough Airshow, pilots faced extreme heat conditions with temperatures reaching up to 40 degrees Celsius (100 degrees Fahrenheit), significantly impacting their performance during their flying displays (Boatman, 2022). Additionally, the use of protective gear, such as parachutes and G-suits, can exacerbate heat stress by trapping heat and moisture, further increasing body temperature and discomfort (Hancock & Vasmatzidis, 2003).

Conversely, exposure to cold temperatures can lead to hypothermia,[15] where the body's core temperature drops below normal. Hypothermia affects neuromuscular functions, leading to decreased manual dexterity, slower reaction times, and impaired coordination (Parsons, 1999). Cheung and Sleivert (2004) explored the physiological responses to cold exposure, noting impairments in neuromuscular function and coordination. These effects are critical for display pilots who rely on precise control inputs and rapid responses to dynamic flight conditions. Cognitive functions are also impaired in wintry conditions, with reduced mental agility and increased susceptibility to errors (Taylor et al., 2016; Falla et al., 2021). Usually, the preparation period for display pilots in the Northern Hemisphere takes place during the winter and spring seasons, with weather conditions that could be suboptimal and characterized mainly by low temperatures. In these conditions, pilots could experience reduced manual dexterity due to numbness in their fingers, slower reaction times, and increased stiffness in their movements, making precise control of the aircraft more challenging and posing a significant risk during high-speed, low-altitude maneuvers.

To mitigate the effects of thermal stress on display pilots' performance, numerous strategies can be applied. First, proper hydration and nutrition are essential to maintain physiological balance. Ensuring that pilots drink adequate water and consume balanced meals helps to regulate their body temperature and energy levels, especially during physically demanding performances. For instance, drinking electrolyte solutions can prevent dehydration, while eating meals rich in proteins and complex carbohydrates can provide sustained energy.

Next, wearing temperature-appropriate clothing and using cockpit environmental controls can help regulate body temperature. Pilots can wear moisture-wicking fabrics to stay cool in hot conditions or thermal layers to retain heat in colder environments. Additionally, using cockpit ventilation systems or personal cooling devices can further assist in maintaining a comfortable temperature. For example, pilots might use portable fans or cooling vests in hot climates to prevent overheating.

Lastly, training programs that simulate extreme temperature conditions can also prepare pilots for the upcoming display season. By practicing in controlled environments that mimic the heat or cold they may encounter, pilots can acclimatize their bodies and develop strategies to cope with thermal stress. This might include training in high-heat chambers or cold-weather gear exercises, enabling them to build tolerance and improve performance under various thermal conditions.

Impact of Testosterone Levels

Testosterone, a hormone primarily produced in men and in smaller amounts in women, plays a significant role in regulating various physiological and psychological processes (Gouchie & Kimura, 1991; Leedy & Wilson, 1985; Riachy et al., 2020). It impacts human performance in numerous ways, influencing physical capabilities, cognitive functions, mood, and behavior (Archer, 2006; Zitzmann, 2020). For air show performers, optimizing testosterone levels can enhance various aspects of their performance, from physical competence to mental acuity.

Barker (2020) identified the effect of testosterone and other biochemicals in the human body on the performance of display pilots. His research highlight how hormonal balances, particularly levels of testosterone, significantly impact both the physical and cognitive aspects of pilot performance. Testosterone, in particular, was shown to influence muscle mass, strength, and endurance, which are critical for enduring the physical demands of aerobatic maneuvers. Moreover, Barker's findings indicate that testosterone levels affect cognitive functions such as memory, attention, and decision making, all of which are crucial for executing precise and complex flight routines.

In addition to testosterone, Barker's study examines the roles of other biochemicals, such as cortisol and adrenaline. Cortisol, often referred to as the stress hormone, was found to have a complex relationship with testosterone (Hloogeveen & Zonderland, 1996; Leedy & Wilson, 1985). Elevated cortisol levels, often resulting from chronic stress, can suppress testosterone production, leading to reduced physical performance and cognitive function. Conversely, balanced cortisol levels can help maintain testosterone levels and support optimal performance. Adrenaline, another key chemical, was identified as a critical factor in enhancing alertness and reaction times during high-stress situations, although its effects are transient and must be managed to prevent long-term stress-related issues (Krahenbuhl, 1975).

Conceptual Model: Resilient Performance Model (RPM)

In aviation, the angle of attack (AoA) is a critical factor in determining lift, with the critical AoA, beyond which lift dramatically decreases resulting in a stall (Anderson, 2011). Analogously, an air show performer's experienced stress can either enhance or degrade their performance. Just as exceeding the critical AoA can lead to a loss of lift, surpassing a critical stress level can lead to a sharp decline in performance, risking mental and physical breakdowns. Additionally, the duration of exposure to stress plays a significant role, as prolonged stress can further exacerbate performance degradation. Understanding these parallels, the RPM is conceptualized and aims to highlight the need for air show performers to maintain superior performance levels while effectively managing and reducing stress over time.

Theoretical Background

To fully understand the RPM, it is essential to provide a theoretical background that encompasses both fundamental aerodynamics and human performance principles related to stress.

Angle of Attack (AoA) and Lift

The AoA is a fundamental concept in aerodynamics and is crucial for understanding how lift is generated by an airfoil. The lift force (L) produced by an airfoil can be described by the lift equation:

$$L = C_L \cdot \rho\, v^2 S,$$

where L is the lift force, C_L is the coefficient of lift (which varies with the AoA), ρ is the air density, v is the velocity of the air relative to the airfoil, and S is the wing area.

The coefficient of lift (C_L) is a dimensionless number that represents the lift characteristics of an airfoil at a given AoA. It typically increases with the AoA up to a certain point, known as the critical AoA (Figure 7.3). The critical AoA is the point at which the C_L reaches its maximum value. Beyond this point, any further increase

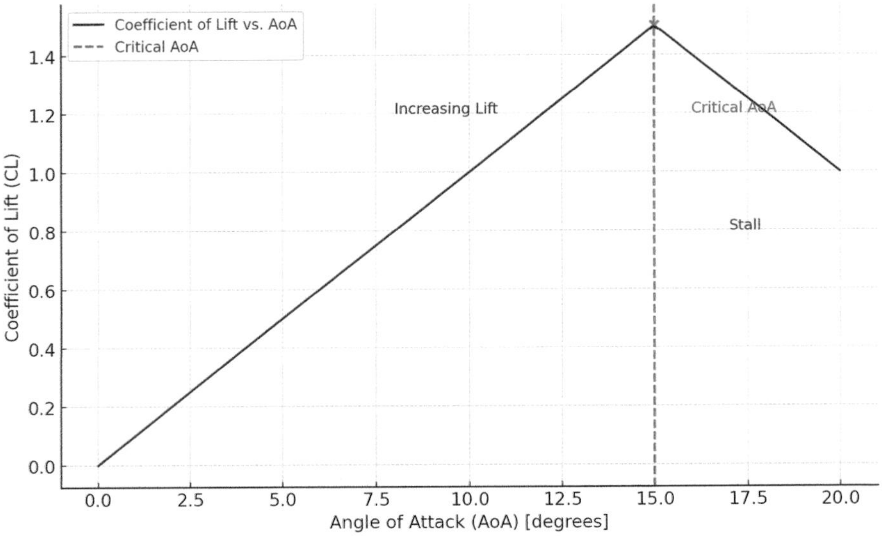

FIGURE 7.3 Example of the relationship between lift and Angle of Attack.

in AoA leads to a rapid decrease in lift, causing the airfoil to stall. This phenomenon occurs due to the airflow separating from the upper surface of the wing, resulting in a drastic loss of lift (Anderson, 2011).

Figure 7.3 illustrates an example of the relationship between the AoA and the C_L. The curve shows how the C_L increases with the AoA up to the critical AoA, marked by the dashed line. At the critical AoA of 15 degrees, the C_L reaches its maximum value of 1.5. Beyond this point, further increases in AoA lead to a decrease in lift, indicating a stall.

Stress and Performance

An effort to capture the relationship between stress and performance is encapsulated by the Yerkes-Dodson Law (Yerkes & Dodson, 1908), which posits that performance increases with stress to an optimal point, after which it declines. According to Yerkes and Dodson, there is an inverted U-shaped relationship between stress and performance (Figure 7.4), suggesting that moderate levels of stress can enhance performance. However, excessively high stress levels can overwhelm an individual's capacity to cope, leading to anxiety, errors, and a decline in performance.

The visual representation in Figure 7.4 illustrates the Yerkes-Dodson Law, depicting the relationship between stress levels and performance. The inverted U-shaped curve demonstrates that performance increases with stress up to an optimal point, beyond which performance declines or might be maintained for a brief period of time, depending on the individual and the source of stress. The dashed line marks the critical stress level where stress exceeds the optimal point, leading to performance degradation.

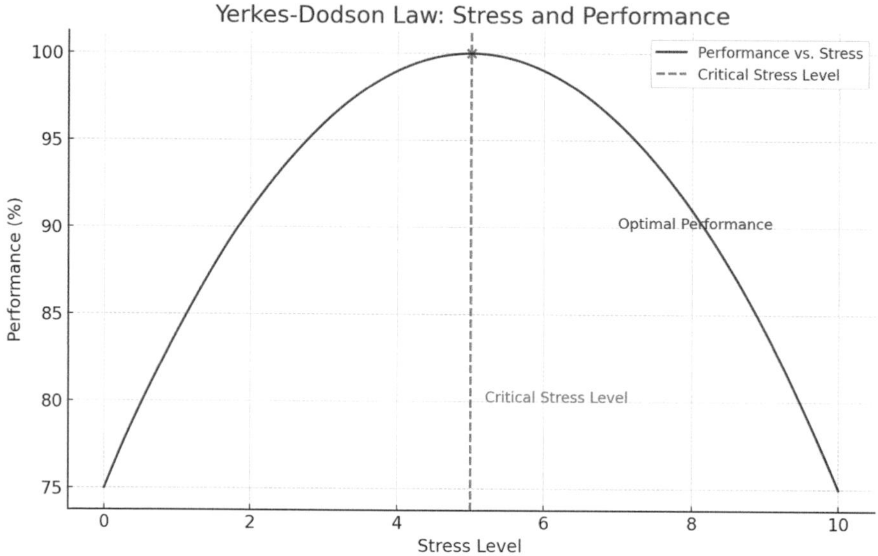

FIGURE 7.4 Yerkes-Dodson law representation.

When stress exceeds the optimal level, it reaches a critical stress level where performance degradation occurs. Beyond this stress level, individuals experience diminished cognitive processing capacity, impaired decision making, and an increased likelihood of mistakes. This critical threshold varies among individuals and can be influenced by factors such as experience and training (Teigen, 1994).

Prolonged exposure to high stress aggravates performance decline over time, while chronic stress can lead to burnout, fatigue, and various mental health issues, fundamentally undermining an individual's capacity to perform at optimal levels. The physiological response to prolonged stress involves the release of cortisol and other stress hormones, which can have deleterious effects on cognitive function, physical health, and overall performance (McEwen & Stellar, 1993). In addition, LeBlanc (2009) emphasizes that acute stress can disrupt cognitive processes essential for performance, particularly in high-stakes environments such as health professions, where quick thinking and precise actions are crucial. Furthermore, Conway et al. (2013) provide a mechanistic account of how stress-induced performance degradation occurs, illustrating the intricate interplay between stress, cognitive load, and performance outcomes. They highlight the necessity of understanding these mechanisms to develop effective interventions that mitigate the adverse effects of stress on performance.

Table 7.3 provides a comparative analysis of aircraft and human performance based on the AoA–lift and stress–performance relationships.

The RPM: Balancing Performance and Stress over Time

By drawing parallels between the AoA–lift relationship in aerodynamics and the stress–performance relationship in human factors, we can better understand and

TABLE 7.3
Comparison of Aircraft and Human Performance

Parameter	Aerodynamics (AoA and Lift)	Human Performance (Stress and Performance)
Optimal Range	Moderate AoA for maximum lift	Moderate stress for optimal performance.
Critical Threshold	Critical AoA leading to stall	Critical stress level leading to performance drop.
Performance Degradation	Post-stall: loss of lift	Over-stressed: mental/physical breakdown.
Time Factor	Not available	Extended time under high stress exacerbates the decline.

manage the performance of air show performers. Just as exceeding the critical AoA leads to a stall, surpassing the critical stress level leads to a significant drop in performance, especially over prolonged periods, necessitating effective stress and time management strategies.

Since display pilots need to achieve peak performance at all times, it is essential to accustom themselves to stress and manage to lower its effects on their performance. It is even better to start performing at higher stress levels within a certain time frame within a controlled training environment after appropriate training.

With the RPM, a framework is suggested that expands on the Yerkes-Dodson Law to include the *time factor*, highlighting the need for air show performers to *maintain* superior performance levels while effectively managing and *reducing* stress over time. This model emphasizes the importance of resilience, stress management techniques, and continuous improvement to sustain peak performance without the detrimental effects of prolonged high stress.

Achieving optimal performance involves maintaining performers within an optimal stress zone to avoid reaching the *Critical-Stress Threshold* (*CST*) that leads to performance decline. Building resilience through physical conditioning, mental fortitude, and emotional stability is crucial. Regular training in high-pressure scenarios helps performers adapt to stress and maintain composure during actual performances. Implementing effective stress management techniques, such as mindfulness meditation, deep-breathing exercises, and visualization, ensures that air show performers are prepared and remain focused (Karachalios, 2022; Karachalios & Adjekum, 2023). These practices are essential for quickly returning to the optimal stress zone when stress levels begin to rise even as soon as they release the brakes for takeoff.

Additionally, continuous monitoring of stress levels and performance metrics allows for real-time adjustments. Utilizing wearable technology and biofeedback devices to track physiological indicators of stress (Hickey et al., 2021) provides immediate feedback to performers and their coaches.

Figure 7.5 illustrates the RPM, demonstrating how air show performers can maintain peak performance at 100% while managing and reducing their stress levels over time.

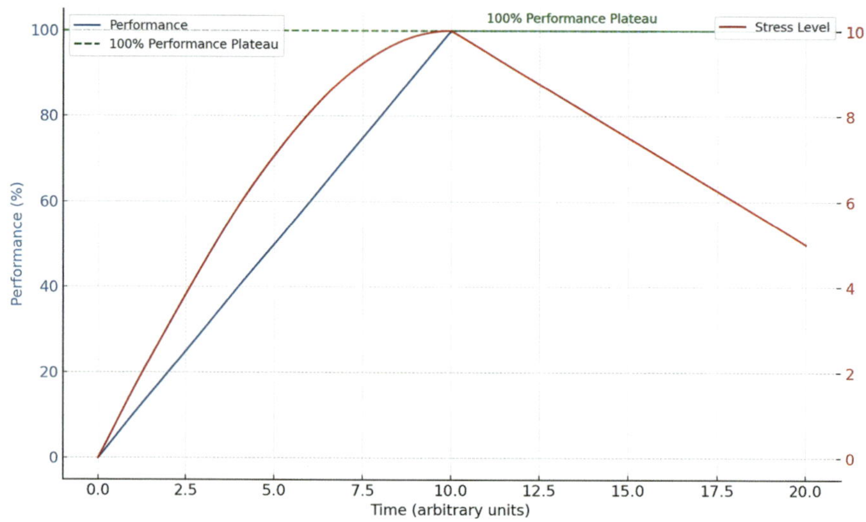

FIGURE 7.5 Resilient Performance Model (RPM) and the Γ-shaped performance curve.

The *performance curve*, depicted by the blue line, illustrates how performers can achieve and maintain 100% performance. Initially, as time progresses, performance steadily increases, reaching the peak performance level around the 10-unit mark. Once 100% performance is reached, it is maintained as a plateau throughout the remaining period, visually resembling the Greek letter gamma "Γ."

The *stress level curve*, shown by the red line, initially increases, peaking simultaneously with the performance curve when it reaches 100%. After hitting the peak stress level, stress begins to decrease slightly, even as performance is maintained at 100%, in a Γ-shape (*gamma*-shape). This reduction in stress can be attributed to the effective application of stress management techniques and resilience building.

Critical points and the plateau are highlighted in the graph. The critical performance point is indicated by the intersection of the blue and green lines at 100%, marking the maximum performance level. The dashed line represents the *100% performance plateau*, underscoring that performance can be sustained at its peak even while stress levels are effectively managed and reduced.

The Critical-Stress Threshold (CST)

The CST is a pivotal concept within the RPM that is used to understand and manage the stress–performance relationship among air show performers. Similar to the critical AoA in aerodynamics, where exceeding a specific angle results in a rapid decline in lift, the CST represents the point at which a performer's stress level exceeds their optimal functioning range, leading to significant performance degradation.

For air show performers, understanding the CST is crucial for several reasons. First, it allows for the identification of stress levels that can be beneficial versus those that are detrimental. Just as the AoA can be optimized for maximum lift without stalling by extending the flaps, stress can be managed to enhance performance without causing burnout or mental breakdowns. Ultimately, display pilots need to be aware of their comfort zones at all times.

Identifying the CST involves a combination of physiological, psychological, and performance metrics. These metrics include heart rate variability, cortisol levels, self-reported stress levels, and cognitive performance indicators. By monitoring these parameters, it is possible to determine when a performer is approaching or exceeding their CST (Holm et al., 1996; Crocker et al., 1988; Perna et al., 2003). Physiological indicators such as elevated heart rates, changes in heart rate variability, and increased cortisol levels can indicate rising stress levels. Wearable technology can continuously monitor these indicators, providing real-time data. Psychological indicators, including self-reported stress assessments, mood and anxiety scales, can help gauge a performer's mental state and identify trends, and potential issues, before they become critical. Furthermore, performance indicators such as cognitive tasks, reaction time tests, and performance under simulated high-stress conditions can provide insight into how stress impacts a performer's abilities (Hanton et al., 2004; Clark and Williamon, 2011).

Managing the CST

Once the CST is identified, several strategies can be employed to manage and increase this threshold, thereby enhancing performance and resilience.

Mental resilience training techniques, such as Cognitive Behavioral Therapy (CBT) (Turner & Napolitano, 2010) and Stress Inoculation Training (SIT) (Meichenbaum, 1985), can help performers reframe negative thoughts, build coping mechanisms and gradually increase their tolerance to stress (Saunders et al., 1996). Incorporating mindfulness practices such as mindfulness-based stress reduction (MBSR) (Darses et al., 2023; Meland et al., 2015) can help performers stay present, manage stress, and maintain focus. Overall, these practices can improve emotional regulation and reduce anxiety (Kabat-Zinn, 1990), pushing the CST further.

Additionally, maintaining high levels of physical fitness through aerobic and anaerobic conditioning can improve overall stress resilience. A well-conditioned body can better handle the physical demands of high-G maneuvers and the associated stress, thus supporting a higher CST (Crocker et al., 1988; Perna et al., 2003). Therefore, continuous monitoring of stress indicators and performance metrics is important for maintaining awareness of one's stress level compared to one's CST (Mesagno & Mullane-Grant, 2010).

Practical Implications

The RPM has several practical applications for enhancing safety and ensuring the top-performance of air show performers. Implementing the RPM involves key *resilient* strategies for optimizing performance while ensuring safety.

Designing well-thought display profiles is essential for balancing showmanship with safety. Firstly, by ensuring that the complexity of maneuvers matches the skill level of the performers which involves a gradual progression to more difficult routines as performers gain experience and manage the *flying-in-front-of-a-crowd* stress with confidence. Then, the display profile needs to be structured in a manner that there's an alternation of intensity in the maneuvers, having high-intensity maneuvers with less demanding ones to interchange, allowing performers to recover and maintain optimal performance levels. Balancing aerobatic maneuvers that involve high G-forces, rolls, and prolonged periods of high-G or negative-G environments with less physically and mentally demanding maneuvers helps prevent fatigue. Limiting the duration of high-G or negative-G maneuvers further prevents physical strain and mental fatigue.

Continuously gathering feedback from performers and using this information to adjust and refine display profiles is equally essential. This iterative process helps ensure that the profiles remain appealing to the crowd yet manageable for the display pilot. Therefore, incorporating safety margins in the design of display profiles is critical, including minimum altitudes, safe distances from the audience, and buffers for error recovery.

Furthermore, integrating clear and well-practiced emergency procedures into display profiles is crucial so that performers can be trained to execute these procedures effectively under stress to minimize risks during unexpected situations.

Understanding the CST is vital for training air show performers to ensure they maintain stress within optimal levels. Regularly assessing performers' stress levels through psychological evaluations and physiological monitoring helps identify when their stress is approaching critical levels, and this needs to be addressed. Display pilots under training might require a break for a day to reset their stress levels and assist their physiological and psychological adaptation to the newly exposed low-level aerobatic environment. Therefore, it is key to tailor training programs to the individual needs of performers, considering their unique stress responses. For this purpose, coping mechanisms can include training sessions that replicate high-stress scenarios in a controlled environment, such as a simulator or a two-seater aircraft at higher altitudes, allowing performers to practice stress management techniques. Part of the training to self-recognize one's CST is to incorporate a variety of stress-reduction techniques in ground training, such as deep-breathing exercises and visualization, which can help them quickly reduce stress levels before, during, and after their flying display performances.

Example of RPM Application

Consider an air show performer who undergoes rigorous training to build physical and mental resilience. Initially, as the performer starts their display training, stress levels increase due to the demanding nature of the activities. With consistent training, stress management techniques, and continuous monitoring, the performer gradually reaches peak performance.

Once the performer achieves 100% performance, they maintain this level through regular practice. Meanwhile, the stress levels, having peaked, begin to decrease due to the performer's improved ability to handle stress, relaxation techniques, and adequate recovery periods. This balance allows the performer to sustain resiliently an optimal performance without succumbing to the negative effects of stress.

The RPM consequently provides a comprehensive framework for air show performers to achieve and sustain peak performance while effectively managing and reducing their stress levels over time.

Case Study: Enhancing Air Show Performers' Mental and Physical Strength Training Inspired by Boxers

Mindfulness and presence, crucial for optimal performance under normal conditions (one G), are predominantly studied within controlled environments with minimal physical stress (Kabat-Zinn, 1990). This scenario applies well to air show performers before takeoff when they can manage external factors and maintain standard heart rate levels without extra stressors. However, once airborne, the dynamics shift significantly, making all the learned methods of mental preparedness quite irrelevant.

During performances, display pilots are exposed to a highly dynamic and physically demanding environment, performing continuous maneuvers such as rolls, turns, loops, and spins at low altitudes in front of an audience. These maneuvers significantly impact their physiological state, inducing extreme forces on the body, rapid head movements, and intense control manipulations under varying gravitational forces. Such conditions can escalate stress, lead to disorientation, and impair decision making capabilities due to the interplay of positive and negative G-forces, which influence cerebral blood flow, as discussed earlier, and balance mechanisms in the ears (Kumar et al., 2020). For instance, a fast jet solo display pilot performing a high-G turn may experience up to nine Gs for several seconds, which can lead to visual impairment (grayout), loss of consciousness (G-LOC), and significant cognitive load as they try to maintain situational awareness (Antunano, 2000).

Analogous Training Insights from Boxing

The training and coping strategies of heavyweight boxers provide valuable insights for developing effective programs for air show performers. Boxers endure significant physical impacts while needing to think strategically about their next moves, maintaining composure and offensive spirit even under physical distress (Wacquant, 2006). Their training emphasizes mental clarity and strategic thinking even while in a so-called *survival mode*. For example, during a match, a boxer must remain calm, anticipate the opponent's moves, and execute the trained techniques despite being under significant physical and psychological stress (Smith, 2006).

Like boxers, display pilots can benefit from enhanced mental resilience training, involving training sessions that replicate the physical and cognitive demands of aerobatic performances, teaching pilots to maintain composure and clarity in decision making under stress (Lazarus & Folkman, 1984).

Display pilots should also engage in physical conditioning similar to boxers, preparing to endure and adapt to extreme G-forces. Strength training, cardiovascular improvement, and overall endurance exercises will support their ability to handle the demanding needs of low-altitude aerobatic maneuvers (McGill, 2007). For example, air show performers could undertake resistance training to strengthen the muscles involved in the AGSM, which is critical for maintaining consciousness during high-G maneuvers (Green & Muir, 2002).

Cognitive flexibility is another key area. Boxers are trained to quickly switch between defensive and offensive strategies under duress (Ottoboni et al., 2015). Pilots can similarly be trained in cognitive flexibility, learning to adjust their display profile based on real-time feedback during their flights. This training could involve scenarios where pilots must rapidly alter flight paths or respond to unexpected environmental changes, thereby enhancing their ability to make swift, accurate decisions under pressure.

Lastly, adapting traditional mindfulness practices to high-stress environments can help pilots maintain focus and awareness during physically disorienting performances, a technique that has shown effectiveness in other high-performance domains (Grossman et al., 2004). Techniques such as mindfulness meditation, controlled breathing, and body scanning can be incorporated into pilots' training regimens to enhance their ability to remain calm and focused during the air show weekend.

Case Study: Effect of COVID-19 on Air Show Performers Currency and Proficiency

The COVID-19 pandemic (World Health Organization, 2025) had a profound and multifaceted impact on the aviation industry, affecting everything from commercial flights to the niche sector of air show performers. The global response to the pandemic resulted in unprecedented travel restrictions, causing a significant reduction in air transport mobility, which had far-reaching consequences on the industry's operational and economic stability.

Dube et al. (2021) discuss the overall prospects for recovery in the aviation industry, highlighting how the drastic reduction in flights led to substantial financial losses and operational challenges. Their study points out that the recovery process will be gradual and requires robust strategies to address the long-term impacts of the pandemic. Similarly, Nižetić (2020) provides a detailed analysis of how the pandemic influenced air transport mobility, energy consumption, and environmental aspects. While there was a temporary reduction in environmental pollutants due to decreased air travel, the overall impact on the industry was negative, with significant disruptions to operations and energy management.

Focusing on air passenger travel, Li and Wang (2024) present empirical findings that indicate a dramatic decline in passenger numbers and a shift in travel behaviors, with increased apprehensions about flying. This shift not only affected commercial airlines but also had a ripple effect on air show performers who rely on the broader aviation infrastructure for their events. In parallel, Paraschi et al. (2024) examine the crisis management strategies employed by Greek airlines, emphasizing the importance of adaptive measures to cope with the sudden downturn. Airlines had to innovate and implement new safety protocols to regain passenger trust and ensure operational viability, a scenario mirrored in the air show industry where performers had to adapt to new safety and health guidelines.

The psychological impact of the pandemic on individuals within the aviation sector was significant. Khudaykulov et al. (2024) explore the effects of job insecurity and fear of COVID-19 on mental health in China, noting increased levels of depression and anxiety. These findings are relevant to air show performers who faced similar uncertainties and stress due to the cancellation of events and loss of income. In addition, Kubatko et al. (2023) examine the pandemic's impact on global food system resilience, drawing parallels to how the aviation sector, including air shows, had to develop resilience strategies to cope with the disruptions.

Similarly, Martin et al. (2023) investigate the impact of the pandemic on burnout and stress among US nurses, which emphasizes the broader implications of prolonged stress and workload on performance and mental health. This comparison is pertinent as it reflects the experiences of air show performers who also faced heightened stress due to the pandemic's disruptions.

Overall, the COVID-19 pandemic disrupted the aviation industry in numerous ways, affecting both commercial operations and specialized sectors like air shows. The industry had to adopt innovative solutions and adaptive strategies to address the immediate challenges and pave the way for recovery (Karachalios, 2022). The experiences of air show performers highlight the need for resilience, mental health support, and continuous adaptation to ensure safety and proficiency in a post-pandemic world.

Impact of COVID-19 on Air Show Performers

The cancellation of air shows and restrictions on gatherings during the COVID-19 pandemic (AirshowStuff, 2020) resulted in significantly fewer opportunities for display pilots to engage in regular flying practice and event participation. This disruption posed a serious threat of skill decay (Olaganathan & Amihan, 2021), where critical flying abilities could diminish without the consistent reinforcement typically provided by routine performances and training sessions. Air show performers depend on regular practice to maintain their proficiency, G-tolerance, and reaction times, all of which are essential for executing advanced aerobatic maneuvers safely. Without these opportunities, there is a heightened risk of pilots losing their sharpness and precision, making it more challenging to perform at high levels when air shows eventually resume.

Additionally, lockdowns and social distancing measures imposed further constraints by limiting access to training facilities. Training centers, flight schools, and simulation facilities often had to close or operate at reduced capacity to comply with health guidelines, significantly curtailing pilots' access to essential resources. This restricted availability made it challenging for pilots to participate in simulator sessions and other forms of training (Butuk & Škvareková, 2022).

Mitigation Measures to Maintain Currency and Proficiency

To combat *skill atrophy* during periods of limited flying opportunities, such as those experienced during the COVID-19 pandemic, display pilots and their organizations implemented several strategies that demonstrated the air show community's robust resilient-culture (Karachalios, 2022):

Many organizations transitioned to virtual platforms to continue training and development. Webinars, virtual workshops, and online courses helped pilots to engage with theoretical aspects of their craft, from aerodynamics and new maneuver techniques to safety protocols. The ICAS (ICAS, n.d.) and the European Airshow Council (EAC, 2021) hosted a series of webinars focusing on topics ranging from emergency maneuver training to updates in aviation regulations. These sessions helped pilots stay informed and engaged with the latest developments in aerial display techniques.

Where possible, pilots increased their use of flight simulators to maintain proficiency. Simulators provided a safe, socially distanced way to hone skills and practice routines without needing to leave the ground or violate health guidelines.

Regulatory bodies in some regions temporarily adjusted the standards for currency due to the impact of the pandemic (FAA, 2020). These adjustments were designed to provide pilots with more leeway in meeting currency requirements without compromising overall safety.

The air show community rallied together, sharing resources and best practices through online platforms and social media (EAC, 2021). Veteran pilots and trainers often provided mentorship through virtual meetings, helping less experienced pilots navigate the challenges posed by the pandemic.

In general, the COVID-19 pandemic tested the resilience and adaptability of the air show community. Nevertheless, by leveraging technology, i.e., video conferencing and social media, adjusting regulatory requirements, and supporting each other, display pilots managed to maintain their currency and proficiency despite significant hurdles. These efforts ensured that once restrictions were lifted, pilots were ready to return to the air show business, performing safely and effectively for enthusiastic and eager for action audiences.

NOTES

1 DA Category: Refers to Display Authorization, indicating that the practice must conform to specific authorized parameters that ensure the pilot's preparedness and compliance with safety standards.

Aerobatic Skill Levels: The requirements hold uniformly across all skill levels from standard to unlimited, emphasizing rigorous practice irrespective of proficiency, to maintain high safety and performance standards (UK CAA, 2025).

2 The Icarian illusion of invulnerability refers to a cognitive bias where individuals overestimate their safety and underestimate risks, akin to the myth of Icarus who ignored warnings and flew too close to the sun, ultimately leading to his downfall. Also known as the Dunning-Kruger effect (Dunning, 2011).

3 "Fight instead of fly" refers to the instinctual reaction to confront and combat a threat or stressor directly, rather than fleeing or escaping from it, often associated with the body's acute stress response (Kozlowska et al., 2015).

4 An electroencephalogram (EEG) is a test that measures electrical activity in the brain (Mayo Clinic, 2025).

5 Cognitive dissonance is the psychological discomfort experienced when holding two or more conflicting beliefs, values, or attitudes simultaneously (Harmon-Jones & Mills, 2019).

6 A go/no-go decision is a critical point in a process where a decision is made to either proceed with or halt the project based on predefined criteria or conditions (Kamissoko et al., 2023).

7 *Perfection-itis* is a paraphrase of the noun perfectionism defined as "a refusal to accept any standard short of perfect" (Merriam-Webster, 2025c).
8 In display flying, energy gates are critical points in a maneuver where pilots must precisely manage the aircraft's kinetic and potential energy to maintain control and achieve optimal performance.
9 In display flying, energy points are specific locations within a maneuver where pilots must adjust the aircraft's energy state, balancing speed and altitude, to ensure smooth and precise execution of aerobatic maneuvers.
10 Hypoglycemia is "an abnormally low concentration of sugar in the blood" (Merriam-Webster, 2025b).
11 The glycemic index is a measure that ranks foods based on how quickly and significantly they raise blood sugar levels after consumption. High glycemic index foods like white bread cause rapid blood sugar spikes, moderate glycemic index foods like brown rice have a more gradual effect, and low glycemic index foods like lentils lead to slow, sustained blood sugar increases (Foster-Powell et al., 2002).
12 Hypoxia is a condition in which the body or a region of the body is deprived of adequate oxygen supply at the tissue level (Shaw et al., 2021).
13 Thermal stress is the strain on the body caused by exposure to extreme temperatures, affecting physiological functions and potentially impacting performance and health (Enander & Hygge, 1990).
14 Hyperthermia refers to the state of having an abnormally high body temperature, which can occur naturally or be intentionally induced for therapeutic reasons (Merriam-Webster, 2025a).
15 Hypothermia is a condition where the body's core temperature drops below the required temperature for normal metabolism and body functions, defined as 35.0 degrees Celsius (95.0 degrees Fahrenheit) (Crawford, 1954).

REFERENCES

Adler, R. F., & Benbunan-Fich, R. (2012). Juggling on a high wire: Multitasking effects on performance. *International Journal of Human-Computer Studies, 70*(2), 156–168. https://doi.org/10.1016/j.ijhcs.2011.10.003

Air Accidents Investigation Branch (AAIB). (2017). *Report on the Accident to Hawker Hunter T7, G-BXFI Near Shoreham Airport on 22 August 2015 (Aircraft Accident Report No. 1/2017).* Department of Transport, United Kingdom. https://assets.publishing.service.gov.uk/media/58b9247740f0b67ec80000fc/AAR_1-2017_G-BXFI.pdf

Aircraft Owners and Pilots Association. (2016). *Currency vs. proficiency.* AOPA. www.aopa.org/training-and-safety/active-pilots/safety-and-technique/currency-vs-proficiency

AirshowStuff. (2020, March 12). Numerous airshows cancel in the face of coronavirus pandemic – Here is a full list. *AirshowStuff.* https://airshowstuff.com/v4/2020/numerous-airshows-cancel-in-the-face-of-coronavirus-pandemic-here-is-a-full-list/

Allen, M. S., Jones, M., McCarthy, P. J., Sheehan-Mansfield, S., & Sheffield, D. (2013). Emotions correlate with perceived mental effort and concentration disruption in adult sport performers. *European Journal of Sport Science, 13*(6), 697–706. https://doi.org/10.1080/17461391.2013.771381

Anderson, J. D. (2011). *Fundamentals of aerodynamics.* McGraw-Hill Education.

Antunano, M. J. (2000). *The effects of G-forces on the human body.* Federal Aviation Administration Office of Aerospace Medicine.

Archer, J. (2006). Testosterone and human aggression: An evaluation of the challenge hypothesis. *Neuroscience & Biobehavioral Reviews, 30*(3), 319–345.

Arthur, W., Jr., Bennett, W., Jr., Stanush, P. L., & McNelly, T. L. (1998). *Factors that influence skill decay and retention: A quantitative review and analysis. Human Performance, 11*(1), 57–101. https://doi.org/10.1207/s15327043hup1101_3

Baines Simmons. (2024, April 23). *Human Factors in Air Displays* [Video]. YouTube. www.youtube.com/watch?v=ybDy0Rb7z_8

Balldin, U. I. (1986). Factors influencing G-tolerance. *Clinical Physiology, 6*(3), 209–219. https://doi.org/10.1111/j.1475-097X.1986.tb00620.x

Banks, V. A., Plant, K. L., & Stanton, N. A. (2020). Leaps and shunts: Designing pilot decision aids on the flight deck using Rasmussen's ladders. In R. Charles & D. Golightly (Eds.), *Contemporary ergonomics and human factors 2020* (pp. 1–8). Chartered Institute of Ergonomics and Human Factors.

Barker, D. (2003). *Zero error margin: Airshow display flying analysed*. Freeworld Publications.

Barker, D. (2020). *Anatomy of airshow accidents* (1st ed., Vol. 1). International Council of Air Shows, International Air Show Safety Team.

Bateman, W. A., Jacobs, I., & Buick, F. (2006). Physical conditioning to enhance +Gz tolerance: Issues and current understanding. *Aviation, Space, and Environmental Medicine, 77*(6), 573–580.

Besch, E. L., Werchan, P. M., Wiegman, J. F., Nesthus, T. E., & Shahed, A. R. (1994). Effect of hypoxia and hyperoxia on human +Gz duration tolerance. *Journal of Applied Physiology, 76*(4), 1693. https://doi.org/10.1152/jappl.1994.76.4.1693

Boatman, J. (2022, July 15). Farnborough set to come in hot—And cheap? *FLYING Magazine*. www.flyingmag.com/farnborough-set-to-come-in-hot-and-cheap/

Bond, C. F., Jr., & Titus, L. J. (1983). Social facilitation: A meta-analysis of 241 studies. *Psychological Bulletin, 94*(2), 265–292. https://doi.org/10.1037/0033-2909.94.2.265

British Air Display Association. (n.d.). *About us*. Retrieved April 20, 2025, from www.bada-uk.com/2011/01/about-us/

Butuk, Y., & Škvareková, I. (2022). *Impact of COVID-19 pandemic on pilot training*. University of Žilina. www.drepo.uniza.sk/xmlui/handle/hdluniza/777

Campbell, R. D., & Bagshaw, M. (2002). *Human performance and limitations in aviation* (3rd ed.). Blackwell Publishing.

Causse, M., Peysakhovich, V., & Fabre, E. F. (2016). High working memory load impairs language processing during a simulated piloting task: An ERP and pupillometry study. *Frontiers in Human Neuroscience, 10*, 240. https://doi.org/10.3389/fnhum.2016.00240

Cheung, S. S., & Sleivert, G. G. (2004). Multiple triggers for hyperthermic fatigue and exhaustion. *Exercise and Sport Sciences Reviews, 32*(3), 100–106.

Childs, J. M., & Spears, W. D. (1986). Flight-skill decay and recurrent training. *Perceptual and Motor Skills, 62*(1), 235–242. https://doi.org/10.2466/pms.1986.62.1.235

Clark, M., & Mach, N. (2016). Nutritional strategies of high performance athletes to support training and minimize performance decrements. *Journal of Sports Sciences, 34*(8), 687–697.

Clark, T., & Williamon, A. (2011). Evaluation of a mental skills training program for conservatoire students. *Psychology of Music, 39*(5), 524–541.

Convertino, V. A. (1998). High sustained +Gz acceleration: physiological adaptation to high-G tolerance. *Journal of Gravitational Physiology, 5*(1), P51–P54.

Conway, D., Dick, I., Li, Z., Wang, Y., & Chen, F. (2013). The effect of stress on cognitive load measurement. In P. Kotzé, G. Marsden, G. Lindgaard, J. Wesson, & M. Winckler (Eds.), *Human-Computer Interaction – INTERACT 2013* (pp. 659–666). Springer. https://doi.org/10.1007/978-3-642-40498-6_58

Crawford, J. H. (1954). Hypothermia. *British Medical Journal, 1*, 336–336.

Crocker, P. R. E., Alderman, R. B., & Smith, F. M. R. (1988). Cognitive-affective stress management training with high performance youth volleyball players: Effects on affect, cognition, and performance. *Journal of Sport and Exercise Psychology, 10*(4), 448–452.

Croft, R. J., Kölegård, R., Tribukait, A., Taylor, N. A. S., & Eiken, O. (2021). Effects of acceleration-induced reductions in retinal and cerebral oxygenation on human performance. *Aerospace Medicine and Human Performance, 92*(2), 75–82. https://doi.org/10.3357/AMHP.5731.2021

Dahlstrom, N., Nahlinder, S., Wilson, G. F., & Svensson, E. (2011). Recording of psychophysiological data during aerobatic training. *The International Journal of Aviation Psychology, 21*(2), 105–122. https://doi.org/10.1080/10508414.2011.556443

Darses, F., Bernier, M., Berthelot, V., Fornette, M.-P., Launay, Y., Dozias, B., Chastres, V., & Fournier, J. (2023). Effects of mindfulness training on decision making in critical and high-demand situations: A pilot study in combat aviation. *Safety Science, 166*, 106204. https://doi.org/10.1016/j.ssci.2023.106204

David, T., Clarke, R. J., Uzieblo, A., Panayiotou, D., & Richardson, T. S. (2022). Enabling the automated assessment of precision aerobatic manoeuvres. *AIAA SCITECH 2022 Forum*. https://doi.org/10.2514/6.2022-1287

Dekker, S. W. A. (2003). *Human factors in aviation—A natural history*. Human Factors in Aviation.

Dempsey, J., & Wagner, P. (1999). Exercise-induced arterial hypoxemia. *Journal of Applied Physiology, 87*(6), 1997–2385. https://doi.org/10.1152/jappl.1999.87.6.1997

Driskell, J. E., & Salas, E. (Eds.). (1996). *Stress and human performance* (1st ed.). Psychology Press. https://doi.org/10.4324/9780203772904

Driskell, J. E., Copper, C., & Moran, A. P. (1994). Does mental practice enhance performance? *Journal of Applied Psychology, 79*(4), 481–492. https://doi.org/10.1037/0021-9010.79.4.481

Dube, K., Nhamo, G., & Chikodzi, D. (2021). COVID-19 pandemic and prospects for recovery of the global aviation industry. *Journal of Air Transport Management, 92*, 102022. https://doi.org/10.1016/j.jairtraman.2021.102022

Dunning, D. (2011). The Dunning–Kruger effect: On being ignorant of one's own ignorance. In M. P. Zanna & J. M. Olson (Eds.), *Advances in Experimental Social Psychology* (Vol. 44, pp. 247–296). Academic Press. https://doi.org/10.1016/B978-0-12-385522-0.00005-6

Ebbatson, M., Harris, D., Jarvis, S., & Inns, A. (2010). *The relationship between manual handling performance and recent flying experience in air transport pilots. Ergonomics, 53*(2), 268–277.

Edvardsen, A., Akerø, A., Christensen, C. C., Ryg, M., & Skjønsberg, O. H. (2012). Air travel and chronic obstructive pulmonary disease: a new algorithm for pre-flight evaluation. *Thorax, 67*(11), 964–969. https://doi.org/10.1136/thoraxjnl-2012-201855

Enander, A. E., & Hygge, S. (1990). Thermal stress and human performance. *Scandinavian Journal of Work, Environment & Health, 16*(Suppl 1), 44–50. https://doi.org/10.5271/sjweh.1823

Endsley, M. R. (1995). Toward a theory of situation awareness in dynamic systems. *Human Factors, 37*(1), 32–64. https://doi.org/10.1518/001872095779049543

Ercan, E., & Gunduz, S. H. (2020). The effects of acceleration forces on cognitive functions. *Microgravity Science and Technology, 32*(4), 681–686. https://doi.org/10.1007/s12217-020-09793-0

European Airshow Council. (n.d.). *About*. Retrieved April 20, 2025, from www.europeanairshow.org/about

Falla, M., Micarelli, A., Hüfner, K., & Strapazzon, G. (2021). The effect of cold exposure on cognitive performance in healthy adults: A systematic review. *International Journal of Environmental Research and Public Health, 18*(18), 9725. https://doi.org/10.3390/ijerph18189725

Federal Aviation Administration (FAA). (2018, April 30). *AC 61-98D – Currency requirements and guidance for the flight review and instrument proficiency check*. Federal Aviation Administration. www.faa.gov/documentLibrary/media/Advisory_Circular/AC_61-98D.pdf

Federal Aviation Administration (FAA). (2020, November 19). *Special Federal Aviation Regulation (SFAR) COVID*. www.faa.gov/coronavirus/sfar-covid-faqs

Federal Aviation Administration (FAA). (2023). *Order 8900.1, Volume 5, Chapter 9, Section 1: Issue/Renew/Reevaluate/Rescind a Statement of Aerobatic Competency (SAC)*. Dynamic Regulatory System. https://drs.faa.gov/browse/excelExternalWindow/DRSDOCID100875355920230912130420.0001

Fong, K. L., & Fan, S. W. (1997). An overview of the physiological effects of sustained high +Gz forces on human beings. *Annals of the Academy of Medicine, Singapore, 26*(1), 94–103.

Fornette, M., Bardel, M., Lefrancois, C., Fradin, J., Massioui, F., & Amalberti, R. (2012). Cognitive-adaptation training for improving performance and stress management of Air Force pilots. *The International Journal of Aviation Psychology, 22*, 203–223. https://doi.org/10.1080/10508414.2012.689208

Foster-Powell, K., Holt, S. H. A., & Brand-Miller, J. C. (2002). International table of glycemic index and glycemic load values: 2002. *The American Journal of Clinical Nutrition, 76*(1), 5–56. https://doi.org/10.1093/ajcn/76.1.5

Fusco, P. (2018). *The crowd pleasers: A history of airshow misfortunes from 1910 to the present*. Skyhorse Publishing.

Gannon, M. C., & Nuttall, F. Q. (2004). Effect of a high-protein, low-carbohydrate diet on blood glucose control in people with type 2 diabetes. *Diabetes, 53*(9), 2375–2382. https://doi.org/10.2337/diabetes.53.9.2375

Gibb, R., Gray, R., & Scharff, L. (2011). *Aviation visual perception: Research, misperception, and mishaps*. Ashgate Publishing.

Gillingham, K. (1987). G-tolerance standards for aircrew training and selection. *Aviation, Space, and Environmental Medicine, 58*(10), 1024–1026.

Gouchie, C., & Kimura, D. (1991). The relationship between testosterone levels and cognitive ability patterns. *Psychoneuroendocrinology, 16*(4), 323–334. https://doi.org/10.1016/0306-4530(91)90018-O

Green, N. D. C. (2018). Cognitive impairment during G-force exposure. *Journal of Cognitive Neuroscience, 30*(5), 698–707. https://doi.org/10.1162/jocn_a_01275

Green, N. D. C., & Muir, H. C. (2002). The role of anti-G suits in modern aviation. *Ergonomics, 45*(7), 493–508.

Grossman, P., Niemann, L., Schmidt, S., & Walach, H. (2004). Mindfulness-based stress reduction and health benefits. *Journal of Psychosomatic Research, 57*(1), 35–43.

Guézennec, C.-Y., Louisy, F., Portier, H., Laude, D., Chapuis, B., & Plésant, J. (2001). Effects of aerobatics flight on oxygen consumption and heart rate control: Influence on autonomic cardiovascular regulation during recovery. *European Journal of Applied Physiology, 84*(6), 562–568. https://doi.org/10.1007/s004210100381

Gultepe, I. (2023). A review on weather impact on aviation operations: Visibility, wind, precipitation, icing. *Journal of Airline Operations and Aviation Management*. https://doi.org/10.56801/jaoam.v2i1.1

Habay, J., Van Cutsem, J., Verschueren, J., De Bock, S., Proost, M., De Wachter, J., Tassignon, B., Meeusen, R., & Roelands, B. (2021). Mental fatigue and sport-specific psychomotor performance: A systematic review. *Sports Medicine, 51*(7), 1527–1548. https://doi.org/10.1007/s40279-021-01429-6

Hafen, B. Q., & Sharma, M. (2018). Oxygen saturation. *PubMed*. https://pubmed.ncbi.nlm.nih.gov/30247849/

Hales, B. M., & Pronovost, P. J. (2006). The checklist—a tool for error management and performance improvement. *Journal of Critical Care, 21*(3), 231–235. https://doi.org/10.1016/j.jcrc.2006.06.002

Hancock, P. A., & Vasmatzidis, I. (2003). Effects of heat stress on cognitive performance: The current state of knowledge. *International Journal of Hyperthermia, 19*(3), 355–372.

Hanton, S., Fletcher, D., & Coughlan, G. (2004). Stress in elite sport performers: A comparative study of competitive and organizational stressors. *Journal of Sports Sciences, 23*(10), 1129–1141.

Hardy L. (1992). Psychological stress, performance, and injury in sport. *British medical bulletin, 48*(3), 615–629. https://doi.org/10.1093/oxfordjournals.bmb.a072567

Harmon-Jones, E., & Mills, J. (2019). An introduction to cognitive dissonance theory and an overview of current perspectives on the theory. In E. Harmon-Jones (Ed.), *Cognitive dissonance: Reexamining a pivotal theory in psychology* (2nd ed., pp. 3–24). American Psychological Association. https://doi.org/10.1037/0000135-001

Harris, D. (2011). *Human performance on the flight deck*. Routledge. https://doi.org/10.1201/9781315252988

Haslbeck, A., & Hoermann, H. J. (2016). Flying the Needles: Flight Deck Automation Erodes Fine-Motor Flying Skills Among Airline Pilots. *Human factors*, *58*(4), 533–545. https://doi.org/10.1177/0018720816640394

Hickey, B. A., Chalmers, T., Newton, P., Lin, C. T., Sibbritt, D., McLachlan, C. S., Clifton-Bligh, R., Morley, J., & Lal, S. (2021). Smart devices and wearable technologies to detect and monitor mental health conditions and stress: A systematic review. *Sensors, 21*(10), 3461. https://doi.org/10.3390/s21103461

Hidalgo-Muñoz, A. R., Matton, N., Causse, M., Rouillard, Y., & El-Yagoubi, R. (2018). Cardiovascular correlates of emotional state, cognitive workload and time-on-task effect during a realistic flight simulation. *International Journal of Psychophysiology, 128*, 62–69. https://doi.org/10.1016/j.ijpsycho.2018.04.002

Hloogeveen, A. R., & Zonderland, M. L. (1996). Relationships between testosterone, cortisol and performance in professional cyclists. *International Journal of Sports Medicine, 17*(6), 423–428. https://doi.org/10.1055/s-2007-972872

Holm, J. E., Holroyd, K. A., & Keefe, F. J. (1996). A meta-analysis of psychosocial interventions for chronic pain. *Health Psychology, 15*(1), 49.

Hormeño-Holgado, A. J., & Clemente-Suárez, V. J. (2019). Effect of different combat jet manoeuvres in the psychophysiological response of professional pilots. *Physiology & Behavior, 208*, 112559. https://doi.org/10.1016/j.physbeh.2019.112559

Humphreys, S., Deyermond, R., Bali, I., Stevenson, M., & Fee, J. P. (2005). The effect of high altitude commercial air travel on oxygen saturation. *Anaesthesia*, *60*(5), 458–460. https://doi.org/10.1111/j.1365-2044.2005.04124.x

Hyten, C., & Ludwig, T. D. (2017). Complacency in process safety: A behavior analysis toward prevention strategies. *Journal of Organizational Behavior Management, 37*(3–4), 240–260. https://doi.org/10.1080/01608061.2017.1341860

International Council of Air Shows. (n.d.). *Home*. Retrieved April 20, 2025, from https://airshows.aero

International Council of Air Shows (ICAS). (2016). *Air show performers safety manual*. ICAS Publications.

International Council of Air Shows. (2019). *Aerobatic Competency Evaluation (ACE) Manual* (Rev. 9.0). https://airshows.aero/ViewDoc/3129

Jensen, R.S. (1995). *Pilot judgment and crew resource management* (1st ed.). Routledge. https://doi.org/10.4324/9781315246796

Johannigman, J., Gerlach, T., Cox, D., Juhasz, J., Britton, T., Elterman, J., Rodriquez, D. Jr., Blakeman, T., & Branson, R. (2015). Hypoxemia during aeromedical evacuation of the walking wounded. *Journal of Trauma and Acute Care Surgery, 79*(4 Suppl 2), S216–S220. https://doi.org/10.1097/TA.0000000000000736

Johns Hopkins University. (2018, February 10). The mind of a medalist: How the brain can lead to Olympic gold. *Neuroscience News*. https://neurosciencenews.com/olympic-gold-mind-8471/

Johnson, O. C., Carlson, J. B., Veverka, D. V., & Self, B. P. (2007). Effects of low blood glucose on pilot performance. *Bios, 78*(3), 95–100.

Kabat-Zinn, J. (1990). *Full catastrophe living: Using the wisdom of your body and mind to face stress, pain, and illness*. Delta.

Kahneman, D. (1973). *Attention and effort*. Prentice-Hall.

Kamissoko, D., Gourc, D., Marmier, F., & Clément, A. (2023). A go/no-go decision-making model based on risk and multi-criteria techniques for project selection. *International Journal of Decision Support System Technology, 15*(2), 1–21. https://doi.org/10.4018/IJDSST.315641

Karachalios, M. (2022). *An evaluation of the relationships between resilient safety culture, safety risk parameters, and mindfulness in the international air show community* [PhD dissertation, University of North Dakota]. https://commons.und.edu/theses/4267/

Karachalios, M., & Adjekum, D. K. (2023). *Air show performers: Safety, risk management, and psychological factors*. CRC Press. https://doi.org/10.1201/9781003431879

Karn, S. N. (2022). A characterization of the effects of the anti-G straining maneuver on pilot breathing (Master's thesis, Case Western Reserve University).

Kasture, S., Sharma, M., & Nataraja, M. S. (2020). Correlation of age, height, and gender with +Gz tolerance among healthy Indian participants. *Indian Journal of Aerospace Medicine, 64*(1), 14–17. https://doi.org/10.25259/IJASM_13_2020

Keller, J., Mendonca, F., Cutter, J., Suckow, M., & Dillman, B. (2020). Justification and development of competencies to transform a collegiate aviation flight program. *The Journal of Competency-Based Education*, *5*(3), e01216.

Kenny, D. T., Davis, P., & Oates, J. (2004). Music performance anxiety and occupational stress amongst opera chorus artists and their relationship with state and trait anxiety and perfectionism. *Journal of Anxiety Disorders, 18*(6), 757–777. https://doi.org/10.1016/j.janxdis.2003.09.004

Kerksick, C. M., Wilborn, C. D., Roberts, M. D., Smith-Ryan, A. E., Kleiner, S. M., Jäger, R., & Kreider, R. B. (2018). ISSN exercise & sports nutrition review update: Research & recommendations. *Journal of the International Society of Sports Nutrition, 15*(1), 1–57.

Khudaykulov, A., Zheng, C., Obrenovic, B., Godinic, D., Alsharif, H. Z. H., & Jakhongirov, I. (2024). The fear of COVID-19 and job insecurity impact on depression and anxiety: An empirical study in China in the COVID-19 pandemic aftermath. *Current Psychology, 43*, 8471–8484. https://doi.org/10.1007/s12144-022-02883-9

Kilduff, L. P., Finn, C. V., Baker, J. S., Cook, C. J., & West, D. J. (2013). Preconditioning strategies to enhance physical performance on the day of competition. *International Journal of Sports Physiology and Performance, 8*(6), 677–681. https://doi.org/10.1123/ijspp.8.6.677

Kirwan, J. P., Cyr-Campbell, D., Campbell, W. W., Scheiber, J., & Evans, W. J. (2001). Effects of moderate and high glycemic index meals on metabolism and exercise performance. *Metabolism: clinical and experimental, 50*(7), 849–855. https://doi.org/10.1053/meta.2001.24191

Klostermann, A., Oprins, E. A. P. B., & van den Bosch, K. (2022). Factors influencing attenuating skill decay in high-risk industries: A scoping review. *Safety, 8*(2), 22. https://doi.org/10.3390/safety8020022

Koch, I., Poljac, E., Müller, H., & Kiesel, A. (2018). Cognitive structure, flexibility, and plasticity in human multitasking: An integrative review of dual-task and task-switching research. *Psychological Bulletin, 144*(6), 557–583. https://doi.org/10.1037/bul0000144

Koglbauer, I. (2015). Training for prediction and management of complex and dynamic flight situations. *Procedia - Social and Behavioral Sciences, 209*, 268–276. https://doi.org/10.1016/j.sbspro.2015.11.232

Kourtidou-Papadeli, C., Papadelis, C., Koutsonikolas, D., Boutzioukas, S., Styliadis, C., & Guiba-Tziampiri, O. (2008). High altitude cognitive performance and COPD interaction. *Hippokratia, 12*(Suppl 1), 84–90.

Kozlowska, K., Walker, P., McLean, L., & Carrive, P. (2015). Fear and the defense cascade: Clinical implications and management. *Harvard Review of Psychiatry, 23*(4), 263–287. https://doi.org/10.1097/HRP.0000000000000065

Krahenbuhl, G. S. (1975). Adrenaline, arousal and sport. *The American Journal of Sports Medicine, 3*(3), 117–121. https://doi.org/10.1177/036354657500300304

Kubatko, O., Merritt, R., Duane, S., & Piven, V. (2023). The impact of the COVID-19 pandemic on global food system resilience. *Mechanism of an Economic Regulation, 3*(1), 56–65. https://mer-journal.sumy.ua/index.php/journal/article/view/144

Kumar, A. (2023). Effect of age and flying experience on heart rate response of fighter aircrew during high-G exposure in the high-performance human centrifuge. *Indian Journal of Aerospace Medicine, 67*(1), 2–7. https://doi.org/10.25259/IJASM_62_2020

Kumar, A., Nataraja, M. S., & Sharma, V. (2020). Analysis of G-induced loss of consciousness (G-LOC) and almost loss of consciousness (A-LOC) incidences in high-performance human centrifuge at Institute of Aerospace Medicine Indian Air Force. *Indian Journal of Aerospace Medicine, 63*(2), 53–60.

Landman, A., de Vries, D., & Binsch, O. (2022). *Retention of military combat lifesaving skills during six months following classroom-style and individualized-style initial training. Military Psychology, 35*(6), 590–602. https://doi.org/10.1080/08995605.2022.2144034

Lazarus, R. S., & Folkman, S. (1984). *Stress, appraisal, and coping*. Springer Publishing Company.

LeBlanc, V. R. (2009). The effects of acute stress on performance: Implications for health professions education. *Academic Medicine, 84*(10 Suppl), S25–S33. https://doi.org/10.1097/ACM.0b013e3181b37b8f

Leedy, M. G., & Wilson, M. S. (1985). Testosterone and cortisol levels in crewmen of U.S. Air Force fighter and cargo planes. *Psychosomatic Medicine, 47*(4), 333. https://journals.lww.com/psychosomaticmedicine/abstract/1985/07000/Testosterone_and_cortisol_levels_in_crewmen_of.3.aspx

Li, Y., & Wang, J. (2024). The impact of COVID-19 pandemic on air passenger travel: A focus on empirical findings. *Transport Reviews, 44*(2), 461–483. https://doi.org/10.1080/01441647.2023.2244165

Lieberman, H. R. (2003). Nutrition, brain function and cognitive performance. *Appetite, 40*(3), 245–254.

Lobo, J. L., Del Ser, J., Moravek, Z., De Simone, F., Presta, R., & Collina, S. (2016). Cognitive workload classification using eye-tracking and EEG data. In *Proceedings of the*

International Conference on Human-Computer Interaction in Aerospace (HCI-Aero '16) (Article No. 16). Association for Computing Machinery. https://doi.org/10.1145/2950112.2964585

Mahlfeld, W., Hasse, C., Grasshoff, D., & Bruder, C. (2011). The effect of complacency potential on human operators' monitoring behaviour in aviation. In D. de Waard, N. Gérad, L. Onnasch, & D. Manzey (Eds.), *Human Centred Automation* (pp. 133–144). Shaker Publishing.

Malouf, R., & Brust, J. C. (1985). Hypoglycemia: causes, neurological manifestations, and outcome. *Annals of Neurology, 17*(5), 421–430. https://doi.org/10.1002/ana.410170502

Mandrick, K., Peysakhovich, V., Rémy, F., Lepron, E., & Causse, M. (2016). Neural and psychophysiological correlates of human performance under stress and high mental workload. *Biological Psychology, 121*, 62–73. https://doi.org/10.1016/j.biopsycho.2016.10.002

Mantantzis, K., Schlaghecken, F., Sünram-Lea, S. I., & Maylor, E. A. (2019). Sugar rush or sugar crash? A meta-analysis of carbohydrate effects on mood. *Neuroscience & Biobehavioral Reviews, 101*, 45–67. https://doi.org/10.1016/j.neubiorev.2019.03.016

Marcora, S. M., Staiano, W., & Manning, V. (2009). Mental fatigue impairs physical performance in humans. *Journal of Applied Physiology, 106*(3), 857–864. https://doi.org/10.1152/japplphysiol.91324.2008

Mark, J. A., Kraft, A. E., Ziegler, M. D., & Ayaz, H. (2022). Neuroadaptive training via fNIRS in flight simulators. *Frontiers in Neuroergonomics, 3*, Article 820523. https://doi.org/10.3389/fnrgo.2022.820523

Martin, B., Kaminski-Ozturk, N., O'Hara, C., & Smiley, R. (2023). Examining the impact of the COVID-19 pandemic on burnout and stress among U.S. nurses. *Journal of Nursing Regulation, 14*(1), 4–12. https://doi.org/10.1016/S2155-8256(23)00063-7

Mayo Clinic. (2025). *EEG (electroencephalogram)*. www.mayoclinic.org/tests-procedures/eeg/about/pac-20393875

McEwen, B. S., & Stellar, E. (1993). Stress and the individual: Mechanisms leading to disease. *Archives of Internal Medicine, 153*(18), 2093–2101.

McGill, S. M. (2007). *Low back disorders: Evidence-based prevention and rehabilitation*. Human Kinetics.

McLaughlin, A. C., & Byrne, V. E. (2020). A fundamental cognitive taxonomy for cognition aids. *Human Factors, 62*(6), 865–873. https://doi.org/10.1177/0018720820920099

Meichenbaum, D. (1985). *Stress inoculation training*. https://doi.org/10.4324/9781315748931-10

Meland, A., Fonne, V., Wagstaff, A., & Pensgaard, A. M. (2015). Mindfulness-based mental training in a high-performance combat aviation population: A one-year intervention study and two-year follow-up. *International Journal of Aviation Psychology, 25*(1), 48–61. https://doi.org/10.1080/10508414.2015.995572

Merriam-Webster. (2025a). *Hyperthermia*. In *Merriam-Webster.com dictionary*. Retrieved April 20, 2025, from www.merriam-webster.com/dictionary/hyperthermia

Merriam-Webster. (2025b). *Hypoglycemia*. In *Merriam-Webster.com dictionary*. Retrieved April 20, 2025, from www.merriam-webster.com/dictionary/hypoglycemia

Merriam-Webster. (2025c). *Perfectionism*. In *Merriam-Webster.com dictionary*. Retrieved April 20, 2025, from www.merriam-webster.com/dictionary/perfectionism

Mesagno, C., & Mullane-Grant, T. (2010). A comparison of different pre-performance routines as possible choking interventions. *Journal of Applied Sport Psychology, 22*, 343–360. https://doi.org/10.1080/10413200.2010.491780

Mohanavelu, K., Poonguzhali, S., Ravi, D., Singh, P. K., Mahajabin, M., Ramachandran, K., Singh, U. K., & Jayaraman, S. (2020). Cognitive workload analysis of fighter aircraft pilots in flight simulator environment. *Defence Science Journal, 70*(2), 131–139.

Moitzi, A. M., & König, D. (2023). Longer-Term Effects of the Glycaemic Index on Substrate Metabolism and Performance in Endurance Athletes. *Nutrients, 15*(13), 3028. https://doi.org/10.3390/nu15133028

Moray, N., & Inagaki, T. (2000). Attention and complacency. *Theoretical Issues in Ergonomics Science, 1*(4), 354–365. https://doi.org/10.1080/14639220052399159

Morris, C. H., & Leung, Y. K. (2006). Pilot mental workload: How well do pilots really perform? *Ergonomics, 49*(15), 1581–1596. https://doi.org/10.1080/00140130600857987

Morrison, J. G., Forster, E. M., Hitchcock, E. M., Barba, C. A., Santarelli, T. P., & Scerbo, M. W. (1994). Cumulative effects of +Gz on cognitive performance. *Proceedings of the Human Factors and Ergonomics Society Annual Meeting, 38*(1), 46–50. https://doi.org/10.1177/154193129403800108

Murray, C. (2007). Mental preparation for flying: The role of mental practice in pilot training. *Aviation Psychology and Applied Human Factors, 1*(1), 45–52.

Neff, P. (2022). The five hazardous attitudes, a subset of complacency. *International Journal of Aviation, Aeronautics, and Aerospace, 9*(1). https://doi.org/10.15394/ijaaa.2022.1677

Newman, D. (2015). *High G flight: Physiological effects and countermeasures* (1st ed.). Routledge. https://doi.org/10.4324/9781315586779

Nižetić, S. (2020). Impact of coronavirus (COVID-19) pandemic on air transport mobility, energy, and environment: A case study. *International Journal of Energy Research, 44*(13), 10953–10961. https://doi.org/10.1002/er.5706

Olaganathan, R., & Amihan, R. A. H. (2021). *Impact of COVID-19 on pilot proficiency – A risk analysis. Global Journal of Engineering and Technology Advances, 6*(3), 1–9. https://doi.org/10.30574/gjeta.2021.6.3.0023

Orasanu, J. M., Martin, L., & Davison, J. (2001). Cognitive and contextual factors in aviation accidents: Decision errors. In E. Salas & G. Klein (Eds.), *Linking expertise and naturalistic decision making* (pp. 209–225). Mahwah, NJ: Lawrence Erlbaum Associates.

Ossard, G., Clère, J. M., Kerguelen, M., Melchior, F., & Seylaz, J. (1994). Response of human cerebral blood flow to +Gz accelerations. *Journal of Applied Physiology, 76*(5), 2114–2118. https://doi.org/10.1152/jappl.1994.76.5.2114

Ottoboni, G., Russo, G., & Tessari, A. (2015). What boxing-related stimuli reveal about response behaviour. *Journal of Sports Sciences, 33*, 1019–1027. https://doi.org/10.1080/02640414.2014.977939

Paraschi, E. P., Panagopoulos, A., & Fragkogianni, M. (2024). COVID-19 crisis management in Greek airlines. *Journal of the Air Transport Research Society, 3*, 100032. https://doi.org/10.1016/j.jatrs.2024.100032

Parsons, K. (1999). *Human thermal environments: The effects of hot, moderate, and cold environments on human health, comfort and performance.* https://doi.org/10.1201/9780203302620

Perna, F., Antoni, M., Baum, A., Gordon, P., & Schneiderman, N. (2003). Cognitive behavioral stress management effects on injury and illness among competitive athletes: A randomized clinical trial. *Annals of Behavioral Medicine, 25*, 66–73. https://doi.org/10.1207/S15324796ABM2501_09

Pilcher, J., Nadler, E., & Busch, C. (2002). Effects of hot and cold temperature exposure on performance: A meta-analytic review. *Ergonomics, 45*, 682–698. https://doi.org/10.1080/00140130210158419

Pooja, K., Kumar, A., & Dinakar, S. (2022). A case of low G-tolerance: An approach to aeromedical evaluation and disposal. *Indian Journal of Aerospace Medicine, 66*(1), 39–42. https://doi.org/10.25259/IJASM_11_2022

Post, J., Noordzij, L., Kam, M., Blauw, G., Cohen, A., & Gerven, J. (2002). Evaluation of tests of central nervous system performance after hypoxemia for a model for cognitive impairment. *Journal of Psychopharmacology, 16*, 337–343. https://doi.org/10.1177/026988110201600408

Radstake, W. E., Jillings, S., Laureys, S., Demertzi, A., Sunaert, S., Van Ombergen, A., & Wuyts, F. L. (2023). Neuroplasticity in F16 fighter jet pilots. *Frontiers in Physiology, 14*, 1082166. https://doi.org/10.3389/fphys.2023.1082166

Rangel, M. V. D. S., de Sá, G. B., Farinatti, P., & Borges, J. P. (2023). Neuro-cardiovascular responses to sympathetic stimulation in fighter pilots. *Aerospace Medicine and Human Performance, 94*(10), 761–769. https://doi.org/10.3357/AMHP.6223.2023

Repetitio est mater studiorum. (n.d.). In *Proverbius.net*. Retrieved from www.proverbius.net/en/latin-proverbs/repetitio-est-mater-studiorum

Riachy, R., McKinney, K., & Tuvdendorj, D. R. (2020). Various factors may modulate the effect of exercise on testosterone levels in men. *Journal of Functional Morphology and Kinesiology, 5*(4), 81. https://doi.org/10.3390/jfmk5040081

Rickards, C. A., & Newman, D. G. (2005). G-induced visual and cognitive disturbances in a survey of 65 operational fighter pilots. *Aviation, Space, and Environmental Medicine, 76*(5), 496–500.

Risko, E. F., & Gilbert, S. J. (2016). Cognitive offloading. *Trends in Cognitive Sciences, 20*(9), 676–688. https://doi.org/10.1016/j.tics.2016.07.002

Rodríguez-Roisin, R., & Roca, J. (2005). Mechanisms of hypoxemia. *Intensive Care Medicine, 31*(8), 1017–1019. https://doi.org/10.1007/s00134-005-2678-1

Roland, D. (1994). How professional performers manage performance anxiety. *Research Studies in Music Education, 2*(1), 25–35. https://doi.org/10.1177/1321103X9400200105

Roza, D. (2024, April 15). F-16 pilots use new sensors to max out human performance. *Air & Space Forces Magazine*. www.airandspaceforces.com/f-16-pilot-human-performance-sensor/

Ruthruff, E., Remington, R. W., & Johnston, J. C. (2001). Switching between simple cognitive tasks: The interaction of top-down and bottom-up factors. *Journal of Experimental Psychology: Human Perception and Performance, 27*(6), 1404–1419. https://doi.org/10.1037/0096-1523.27.6.1404

Saastamoinen, K., & Maunula, K. (2021). Usefulness of flight simulator as a part of military pilots training–case study: Grob G 115E. *Procedia Computer Science, 192*, 1670–1676. https://doi.org/10.1016/j.procs.2021.08.171

Salas, E., & Burke, C. S. (2002). Simulation for training is effective when … *BMJ Quality & Safety, 11*(2), 119–120.

Samardzić, R., Milovanović, A., & Pavlović, S. (2017). Neurological implications of high G-force exposure. *Neuroscience Letters, 650*, 89–95. https://doi.org/10.1016/j.neulet.2017.04.046

Samuel, J., & Franklin, C. (2008). Hypoxemia and hypoxia. In J. A. Myers, K. W. Millikan, & T. J. Saclarides (Eds.), *Common surgical diseases: An algorithmic approach to problem solving* (pp. 391–394). Springer. https://doi.org/10.1007/978-0-387-75246-4_97

Sandi, C. (2013). Stress and cognition. *Wiley Interdisciplinary Reviews: Cognitive Science, 4*(3), 245–261. https://doi.org/10.1002/wcs.1222

Santos, M., Vollam, S., Pimentel, M. A. F., Areia, C., Young, L., Roman, C., Ede, J., Piper, P., King, E., Harford, M., Shah, A., Gustafson, O., Tarassenko, L., & Watkinson, P. (2022). The use of wearable pulse oximeters in the prompt detection of hypoxemia and during movement: Diagnostic accuracy study. *Journal of Medical Internet Research, 24*(2), e28890. https://doi.org/10.2196/28890

Sarkar, M., Niranjan, N., & Banyal, P. K. (2017). Mechanisms of hypoxemia. *Lung India, 34*(1), 47. https://doi.org/10.4103/0970-2113.197116

Sarter, N., & Woods, D. (1995). How in the world did we ever get into that mode? Mode error and awareness in supervisory control. *Human Factors: The Journal of Human Factors and Ergonomics Society, 37*(1), 5–19. https://doi.org/10.1518/001872095779049516

Saunders, T., Driskell, J., Johnston, J., & Salas, E. (1996). The effect of stress inoculation training on anxiety and performance. *Journal of Occupational Health Psychology, 1*(2), 170–186. https://doi.org/10.1037//1076-8998.1.2.170

Scofield, J. E., Koenig, L. N., & Radvansky, G. A. (2020). *Repetition and recency in task performance: Behavioral and electrophysiological evidence for enhanced recognition memory. Journal of Applied Cognitive Psychology, 36*(4), 589–602.

Shaw, D. M., Cabre, G., & Gant, N. (2021). Hypoxic hypoxia and brain function in military aviation: Basic physiology and applied perspectives. *Frontiers in Physiology, 12*, 665821. https://doi.org/10.3389/fphys.2021.665821

Shender, B., Forster, E., Hrebien, L., Ryoo, H., & Cammarota, J. (2003). Acceleration-induced near-loss of consciousness: The "A-LOC" syndrome. *Aviation, Space, and Environmental Medicine, 74*(10), 1021–1028.

Sikora, I., Hari, P., & Hanusch, A. (2020). Mitigating loss of control in flight (LOC-I) through human factors and Crew Resource Management techniques. *Aviation Psychology and Applied Human Factors, 10*(3), 155–61.

Svensson, E., Angelborg-Thanderz, M., Sjöberg, L., & Olsson, S. (1997). Information complexity–mental workload and performance in combat aircraft. *Ergonomics, 40*(3), 362–380. https://doi.org/10.1080/001401397188206

Taylor, L., Watkins, S. L., Marshall, H., Dascombe, B. J., & Foster, J. (2016). The impact of different environmental conditions on cognitive function: A focused review. *Frontiers in Physiology, 6*, 372. https://doi.org/10.3389/fphys.2015.00372

Teigen, K. (1994). Yerkes-Dodson: A law for all seasons. *Theory & Psychology, 4*, 525–547. https://doi.org/10.1177/0959354394044004

Turner, R., & Swearer Napolitano, S. M. (2010). Cognitive behavioral therapy (CBT). In C. S. Clauss-Ehlers (Ed.), *Encyclopedia of Cross-Cultural School Psychology* (pp. 226–229). Springer. https://doi.org/10.1007/978-0-387-71799-9_75

UK Civil Aviation Authority. (2018). *CAP 1694: Human factors in air displays: Transfer of behaviours and error path study*. UK Civil Aviation Authority. www.caa.co.uk/our-work/publications/documents/content/cap1694/

UK Civil Aviation Authority. (2025). *CAP 1724: Flying display pilot authorisation and evaluation: Requirements and guidance* (Edition 7). www.caa.co.uk/our-work/publications/documents/content/cap1724/

U.S. Air Force (USAF). (2022, March 18). *Air Force Manual 11-246V1: Air Force Heritage Flight Program and Single-Ship Demonstrations* (AFMAN 11-246V1). www.acc.af.mil/Portals/92/11-246V1%20AFMAN%20%2818Mar22%29.pdf

Wacquant, L. J. (2006). *Body & soul: Notebooks of an apprentice boxer*. Oxford University Press.

Whitley, P. (1997). Pilot performance of the anti-G straining maneuver: Respiratory demands and breathing system effects. *Aviation, Space, and Environmental Medicine, 68*(4), 312–316.

Wickens, C. D. (2002). Situation awareness and workload in aviation. *Current Directions in Psychological Science, 11*(4), 128–133. https://doi.org/10.1111/1467-8721.00184

Wilson, J. H. (1993). *The value of currency in professional practice. Journal of Professional Development, 12*(3), 45–52.

World Health Organization. (2025). *Coronavirus disease (COVID-19) pandemic*. www.who.int/emergencies/diseases/novel-coronavirus-2019

Wright, H. W. (1973). Retention of flying skills and refresher training requirements: Effects of Nonflying and Proficiency Flying. *Aerospace Medicine, 44*(5), 509–512.

Wylie, A. (2024, March 22). Helmet-based pilot monitoring for real-time health tracking. *Defense Advancement*. www.defenseadvancement.com/news/helmet-based-pilot-monitoring-for-real-time-health-tracking/

Yerkes, R. M., & Dodson, J. D. (1908). The relation of strength of stimulus to rapidity of habit-formation. *Journal of Comparative Neurology and Psychology, 18*(5), 459–482.

Yoshie, M., Kudo, K., Murakoshi, T., & Ohtsuki, T. (2009). Music performance anxiety in skilled pianists: Effects of social-evaluative performance situation on subjective, autonomic, and electromyographic reactions. *Experimental Brain Research, 199*(2), 117–126. https://doi.org/10.1007/s00221-009-1979-y

Zhang, Y., & Zhang, W. (2022). A kind of psychological immunity training for pilots. *Proceedings of the 2022 International Conference on Education, Philosophy and Social Sciences (ICEPSS 2022)*. Madison Academic Press. https://madison-proceedings.com/index.php/aehssr/article/view/306

Zhu, W., Zhang, C., Liu, C., Yuan, J., Li, X., Wang, Y., & Jiang, C. (2023). Recognition of pilot mental workload in the simulation operation of carrier-based aircraft using the portable EEG. In *Proceedings of the 2023 3rd International Conference on Human Machine Interaction (ICHMI '23)* (pp. 43–49). Association for Computing Machinery. https://doi.org/10.1145/3604383.3604391

Zitzmann, M. (2020). Testosterone, mood, behaviour and quality of life. *Andrology, 8*(6), 1598–1605. https://doi.org/10.1111/andr.12867

FURTHER READINGS

Cox, M. (n.d.). *Human factors in flying displays – Practical guidance for display pilots & flying display directors* (p. 27). National Air Traffic Services (NATS), United Kingdom.

Dismukes, R. K., Berman, B. A., & Loukopoulos, L. D. (2015). *The limits of expertise: Rethinking pilot error and the causes of airline accidents*. Routledge.

Rumbold, J. L., Fletcher, D., & Daniels, K. (2020). A systematic review of stress management interventions with sport performers. *Journal of Sport and Exercise Psychology, 42*(4), 389–406.

Smith, M. S. (2006). Psychological preparation of professional boxers. *Journal of Sports Science & Medicine, 5*(4), 59–63.

U.S. Air Force (USAF). (2022, March 18). *Air Force Manual 11-246V1: Air Force Heritage Flight Program and Single-Ship Demonstrations* (AFMAN 11-246V1). www.acc.af.mil/Portals/92/11-246V1%20AFMAN%20%2818Mar22%29.pdf

Zitzmann, M., & Nieschlag, E. (2001). Testosterone levels in healthy men and the relation to behavioral and physical characteristics: Facts and constructs. *European Journal of Endocrinology, 144*(3), 183–197.

8 Mentorship

Manolis Karachalios

INTRODUCTION

Mentoring, as defined by Kram (1988), is a developmental partnership wherein a more experienced individual guides a less experienced person. This relationship is crucial in training air show performers, providing essential guidance, insights, and practical experience necessary for display pilots to excel. This mentoring process is not limited to the training season but extends throughout the years, facilitated by the strong family culture within the air show community. This enduring connection allows air show performers and their mentors to continuously exchange experiences and support each other's growth.

Display pilot instructors, acting as mentors, take on a sense of care and responsibility, not only offering valuable techniques and insights to their trainees but also sharing extensive knowledge, skills, and expertise. Mentors offer tailored feedback and constructive criticism, refining trainees' skills and enhancing their performance (Ragins & Kram, 2007). However, this mentorship extends beyond formal training sessions, with display pilots readily offering their guidance and support to peers whenever requested. The collaborative and supportive environment within the air show community ensures that mentorship is always available, fostering a culture of continuous learning and mutual support.

Both safety and professionalism are critical aspects of mentoring. Mentors instill a deep respect for safety, teaching trainees to balance performance with risk management and make sound decisions under pressure. They also exemplify the professionalism and discipline required of display pilots, including composure, adherence to protocols, and integrity (Eby et al., 2008).

Ultimately, mentoring is indispensable in the careers of display pilots, both in receiving guidance and in passing it on to new generations of air show performers. This continuous cycle of mentorship ensures the perpetuation of skills, knowledge, and professional values within the air show community.

DOI: 10.1201/9781003504801-11

IMPORTANCE OF MENTORSHIP IN INITIAL TRAINING FOR NEW DISPLAY PILOTS

The importance of mentorship in the initial training of new display pilots cannot be overstated. During this critical phase, mentorship provides the foundational support that inexperienced display pilots need to transition from theoretical knowledge to practical application. In the following section, an analysis of mentorship practices within the UK Civil Aviation Authority (CAA) and the USA Federal Aviation Administration (FAA) will be conducted, followed by a comparative evaluation of these two approaches.

UK CAA

According to CAP 1724 (UK Civil Aviation Authority, 2025), a pilot wishing to apply for a Display Authorisation (DA) must first establish contact with a Display Authorisation Evaluator (DAE) who holds a current authorization relevant to the skills and categories in which the pilot wishes to be evaluated and is responsible for assessing the applicant's suitability for display flying.

Pre-Evaluation Mentoring

The pre-evaluation mentoring phase, crucial for both the DAE and the applicant, aims to establish the applicant's motivation and commitment. The DAE must gauge the applicant's dedication and ensure they understand the demands and responsibilities of display flying. As per CAP 1724, human factors, such as motivation, attitude, and interpersonal skills, are critical during this initial assessment. Building early rapport during this phase forms the basis for ongoing mentorship, guidance, and support, which are essential for the safety of UK display flying. Additionally, pre-evaluation mentoring provides an opportunity for the DAE to assist the applicant with their DA application process, offering support and advice to navigate the requirements.

Initial Assessment

The initial assessment, ideally conducted face-to-face, involves a comprehensive evaluation by the DAE mentor. The DAE must explain in detail what is required of a DA holder and consider several key areas:

- Experience: The applicant's flying experience is scrutinized to ensure they have the requisite skills and hours.
- Documentation: The DAE must verify the applicant's pilot license, medical certificate, and Certificate of Renewal (if applicable).
- Motivation: Open questioning is used to understand the applicant's motivation for display flying. The DAE must determine if the applicant's desire is to proficiently display the aircraft safely rather than seeking personal attention or taking unnecessary risks.
- Commitment and Training: The DAE and the applicant must agree on the necessary steps to achieve the required standard for DA evaluation, including a realistic time limit. This involves a significant commitment in terms of time, training, and financial resources, which must be clearly understood and agreed upon by the applicant.

To further assess and understand the applicant's flying ability, the guidelines encourage DAEs to fly with candidates. These assessment flights need not be of the exact type of aircraft offered for display, especially if they are single-seat aircraft. This practical evaluation helps the DAE gauge the pilot's competence and readiness for display flying.

Notification and Training

Once the DAE is satisfied that the applicant is suitable for training, they must notify the CAA so that the training and mentoring process can officially begin. It is important to note that the CAA ultimately issues the DA based on the DAE's recommendation. Therefore, if the DAE has any doubts about the applicant's suitability, they are advised to contact the CAA for further guidance.

USA FAA

The FAA's (2022) guidelines for mentorship in the context of applying for an initial SAC (Statement of Aerobatic Competency) card highlight the importance of structured mentorship in the training and evaluation of display pilots. Here is an analysis of the key points from the FAA's guidelines:

Initial Applicant Requirements

The FAA requires that an applicant for an initial SAC card must provide the name and contact information of their trainer or mentor. This requirement ensures that the applicant has received guidance and support from an experienced professional during their preparation. It also allows the FAA to verify the mentorship and training history of the applicant, adding an extra layer of accountability and quality assurance.

Evaluation Application Process

After identifying and contacting a qualified Aerobatic Competency Evaluator (ACE), the applicant and ACE must agree on the date and location of the evaluation. The applicant is then required to notify the International Council of Air Shows (ICAS) with the name and contact information of the person(s) who mentored them during their preparation, ensuring that ICAS is aware of the mentorship structure supporting the applicant.

Role of the Recommending ACE/Aerobatic Expert

The recommending ACE or aerobatic expert has several responsibilities in addition to covering all areas of the evaluation (International Council of Air Shows (ICAS), 2019):

- Feedback/Mentoring for Showmanship: The ACE provides feedback and mentoring on showmanship, which is crucial for the performance aspect of aerobatics. This includes coaching on how to present maneuvers effectively and engage the audience, ensuring that the performance is both safe and entertaining.
- Safety Considerations: Safety is a primary concern when flying a display. The ACE mentors the applicant on safety considerations, ensuring that the applicant understands and adheres to safety protocols. This mentoring helps to minimize risks and enhance the overall safety of air show performances.

TABLE 8.1
Comparison of FAA and UK CAA Mentorship Approaches

Aspect	USA FAA Approach	UK CAA Approach
Initial Contact and Mentorship Verification	Requires applicants to provide the name and contact information of their trainer or mentor.	Requires pilots to establish contact with a DAE and agree to be mentored.
Mentorship During Preparation	Applicants must notify ICAS of the mentor(s) who guided them in preparation for the evaluation.	Includes pre-evaluation mentoring to assess motivation and commitment and ensure understanding of display flying requirements.
Roles and Responsibilities of the Mentor/ Evaluator	Recommending ACE provides feedback and mentoring on showmanship and safety considerations.	DAE conducts a comprehensive initial assessment covering experience, documentation, motivation, and commitment. Encouraged to conduct assessment flights during the pre-evaluation phase.
Safety and Professionalism	ACE mentors applicants on safety protocols and risk management to ensure safe performances.	Emphasizes safety and professionalism through comprehensive assessment and ongoing mentorship.
Evaluation Process	Requires agreement on the evaluation date and location with the ACE and notification to ICAS.	DAE must notify the CAA once satisfied with the applicant's suitability for training and make a recommendation.

Comparison of FAA and UK CAA Mentorship Approaches

The USA FAA and UK CAA mentorship approaches share several similarities, including the formal establishment of mentorship relationships, emphasis on preparation, and prioritization of safety and professionalism, as shown in Table 8.1. However, the FAA focuses more on the administrative process, safety, and showmanship while engaging the ICAS during the process, whereas the CAA provides a holistic assessment framework, emphasizing motivation and comprehensive pilot evaluation.

LIFETIME MENTORSHIP FOR AIR SHOW PERFORMERS

Continuous mentoring is essential beyond initial training and evaluations. Ongoing mentorship helps pilots refine their skills, stay updated with best practices, and receive timely feedback on their performance. This continuous support is crucial for pilots to maintain high standards of competence and safety throughout their careers.

For instance, Zynda (2023) emphasizes the significance of a structured aviation mentorship program in ensuring the sustained development and safety of pilots.

Similarly, Cordova et al. (2024) highlight the positive outcomes from a well-designed pilot development mentor program, showing that mentorship leads to better skill refinement and adherence to safety protocols. Additionally, Anca (2020) supports this by discussing effective mentor behaviors in aviation and the positive impact of continuous guidance on a pilot's career.

Moreover, structured support systems within larger organizations, such as military aerobatic teams, inherently provide an environment where pilots can regularly interact with experienced mentors. These structures are crucial for fostering a culture of safety and continuous improvement. Larger organizations offer regular check-ins, evaluations, and feedback sessions, which are vital for pilots' professional development and safety (Karachalios, 2022). On the other hand, for independent pilots or those outside large organizations, developing formal mentorship programs can bridge the gap. Barker (2003) notes that these programs should facilitate regular check-ins, evaluations, and feedback sessions to ensure continuous professional development, providing the necessary support and oversight that independent pilots might otherwise miss.

In addition, effective supervision is integral to a pilot's career, especially in high-risk areas like air displays. Regular oversight by experienced air show performers, such as DAEs, ACEs, or team leaders, ensures that display pilots adhere to safety standards and can maintain peak performance during air shows. This supervision is critical in identifying and correcting any issues before they lead to accidents. Goodwyn et al. (2022) discuss how peer coaching among pilots can champion well-being and mitigate hazardous attitudes, underscoring the necessity of ongoing mentorship and oversight.

Furthermore, Newman (2021) provides insights into factors contributing to accidents during aerobatic flight operations, underscoring the importance of experienced supervision in mitigating risks. Promoting a resilient safety culture where mentorship and supervision are valued can lead to better compliance with safety protocols and a proactive approach to risk management (Karachalios, 2022). This culture should encourage open communication, reporting of near misses, and collaborative problem-solving (Karachalios, 2022; Karachalios & Adjekum, 2023).

Finally, Adjekum and Tous (2020) explore the relationship between organizational management factors and resilient safety culture in aviation programs, highlighting the importance of safety management systems (SMSs) in fostering such a culture. Additionally, Johnson et al. (2023) emphasize the need for creating a mentoring culture within training programs, which can significantly enhance safety and performance standards in aviation.

MENTORSHIP IN THE CONTEXT OF AIR DISPLAY ACCIDENTS

The analysis of mentorship in the context of air display accidents, as described in the CAP 1724 (UK CAA, 2022) document, emphasizes the critical role of mentorship and supervision in ensuring pilot competence and safety. For instance, the 2014 OV-10B Bronco accident report indicates that the pilot's display was not adequately evaluated beyond the initial authorization process, and there was a lack of continuous mentoring (UK Air Accidents Investigation Branch, 2014), highlighting that this lack of oversight might have contributed to the accident. Similarly, in the case of the 2023 Extra EA300/S accident, the report highlights issues of insufficient supervision (Safety Investigation & Safety Investigation Authority of Finland (SIAF), 2024).

CAP 1724 also points out the challenges faced by pilots who are not part of a larger organization due to limited access to structured mentoring and supervision. This isolation can result in missed opportunities for feedback and improvement, which are critical for maintaining safety standards.

Case Study for Lack of Continuous Mentorship: Fairchild Air Force Base B-52 Crash

The Fairchild Air Force Base B-52 crash on June 24, 1994 (Aviation Safety Network (ASN), 2024) is a poignant case study of how failed mentorship can influence the safety and outcomes of air display activities. This tragic incident, analyzed in Tony Kern's (1995) "Darker Shades of Blue: A Case Study of Failed Leadership," highlights the significant role that inadequate mentorship and poor leadership can play in aviation accidents.

The crash involved a B-52H aircraft, known as *Czar 52*, which was performing a practice session for an upcoming air show, which crashed while attempting a tight, low-altitude 360-degree turn, resulting in the death of all four crew members.

According to Kern, mentorship, or the lack thereof, was a critical factor in this accident. The analysis reveals that the captain, despite being a skilled pilot, had a long history of violating air discipline and regulations. Unfortunately, his behavior was not adequately addressed by his superiors, and there was a significant lack of effective mentorship and supervision.

A Contributing Factor

The captain's career at Fairchild AFB was marked by repeated breaches of flight discipline. Despite his reputation as an excellent pilot, his actions often violated regulations and safety protocols. The failure to provide continuous and effective mentorship allowed these dangerous behaviors to persist, as his senior officers failed to consistently mentor and supervise, allowing him to believe he could operate outside established norms without consequence (Kern, 1995).

Moreover, the command climate at Fairchild AFB did not support a culture of compliance and safety. Leaders at multiple levels failed to enforce standards and address the captain's violations effectively. For example, despite numerous red flags and recommendations to ground the captain, the leadership did not act decisively to curb his risky behavior (Kern, 1995).

Contagious Impact

The lack of proper mentorship and leadership had a cascading effect on junior officers and crew members. The captain's behavior set a poor example, which some junior aviators attempted to emulate, leading to near-disastrous results. Consequently, the absence of a robust mentorship framework meant that younger pilots did not receive the guidance needed to develop safe flying practices, ultimately allowing a culture of unsafe behavior and eroding trust in leadership (Kern, 1995).

THE ROLE OF MENTORSHIP IN SUSTAINING EXCELLENCE IN MILITARY DEMONSTRATION TEAMS

Mentorship acts as the lifeblood of excellence within aerobatic teams, infusing new pilots with the skills and resilience needed to soar to perfection. The initial training phase is a trial where raw talent is forged into precision and composure, with constructive criticism sharpening flying skills and unwavering emotional support fortifying the resilience essential for the intense display season ahead. This dynamic vividly comes to life in elite teams like the Red Arrows of the Royal Air Force (RAF) and the Canadian Forces (CF) Snowbirds, as well as solo display teams like the French Rafale Solo Display. Here, experienced pilots take new recruits under their wings, meticulously preparing them both technically and emotionally for the high-stakes drama of the air show season.

Case Study 1. Mentorship in Aerobatic Teams: Red Arrows and Snowbirds

The Red Arrows embark on a grueling, lengthy training session before the display season begins. New team members face a daunting selection process, emerging only to be seamlessly woven into the team through a meticulously phased training approach. Experienced team members provide continuous feedback and support, ensuring that new pilots are fully integrated into the team's high standards. This integration helps new members align with the team's culture of excellence, where each member supports the others. This culture is reinforced through regular team-building activities and a focus on mutual respect and trust (RAF, 2024).

Furthermore, the Snowbirds place paramount importance on the emotional and mental fortitude of their pilots. Resources for stress management, a nurturing team atmosphere, and resilience-building exercises ensure that each pilot can perform with unflinching confidence under the searing pressure of the spotlight. Training flights are meticulously planned and executed, with pilots providing feedback and guidance to new team members (Royal Canadian Air Force, 2021.

Case Study 2. The Role of Mentorship in Solo Display Teams: The French Air Force and Space Rafale Solo Display Example

In the French Air Force and Space Rafale Solo Display team, mentorship plays a critical role in sustaining excellence and ensuring safety in solo display performances. This mentorship model is a torch-passing ceremony, where the previous year's display pilot becomes the sage and guide for the incoming talent. This system ensures a seamless handover of invaluable knowledge, hard-earned experience, and unwavering emotional support, preserving the standards of awe-inspiring performances (Ministère des Armées, 2024).

For solo pilots, such as those in the Rafale Solo Display team, the mentorship is highly personalized. This approach focuses on individual skill refinement and resilience building. The outgoing display pilot shares practical considerations that can enhance performance and safety. These considerations might include specific techniques for handling the aircraft, optimizing display maneuvers, and managing the aircraft's limits under different conditions. For instance, the mentor might share considerations on how to perform a precise high-G turn under strong crosswinds or how to manage the jet's energy during a vertical dive to avoid a controlled flight into the terrain. The mentor also provides insights into potential hazards that are not documented or obvious, such as environmental factors like wind shear at specific air show locations or technical quirks of the aircraft relevant to the low-level air show environment, that the new display pilot needs to be aware of.

Emotional resilience is equally crucial for solo display pilots who face significant pressure to perform flawlessly. The relationship between the outgoing and incoming pilots fosters a keen sense of rapport, which helps the new pilot build confidence. The mentor provides moral and psychological support, helping the new pilot manage stress and maintain focus. The mentor is also the air show coordinator, allowing the display pilot to concentrate on his primary task of flying with precision and safety in his jet. This support system is vital for preparing the new pilot for the unique pressures of solo displays, such as handling media attention.

Comparative Analysis in Mentorship Approaches for Solo and Formation Display Pilots

The mentorship approaches for solo and formation display pilots, though distinct, share the common goal of sustaining high standards of performance. For solo pilots, the mentorship is highly personalized, focusing on individual skill refinement and resilience building. The direct transfer of knowledge and one-on-one support help solo display pilots navigate the unique pressures of performing alone. In the Rafale Solo Display team, the one-on-one mentorship ensures that the new display pilot receives personalized guidance tailored to their specific needs.

In contrast, mentorship for formation pilots is inherently collaborative. It emphasizes team coordination, collective performance, and effective communication. Formation pilots benefit from a support system that fosters teamwork and mutual trust, which is essential for executing complex group maneuvers. The mentorship programs in teams like the Red Arrows and Snowbirds ensure that new pilots are seamlessly integrated into the team, learning to work in perfect harmony with their peers.

APPLICATIONS OF MENTORSHIP STRATEGIES FROM DISPLAY PILOTS TO HIGH-PERFORMANCE TEAMS IN VARIOUS INDUSTRIES

The mentorship strategies employed by display pilots, renowned for their breathtaking aerial maneuvers, can be effectively applied to other high-risk, high-performance, and high-reliability (HRO) industries. These strategies enhance skills, foster resilience, and ensure performance excellence. Fields, such as healthcare, nuclear power, firefighting, corporate leadership, elite sports, military special operations, and software development, can greatly benefit from these approaches.

In healthcare, mentorship serves as the lifeline that ensures the seamless transfer of critical skills and knowledge from seasoned professionals to new practitioners. Imagine a veteran surgeon guiding a novice through the intricacies of a complex procedure, much like a display pilot mentoring a new team member on precision maneuvers. Atul Gawande (2011), in "The Checklist Manifesto," illustrates how mentorship and structured checklists improve surgical outcomes. High-stress situations, where split-second decisions are crucial, benefit from mentorship programs that mirror the emotional and psychological support seen in aerobatic teams. For example, the Center for Professionalism and Peer Support at Brigham and Women's Hospital redesigned the peer support program to provide one-on-one peer support and significantly reduced burnout and improved job satisfaction among clinicians, fostering a supportive environment that enhances both performance and mental well-being (Shapiro & Galowitz, 2016).

In the nuclear power industry, where precision and coordination are paramount, mentorship can emulate the meticulous approach of the Red Arrows. Experienced engineers guide new employees through the labyrinth of plant operations and safety protocols, ensuring team synchronization similar to the formation of planes moving in harmony. The Palo Verde Nuclear Generating Station's mentorship program emphasizes knowledge transfer and safety protocols (Reason, 2016). Here, seasoned engineers offer guidance on handling high-pressure situations, akin to display pilots supporting each other during complex maneuvers, which is crucial for maintaining composure and precision in a high-risk environment.

Firefighting demands exceptional skill and safety, and mentorship strategies from aerobatic teams, such as detailed debriefing sessions and constructive feedback, can be seamlessly integrated into training programs. The Fire Department of New York (FDNY) has a comprehensive mentorship program where veteran firefighters mentor new recruits, sharing practical knowledge about fire dynamics, rescue techniques, and safety measures (Salka, 2005). Firefighting, much like formation flying, relies heavily on teamwork and trust. Therefore, mentorship programs forge strong bonds among team members, ensuring effective cooperation during emergencies, mirroring the Red Arrows' emphasis on mutual respect and trust.

Furthermore, in corporate leadership, mentorship is vital for transferring knowledge and emotional skills from experienced executives to emerging leaders. Visualize a senior executive mentoring a junior leader on strategic decision making and crisis management, much like a display pilot sharing tips for optimizing maneuvers. General Electric's (GE's) leadership development program exemplifies this approach, fostering continuous improvement and innovation (Deschamps, 2009). Such an approach to mentorship from GE's leadership also cultivates emotional intelligence, helping leaders build confidence and resilience, ultimately enabling them to handle high-pressure scenarios and maintain composure (Goleman, 2020).

Additionally, elite sports, such as basketball, rely on mentorship to refine skills and enhance performance (Fraina & Hodge, 2020). Picture an experienced athlete mentoring a younger teammate, providing detailed feedback on free-throw techniques and mental preparation. The relationship between Michael Jordan and Kobe Bryant is a classic example of such a type of mentorship (Tucker, 2020), akin to a display pilot offering constructive criticism to improve maneuvers. Moreover, elite sports teams require precise coordination and collective performance, and mentorship strategies emphasizing teamwork and synchronization similar to those used by the Red Arrows can be effectively applied. Veteran players guide new members in integrating seamlessly into the team, ensuring alignment with team ethos, strategies, and goals (Lyle, 2002).

Moreover, in military operations, mentorship is crucial for transferring tactical knowledge and skills from seasoned soldiers to new recruits. Envision a veteran soldier mentoring a new recruit on combat tactics and mission planning, much like a display pilot sharing expertise on aircraft handling. The U.S. Navy SEALs employ a mentorship program where experienced operators train newcomers in critical skills necessary for mission success in high-stress environments requiring significant emotional and psychological support (Kram & Hall, 1989; Wong et al., 2003).

Lastly, in software development, mentorship programs are vital for continuous learning and skill enhancement. Google's Engineering Residency program exemplifies this effort, fostering a culture of continuous learning and innovation (Hunt & Thomas, 1999), where experienced developers help new team members integrate seamlessly and contribute effectively, maintaining high standards of performance and innovation, much like an aerobatic team executing complex aerial stunts.

REFERENCES

Adjekum, D. K., & Tous, M. F. (2020). Assessing the relationship between organizational management factors and a resilient safety culture in a collegiate aviation program with Safety Management Systems (SMS). *Safety Science, 131*, 104909.

Anca, J. (2020). *Mentor behaviours: A field study of good practice mentoring behaviours in Australian aviation* [Doctoral dissertation, Swinburne University of Technology]. Swinburne Research Bank. https://researchbank.swinburne.edu.au/file/d49c7e77-d6ee-4293-9bec-5bb422477829/1/jose_anca_thesis.pdf

Aviation Safety Network (ASN). (2024). Accident Boeing B-52H-170-BW Stratofortress 61-0026, Friday 24 June 1994. *Aviation Safety Network*. https://asn.flightsafety.org/wikibase/17433

Barker, D. (2003). *Zero error margin: Airshow display flying analysed*. Freeworld Publications.

Cordova, C. G., Parsons-Daisley, B., & Fox, E. (2024). High-flying results: A case study in a pilot development mentor program. *Journal of Aviation/Aerospace Education & Research, 33*(3), 3.

Deschamps, J. P. (2009). *Innovation leaders: How senior executives stimulate, steer and sustain innovation*. John Wiley & Sons.

Eby, L., Allen, T., Evans, S., Ng, T., & DuBois, D. (2008). Does mentoring matter? A multidisciplinary meta-analysis comparing mentored and non-mentored individuals. *Journal of Vocational Behavior, 72*(2), 254–267. https://doi.org/10.1016/J.JVB.2007.04.005

Federal Aviation Administration (FAA). (2022). *Order 8900.1A – Flight standards information management system*. Federal Aviation Administration (FAA). www.faa.gov/regulations_policies/orders_notices/index.cfm/go/document.information/documentID/1041430

Fraina, M., & Hodge, S. (2020). Mentoring relationships among athletes, coaches, and athletic administrators: A literature review. *Journal for the Study of Sports and Athletes in Education, 14*, 140–164. https://doi.org/10.1080/19357397.2020.1768033

Gawande, A. (2011). *The checklist manifesto: How to get things right*. Metropolitan Books.

Goleman, D. (2020). *Emotional intelligence: Why it can matter more than IQ*. Bloomsbury Publishing.

Goodwyn, N., Beech, N., Garvey, B., Gold, J., Gulliford, R., Auty, T., Sajjadi, A., Arrigoni, A., Mahtab, N., Jones, S., & Beech, S. (2022). Flying high: Pilot peer coaching to champion well-being and mitigate hazardous attitudes. *European Journal of Training and Development*. https://doi.org/10.1108/ejtd-09-2021-0136

Hunt, A., & Thomas, D. (1999). *The pragmatic programmer: From journeyman to master*. Addison-Wesley Professional.

International Council of Air Shows. (2019). *ACE manual—Revision 9*. International Council of Air Shows.

Johnson, W. B., Long, S., Smith, D. G., & Griffin, K. A. (2023). Creating a mentoring culture in graduate training programs. *Training and Education in Professional Psychology, 17*(1), 63.

Karachalios, M. (2022). *An evaluation of the relationships between resilient safety culture, safety risk parameters, and mindfulness in the international air show community* [PhD dissertation, University of North Dakota]. https://commons.und.edu/theses/4267/

Karachalios, M., & Adjekum, D. K. (2023). *Air show performers: safety, risk management, and psychological factors*. CRC Press. https://doi.org/10.1201/9781003431879

Kern, T. (1995). Darker shades of blue: A case study of failed leadership. L200, *Developing Leaders and Organizations*, by CGSC/Department of Command and Leadership, 321–342.

Kram, K. E. (1988). *Mentoring at work: Developmental relationships in organizational life*. University Press of America.

Kram, K., & Hall, D. (1989). Mentoring as an antidote to stress during corporate trauma. *Human Resource Management, 28*, 493–510. https://doi.org/10.1002/HRM.3930280405

Lyle, J. (2002). *Sports coaching concepts: A framework for coaches' behaviour*. Routledge.

Ministère des Armées. (2024, June 28). *Voltige aérienne et Rafale Solo Display: une même mission*. www.defense.gouv.fr/air/actualites/voltige-aerienne-rafale-solo-display-mission

Newman, D. G. (2021). Factors contributing to accidents during aerobatic flight operations. *Aerospace Medicine and Human Performance, 92*(8), 612–618.

Ragins, B. R., & Kram, K. E. (Eds.). (2007). *The handbook of mentoring at work: Theory, research, and practice*. Sage Publications.

Reason, J. (2016). *Managing the risks of organizational accidents*. Routledge.

Royal Air Force, United Kingdom. (2024). Red Arrows, Royal Air Force. *Red Arrows, Royal Air Force*. www.raf.mod.uk/display-teams/red-arrows/

Royal Canadian Air Force. (2021, July 27). *Snowbirds – Royal Canadian Air Force*. Government of Canada. www.canada.ca/en/air-force/services/showcasing/snowbirds.html

Safety Investigation & Safety Investigation Authority of Finland (SIAF). (2024). *Aviation accident at Selänpää Aerodrome on July 16, 2023 (Accident Investigation L2023-02; p. 32)*. https://turvallisuustutkinta.fi/material/sites/otkes/otkes/mn06fy4o2/L2023-02_Kouvola_investigation_report.pdf#page=2.08

Salka, J. (2005). *First in, last out: Leadership lessons from the New York Fire Department*. Penguin.

Shapiro, J., & Galowitz, P. (2016). Peer support for clinicians: A programmatic approach. *Academic Medicine: Journal of the Association of American Medical Colleges, 91*(9), 1200–1204. https://doi.org/10.1097/ACM.0000000000001297

Tucker, H. (2020, February 24). Michael Jordan remembers his "little brother" Kobe Bryant. *USA TODAY*. www.usatoday.com/story/sports/nba/2020/02/24/michael-jordan-speech-kobe-bryant-memorial/4860574002/

UK Air Accidents Investigation Branch. (2014). *AAIB Bulletin 1/2014 (Accident Investigation 1/2014; p. 19)*. UK Air Accidents Investigation Branch. https://assets.publishing.service.gov.uk/media/5422f8beed915d13710006f9/Rockwell_OV-10B_Bronco_G-BZGK_01-14.pdf

UK Civil Aviation Authority. (2025, March 10). *CAP 1724: Flying display pilot authorisation and evaluation: Requirements and guidance* (Edition 7). www.caa.co.uk/our-work/publications/documents/content/cap1724/

Wong, L., Bliese, P., & McGurk, D. (2003). Military leadership: A context specific review. *Leadership Quarterly, 14*, 657–692. https://doi.org/10.1016/J.LEAQUA.2003.08.001

Zynda, J. (2023). *Aviation mentorship program* (Honors thesis, Western Michigan University). ScholarWorks at WMU. https://scholarworks.wmich.edu/honors_theses/3675/

9 Pre-Show Preparation

Manolis Karachalios

INTRODUCTION

Air show preparation is a *Sisyphean*[1] yet exact process that begins immediately after the conclusion of the previous display season, exemplifying the solid dedication of air show performers to excellence and safety. The timing of the pre-display season varies by hemisphere: in the Northern Hemisphere, it typically spans from November to March, while in the Southern Hemisphere, it extends from May to September. These periods have been established based on climatic conditions conducive to outdoor exhibitions, ensuring that performers can fly their displays and the audience can watch in optimal weather (Salas & Maurino, 2010).

In preparing for air shows, the "*5M*" model is employed as a foundational framework, although it is traditionally associated with flight safety or the Human Factors Analysis and Classification System (HFACS) (Wiegmann & Shappell, 2003). By leveraging the 5Ms—huMan, Machine, Medium, Mission, and Management—air show performers adopt a thorough and systematic approach to their preparation strategies. This methodical application not only enhances the safety of each performance but also optimizes performance quality and showmanship, reflecting the depth of planning and commitment required in the high-risk environment of air shows (Federal Aviation Administration [FAA], 2023).

The preparation begins with the *huMan* element, where performers focus primarily on flight training to refine their aerobatic maneuvers and ensure peak aerial performance. Physical conditioning is also a significant area of emphasis, preparing the body to withstand the physical demands of high G-force maneuvers, as well as the possible effects of accumulative fatigue during prolonged display seasons (Newman, 2016). Additionally, psychological training is included to develop mental resilience and effective stress management techniques, which are vital for maintaining concentration and performance under the pressure of flying displays in front of an audience (Helmreich et al., 2017).

The *Machine* component involves rigorous maintenance and potential modifications of aircraft to meet specific display requirements, such as stunt flying display or special effects like pyrotechnics and ribbon cutting.[2] This ensures that the aircraft is not only safe but also equipped to enhance the spectacle of each performance (FAA, 2016).

DOI: 10.1201/9781003504801-12

Pre-show preparation also includes consideration of the *Medium*, referring to readiness for various environmental and weather conditions that might affect the display. This aspect of preparation ensures that performers can adapt their routines to any changes in external conditions, thereby maintaining safety and performance accuracy (Karachalios & Adjekum, 2023).

The *Mission* aspect involves media training and effective communication strategies to promote the flying display effectively. This training helps performers engage with their audience and media partners, enhancing the visibility and appeal of the event (Flin & O'Connor, 2017).

Lastly, *Management* involves meticulous planning of finances and time. Securing sponsorships and managing budgets are crucial for the financial sustainability of, especially, civilian air show performers. Time management is equally important, involving the scheduling of appearances and managing invitations according to a tightly coordinated annual schedule (Karachalios, 2022).

With the *5Ms* in mind, this Chapter examines the required preparation for an air show that begins with detailed *pre-arrival* planning, including logistical setups, training schedules, and the establishment of safety protocols, all tailored to the specific venue and expected weather conditions. Following this, the focus shifts to *post-arrival* activities, where the real-time application of these plans takes place. This includes finalizing the setup of equipment in the air show venue, conducting rehearsals, and engaging in community and media sessions.

Each section of this Chapter highlights how a systematic approach to both *pre-arrival* and *post-arrival* preparations enables air show performers to meet the demands of the show and exceed audience expectations, ensuring every performance is *precise*, *entertaining*, and ultimately *safe*.

PRE-ARRIVAL PLANNING AND PREPARATIONS

Display pilots engage in extensive preparations well in advance to ensure both seamless and safe performances. This process transcends standard protocols, reflecting a deeply ingrained practice of excellence in planning and operational foresight. Air shows are typically organized at airports to facilitate all the necessary operations, including both static and dynamic displays of aircraft, offering comprehensive facilities that accommodate everything from taxiing to takeoff and landing operations safely and effectively. However, air show sites are not limited to airports alone. Seaside air show events have become increasingly popular, offering unlimited access to audiences while combining other side events like musical concerts (Bournemouth Air Festival, 2024; Office du tourisme du Touquet, 2024; Pacific Airshow, 2024). In these scenarios, air show participants take off from a nearby airport, fly to the show site, perform their display, and return to the departure airport. Such operations add complexity to preparation needs, requiring additional considerations, such as fuel management, communication with multiple air traffic control (ATC) units, and weather conditions at various points of the route.

For events where the air show takes place at the same airport where the aircraft is parked, the planning phase begins with the inbound flight to the air show site. This involves in-depth preparation for fuel, weather, and flight route (Hajek, 2015; Karachalios, 2022). Pilots carefully outline their route, considering not only the required fuel but also potential diversion airfields while accounting for weather conditions en route and at the destination airport. They ensure a sufficient fuel supply to cover unforeseen delays due to ATC restrictions or weather-induced deviations. Military fast jets with air-to-air refueling capability often fly without external fuel tanks to preserve the aesthetic of their paint job, relying on airborne tankers for support during transit (see Figure 9.1). Additionally, pilots acquire scaled maps and charts, focusing on detailed graphical depictions of the venue and surrounding areas to pinpoint critical visual references for orientation during the display. This helps them understand the terrain at the display area and its potential effects on their display profile and altitudes.

As in all aspects of aviation, timing and communication are essential (Krivonos, 2007; Üzülmez & Yilmaz, 2017), but for air show pre-arrival planning and preparation become vital (Barker, 2003). Air show performers aim to arrive at the pre-planned time on the arrival day or, on the air show day, punctually for their displays, since accurate timing preserves the schedule and enhances the event's flow. Therefore, performers proactively communicate with event organizers and air bosses about any potential delays.

FIGURE 9.1 The Thunderbirds receive aerial refueling from a KC-135 inbound to an air show, 2017. (Copyright: U.S. Air Force photo by Tech. Sgt. Christopher Boitz.)

Furthermore, logistics play a substantial role in pre-show preparation. Pilots maintain an accurate log of their arrival and departure times, including planned stops for meals or fuel and the locations for aircraft parking and necessary ground equipment, ensuring no detail is overlooked. On arrival day, they usually conduct aerial reconnaissance of the crowd line and designated display areas, noting any zones to avoid and any obstacles affecting their flying display, which is crucial for maintaining safety margins. As soon as they arrive at the air show site, they meet the Flying Display Director or any other member of the flying control committee to confirm safety briefing and rehearsal times to seamlessly integrate them into their schedules, ensuring that preparation flows smoothly. Recognizing the importance of adequate rest, air show performers arrange accommodations well in advance with the event organizers to avoid any last-minute issues.

As an essential part of their logistical preparation, air show performers ensure they have all the necessary equipment and spare parts before arrival to prevent potential disruptions on the day of the display. Therefore, the availability of ground and engineering support by the event organizer is confirmed well in advance. Performers also check for specific needs, such as the correct type of fuel, engine oil, and smoke oil, ensuring all logistical elements are in place.

Compliance with legal and safety standards is imperative. Every performer diligently adheres to national safety regulations relevant to the event's location, and they stay informed by the air boss about any temporary restrictions or adjustments specific to the air show, such as altitude limitations and noise abatement procedures.[3] Before departure to the air show site, performers verify that all personal and aircraft documents are current and meet the necessary legal standards, including licenses, registrations, and required approvals.

Ultimately, air show organizers continuously assist air show performers in coordinating and accommodating their needs prior to the air show weekend, especially with the use of technology. Besides emails, phone calls, and text messages, common practices include using electronic questionnaires to gather required data and needs from participants, providing cloud-based access to air show-related documents, having video conferences, and maintaining social media chat groups that offer real-time updates on critical information.

POST-ARRIVAL PREPARATIONS

Upon their arrival at the air show venue, air show performers are excited to meet the organizers, immerse themselves in the festive environment, and become part of the air show family. However, this excitement is balanced with a specific and strict preparation plan for the air show that leaves them no room to relax.

Much like elite athletes, air show performers must navigate a range of environmental and venue-related factors that can significantly impact their performance. Research has shown that such factors play a critical role in the outcomes for elite male track athletes, affecting their ability to compete at their best (Hollings et al., 2012). These factors include altitude, temperature, and the specific layout of the competition area, all of which require athletes to adapt their strategies and conditioning. Similarly, air show performers must adjust to the unique conditions of each venue, from varying altitudes and temperatures to different layouts and spectator configurations. The importance of acclimatizing to these conditions cannot be overstated, as it ensures that performers can execute their routines with precision and safety.

In addition to environmental adjustments, the rigors of international travel pose significant challenges. Elite athletes must maintain peak physical condition despite the stresses of frequent travel (Pipe, 2011). This includes managing jet lag, dietary changes, and maintaining a training regimen while on the move. Air show performers face similar demands, as they often travel long distances to participate in aerial events. This requires them to stay physically and mentally fit to perform their display profiles safely and consistency in daily routines helps them mitigate the fatigue and disruption caused by travel.

Furthermore, organizational stress is another aspect shared by elite athletes and air show performers. The need to coordinate with event organizers, manage technical issues, and ensure adherence to safety protocols can create a high-pressure environment. Woodman and Hardy (2001) have highlighted the impact of organizational stress on elite athletes, noting that factors such as logistical challenges, team dynamics, and performance expectations can significantly affect an athlete's mental state and performance. For air show performers, effective delegation of responsibilities and thorough planning are crucial in managing these stresses and ensuring smooth operations. By clearly defining roles and establishing detailed plans, performers can reduce uncertainty and focus on their performance.

Finally, the ability to perform in extreme climatic conditions is a challenge for both elite athletes and air show performers. Just as elite female athletes must adapt to compete in harsh weather conditions during world championship events, as suggested by Zeiler and Shipway (2020), air show performers must be prepared to execute their routines flawlessly, without the weather affecting their performance. This involves not only physical preparation but also mental resilience to stay focused under challenging weather conditions, maintain peak performance levels, and ensure safety.

Physical Acclimatization

Air shows, with their high-stakes nature, often take place far from a performer's home base, necessitating meticulous travel and preparation planning to ensure that by the time of the event, pilots are fully acclimated both physically and psychologically and their aircraft is ready. Maintaining peak performance during the air show weekend requires a holistic approach to managing human factors, such as fatigue, sleep, and jet lag, while adjusting to varying environmental conditions, including venues at higher elevations.

Air show performers, much like elite athletes, must undergo careful physical acclimatization to maintain peak performance and ensure safety. This process involves adjusting to new time zones, climates, and environmental conditions, which can significantly impact their physiological and psychological readiness.

One major aspect of physical acclimatization is managing *jet lag*, a common issue faced by both air show performers and athletes who travel across time zones. Jet lag disrupts the body's internal clock, leading to fatigue, sleep disturbances, and impaired cognitive and physical performance (Caldwell, 2005; Caldwell et al., 2009). Lee and Galvez (2012) suggest that athletes must employ strategies such as controlled light exposure, sleep hygiene practices, and gradual adaptation to new time zones to mitigate the effects of jet lag. Similarly, air show performers must adopt these techniques to maintain alertness and precision during their routines.

The influence of *circadian rhythms* on athletic performance has been extensively studied. Manfredini et al. (1998) discuss how circadian rhythms affect various physiological processes, including energy levels, reaction times, and muscle strength. Disruptions to these rhythms, often caused by travel, can hinder performance. For air show performers, aligning their routines with their natural circadian rhythms as closely as possible helps maintain optimal performance levels.

Furthermore, *fatigue* is another critical factor that can affect performance. In aviation, fatigue is a well-documented issue, impacting pilot performance and safety (Gaines et al, 2020). Bendak and Rashid (2020) conducted a systematic review that emphasizes the importance of managing fatigue through adequate rest, proper scheduling, and fatigue risk management systems. For air show performers, maintaining consistent daily routines and ensuring sufficient rest is vital to counteract the physical and mental demands of their profession.

Managing *travel fatigue* and *jet lag* effectively is essential for maintaining performance. Rensburg et al. (2020) offer practical tips for athletes, such as staying hydrated, avoiding caffeine and alcohol before bedtime, and using strategic naps. These strategies are equally applicable to air show performers who need to stay sharp and focused despite the rigors of travel.

Proper *sleep hygiene* is a crucial factor for recovery and performance. Fowler et al. (2020) found that good sleep hygiene practices, such as maintaining a regular sleep schedule and managing light exposure, can significantly improve performance after long-haul travel. Similarly, air show performers must prioritize sleep hygiene to ensure they are well-rested and alert for their displays (Karachalios, 2022).

Environmental conditions can also pose challenges. Loat and Rhodes (1989) and Zhou et al. (2021) explore how factors like altitude, humidity, and temperature can affect human performance. For air show performers, adapting to different environmental conditions at each venue is crucial. This might involve acclimatizing to cope with extreme temperatures, adjusting to varying humidity levels, or higher venue elevation. Specifically, heat stress can be a significant issue for performers operating in hot environments (Goswami & Sharma, 2007). Guy et al. (2015) discuss how elite athletes adapt to hot conditions through strategies like acclimatization, hydration, and cooling techniques. Air show performers can benefit from similar strategies to prevent heat-related performance decrements.

High altitude and G-forces also pose unique physiological challenges. Linakis et al. (2017) review the potential differences in chemical exposures for fighter pilots resulting from high altitude and G-forces. In addition, research by Balldin (1986) and Besch et al. (1994) highlights the importance of understanding how factors like hypoxia and hyperoxia affect G-tolerance. Display pilots must be well-acquainted with their limits and utilize countermeasures such as specialized G-suits and Anti-G Straining Maneuvers to prevent G-Induced Loss of Consciousness (G-LOC).

Consistency in Daily Routines

To mitigate these challenges, pilots adhere to a strict regimen involving rigorous physical training and consistent eating habits. Physical training routines often include cardiovascular exercises to enhance stamina and strength training to cope with the physical demands of maneuvering the aircraft under intense conditions, as discussed in Chapter 5. An example of this is how members of the USN Blue Angels undergo intensive physical training to endure the high G-forces experienced during their aerobatic maneuvers (Gabriel, 2018; Prime Video, 2024).

It is also important for display pilots to maintain their existing daily routines, including dietary habits. For instance, a display pilot might follow a high-protein, low-carb diet to sustain energy levels and avoid gastrointestinal issues that could arise from eating unfamiliar foods (Gargano et al., 2021). Therefore, they often bring their own food to air shows to ensure their nutritional intake remains consistent, familiar, and uncontaminated (Karachalios, 2022). For instance, during international air shows, teams from different countries might import staple foods to prevent digestive issues that could arise from unfamiliar cuisine or to adhere to their religious dietary restrictions, such as halal[4] food.

Delegation of Responsibilities

Delegation of responsibilities is also crucial, especially during the air show weekend. Tasks related to aircraft maintenance, media interactions, and detailed flight planning are assigned to other team members. This division of labor is evident in military aerobatic teams like the Royal Air Force's Red Arrows, where each member has a clearly defined role supporting the pilots, allowing them to concentrate solely on their performance (Royal Air Force (RAF), 2024). For example, while the pilots focus on preparing for their formation flying and getting adequate rest, a dedicated team of engineers, the Blues, ensure the aircraft's mechanical systems are in optimal condition, while media officers manage public relations and schedule interviews and appearances.

Stress Management Techniques

Effective stress management is vital for optimal performance. By creating a performance *bubble* before their display, pilots isolate themselves to mentally prepare and reduce stress (Barker, 2003; Karachalios, 2022; Karachalios & Adjekum, 2023). This might involve sitting quietly in a designated area away from the busy show environment or using noise-canceling headphones to listen to calming music or focus-enhancing sounds. For Patty Wagstaff (2018), a legendary figure in the air show world, the *bubble* is sometimes her car, where she can remain cool and free from distractions. This allows the necessary *compartmentalization*, as Wagstaff suggests, and consequently, the mental focus for air show performers before their display, especially if they need to wait for prolonged periods after attending the safety briefing and while awaiting their show time.

Environmental Assessments

Environmental assessments and proactive decision making are integral to the post-arrival process at an air show. Display pilots assess weather conditions well in advance to determine if it is safe to perform. They avoid flying under marginal conditions that could compromise safety or performance. For example, if the forecasted crosswind exceeds operational thresholds, a display pilot will delay their performance. However, in the case of a strong tailwind, they may coordinate with the air boss to change the takeoff runway and successively adjust their display flow to maintain a consistent sequence of maneuvers.

Pre-Show Checklist and Planning

A pre-show checklist is essential to ensure nothing is overlooked. Pilots review every detail, from the technical aspects of the aircraft to the sequence of the show. They also prepare a display card that can be used as a quick reference guide that outlines every element of their performance, including timing for takeoff and landing, frequencies for communication with the air boss and ATC, weather conditions at the air show venue, and the alternate airfields, and procedures for critical emergencies.

For example, the Zeus Demo Team of the Hellenic Air Force employs a memory rule of thumb before lining up for takeoff. The demo pilots communicate with their safety observer to confirm that the critical systems of the aircraft are set for display. This includes ensuring that the ejection seat is armed, the video recording system is on, the flight control system is properly functioning, and the radar altimeter[5] is serviceable. Most critically, they ensure that the barometric altimeter is set to zero,[6] using the airfield's elevation as the main reference point for all maneuvers.

Air Show Safety Briefing

The ultimate key to the safe conduct of an air show is the air show safety briefing, prepared and delivered by the Flying Display Director (FDD) before each display day of the air show weekend (Karachalios, 2022; UK Civil Aviation Authority [UK CAA], 2025). This briefing covers crucial aspects of safety, such as weather limits, safe separations from the crowd, display lines, and the time sequence of display events. Attendance is mandatory for all air show participants, including pilots and support members such as ATC and weather personnel.

During the briefing, clear communication protocols with ATC are established to minimize unnecessary communications during critical phases of the performance.

Additionally, to avoid distractions and ensure a sterile Showbox, discrete frequencies are allocated for the display. If display pilots feel the radio becomes busy for no reason, they may request radio silence[7] on the radio from ATC by either calling "*all stations stop transmitting*" or a simple "*all stations standby*" to ensure they are not distracted by radio calls during their routine.

But most importantly, the safety briefing is an excellent opportunity to meet all members of the air show team and the air boss/FDD. It sets the tone for the main priorities of the weekend to run smoothly: expecting excellence and maintaining safety.

HUMAN FACTORS, EXCELLENCE, AND SAFETY IN PREPARATION FOR HIGH-PERFORMANCE EVENTS

The preparation strategies employed by air show performers can be closely analyzed and expanded upon through the lenses of excellence management, safety management, and human factors. By drawing parallels with other high-performance disciplines like MotoGP®, Formula 1®, and NBA® basketball, we can gain a deeper understanding of these practices.

Excellence Management

Excellence management in air shows involves careful planning and continuous improvement, akin to the strategies used by high-performance sports teams. Formula 1® teams, for example, constantly analyze race data to refine strategies and improve vehicle performance. Similarly, air show performers review their flight data and debrief thoroughly after each show to identify areas for improvement. This continuous loop of performance review and enhancement is also seen in MotoGP®, where riders and teams adjust their approaches based on track conditions and previous race outcomes.

Safety Management

Safety management is perhaps the most critical element, given the high risks involved in air shows. This includes rigorous adherence to safety protocols, emergency response planning, and regular maintenance checks—practices shared with other high-risk sports. For instance, MotoGP® riders rely on advanced protective gear and strictly enforced safety regulations to minimize risks during races. Lippi et al. (2007) discuss changes in technical regulations and drivers' safety in top-class motor sports, highlighting the impact of regulatory advancements on reducing injuries. The importance of helmets in preventing injuries among motorcycle riders is similarly emphasized by Liu et al. (2008) in their systematic review.

Similarly, Formula 1® has evolved with enhanced safety measures such as the HALO system, designed to protect drivers from head injuries during crashes. Frömmig (2023) provides an in-depth analysis of safety in motorsport, while Kandula and Anima (2019) explore the effectiveness of the HALO system in Formula 1® through impact simulations. Mellor (2002) and Savage (2010) examine the role of impact testing and the development of penetration resistance in Formula 1® cars, emphasizing the continuous evolution of safety features in the sport. These studies collectively underline the critical nature of safety management across various high-risk activities.

In air shows, the safety protocols extend to ensuring aircraft are maintained to the highest standards and that emergency procedures are well-practiced and effective. This comprehensive approach to safety management is crucial in preventing accidents and ensuring the well-being of participants and spectators alike.

Human Factors

Human factors play a crucial role in ensuring the safety and performance of aviators (UK Civil Aviation Authority, 2023), particularly air show performers. This encompasses psychological readiness, physical fitness, and cognitive abilities, which are similarly crucial for athletes in MotoGP®, Formula 1®, and the NBA®.

Psychological readiness involves mental toughness and the use of psychological strategies to cope with high-pressure situations, as discussed by Crust and Azadi (2010). This mental resilience is critical for air show performers who must remain focused and composed while executing complex aerial maneuvers. Furthermore, Lubitskaya and Serdyukov (2020) highlight the importance of motivation in the psychological readiness of athletes, which is equally applicable to air show performers who must stay motivated and mentally prepared for their performances.

Physical fitness is another essential component, as elite motor racing drivers demonstrate through their high levels of VO2 peak,[8] body composition, and neck strength (McKnight et al., 2019). Similarly, air show performers require peak physical fitness to handle the physical demands of high-G maneuvers and extended periods of flying. The systematic review by Mancha-Triguero et al. (2019) on physical fitness in basketball players underscores the importance of maintaining optimal fitness levels, a principle that extends to air show performers. Xiao et al. (2021) also discuss the benefits of functional training in enhancing athletes' physical fitness, which can be applied to the training regimens of air show performers.

Cognitive abilities are equally critical, as air show performers must process vast amounts of information quickly and accurately. Birrer and Morgan (2010) emphasize the role of psychological skills training in enhancing performance in high-intensity sports, which can be adapted to the training of air show performers. Similarly, Cook and Khan (2001) discuss the importance of cognitive skills in motorsports, highlighting the need for sharp decision making abilities under pressure. The research by Land and Tatler (2001) and Pammer and Blink (2018) on visual processing in expert drivers provides insights into how visual strategies and expert cognitive processing can enhance performance, which can be relevant for air show performers who rely heavily on visual cues and rapid decision making.

For example, just as a Formula 1® driver must have the mental resilience to make split-second decisions at high speeds, air show performers must maintain intense focus to execute precise maneuvers while managing the stresses of high-G display flying at low altitudes (Cook & Khan, 2001). Moreover, NBA® players' ability to remain mentally sharp throughout a game under physical exhaustion parallels the endurance air show performers need during long displays or more than one flight (Cao et al., 2010). Last, the ability to manage pressure, as discussed by Maher et al. (2018) and Wolch et al. (2020), is critical for both basketball players and air show performers, who must perform flawlessly under high-pressure conditions.

ULTIMATE PRIORITIES DURING DISPLAY FLYING: SAFETY FIRST

Stay Away From the Crowd, Away From the Ground, Recover.
or
Save the Crowd, Save the Pilot, Save the Jet.

The distinction between a successful air show display and a disaster hinges on maintaining clear priorities. These priorities are encapsulated in two essential mantras that guide display pilots: "*Stay Away From the Crowd, Away From the Ground, Recover*" and "*Save the Crowd, Save the Pilot, Save the Jet.*" Both emphasize a hierarchy of safety that is crucial for safe display flying. The first mantra points out the importance of spatial awareness and altitude management to avoid endangering primarily the audience and then oneself, while the second highlights the imperative of prioritizing human life and safety over the aircraft itself. These guiding principles ensure that pilots remain mindful of the most critical aspects of safety during their performances.

Stay Away from the Crowd, Away from the Ground, Recover!

This mantra accentuates the fundamental aspects of spatial awareness and recovery from unplanned conditions during air show performances, inspired by Weick and Sutcliffe's (2007) principles of resilient performance. Each component highlights a critical aspect of flight safety:

Stay Away from the Crowd
Ensuring the safety of the spectators is the top priority; display pilots perform while maintaining a safe distance from the crowd to prevent any risk of accidents that could endanger lives on the ground. This rule is strictly enforced by air show organizers and regulatory bodies and is a non-negotiable aspect of display flying.

Away from the Ground

Maintaining adequate altitude is essential for providing a safety buffer that allows display pilots to recover from unexpected situations. This principle helps prevent a collision with the ground or any obstacles, which can be catastrophic.

Recover

Immediate recovery from any deviation or loss of control is crucial. Air show performers practice recognizing when a maneuver is not going as planned and must initiate recovery procedures without hesitation, an ability that ensures the pilot's safety and the integrity of the performance.

Save the Crowd, Save the Pilot, Save the Jet

This hierarchy of priorities reinforces the overarching principle of human safety more than anything else, followed by the preservation of the aircraft:

Save the Crowd

The primary concern during any display flight is the safety of the spectators. All maneuvers are planned and executed with this priority in mind, ensuring that the crowd is never put at risk.

Save the Pilot

The next priority is the safety of the pilot. In the event of an emergency, the pilot's well-being takes precedence. Training programs emphasize quick decision making skills and effective use of safety equipment, such as ejection seats and parachutes, to protect the pilot in critical situations.

Save the Jet

While preserving the aircraft is important, it is secondary to ensuring the safety of the crowd and the pilot. Pilots are trained to make tough decisions, including abandoning the aircraft, if necessary, to protect lives.

CASE STUDY: BENCHMARKING EXCELLENCE—LESSONS FROM THE BLUE ANGELS ON PRE-SHOW PREPARATION

The USN Blue Angels' approach to pre-show preparation offers a robust model of meticulous planning and execution, reflective of their high standards in precision flying and operational excellence. Central to their preparation is the detailed Support Manual (U.S. Navy, 2024a), a culmination of 78 years of air show experience, which dictates every aspect of their operations, from maintenance to public engagements (see Figure 9.2).

Key Components of the Blue Angels' Preparation

The Blue Angels' support manual (U.S. Navy, 2024a) is a comprehensive guide covering key aspects of their operations and expected preparations and support activities needed by the air show organizers.

FIGURE 9.2 Fat Albert, Blue Angels' C-130, unloading equipment in preparation for an air show. (Copyright: San Diego Air and Space Museum Archive (SDASM).)

Beyond just flying operations, in their support manual, the Blue Angels emphasize community outreach and effective media engagement as part of their mission. Detailed plans for community activities and interactions are crafted well in advance, underlining their role as ambassadors of the Navy and Marine Corps.

It needs to be acknowledged that while their operations are highly structured, the Blue Angels' team is trained to be flexible and responsive to on-the-ground realities, adjusting their plans as needed to accommodate unforeseen circumstances or better opportunities for community engagement.

Chapter III of the Manual, titled "Personnel Support," can be overwhelming for first-time event organizers. It outlines the extensive support required from local organizers, akin to the logistical needs of a touring rock band. Before the season begins, team representatives conduct winter visits to prospective sites to assess feasibility and discuss requirements, such as hangar space, fuel supplies, and local accommodations. Organizers must provide between 65 and 85 hotel rooms, ideally within a 30-minute drive from the airfield, ensuring comfort and convenience for the team. A variety of vehicles are needed to support the team's logistics, including cars and vans for personnel and equipment transport. Access to a local physician and nearby hospital information is crucial for emergencies, along with provisions for routine medical needs. High-quality gym access is mandated, with strength training equipment and suitable hours to maintain the physical conditioning required for performance (Simenz et al., 2005). Organizers should provide meals and beverages to support the team during busy show periods, ensuring they remain energized and focused. Additionally, reserved seating for Blue Angels' friends and family, along with managing passes, is essential for smooth operations and security during events.

This detailed support structure of the Manual is vital for maintaining the team's high performance throughout the 2024 display season, during which they were planned to perform 62 flight demonstrations at 32 separate locations across the United States (U.S. Navy, 2024b). This demanding schedule, spread over eight months, highlights the necessity for meticulous planning and systematic coordination. These efforts are essential not only for demonstrating the excellence and teamwork of the US Navy and Marine Corps aviation but also for ensuring the safety of both the community and the Blue Angels team itself.

Lessons Learned

The Blue Angels' methodical approach serves as a benchmark for other performance teams, both in the military and civilian sectors. Their emphasis on rigorous preparation, detailed procedural adherence, and proactive community engagement can be adapted to enhance the operational effectiveness and public impact of other aerobatic teams. Their use of a comprehensive support manual and readiness to integrate technological tools also provide valuable lessons in operational efficiency and adaptability.

However, this detailed support manual might be overwhelming for civilian single-aircraft air show performers. While the extensive preparations of the Blue Angels set a high standard, sections of their support manual can be used in a simplified yet constructive method by air show performers and organizers. This systematic approach to event preparation can help manage the administrative burden that often overwhelms display pilots. By adopting these practices, even smaller teams can benefit from a structured and efficient preparation process.

The Blue Angels' preparation for air show performances involves an extensive setup that is significantly more resource-intensive than the flying training and preparation of the display pilots themselves. The coordination involved in preparing for the team's arrival at an air show site requires a substantial number of personnel, reinforcing the immense value of teamwork. This intricate orchestration involves not only the pilots but also numerous support staff who manage everything from equipment maintenance to public relations. It demonstrates a complex, collaborative effort crucial for the smooth execution of each performance, highlighting the importance of each team member to the overall success of the show. Moreover, such an approach stresses that the effectiveness of the team extends far beyond the individual capabilities of its pilots, emphasizing the critical role of support personnel in achieving excellence in air show performances.

NOTES

1 Of or relating to Sisyphus, in Greek mythology a king of Corinth who was condemned in Hades endlessly to roll a heavy stone up a hill only for it to roll down again as he reached the top; resembling Sisyphus or that of Sisyphus; endless, laborious, and ineffective.

(Oxford English Dictionary, 2024b)

2 The ribbon cut is a precision flying maneuver where a pilot slices through a ribbon strung between two poles in inverted flight using the aircraft's propeller or rudder, or in a knife-edge attitude using the aircraft's wing (International Council of Air Shows (ICAS), 2016).
3 Noise abatement procedures are specific flight operation techniques and regulations designed to minimize aircraft noise over populated areas, typically involving modified flight paths, altitude restrictions, reduced engine thrust, and designated takeoff and landing patterns (Erkelens, 2000).
4 Halal foods are conforming to Islamic dietary laws (Oxford English Dictionary, 2024a).
5 Radar altimeter measures the height that the aircraft is flying over the ground or water.
6 Barometric altitude, when set to zero, refers to the condition where an aircraft's altimeter is adjusted to read zero at a specific reference point, usually the ground level at the air show airport. This setting, known as QFE, allows the pilot to read the height above that specific point, also referred to as show center or display datum, aiding in the flying display by providing a direct indication of altitude relative to the airfield.
7 Radio silence is a procedure used to keep the frequency channel exclusively available to the aircraft or formation displaying, ensuring clear and uninterrupted communication with the air boss/FDD, the safety observer, and within the formation (SKYbrary Aviation Safety, 2024; UK CAA, 2020).
8 VO2 (oxygen consumption) is a measure of the volume of oxygen that an individual can utilize during intense exercise, reflecting their aerobic capacity and cardiovascular fitness (Mancini et al., 2000).

REFERENCES

Balldin, U. I. (1986). Factors influencing G-tolerance. *Clinical Physiology, 6*(3), 209–219. https://doi.org/10.1111/j.1475-097X.1986.tb00620.x

Barker, D. (2003). *Zero error margin: Airshow display flying analysed*. Freeworld Publications.

Bendak, S., & Rashid, H. (2020). Fatigue in aviation: A systematic review of the literature. *International Journal of Industrial Ergonomics, 76*, 102928. https://doi.org/10.1016/j.ergon.2020.102928

Besch, E. L., Werchan, P. M., Wiegman, J. F., Nesthus, T. E., & Shahed, A. R. (1994). Effect of hypoxia and hyperoxia on human +Gz duration tolerance. *Journal of Applied Physiology, 76*(4), 1693–1700. https://doi.org/10.1152/jappl.1994.76.4.1693

Birrer, D., & Morgan, G. (2010). Psychological skills training as a way to enhance an athlete's performance in high-intensity sports. *Scandinavian Journal of Medicine & Science in Sports, 20*. https://doi.org/10.1111/j.1600-0838.2010.01188.x

Bournemouth Air Festival. (2024). *Bournemouth Air Festival*. Bournemouth Air Festival. https://bournemouthair.co.uk/

Caldwell, J. (2005). Fatigue in aviation. *Travel Medicine and Infectious Disease, 3*(2), 85–96. https://doi.org/10.1016/J.TMAID.2004.07.008

Caldwell, J. A., Mallis, M. M., Caldwell, J. L., Paul, M. A., Miller, J. C., & Neri, D. F. (2009). Fatigue countermeasures in aviation. *Aviation, Space, and Environmental Medicine, 80*(1), 29–59.

Cao, Z., Price, J., & Stone, D. (2010). Performance under pressure in the NBA. *Journal of Sports Economics, 12*, 231–252. https://doi.org/10.1177/1527002511404785

Cook, J., & Khan, K. (2001). Performance in motor sports. *British Journal of Sports Medicine, 35*, 290–291. https://doi.org/10.1136/bjsm.35.5.290

Crust, L., & Azadi, K. (2010). Mental toughness and athletes' use of psychological strategies. *European Journal of Sport Science, 10*, 43–51. https://doi.org/10.1080/17461390903049972

Erkelens, L. (2000, August). Research into new noise abatement procedures for the 21st century. In *AIAA Guidance, Navigation, and Control Conference and Exhibit* (p. 4474).

Federal Aviation Administration. (2023). *Pilot's handbook of aeronautical knowledge* (FAA-H-8083-25C). www.faa.gov/regulations_policies/handbooks_manuals/aviation/faa-h-8083-25c.pdf

Flin, R., & O'Connor, P. (2017). *Safety at the sharp end: A guide to non-technical skills.* CRC Press.

Fowler, P., Knez, W., Thornton, H., Sargent, C., Mendham, A., Crowcroft, S., Miller, J., Halson, S., & Duffield, R. (2020). Sleep hygiene and light exposure can improve performance following long-haul air travel. *International Journal of Sports Physiology and Performance*, 1–10. https://doi.org/10.1123/ijspp.2019-0931

Frömmig, L. (2023). Safety in motorsport. In *Basic course in race car technology: Introduction to the interaction of tires, chassis, aerodynamics, differential locks and frame* (pp. 429–454). Springer Fachmedien Wiesbaden.

Gabriel, M. N. (2018, April 19). It takes intensive training, perseverance to become a member of U.S. Navy Blue Angels. *Treasure Coast.* www.tcpalm.com/story/entertainment/whattodoin772/2018/04/18/takes-intensive-training-perseverance-become-member-u-s-navy-blue-angels/526397002/

Gaines, A. R., Morris, M. B., & Gunzelmann, G. (2020). Fatigue-related aviation mishaps. *Aerospace Medicine and Human Performance, 91*(5), 440–447.

Gargano, D., Appanna, R., Santonicola, A., Bartolomeis, F., Stellato, C., Cianferoni, A., Casolaro, V., & Iovino, P. (2021). Food allergy and intolerance: A narrative review on nutritional concerns. *Nutrients, 13*. https://doi.org/10.3390/nu13051638

Goswami, P., & Sharma, S. (2007). Evaluation of pre takeoff heat stress in a modern fighter cockpit: A field study. *Indian Journal of Aerospace Medicine, 51*(2), 54–64.

Guy, J., Deakin, G., Edwards, A., Miller, C., & Pyne, D. (2015). Adaptation to hot environmental conditions: An exploration of the performance basis, procedures and future directions to optimise opportunities for elite athletes. *Sports Medicine, 45*, 303–311. https://doi.org/10.1007/s40279-014-0277-4

Hajek, D. (2015, August 30). An aerobatics pilot spins (and rolls, and loops) a career from a crash. *NPR.* www.npr.org/2015/08/30/433515500/an-aerobatics-pilot-spins-and-rolls-and-loops-a-career-from-a-crash

Helmreich, R. L., Merritt, A. C., & Wilhelm, J. A. (2017). The evolution of crew resource management training in commercial aviation. In *Human error in aviation* (pp. 275–288). Routledge.

Hollings, S., Hopkins, W., & Hume, P. (2012). Environmental and venue-related factors affecting the performance of elite male track athletes. *European Journal of Sport Science, 12*, 201–206. https://doi.org/10.1080/17461391.2011.552640

International Council of Air Shows (ICAS). (2016). *Air show performers safety manual* (p. 27). ICAS. www.airshows.aero/GetDoc/3444

Kandula, R., & Anima, C. (2019). High-speed savior: Exploring Formula One's HALO frame using impact simulations. *International Journal of Science and Research (IJSR), 8*. https://doi.org/10.21275/SR24119231329

Karachalios, M. (2022). *An evaluation of the relationships between resilient safety culture, safety risk parameters, and mindfulness in the international air show community* [PhD dissertation, University of North Dakota]. https://commons.und.edu/theses/4267/

Karachalios, M., & Adjekum, D. K. (2023). *Air show performers: Safety, risk management, and psychological factors.* CRC Press. https://doi.org/10.1201/9781003431879

Krivonos, P. D. (2007, June). *Communication in aviation safety: Lessons learned and lessons required.* In R. Kennedy (Ed.), *Proceedings of the Regional Seminar of the Australia and New Zealand Societies of Air Safety Investigators* (Vol. 4, pp. 1–35). Australia and New Zealand Societies of Air Safety Investigators.

Land, M., & Tatler, B. (2001). Steering with the head: The visual strategy of a racing driver. *Current Biology, 11*, 1215–1220. https://doi.org/10.1016/S0960-9822(01)00351-7

Lee, A., & Galvez, J. (2012). Jet lag in athletes. *Sports Health, 4*, 211–216. https://doi.org/10.1177/1941738112442340

Linakis, M. W., Job, K. M., Liu, X., Collingwood, S. C., Pangburn, H. A., Ott, D. K., & Sherwin, C. M. T. (2017). Riding (high) into the danger zone: A review of potential differences in chemical exposures in fighter pilots resulting from high altitude and G-forces. *Expert Opinion on Drug Metabolism & Toxicology, 13*(9), 925–934. https://doi.org/10.1080/17425255.2017.1360867

Lippi, G., Salvagno, G., Franchini, M., & Guidi, G. (2007). Changes in technical regulations and drivers' safety in top-class motor sports. *British Journal of Sports Medicine, 41*, 922–925. https://doi.org/10.1136/bjsm.2007.038778

Liu, B., Ivers, R., Norton, R., Boufous, S., Blows, S., & Lo, S. (2008). Helmets for preventing injury in motorcycle riders. *The Cochrane Database of Systematic Reviews, 1*, CD004333. https://doi.org/10.1002/14651858.CD004333.pub3

Loat, C., & Rhodes, E. (1989). Jet-lag and human performance. *Sports Medicine, 8*, 226–238. https://doi.org/10.2165/00007256-198908040-00003

Lubitskaya, O., & Serdyukov, E. (2020). Motivation in the structure of psychological readiness of athletes to competition. *The Humanities and Social Studies in the Far East*. https://doi.org/10.31079/1992-2868-2020-17-4-132-138

Maher, R., Marchant, D., Morris, T., & Fazel, F. (2018). Managing pressure at the free-throw line: Perceptions of elite basketball players. *International Journal of Sport and Exercise Psychology, 18*, 420–436. https://doi.org/10.1080/1612197X.2018.1536159

Mancha-Triguero, D., García-Rubio, J., Calleja-González, J., & Ibáñez, S. (2019). Physical fitness in basketball players: A systematic review. *The Journal of Sports Medicine and Physical Fitness, 59*(9), 1513–1525. https://doi.org/10.23736/S0022-4707.19.09180-1

Mancini, D., LeJemtel, T., & Aaronson, K. (2000). Peak Vo2. *Circulation, 101*(10), 1080–1082. https://doi.org/10.1161/01.CIR.101.10.1080

Manfredini, R., Manfredini, F., Fersini, C., & Conconi, F. (1998). Circadian rhythms, athletic performance, and jet lag. *British Journal of Sports Medicine, 32*, 101–106. https://doi.org/10.1136/bjsm.32.2.101

McKnight, P., Bennett, L., Malvern, J., & Ferguson, D. (2019). VO2peak, body composition, and neck strength of elite motor racing drivers. *Medicine & Science in Sports & Exercise*. https://doi.org/10.1249/MSS.0000000000001961

Mellor, A. N. (2002). Impact testing in Formula One. *International Journal of Crashworthiness, 7*(4), 475–486. doi:10.1533/cras.2002.0231

Newman, D. (2016). *High G flight: Physiological effects and countermeasures*. Routledge. https://doi.org/10.4324/9781315586779

Office du tourisme du Touquet. (2024). *Meeting aérien—Le Touquet air show*. Office du Tourisme du Touquet. www.letouquet.com/agenda/meeting-aerien-le-touquet-air-show/

Oxford English Dictionary. (2024a). Halal. *Oxford English Dictionary*. www.oed.com/dictionary/halal_adj

Oxford English Dictionary. (2024b). Sisyphean. In *Oxford English Dictionary*. Oxford English Dictionary. www.oed.com/dictionary/sisyphean_adj

Pacific Airshow. (2024). *Pacific Airshow*. Pacific Airshow. www.pacificairshow.com

Pammer, K., & Blink, C. (2018). Visual processing in expert drivers: What makes expert drivers expert? *Transportation Research Part F: Traffic Psychology and Behaviour, 55*, 353–364.

Pipe, A. (2011). International travel and the elite athlete. *Clinical Journal of Sport Medicine: Official Journal of the Canadian Academy of Sport Medicine, 21*(1), 62–6. https://doi.org/10.1097/JSM.0b013e318205dfc9

Prime Video (Director). (2024, March 20). *The Blue Angels* [Video recording]. www.youtube.com/watch?v=f36UTLWPL6I

Rensburg, D., Fowler, P., & Racinais, S. (2020). Practical tips to manage travel fatigue and jet lag in athletes. *British Journal of Sports Medicine, 55*, 821–822. https://doi.org/10.1136/bjsports-2020-103163

Royal Air Force (RAF). (2024). *Red Arrows, Royal Air Force*. Royal Air Force. www.raf.mod.uk/display-teams/red-arrows/

Salas, E., & Maurino, D. (Eds.). (2010). *Human factors in aviation*. Academic Press.

Savage, G. (2010). Development of penetration resistance in the survival cell of a Formula 1 racing car. *Engineering Failure Analysis, 17*(1), 116–127.

Simenz, C., Dugan, C., & Ebben, W. (2005). Strength and conditioning practices of national basketball association strength and conditioning coaches. *Journal of Strength and Conditioning Research, 19*, 495–504. https://doi.org/10.1519/15264.1

SKYbrary Aviation Safety. (2024). Radio silence. *SKYbrary Aviation Safety*. https://skybrary.aero/articles/radio-silence

UK Civil Aviation Authority (UK CAA). (2020, November 26). *Radiotelephony Manual, CAP 413* (Version 23.1). UK Civil Aviation Authority. www.caa.co.uk/publication/download/18165

UK Civil Aviation Authority. (2023, February 2). *CAP 737: Flight crew human factors handbook* (Edition 2). www.caa.co.uk/our-work/publications/documents/content/cap-737/

UK Civil Aviation Authority. (2025, February 4). *CAP 403: Flying displays and special events – Safety and administrative requirements and guidance* (Edition 22). www.caa.co.uk/our-work/publications/documents/content/cap-403/

U.S. Navy. (2024a). *Blue Angels Support Manual*. www.blueangels.navy.mil/assets/docs/resources/support-manual.pdf

U.S. Navy. (2024b). *U.S. Navy Blue Angels*. U.S. Navy Blue Angels. www.blueangels.navy.mil/

Üzülmez, M., & Yilmaz, A. K. (2017, October). Holistic time management in aviation: Analyzing the causes of delay in Europe airspace. In *3rd International Conference on Economic and Social Impacts of Globalization* (p. 226).

Wagstaff, P. (2018, May 30). Pilots must compartmentalize while in the air. *Plane & Pilot Magazine*. www.planeandpilotmag.com/article/pilots-must-compartmentalize-while-in-the-air/

Weick, K. E., & Sutcliffe, K. M. (2007). *Managing the unexpected: Resilient performance in an age of uncertainty*. Jossey-Bass.

Wiegmann, D. A., & Shappell, S. A. (2003). The human factors analysis and classification system (HFACS). In *A human error approach to aviation accident analysis*. Routledge.

Wolch, N., Arthur-Cameselle, J., Keeler, L., & Suprak, D. (2020). The effects of a brief mindfulness intervention on basketball free-throw shooting performance under pressure. *Journal of Applied Sport Psychology, 33*, 510–526. https://doi.org/10.1080/10413200.2020.1720044

Woodman, T., & Hardy, L. (2001). A case study of organizational stress in elite sport. *Journal of Applied Sport Psychology, 13*, 207–238. https://doi.org/10.1080/104132001753149892

Xiao, W., Soh, K., Wazir, M., Talib, O., Bai, X., Bu, T., Sun, H., Popović, S., Masanovic, B., & Gardašević, J. (2021). Effect of functional training on physical fitness among athletes: A systematic review. *Frontiers in Physiology, 12*. https://doi.org/10.3389/fphys.2021.738878

Zeiler, P., & Shipway, R. (2020). Some like it hot? Elite female athlete perspectives on competing in extreme climatic conditions at world championship sports events. *International Journal of Event and Festival Management*. https://doi.org/10.1108/ijefm-08-2020-0045

Zhou, B., Ding, L., Chen, B., Shi, H., Ao, Y., Xu, R., & Li, Y. (2021). Physiological characteristics and operational performance of pilots in the high temperature and humidity fighter cockpit environments. *Sensors, 21*(17), 5798.

10 Case Studies in Training
Framework for Resilient Excellence

Manolis Karachalios

INTRODUCTION

In the world of display flying, excellence is not stumbled upon by luck. It has been earned through systematic training, detailed-oriented preparation, and a persistent pursuit of improvement. In this Chapter, we explore the training methods of some of the most acclaimed air show performers, both civilian and military, offering a vivid comparative analysis that highlights the varied paths to mastering this aerial art form.

It needs to be noted that the air show community comprises display pilots with exceptional flying skills and outstanding levels of both airmanship and showmanship. Given the vast diversity in training experiences, it would be impossible to study every method each pilot has undergone to achieve the level of excellence required to perform in front of thousands. As discussed in previous chapters, training methods depend on the pilot's background, nationality, aviation regulations, and other factors. To remain within the scope of this publication, we have selected a handful of civilian air show performers based on the available information online about their display flying training, their extensive background, and their recognized achievements: Dario Costa, Mike Goulian, Rob Holland, and Frank van Houten.

Each pilot carves their own path to greatness, influenced by firsthand experiences and career demands. Dario Costa's training, marked by resilience and innovation, stands out; Mike Goulian's precision and discipline are exemplary, while Rob Holland's cutting-edge maneuvers and Frank van Houten's approach to safety paint a picture of diverse expertise.

In the second part of the case studies, we will shift our focus to the structured training programs of military display teams, specifically highlighting the US Air Force Heritage Flight Program (USAFHFP), the USAF Viper West Demonstration Team, and the Hellenic Air Force (HAF). These programs offer a systematic approach to developing exceptional display pilots, providing all the necessary resources to guarantee exceptional flying skills and mindful decision making, with the ultimate goal of promoting excellence and achieving safety.

The USAFHFP, for example, combines the legacy of air combat with modern fighter jets, while the Viper West team hones the agility of its jets and the precision of its pilots. The HAF's program, steeped in tradition, blends consistent training with operational needs.

DOI: 10.1201/9781003504801-13

By weaving together these diverse training methodologies and their outcomes, we aim to uncover the threads of commonality that knit the fabric of excellence in display flying. With a holistic comparative analysis, our aim is to find the strengths of each approach and uncover best practices that can elevate training programs universally. Through these case studies, the reader will gain an intimate understanding of the dedication, discipline, and innovation necessary for a display pilot to reach the level of mastery required and expected by the air show community.

Closing the chapter, and while striving to enrich our analysis, we introduce a comprehensive model for display pilot training: the *EXCELS Model*. This model distills the essence of successful training into six core elements: *Education*, *eXperience*, *Conditioning*, *Emergency preparedness*, *Leadership*, and *Safety*, with the aim to provide a blueprint for aspiring display pilots to achieve the highest levels of performance and safety, ensuring the continued evolution and excellence of the air show community.

CASE STUDIES FOR CIVILIAN AIR SHOW PERFORMERS

Case Study 1. Dario Costa: Mental Preparation

Dario Costa is an Italian professional Red Bull Air Race competitor, stunt pilot, aerobatic performer, and flight instructor. He grew up in Bologna, Italy, and began his aviation journey at the age of 14, performing his first solo flight at 16. Costa studied physics at the University of Bologna and later became Italy's youngest flight and aerobatics instructor (Costa, 2025).

Dario Costa is notable for several achievements, including being the first pilot to fly through a tunnel setting a Guinness World Record in September 2021 for the longest tunnel flown by an airplane (Red Bull Air Race, 2024). Throughout his career, Costa has been dedicated to aerobatic flying, teaching the techniques in Italian flying schools, and competing in the Red Bull Air Race series, while his dedication and contributions to aviation have earned him numerous awards and accolades.

Training Approach and Philosophies

Costa's training regimen is distinguished by its use of advanced simulation tools that mimic real-time aerodynamic changes and environmental challenges (Costa, 2025). This methodology not only enhances his technical skills but also bolsters his confidence to perform under high-stress conditions. His preparation for the Tunnel Pass, even though it did not include any aerobatic maneuvers, required meticulous planning and precision flying within extremely confined space, necessitating reaction times of less than 250 milliseconds to adapt to subtle air movements and spatial limitations (Red Bull Air Race, 2024).

First, he used simulating techniques, not with software but rather with a car itself, to gain the feeling and the visual references while driving at the same speed as when he would be flying his aircraft. He installed cameras on the car at eye-sight to be able to debrief his ride with a high-definition virtual reality (VR) device that allowed him a mental rehearsal unlimited times before the actual day. Then, the majority of his physical training was conducted with the video in the background, having random dots on a screen that he had to touch to train his reaction time while wearing strobe glasses[1] that were creating random flashes simulating the sequencing lights on top and on the sides of the tunnel. Such scientifically backed techniques had the intent to accustom him both mentally and physically to the unique flying environment.

The success of the Tunnel Pass was also a testament to the importance of teamwork. The project involved a collaboration among a team of 40 professionals, including engineers, safety experts, and logistical coordinators, which allowed Costa to obtain scientific support for his endeavor and be thoroughly prepared before the flight. The collective effort was pivotal in developing the specialized training technologies and methods that supported Costa's successful flight (Medalen, 2021).

Beyond his competitive achievements, Costa is deeply committed to the aviation community, actively sharing his knowledge and expertise. At the 2024 European Airshow Council (EAC) Convention, he shared his experiences and engaged in a constructive discussion with the extended air show community (European Airshow Council (EAC), 2024). Beyond that, he actively inspires and educates new generations of aspiring pilots by engaging in social media interaction with his fan base. It is without saying that his dedication to promoting continuous learning and innovative training practices helps foster a culture of safety and excellence in aviation.

Case Study 2. Mike Goulian: Physical Preparation

Michael "Mike" Goulian is a distinguished American aviator renowned for his breathtaking air show performances and his participation in the Red Bull Air Race. He began flying at an early age and quickly rose to prominence by winning the US National Advanced Aerobatic Championship in 1990 and the Unlimited Championship in 1995. Goulian exemplifies a holistic and rigorous approach to display flying training, with training methodologies that incorporate a blend of dedicated physical fitness regimes and detailed scenario-based training exercises, which are crucial for maintaining his edge during the display season (Goulian, 2024).

Understanding the physical demands of aerobatic flying, Goulian commits to a stringent physical training program. To combat intense G-forces during maneuvers, Goulian's fitness routine focuses on enhancing his cardiovascular health and muscular strength, primarily focusing on long-distance cycling, often covering more than 30 miles multiple times a week (Goulian, 2024).

Mental acuity is equally crucial for Goulian. He honed his concentration skills not only through flying but also through golf, a sport that mirrors the focus and precision (Wang et al., 2021) required in aviation. This combination of physical fitness and mental discipline enables Goulian to perform at the highest level in both air shows and races.

Furthermore, his dual role as both a competitor and an instructor enriches Goulian's expertise. Teaching and mentoring allow him to refine his knowledge and skills continually as he explains and demonstrates techniques to student pilots. Meanwhile, competing keeps his skills sharp and provides practical experiences that he can bring back to his instructional practices. This synergy enhances his abilities in both domains, making him a better pilot and instructor pilot.

Case Study 3. Rob Holland: Innovative Showmanship

Rob Holland is an American air show pilot renowned for his innovative aerobatic maneuvers. He earned his pilot's license as a teenager and quickly progressed into aerobatics. With extensive flying experience, Rob has spent more than 20 years performing full-time air shows, winning numerous national and world awards. His airmanship epitomizes the blend of relentless innovation with rigorous discipline in his training methods, while his mastery in performing his aircraft and his approachable demeanor have garnered him a dedicated global fan base (Ultimate Airshows, 2024).

Interestingly, Holland's aviation training experience is broad, expanding beyond flying air shows. Holding a Bachelor of Science degree in aviation/airway management and aviation flight operations, Rob has been committed to lifelong learning and continuous skill enhancement since becoming a pilot in 1992. He has accumulated more than 15,100 hours of flight time in over 180 diverse types of aircraft. Building valuable flight time and experience as a corporate pilot, commuter pilot, banner tower, flight instructor, and ferry pilot, and operating his own aerobatic flight school, Rob holds ratings as an airline transport pilot, certified flight and ground instructor, as well as glider and seaplane ratings (Ultimate Airshows, 2024). Moreover, he remains actively involved in the aerobatic competition and air show community, often participating in workshops and collaborative sessions with other display pilots (International Aerobatic Club (IAC), 2024).

Holland always strives to design routines that are different and engaging, aiming to create a non-stop, action-packed performance that, "*if you blink, you will miss something*," he says (Aero-News Network, 2014). A distinct maneuver that captures his innovative approach to aerobatics is the "*cobra maneuver*,[2]" inspired by Russian techniques but introduced with his MXS aircraft in a dramatic way, complete with smoke and a 90-degree pitch angle while zeroizing forward movement. Despite his extreme-flying approach to aerobatics, Holland places a strong emphasis on safety. Each new maneuver undergoes a rigorous assessment process, beginning with theoretical analysis, followed by trial flights at a higher altitude, and finally, careful integration into live practice sessions. This methodical approach ensures that every aspect of his performance maintains the highest safety standards (Dayton Air Show, 2017).

One of the training tools extensively used by Holland is the use of cameras to record his practices or display sessions. Besides a ground camera recording his display, he has several onboard cameras, including one installed on the top of his helmet capturing his view and his cross-check points, others aiming inside the cockpit, one looking in front of the windscreen, and others installed at the wingtip toward the fuselage. In one of his videos available online (Rob Holland Aerosports, 2024), he uses the video footage to provide a self-critique of his performances, where he examines planned flight parameters compared to actual execution, analyzing the elegance of the maneuvers as well as its smoothness or aggressiveness. This ensures the aircraft not only flies well but also looks good from the outside while it is placed correctly inside the air show box. It is this detailed type of self-critique and pursuit of perfection that provides the required discipline and dedication for excellence in all he does. Most importantly, sharing his critique practice online with his fans is a testament to his effort to promote excellence within the aerobatic and air show community

(Rob Holland Aerosports, 2024). Ultimately, such a humble approach to excellence is truly inspiring for those just starting their aerobatic journey.

Case Study 4. Frank van Houten: Striking the Balance between Commercial and Display Flying

Frank van Houten's passion for aviation began at age 12, leading him to a dual-role career as an air show performer and business jet pilot. His diverse experience spans from performing in the Sukhoi Su-26 aerobatic aircraft at air shows to piloting a state-of-the-art business jet. Frank's dedication to flying is captured in his motto, "*My life is flying*" (Dutch Rush Aerobatics, 2024), which is the drive for his meticulous preparation and innovative performances, as shown in Figure 10.1 while he's flying a sunset show with pyros.

Frank's aviation training is extensive and multifaceted. His journey includes training as a Navy pilot in the Netherlands and as a captain of the Bombardier Global 6000 business jet. This background instills a prominent level of discipline and safety in his display flying. The skills required for business aviation—advanced navigation, system management, superior knowledge of aviation rules and regulations, and decisive problem-solving—translate seamlessly into the split-second timing and flawless execution necessary for aerobatics.

Furthermore, Frank van Houten's dual role as both an air show performer and a civilian business pilot adds another layer to his professional expertise. Business aviation's stringent standards instill a mindset of meticulous attention to detail in flight operations, while aerobatics offers Frank a platform to showcase his creative

FIGURE 10.1 Frank van Houten performing at Athens Flying Week (AFW) International Air Show, 2018. (Copyright: AFW.)

flying techniques. Ultimately, his ability to adapt between aerobatics and business jet flying highlights his broad skill set.

However, like many air show performers involved in commercial flying, he faces the complexity of balancing the roles of air show performer and business jet pilot. This dual engagement requires him to thoroughly maintain his flying currency and proficiency for air shows, requiring an elevated level of adaptability. Frank achieves this balance through adequate preparation and proactive communication with organizers, ensuring nothing surprises him at the last moment (Smoke On … Go, 2021).

Additionally, Frank is deeply involved in the air show community and aviation circles as an inspiring keynote speaker. Drawing from his vast aviation experience, Frank shares insights on transitioning from aerobatics to military flying to business jet life, emphasizing team building and maintaining high morale, and ultimately representing the excellence culture of the air show community.

Possibly due to his engagement with the business jet aviation, one of the main focus areas for Frank van Houten is the integration of professional aerobatics with marketing and branding, which is closely related to his air show training. His expertise in aerobatics ensures the high quality and safety of the performances, which are central to his branding strategy, while the rigorous training and precision required for aerobatic flying contribute to creating captivating shows that serve as effective promotional platforms.

Common Ground: Pursuit of Excellence

A key attribute shared among these top civilian air show performers is their ability to adapt and innovate. Whether it is Rob Holland developing new maneuvers or Dario Costa attempting record-breaking flights under extreme conditions, each pilot continually seeks to innovate, challenging themselves and ultimately avoiding the trap of becoming complacent with their impressive achievements. This drive for innovation not only enhances their own skills but also propels the entire aerobatic community forward, setting new standards and expanding the possibilities of aerial performance.

Another hallmark common to these pilots is their comprehensive approach to preparation. They integrate physical fitness, mental conditioning, and technical skill development into their routines, ensuring that they are prepared for the intense demands of aerobatic flying. This detailed and proactive approach to preparation is crucial not only for executing complex maneuvers safely but also for maintaining superior performance throughout their careers.

Despite the thrill and spectacle of aerobatics, all these pilots prioritize safety. Mike Goulian's commitment to precision highlights universal respect for the risks involved in their profession. By adhering to strict safety protocols and continuously refining safety practices, they help nurture a culture of safety and excellence within the air show community.

Additionally, each pilot also contributes to the growth and sustainability of the air show community through education and mentorship. By sharing their knowledge, whether through flight schools like Rob Hollands' or through collaborative learning environments that Dario Costa engages in, they help cultivate the next generation of pilots. Their efforts ensure that the legacy of excellence in air shows continues, enriched by their experiences and insights.

Finally, pilots like Frank van Houten exemplify the integration of professional aerobatics with marketing and branding strategies, demonstrating the potential of air shows as platforms for broader promotional activities. With this professional approach, they highlight not only their flying skills but also display how they understand and leverage the commercial aspects of their air show performances.

CASE STUDIES FOR MILITARY AIR SHOW PERFORMERS

Case Study 5. Heritage Flight Training

The USAFHFP training syllabus serves as a comprehensive blueprint designed to prepare pilots for the unique demands of performing Heritage Flights, displaying the rich history and evolution of air combat through formation flights of vintage and modern fighter jets (US Air Force (USAF), 2022). These flights are not only a demonstration of the pilots' flying skills but also a living tribute to the history of air combat. The program's rigorous training syllabus is structured in a building block approach to ensure that every pilot is equipped with the knowledge, skills, and sensitivity required to handle both vintage warbirds and modern fighter jets within the same formation, a task that requires an exceptional level of precision and coordination.

The program starts with an intensive ground training segment, which is essential for laying the foundational knowledge necessary for successful heritage flights. This stage covers the historical context of the aircraft, instilling a sense of respect and duty in the pilots. It is complemented by thorough briefings on Federal Aviation Administration (FAA) and USAF regulations, ensuring pilots are well-versed in the legal and safety standards that govern their operations.

Training then progresses into a detailed exploration of formation flying skills. Pilots learn about various formation profiles and the aerodynamic interaction between different aircraft types—knowledge crucial for maintaining safe separation and handling the unique airflow characteristics that affect aircraft behavior during close formation flying. Moreover, the emphasis on standardized communication techniques with and without radio calls is vital for success in this highly coordinated environment.

The flight training segments of the syllabus are methodically structured to incrementally build the pilots' skills. Beginning with basic maneuvers in two-ship formation and observation flights in different types of aircraft, pilots gradually move through more complex formations, concluding in the mastery required for four-ship formations. Each stage is designed to challenge the pilots progressively, ensuring they can build confidence as they master each level of complexity while also facilitating the practical application of theoretical knowledge gained during ground training, bridging the gap between classroom learning and in-flight execution.

Case Study 6. Comprehensive Training Program for USAF Single-ship Air Show Performers

The training program for the USAF single-ship demonstration teams epitomizes a detailed and careful approach, thoroughly crafted to prepare pilots for the challenging demands of air show performances (USAF, 2022). This program is designed not only to ensure operational excellence and safety while flying intense maneuvers in the hostile low-level environment but also to instill confidence and composure in the demonstration pilot while performing in front of an audience.

The training is initiated with an intensive ground training phase, which is crucial for laying the foundation for both theoretical knowledge and practical application. This initial phase equips demonstration pilots with a deep understanding of standard air show procedures, detailed aircraft handling characteristics, and comprehensive safety protocols. Topics such as fuel requirements, emergency procedures, and team management are covered extensively, providing pilots with the operational, safety, and logistical knowledge essential for conducting air show performances.

Following ground training, pilots engage in advanced simulator training sessions. These sessions are critical as they allow pilots to transition theoretical knowledge into practice without the risks associated with live-display flying. Pilots rehearse both high and low show profiles, mastering the techniques needed to execute precise maneuvers and handle emergency scenarios. The simulator is a pivotal tool in this phase, providing a safe environment for pilots to practice emergencies, adverse weather conditions, high-speed dive recoveries,[3] slow-speed handling,[4] and out-of-control situations.[5] This *risk-free* setting is invaluable for refining skills and building confidence, which is essential for the flawless execution of display routines.

The progressive nature of the flight training is designed to incrementally challenge and develop the pilots' capabilities. Starting with observation sorties, upgraded demo pilots first learn by watching their demo instructor pilot perform maneuvers in order to understand the flow and timing of each maneuver within the air show context.

As training progresses, pilots gradually take a more active role, initially from the back seat and then moving to the front seat to directly engage in flying the standard demonstration profiles. These sorties are carefully structured to ensure that pilots develop a robust ability to manage both high and low show elements safely, precisely, but also elegantly. The training advances from performing maneuvers at higher altitudes to executing full profiles at the lower altitudes, which are inherently more demanding.

A significant emphasis is placed on emergencies and maneuver abort procedures throughout the training sorties. Upgraded demonstration pilots are required to execute multiple maneuver aborts to set safety calls from their instructor, ensuring they can react swiftly and appropriately under any circumstances.

The apex of this extensive training program is the solo performance sorties. These sorties represent the final test of a pilot's readiness to participate in air shows, demonstrating mastery over both high and low show profiles. By this stage, demo pilots are expected to exhibit not only technical proficiency but also the ability to seamlessly integrate into the choreographed flow of the air show, all while adhering to stringent safety protocols.

Case Study 7. Adapting Proven Training Systems for Fast Jet Display Excellence, A Personal Account from the Hellenic Air Force

In 2010, my life took a thrilling turn as I was one of the two pilots from the HAF selected to be upgraded into F-16 demonstration pilots with the USAF Viper

West Demonstration Team. This experience actively involved me in the amazing world of air shows, specifically in the world of fast jet flying displays, all from the cockpit of the legendary F-16. The training was far from ordinary, blending flying skills with essential human skills like team management, media relations, and mindfulness.

The First Flight

My first demo flight in the F-16, on the January 7, 2010, was an unforgettable fun ride. Sitting in the rear cockpit with my instructor at the controls, I could not stop laughing out of sheer joy. It felt like the wildest roller coaster ride, with everyone on the ground watching an F-16 maneuvering like they had never seen before.

The Challenges

Every maneuver of the display presented a new challenge to me, from takeoff to landing. The Victory Roll[6] was particularly tough initially. I had to execute a roll while climbing steeply, using the sun for orientation. During one of my first solo training flights, while rolling and climbing through 11,000 feet, the aircraft departed experiencing –3.5 Gs due to roll-coupling[7] (Jahnke, 1998; Teodorescu, 2007; USAF Test Pilot School, 1989). Fortunately, I managed to recover and reposition for the next maneuver, yet well over the crowd line. My instructor later remarked that he had never seen such a reaction from the F-16 before. We watched the HUD video several times and captured several lessons learned that helped me to be more cautious with control inputs while demonstrating the capabilities of the F-16, a jet that, mainly due to the fly-by-wire[8] technology, sometimes is flown by the "*computer*,"[9] not the pilot.

Intensive Training

The training program was condensed into two weeks for each upgraded demo pilot. This meant our instructor flew twice daily, with extensive briefings and debriefings. The compressed schedule demanded intense focus and dedication, pushing me to my physical and psychological limits and ensuring we absorbed every critical lesson.

Yet, flying my first solo display felt like my first solo flight ever. It was a mix of loneliness and immense pride. I felt confident but also humbled by the lessons gained and ready to continue with this incredible path that I was lucky to be given.

Simulator Sessions

The simulator session, although not perfectly accurate in replicating maneuvers and the display profile, was invaluable. It enhanced my training process by practicing my display routine and familiarizing myself with the flight parameters, with the hardest part being memorizing all the parameters, i.e., minimum, maximum, and target airspeeds and altitudes, for each maneuver without relying on my mission card.

Media and Public Affairs

Beyond the formal ground school that included lectures, I had several discussions with my instructor about handling media and public affairs (PA). Beyond other important aspects of PA, I learned that if faced with a challenging question, I should "*bridge*" it with a narrative aligning with the HAF's PA objectives.

Mental Preparation

Before flying my display, whether in training or at actual air shows, I would mentally walk through the entire demo in a closed room, from takeoff to landing. This technique was introduced by my instructor, a completely novel approach for me, and it helped immensely. Though I had done chair flying during my aviation training, this was an entirely different and beneficial experience that helped me to enter into my "*bubble*" prior to flying a display.

Risk Management

Furthermore, the demo training introduced me to the rules and regulations related to air shows—a completely new topic for me. My instructor helped me understand weather limits, such as cloud ceilings and winds, and adapt my profile accordingly. I also learned about the challenges of flying over water, particularly related to depth perception, and the need to make considerable fuel calculations for staged displays away from the departure airfield.

Professional and Personal Growth

This training experience has profoundly shaped me both professionally and personally. It paved the way for my current endeavors, which I would not have considered before being selected for this training, such as authoring a second book about air show performers and serving the esteemed EAC as a board member. None of this would have been possible without the unwavering commitment and support of the HAF's leadership, especially during the first couple of display seasons amid the Greek economic crisis.[10]

A Testament to Adaptive Excellence

Adaptive excellence is focused on the pursuit of superior performance and excellence through the flexible application of established practices to new environments, especially in aviation and high-reliability organizations. It upholds the importance of adapting proven systems to align with different organizational contexts, promoting the adoption of these systems while maintaining respect for local norms and guaranteeing *resilient safety* in operations.

This collaborative training endeavor between the USAF Viper West Demonstration Team and the HAF represents a paradigm of *adaptive excellence*. The USAF Viper West Demonstration Team brought a wealth of experience and a meticulously crafted training program to the table, known for its rigorous standards and exceptional outcomes. With both teams operating the F-16 and upholding a *shared ethos* of *flight discipline* and *airmanship*, the challenge was to adapt the robust training requirements of Viper West to fit the HAF's unique operational context. This process, while challenging, demonstrated the goal of *adaptive excellence*, proving that superior performance and excellence can be achieved without compromises through careful *adaptation* and *respect* for local organizational requirements.

HOLISTIC COMPARATIVE ANALYSIS

While all the training programs covered in the seven case studies emphasize precision, safety, and proficiency, each has unique focuses and methodologies tailored to its specific context, achieving adaptive excellence. A comparative analysis reveals several key commonalities in their training that contribute to excellence in display flying:

- Comprehensive *ground training* provides a solid foundation in regulatory compliance, display flying mechanics, and safety protocols.
- *Advanced simulation* bridges theoretical knowledge with practical application, enabling the safe rehearsal of complex maneuvers and emergency procedures.
- *Progressive flight training* methodically builds confidence and proficiency, from observation to solo performance.
    - It should be noted that for civilian air show performers, participation in *aerobatic competitions* adds an additional layer of self-discipline. It enables continuous critique of the pilots' performances by judges, who then adhere to stringent allowable error margins in their flight profiles.
- *Physical* and *mental conditioning* incorporates fitness routines and resilience techniques to prepare pilots for the high demands of aerobatic flying.
- Emphasis on *safety* ensures pilots can handle unexpected situations confidently through rigorous safety protocols and emergency procedures.
- Integration of *human skills* includes team management, media relations, and public engagement, creating well-rounded and approachable representatives of the air show community and the aviation world in total.
- *Continuous improvement* and *innovation* drive pilots to push the boundaries of display flying excellence, fostering a culture of *perpetual learning* and advancement.

These commonalities demonstrate that a *holistic approach to training* is essential for developing highly skilled, versatile display pilots capable of excelling in the demanding yet exciting environment of air shows and public demonstrations.

THE EXCELS MODEL FOR DISPLAY PILOT TRAINING

Based on the comprehensive analysis of training methods and outcomes discussed in the previous section, the *EXCELS Model* for display pilot training is conceptualized (Figure 10.2), which stands for *Education*, *eXperience*, *Conditioning*, *Emergency preparedness*, *Leadership*, and *Safety*. This model integrates the essential elements identified as critical for achieving excellence in display flying. To understand better, we can imagine a display pilot's journey through this model: first, *Education* instills deep theoretical knowledge; then, *eXperience* builds practical display flying hours; *Conditioning* ensures peak physical fitness; *Emergency preparedness* readies pilots for the unexpected; *Leadership* hones decision making skills; and last, *Safety* ensures

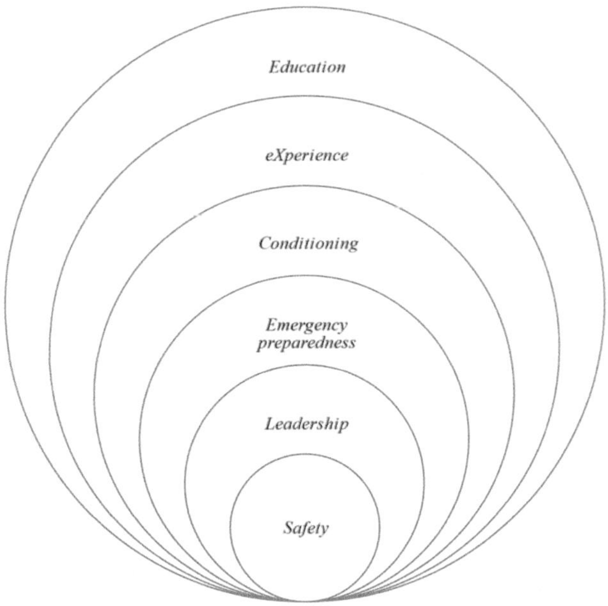

FIGURE 10.2 *EXCELS Model* for display pilot training.

that every maneuver during the display is planned and executed with appropriate risk management.

EDUCATION (GROUND TRAINING)

Education forms the foundational bedrock of the *EXCELS Model*, encompassing thorough ground training that continues throughout the display pilot's career. It includes the following (see also Chapter 5):

- Regulatory Compliance
  This aspect involves detailed instruction on aviation regulations and compliance requirements. Display pilots are trained to understand and adhere to the legal standards that govern their operations, ensuring all flight activities meet the highest safety and regulatory standards crucial for maintaining operational legality and safety (Federal Aviation Administration (FAA), 2022; International Council of Air Shows, 2019b)
- Mechanics of Flight
  Ground training also covers the fundamental mechanics of flight and aerodynamics. Understanding how aircraft operate under various conditions allows pilots to perform maneuvers with greater precision and safety and supports their ability to troubleshoot and respond to in-flight abnormalities effectively (International Council of Air Shows, 2016; Szurovy & Goulian, 1996).

- Safety Protocols
  Display pilots receive rigorous safety training that includes emergency procedures, risk management, and best practices for in-flight and ground operations, ensuring that they are prepared to handle a wide range of potential scenarios, from technical malfunctions to adverse weather conditions, thereby enhancing overall safety (Barker, 2003; European Airshow Council (EAC), 2020; International Council of Air Shows, 2016, 2019a; Karachalios, 2022; Karachalios & Adjekum, 2023)

eXperience (Advanced Simulation and Actual Practice)

eXperience bridges theoretical knowledge with practical application:

- Flight Simulators
  Advanced flight simulators are used to provide display pilots with a safe environment to practice complex maneuvers and emergency responses. These simulators offer a controlled setting where pilots can repeatedly practice and perfect difficult techniques, ensuring precision and accuracy in their performance. Furthermore, simulators replicate real-world conditions, including weather variations, mechanical failures, and other potential hazards, allowing pilots to build automaticity and gain confidence without the risks associated with live training flights.

  The use of flight simulation in aviation has been extensively studied and documented by experts in the field and has been covered in Chapter 5. According to Lee (2017), flight simulators create virtual environments that closely mimic the experience of real flight. These virtual environments are equipped with advanced technology to simulate various flight scenarios, from routine operations to emergency situations. The ability to practice in these simulated environments allows pilots to gain valuable experience and improve their skills without the potential consequences of display parameter deviations.
- Scenario-Based Training
  Real-world practice sessions simulate the conditions pilots will face during actual display performances. These sessions are designed to mirror the dynamic environment of air shows, helping pilots adapt to varying scenarios and refine their techniques under realistic conditions. By closely replicating the challenges of live performances, these practice sessions enable pilots to develop the agility and precision required for display flying, ensuring they can perform safely and effectively during actual events.

  In their comprehensive work on aircrew training, Andrews and O'Neil (2000) highlight the importance of such realistic practice environments. They explain that effective aircrew training programs incorporate scenarios that mimic the conditions of actual missions to build proficiency and confidence.

Conditioning (Physical and Mental)

Conditioning prepares pilots physically and mentally for the demands of aerobatic flying (see also Chapters 5 and 6):

- Fitness Programs
  Display pilots undergo cardiovascular and strength training to build endurance and resilience against high G-forces encountered during aerobatic maneuvers. Maintaining peak physical fitness is essential for ensuring pilots can perform at their best and avoid physical strain or injury during air show flights. Cardiovascular training improves the display pilots' aerobic capacity, enabling them to sustain physical exertion for longer periods, while strength training focuses on enhancing muscle strength, particularly in the core and lower body, to better withstand the intense forces experienced during flight maneuvers.
  In his detailed exploration of the physiological demands of sports training, Hoffman (2014) emphasizes the critical role of physical conditioning in optimizing performance and preventing injury. Hoffman explains that athletes, engage in specialized training regimens that enhance their cardiovascular endurance and muscular strength due to the athletic demands of their performances. This training could be helpful for display pilots helping them manage the physical stresses associated with their activities more effectively.
- Psychological Resilience
  Mental conditioning techniques, such as mindfulness and stress management, are incorporated to help air show performers maintain focus and composure under pressure, enhance cognitive function, reduce anxiety, and improve decision making capabilities (Karachalios & Adjekum, 2023).

Emergency Preparedness (Emphasis on Safety)

Emergency preparedness is crucial for handling unexpected situations:

- Safety Drills
  Regular drills and simulations for emergency procedures ensure pilots are well-prepared to respond to any in-flight emergencies, reinforce the correct responses, and improve reaction times, making pilots more adept at managing crises.
- Risk Management
  Training in proactive risk assessment and mitigation helps display pilots identify potential hazards and implement strategies to minimize risks (Karachalios & Adjekum, 2023).

Leadership (Integration of Soft Skills)

Leadership ensures well-rounded development:

- Team Management
  Training includes leadership and team dynamics to ensure pilots can effectively lead and collaborate with their teams, fostering a cohesive working

environment and improving overall performance, both in the air and on the ground. This comprehensive training is crucial for maintaining the high levels of coordination and mutual trust necessary for successful air show performances.

Mathieu et al. (2015) emphasize the importance of reciprocal team cohesion and its impact on performance, noting that shared leadership and individual competence are key factors. Their research highlights how cohesive teams, where leadership responsibilities are distributed, and members are highly skilled, tend to perform better due to improved communication, trust, and mutual support. This insight is particularly relevant for display teams, where aerobatic team display pilots must rely on each other's abilities and leadership to execute complex maneuvers safely and effectively.

Building on this, Stashevsky and Koslowsky (2006) explore the relationship between leadership team cohesiveness and team performance. They found that strong, cohesive leadership teams are more effective in achieving their goals and maintaining high-performance levels. This study emphasizes the need for display pilots to not only focus on individual skills but also on building strong, cohesive teams. Effective leadership within these teams ensures that all members are aligned with the mission and can collaborate seamlessly during both preparation and performance.

Further complementing these findings, Tung and Chang (2011) examine the effects of empowering leadership on performance in management teams, finding that leaders who empower their team members enhance overall team performance. This approach to leadership fosters a sense of ownership and accountability among team members, encouraging them to contribute their best efforts. For display pilots, empowering leadership means creating an environment where each team member feels valued and capable, leading to more effective collaboration and improved performance during air shows.

- Media Relations

  Display pilots are trained in media interaction and public engagement, skills that are crucial for representing their teams and organizations and positively enhancing the visibility and reputation of the team. This training includes learning how to effectively communicate with the media, engage with the public during events, and utilize social media platforms to promote their activities and achievements.

  According to Dijkmans et al. (2015), effective use of social media can significantly enhance a team's corporate reputation by providing a platform to engage with a broad audience. Their research emphasizes that strategic social media use allows organizations to showcase their expertise, build stronger connections with the public, and manage their public image more effectively. From the perspective of a display pilot, this means not only performing impressive aerial maneuvers but also being adept at sharing their experiences and connecting with fans and followers online.

Yang et al. (2021) further explore the impact of social media use on work efficiency, highlighting the concept of communication visibility. They explain that when team members actively use social media, it increases the transparency of their activities and promotes a culture of openness and engagement. Being visible and communicative on social media platforms can enhance a display team's cohesion and efficiency, as well as improve their individual performance by fostering a supportive and engaged fan base. This visibility also helps build a positive public image and increase the overall impact of their performances.

- Public Engagement
  Since public interaction is a significant component of an air show, techniques for delivering effective public presentations are also taught, ensuring pilots can engage and inspire audiences. This training includes mastering the art of public speaking, understanding audience dynamics, and using multimedia tools to enhance presentations. By becoming proficient in these areas, air show performers can effectively communicate their experiences and knowledge, thereby enhancing the overall impact of their performances.

  Lenz et al. (2015) provide practical strategies for delivering effective lectures, emphasizing the importance of clear communication, engaging content, and interactive elements to maintain audience interest. These principles are directly applicable to display pilots, who must captivate and educate diverse audiences during air shows. By applying these strategies, pilots can deliver presentations that are both informative and engaging, ensuring that their messages resonate with spectators and leave a lasting impression.

  Furthermore, Rossette-Crake (2020) explores the evolution of public speaking in the digital and neoliberal age, highlighting how the integration of digital tools and a focus on performance can enhance public presentations. This perspective is particularly relevant for display pilots, who often use digital media to complement their live presentations. By incorporating videos, photos, and other digital content, air show performers can create a more dynamic and immersive experience for their audiences. Ultimately, understanding the contemporary context of public speaking helps pilots to adapt their presentations to meet the expectations of modern audiences, making their performances more impactful and memorable.

Safety (Innovation and Continuous Learning)

Safety is the core of continuous improvement and innovation (for more information, see Chapters 11, 12, and 13):

- New Maneuvers
  Air show performers are encouraged to develop and practice new aerobatic maneuvers, allowing their skills to remain sharp, preventing themselves from becoming complacent, and allowing for continual improvement and creativity in their performances (Barker, 2003).

- Technological Integration
  Adoption of the latest technologies and techniques is promoted to enhance performance and safety, while pilots use the best tools available to optimize their flying (Karachalios & Adjekum, 2023).
- Peer Learning
  Opportunities for collaborative learning with peers are provided, fostering a culture of continuous learning and knowledge sharing, helping pilots stay informed about best practices and new techniques, and enhancing their overall skill set (Karachalios, 2022).

The *EXCELS Model* for display pilot training provides a framework designed to comprehensively prepare pilots for the demands of air show performances. By integrating *Education*, *eXperience*, *Conditioning*, *Emergency preparedness*, *Leadership*, and *Safety*, this model ensures pilots are not only technically proficient but also mentally resilient, safety-conscious, and effective ambassadors of the aviation industry.

RESILIENT EXCELLENCE AND THE BOUNCING BALL MODEL: THE ENDGAME OF TRAINING

The case studies presented provide a framework not only for performance excellence in aviation but also a blueprint for resilient excellence in high-performance activities, such as elite sports, fighter pilots, and even stock exchange traders, where they seek outstanding levels of performance at all times. For these individuals, a drop in high-level performance is not an option; they must immediately *bounce-up* to their 100% level. For them, to paraphrase the Apollo 13 quote, "*failure is not an option*" (Granath, 2017), "*mediocrity is not an option*," as they do not accept anything less than peak performance. Air show performers have sculptured within them the skills not to *bounce-back* but rather to *bounce-up* whenever their performance has dropped to sub-perfect levels. Their training has given them the mental skills and physical strength to make timely decisions that allow them to find solutions to perform at 100% of the expected performance levels. This doesn't mean that they are burning-out themselves, either mentally or physically; they rather have built up the capacity through training to make the impossible possible, to make what seems *superhuman* to perform with precision and elegance; it is this *resilience* that keeps them performing consistently at the top levels of their display profile. In such a case, they do not just *bounce-back*, but they *bounce-up* to the maximum performance level.

To visualize this, imagine that a display pilot's performance is a *bouncing ball*, similar to the one in Figure 10.3. The pilot holds a *racquet*, representing *resilient excellence*, skillfully hitting the ball to keep it airborne. At takeoff, the pilot starts the game, and with each maneuver, there is a risk that the ball might fall. However, the pilot's adept use of the *racquet* ensures the *ball* bounces higher, maintaining it in flight. If the pilot misses, the ball will not hit the ground because the *safety net* is in place, symbolizing *resilient safety*. This thin *safety net* (*resilient safety*) hovers just above the floor, representing a proactive set of safety measures are in place; it ensures

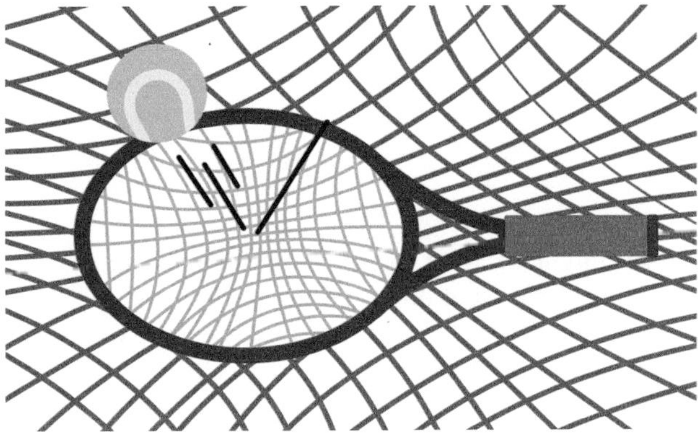

FIGURE 10.3 The *Bouncing Ball Model* for resilient excellence. (Credits: Maria Karachaliou.)

that, even in adverse conditions, the *ball* (*the performance*) remains safe, ready for the pilot to recover with the *racquet* (*resilient excellence*) and continue the flying display without an unsafe outcome.

Nevertheless, a question might be raised about what the most important part of resilience for air show performers is as follows: the excellence one or the one that focuses on safety; the *racquet*, or the *safety net*. Obviously, this is a *rhetorical question* since both aspects of resilience are equally important, not only for air show performers but also for any high-performer: *resilient excellence* is the *precision-insurer* when everything goes as expected, keeping the *ball* in the correct direction, while *resilient safety* is the mechanism that enables the pilot or high-performer to remain *unharmed* when the unexpected happens, and ultimately saving the ball from hitting the ground.

The *bouncing ball model* could have further analysis to investigate what resources are required to build the *safety net*, the *distance* between the net and the racquet, how *high* we *bounce-up* the ball, what exact *spot* on the *racquet* the ball hits, what is the *time* that the ball takes to go *up* and then needs to be hit again, the *time* that it takes to *fall* from the *racquet* to the *safety net*, and of course what is the *frequency* of hitting the ball to keep it at the *highest height*, and the list goes on. Nevertheless, the main focus remains on *resources*, including time and effort, as well as *margins*, both for precision and safety.

In conclusion, this chapter provides insight into the methods that air show performers consistently sustain their performance at high peak, which is not just by looking for excellence. They recognize that excellence in their performance might be *disturbed momentarily* because of external factors or because of manual inputs in the flight controls and throttle, but they have *layered methods* of recovering to the expected peak levels without delay and without letting them lose their flow; this should not be considered as a problematic continuation bias, it is rather a *spring-loaded mentality* of *bouncing-up* their psychological and physical focus to regain excellent levels of performance. For air show performers, "*mediocrity is not an option*"; they never give up if a deviation in their performance has been introduced.

> What we can define as *resilient excellence* is *the ability of an individual or organization to regain superior levels of performance consistently, timely, with minimal effort, as a byproduct of training, experience, conditioning, talent, and preparedness.*

Resilient excellence might be what distinguishes display pilots: the constant seek of excellence while they operate at their optimum mental levels, without additional stress, but with available solutions to their deviations from the expected performance results. They do not want to be mediocre; they are tough critics of themselves, having grand expectations, but they do not allow this demand to harass their elegance of flying, their showmanship, and ultimately their connection with the audience.

NOTES

1 Strobe glasses are eyewear that intermittently obstruct and unblock vision using rapid flashes of light, commonly used in training to improve visual and cognitive processing skills (Carrolla et al., 2021).

2 The cobra maneuver or "Pugachev's Cobra" is an advanced aerobatic move performed by fighter aircraft in which the pilot suddenly raises the aircraft's nose to a near-vertical position before returning to level flight, creating a dramatic deceleration in their forward movement (Skow, 1990).

3 High-speed dive recoveries refer to the techniques and procedures pilots use to safely return an aircraft to level flight after a steep, rapid descent, often involving precise control inputs to manage aerodynamic forces and prevent structural damage or loss of control.

4 Slow-speed handling refers to the techniques and control measures used by pilots to maintain stable and controlled flight at low airspeeds, which is critical for maneuvers such as takeoff, landing, and flying in tight formations or constrained spaces.

5 Out-of-control situations in aviation refer to scenarios where an aircraft deviates from its intended flight path due to loss of aerodynamic control, often requiring immediate corrective actions by the pilot to regain stability.

6 The Victory Roll or Maximum Performance Climb with Rolls is an aerobatic maneuver in which the aircraft, at 330 knots and 300 feet AGL, performs a high-G pull to vertical at show center, executes rapid aileron rolls until reaching 250 knots or 2,500 feet below the waivered airspace, and then recovers by smoothly pulling the nose to the horizon while adjusting power and speed brakes for repositioning (USAF, 2022).

7 Roll-coupling is the divergence experienced during rolling maneuvers, created when a disturbance of about one aircraft axis causes a disturbance about the roll axis (USAF Test Pilot School, 1989).

8 Fly-by-wire technology is a system in which an aircraft's flight controls are operated by electronic signals transmitted from the pilot's inputs to the flight control computers, which then manipulate the control surfaces, replacing traditional mechanical linkages and providing improved handling, stability, and safety (Collinson, 2023).

9 A Digital Flight Control Computer (DFCC) is an advanced avionics system that uses digital processing to manage and control an aircraft's flight surfaces, enhancing precision, stability, and responsiveness by interpreting pilot inputs and environmental data to optimize flight performance and safety (Mackall, 1984).

10 The Greek economic crisis, beginning in late 2009, involved severe government debt and economic recession, leading to austerity measures, international bailouts, and widespread social and political upheaval (Kouretas & Vlamis, 2010).

REFERENCES

Aero-News Network (Director). (2014). *Aero-TV: Ultimate Airshows – In Flight with Rob Holland* [video recording]. www.youtube.com/watch?v=99HesWwSAoA

Andrews, D. H., & O'Neil, H. F. (Eds.). (2000). *Aircrew training and assessment*. CRC Press.

Barker, D. (2003). *Zero error margin: Airshow display flying analysed*. Freeworld Publications.

Carroll, W., Fuller, S., Lawrence, J., Osborne, S., Stallcup, R., Burch, R., Freeman, C., Chander, H., Strawderman, L., Crane, C., Younger, T., Duvall, A., Mock, S., Petway, A., Burgos, B., & Pirolik, A. (2021). Stroboscopic visual training for coaching practitioners: A comprehensive literature review. *International Journal of Kinesiology and Sports Science, 9*(4), 49–59. DOI:10.7575/aiac.ijkss.v.9n.4p.49

Collinson, R. P. G. (2023). Fly-by-wire. In *Introduction to avionics systems* (pp. 115–164). Springer International Publishing.

Costa, D. (2025). *Dario Costa – Official Website*. www.dariocosta.com

Dayton Air Show. (2017). Rob Holland – 2024 CenterPoint Energy Dayton Air Show. https://daytonairshow.com/rob-holland/

Dijkmans, C., Kerkhof, P., & Beukeboom, C. (2015). A stage to engage: Social media use and corporate reputation. *Tourism Management, 47*, 58–67. https://doi.org/10.1016/J.TOURMAN.2014.09.005

Dutch Rush Aerobatics. (2024). Dutch Rush Aerobatics. https://dutchrush.nl/en/

European Airshow Council (EAC). (2020). *Airshow manual (5.1)*. European Airshow Council.

European Airshow Council. (2024). 2024 Speakers. European Airshow Council (EAC). www.europeanairshow.org/2023-speakers

Federal Aviation Administration (FAA). (2022, October 27). *Order 8900.1A – Flight Standards Information Management System*. Federal Aviation Administration (FAA). www.faa.gov/regulations_policies/orders_notices/index.cfm/go/document.information/documentID/1041430

Goulian, M. (2024). Goulian Aerosports. https://mikegoulian.com/

Granath, B. (2017, November). Members of Apollo 13 team reflect on NASA's finest hour. *Spaceport Magazine*, 4(11), 38–40. National Aeronautics and Space Administration. www.nasa.gov/wp-content/uploads/2015/05/spm_nov_2017.pdf

Hoffman, J. (2014). *Physiological aspects of sport training and performance*. Human Kinetics.

International Aerobatic Club (IAC). (2024). Unlimited Team 2024 Bios. IAC. Retrieved from www.iac.org/unlimited-team-2024-bios

International Council of Air Shows. (2016). *Air Show Performers Safety Manual* (p. 27). ICAS. www.airshows.aero/GetDoc/3444

International Council of Air Shows. (2019a). *ICAS air show manual* (7th ed.). International Council of Air Shows. https://airshowdigest.aero/wp-content/uploads/2019/01/ICAS-Air-Show-Manual-7th-Edition.pdf

International Council of Air Shows. (2019b, January 1). *ACE Manual—Revision 9*. International Council of Air Shows.

Jahnke, C. C. (1998). On the roll–coupling instabilities of high–performance aircraft. *Philosophical Transactions of the Royal Society of London. Series A: Mathematical, Physical and Engineering Sciences, 356*(1745), 2223–2239.

Karachalios, M. (2022). *An evaluation of the relationships between resilient safety culture, safety risk parameters, and mindfulness in the international air show community* [PhD dissertation, University of North Dakota]. https://commons.und.edu/theses/4267/

Karachalios, M., & Adjekum, D. K. (2023). *Air show performers: Safety, risk management, and psychological factors*. CRC Press. https://doi.org/10.1201/9781003431879

Kouretas, G., & Vlamis, P. (2010). The Greek crisis: Causes and implications. *Panoeconomicus, 57*, 391–404. https://doi.org/10.2298/PAN1004391K

Lee, A. T. (2017). *Flight simulation: virtual environments in aviation*. Routledge.

Lenz, P., McCallister, J., Luks, A., Le, T., & Fessler, H. (2015). Practical strategies for effective lectures. *Annals of the American Thoracic Society, 12*(4), 561–566. https://doi.org/10.1513/AnnalsATS.201501-024AR

Mackall, D. A. (1984). *AFTI/F-16 digital flight control system experience*. NASA. Langley Research Center NASA Aircraft Controls Research, 1983.

Mathieu, J., Kukenberger, M., D'Innocenzo, L., & Reilly, G. (2015). Modeling reciprocal team cohesion-performance relationships, as impacted by shared leadership and members' competence. *The Journal of Applied Psychology, 100*(3), 713–734. https://doi.org/10.1037/a0038898

Medalen, T. (2021). Dario Costa sets Guinness World Record in Tunnel Pass. Red Bull. Retrieved from www.redbull.com/us-en/tunnel-pass-dario-costa-catalca-turkey

Reason, J. (1997). *Managing the risks of organizational accidents*. Ashgate Publishing.

Red Bull Air Race. (2024). Dario Costa. Red Bull. www.redbull.com/int-en/athlete/dario-costa

Rob Holland Aerosports (Director). (2024). *Critiquing my practice flight* [video recording]. www.youtube.com/watch?v=dIcujao5KIY

Rossette-Crake, F. (2020). 'The new oratory': Public speaking practice in the digital, neoliberal age. *Discourse Studies, 22*(5), 571–589.

Smoke On … Go. (2021). (565) A Smoke On... Go! Aviation Conversation with Frank Van Houten—YouTube. www.youtube.com/watch?v=-4NzJ3OR9u8&t=107s

Stashevsky, S., & Koslowsky, M. (2006). Leadership team cohesiveness and team performance. *International Journal of Manpower, 27*, 63–74. https://doi.org/10.1108/01437720610652844

Szurovy, G., & Goulian, M. (1996). *Advanced aerobatics* (1st ed.). McGraw-Hill Professional.

Teodorescu, B. C. (2007). A crossfeed solution for the roll-coupling problem of high-performance airplanes. *University "Politehnica" of Bucharest Scientific Bulletin, Series D: Mechanical Engineering, 69*(2), 31–44.

Tung, H., & Chang, Y. (2011). Effects of empowering leadership on performance in management team. *Journal of Chinese Human Resource Management, 2*, 43–60. https://doi.org/10.1108/20408001111148720

Ultimate Airshows. (2024). *Rob Holland Ultimate Airshows, Record-Breaking Aerobatic Champion*. Ultimate-Airshows. www.ultimateairshows.com

US Air Force. (2022). Air Force Manual 11-246: Air Force fighter type aircraft demonstrations (A-10, F-16, F-22, F-35) (Vol. 1). USAF. https://static.e-publishing.af.mil/production/1/af_a3/publication/afman11-246v1/afman11-246.pdf

US Air Force Test Pilot School. (1989). Volume II. Flying qualities phase, chapter 9: Roll coupling. In *Volume II. Flying qualities phase* (p. 24). USAF Test Pilot School. https://apps.dtic.mil/sti/tr/pdf/ADA319980.pdf

Wang, K., Frank, C., Tsai, Y., Lin, K., Chen, T., Cheng, M., Huang, C., Hung, T., & Schack, T. (2021). Superior performance in skilled golfers characterized by dynamic neuromotor processes related to attentional focus. *Frontiers in Psychology, 12*. https://doi.org/10.3389/fpsyg.2021.633228

Skow, A. M. (1990). An Analysis of the Su-27 Flight Demonstration at the 1989 Paris Airshow. *SAE Transactions*, *99*, 153–165. www.jstor.org/stable/44472488

Yang, X., Ye, H., & Wang, X. (2021). Social media use and work efficiency: Insights from the theory of communication visibility. *Information & Management, 58*, 103462. https://doi.org/10.1016/J.IM.2021.103462

FURTHER READINGS

Dekker, S. (2017). *The field guide to understanding 'human error'*. CRC Press.

Langer, E. J. (2014). *Mindfulness.* Da Capo Lifelong Books.

Salas, E., Tannenbaum, S. I., Kraiger, K., & Smith-Jentsch, K. A. (2012). The science of training and development in organizations: What matters in practice. *Psychological Science in the Public Interest.*

Zaccaro, S. J., Rittman, A. L., & Marks, M. A. (2001). Team leadership. *The Leadership Quarterly, 12*(4), 451–483.

Part III

Resilient Safety

Manolis Karachalios

INTRODUCTION

The concept of *resilient safety* is crucial in the high-risk, high-stress, and high-reliability environment of air shows, where display pilots perform complex maneuvers in close proximity to the ground, under G, and in tight formations. *Resilient safety* goes beyond traditional active and passive safety measures by emphasizing the ability to anticipate, withstand, recover from, and adapt to various challenges.

One of the most dramatic illustrations of *active safety*[1] in aviation is captured in a declassified video from a US Air Force Arizona Air National Guard F-16 (Aviation Week, 2016). The footage shows the aircraft's Automatic Ground Collision Avoidance

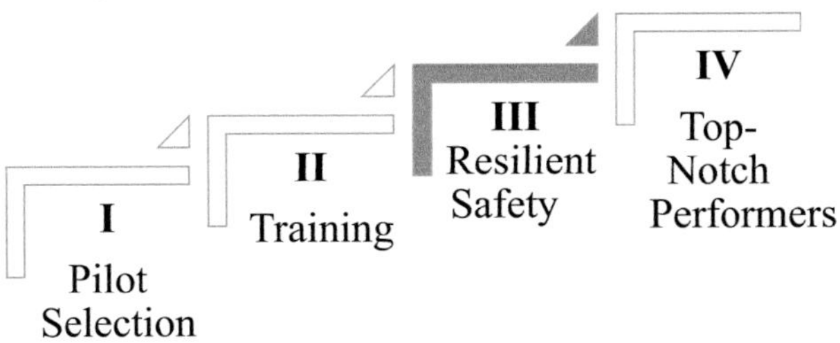

FIGURE PIII.1 Steps to air show excellence: Resilient safety.

DOI: 10.1201/9781003504801-14

System (AGCAS) saving an unconscious pilot from a fatal crash by autonomously pulling the aircraft out of a steep descent. This incident reinforces the critical role of active safety systems like AGCAS in preventing accidents before they occur. The design, integration, and flight testing of AGCAS, as detailed by Swihart et al. (2011a, b), highlight its capability to autonomously avert ground collisions, providing an essential layer of protection for high-performance aircraft. The Air Force Research Laboratory (Air Force Research Laboratory (AFRL), n.d.) further emphasizes the system's success, showcasing how innovations like AGCAS significantly enhance pilot safety and mission success.

While *active safety* systems are designed to prevent incidents, *passive safety*[2] measures are essential when prevention fails. In the context of air shows, passive safety features such as ejection seats become the last line of defense during catastrophic failures. Parate (2022) explores the technology and engineering behind aircraft seat ejection mechanisms, emphasizing their critical role in ensuring pilot survival. Reliable ejection systems can safely extract pilots from life-threatening situations, illustrating the importance of *passive safety* in the overall safety strategy.

> *Resilient safety* is a quality that embodies the capacity of a system, organization, or individual to not only *withstand* (*active safety*) and *recover* (*passive safety*) from adverse events but also to *dynamically adapt* and *grow stronger* in response to risks and uncertainties.

Resilient safety integrates both active and passive safety measures into a broader framework that emphasizes anticipation, recovery, and adaptation. For air shows, this means not only relying on systems like AGCAS and ejection seats but also developing robust training and response strategies. Therefore, display pilots undergo rigorous and regular training to practice responding to various emergency scenarios, including equipment malfunctions and adverse weather conditions.

Furthermore, fostering a *safety culture* that emphasizes continuous learning and improvement is essential for *resilient safety* in air shows. Encouraging display pilots and crew to report any irregularities, near-misses, or safety concerns without fear of retribution allows for continuous assessment and enhancement of safety protocols. This iterative process ensures that the organization can adapt and improve its safety measures, preparing more effectively for future challenges.

In this part of the book, we will study how *resilient safety* principles can be applied to achieve excellence among air show performers. We will explore the key models and practices of *resilient safety*, examining their practical application in the inherently high-risk environment of air shows. Through case studies, expert analyses, and practical recommendations, this part aims to highlight what equips air show performers to excel safely, setting new benchmarks for resilience and excellence in the aviation industry.

Why Resilient Safety?

Resilient safety is essential because it addresses the limitations of traditional safety management approaches, which often focus solely on preventing accidents (active safety) and reducing risks (passive safety). Traditional methods can fall short in complex, dynamic environments where unexpected events and system interdependencies can lead to unforeseen challenges. Consequently, *resilient safety* enhances the ability to cope with and adapt to such disruptions, ensuring continued safe operations (Hollnagel, 2018).

Historical Overview of Resilient Safety

Resilient safety, particularly within high-risk industries like aviation, has undergone significant evolution. The concept of resilience in safety management has been shaped by various academic contributions and practical implementations over time (Adjekum & Fernandez-Tous, 2020; Akselsson et al., 2009; American Psychological Association, 2012; Bergström et al., 2015; Bisbey et al., 2021; Chialastri & Pozzi, 2008; Fasey et al., 2021; Flood & Keegan, 2022; Hollnagel, 2006, 2010, 2014, 2018; Jones et al., 2005; Mowbray, 2011).

The foundation of *resilient safety* began with the concept of high-reliability organizations (HROs), which operate in complex, high-hazard environments with a lower-than-expected number of accidents. This idea was significantly advanced by Karl E. Weick and Kathleen M. Sutcliffe (2011) in their 2001 work. Concurrently, Charles Perrow's (1999) Normal Accident Theory (NAT) suggested that accidents in complex systems are inevitable due to inherent interdependencies.

In the early 2000s, the field of resilience engineering emerged, focusing on enhancing systems' capacities to cope with unexpected events. Erik Hollnagel, David D. Woods, and Nancy Leveson (2006) were pivotal in establishing this field through their edited volume. They introduced the four cornerstones of Resilience Engineering: the ability to *respond*, *monitor*, *learn*, and *anticipate*, which helps organizations prepare for and manage unexpected events, while Hollnagel (2011) further elaborated on these principles.

In aviation, the implementation of safety management systems (SMSs) integrates resilience engineering principles by promoting proactive hazard identification and risk management. This approach is comprehensively documented in the International Civil Aviation Organization's manual (2018). Various case studies, such as those discussed by Dekker (2011), illustrate the practical application of resilience engineering in aviation, highlighting the importance of learning from incidents and fostering a safety culture adaptable to emerging threats.

Contemporary developments in *resilient safety* also address the integration with cyber–physical systems, focusing on vulnerabilities and ensuring robustness against cyber threats. Nancy Leveson's (2016) work provides a comprehensive overview of this integration. Recent research emphasizes adaptive safety management, which includes continuous learning and adaptation as core components of resilient systems, as discussed by Patriarca et al. (2018).

Examples and Applications

Aviation

The successful emergency landing of US Airways Flight 1549 on the Hudson River in 2009 is a prime example of resilient safety. On January 15, 2009, the Airbus A320, piloted by Captain Chesley "*Sully*" Sullenberger and First Officer Jeffrey Skiles, struck a flock of geese shortly after takeoff from LaGuardia Airport in New York City, resulting in a complete loss of engine power (National Transportation Safety Board, 2010). Despite the critical situation, the crew maintained composure and quickly assessed their options. Recognizing that returning to LaGuardia or diverting to another nearby airport was not feasible, Captain Sullenberger decided to execute an unplanned water landing on the Hudson River. The crew's ability to quickly adapt to engine failure and carry out a water landing, a highly challenging maneuver, saved all 155 passengers and crew members on board.

In aviation, continuous training for pilots in handling unexpected situations is a cornerstone of resilient safety. Pilots undergo rigorous and regular simulation training, where they practice responding to various emergency scenarios, including engine failures, severe weather conditions, and other critical in-flight issues. This training ensures that pilots can remain calm and make swift, informed decisions under pressure (Hollnagel, 2018).

Equally important are simulation exercises that often involve the entire flight crew and air traffic control to practice emergency procedures and coordination. These exercises simulate emergency scenarios, providing valuable hands-on experience and helping to identify weaknesses in current protocols. Such a continuous effort to refine procedures allows the aviation industry to enhance its capacity to respond effectively to unexpected events, thereby reinforcing resilient safety (Leveson, 2016).

Furthermore, robust incident reporting systems play a crucial role in fostering resilient safety. These systems encourage pilots and other aviation professionals to report any irregularities, near-misses, or safety concerns without fear of retribution. The data collected from these reports is analyzed to identify trends, potential hazards, and areas for improvement. By learning from past incidents, the aviation industry can implement measures to prevent similar occurrences in the future (Dekker, 2011).

Healthcare

During the COVID-19 pandemic, hospitals had to rapidly adapt to surges in patient numbers, supply shortages, and new treatment protocols. This unprecedented health crisis required healthcare facilities to demonstrate resilient safety principles by quickly and effectively managing the overwhelming demands placed on their systems. Hospitals reconfigured spaces to accommodate the influx of COVID-19-infected patients, converting areas like conference rooms and parking structures into temporary treatment centers (Ranney et al., 2020). They also redirected resources, such as ventilators and personal protective equipment (PPE), to where they were most needed, often innovating to compensate for shortages (Livingston et al., 2020). Implementing new safety measures, such as enhanced infection control protocols and telemedicine services, was critical in minimizing the spread of the virus and ensuring continued care for non-COVID-19 patients (Galea et al., 2020).

Building on these adaptations, developing flexible response plans is a critical component of resilient safety in healthcare. These plans include strategies for managing patient surges, reallocating resources, and adapting to evolving information about the virus and its treatment. Cross-training staff to perform various roles allows hospitals to maintain functionality even when specific departments are overwhelmed. For example, nurses and doctors were trained to support intensive care units (ICUs) despite typically working in different specializations (Shanafelt et al., 2020).

Furthermore, maintaining a culture of learning from incidents and near-misses further strengthens resilience. Throughout the pandemic, healthcare providers continuously reviewed and revised their practices based on emerging data and experiences. This iterative process of learning and adapting ensured that they could better handle subsequent waves of the virus and other unforeseen challenges (Dekker, 2011). By fostering a culture of continuous improvement, healthcare facilities could enhance their preparedness and response strategies, reinforcing their resilient safety frameworks.

Nuclear Power

The response to the Fukushima Daiichi nuclear disaster in 2011 highlighted both the need for and the challenges of resilient safety. On March 11, 2011, a massive earthquake and subsequent tsunami led to the failure of cooling systems at the Fukushima Daiichi Nuclear Power Plant in Japan, resulting in core meltdowns and the release of radioactive materials (Hirano et al., 2012). This disaster points out the importance of having robust, adaptable safety measures in place to manage unforeseen and extreme events.

Effective disaster response and recovery efforts were crucial in mitigating the impact. Emergency responders and plant operators had to work under extreme conditions to stabilize the reactors and prevent further radiation release. Additionally, the Japanese government, in coordination with international agencies, implemented extensive evacuation measures, decontamination efforts, and health monitoring programs for the affected populations. These actions were vital in reducing immediate harm and facilitating long-term recovery.

Moreover, designing redundant safety systems is a crucial aspect of resilient safety in the nuclear industry. Redundancy involves having multiple backup systems in place so that if one system fails, others can take over. At Fukushima, some of the backup generators and cooling systems were compromised by the tsunami, revealing gaps in the plant's safety design. Consequently, nuclear facilities worldwide have since re-evaluated and upgraded their safety systems to ensure better protection against similar events (Hirano et al., 2012).

In addition, conducting regular drills for various emergency scenarios is another critical practice. These drills prepare plant staff to respond quickly and effectively to a range of potential incidents, from equipment malfunctions to natural disasters. By enhancing coordination among different teams and ensuring that all personnel are familiar with emergency procedures, drills play a vital role in disaster preparedness. Post-Fukushima, many nuclear plants have increased the frequency and complexity of their drills to better simulate real-world conditions (Dion-Schwarz et al., 2016).

Ultimately, fostering a safety culture that emphasizes vigilance and continuous improvement is essential for resilient safety (Akselsson et al., 2009). This involves creating an environment where safety is a core value, and all employees are encouraged to report issues, suggest improvements, and stay informed about best practices. Continuous learning from past incidents and near-misses helps organizations to adapt and improve their safety measures (Leveson, 2016). By embracing these principles, the nuclear industry can better safeguard against future disasters and enhance its overall resilience.

Models of Safety

The concept of *resilient safety* has gained significant attention as industries recognize the limitations of traditional safety management approaches in complex, dynamic environments. Traditional methods, often characterized by a reactive focus on preventing accidents and reducing risks, can fall short when dealing with unforeseen challenges and system interdependencies. To address these gaps, various models of resilient safety have been developed. These models emphasize adaptability, continuous improvement, and the ability to maintain functionality under stress. This section explores three influential models: Hollnagel's *Safety-I and Safety-II* (Hollnagel, 2018), Reason's *Swiss Cheese Model* (Reason, 1997), and Woods' *Resilience Engineering* (Woods, 2017), each offering unique perspectives and strategies for enhancing safety and resilience in organizations.

Hollnagel's *Safety-I and Safety-II*

Erik Hollnagel (2018) introduced the concepts of *Safety-I and Safety-II* to distinguish between traditional and modern approaches to safety management.

The *Safety-I* model focuses on preventing things from going wrong by analyzing and eliminating the causes of accidents. It operates on the principle that safety is achieved by reducing the number of adverse outcomes through reactive measures. In this approach, incidents are seen as the result of failures, malfunctions, or human errors. The primary goal is to identify and correct these issues to prevent recurrence. *Safety-I* is essentially about compliance with established rules and procedures, with the assumption that adherence to these standards will minimize risks. Techniques such as root cause analysis, failure mode, and effects analysis (FMEA), as well as fault tree analysis, are commonly used to understand and address the causes of accidents. The focus is on past events and preventing their repetition by improving systems and processes to eliminate identified vulnerabilities. However, *Safety-I* can be limited by its reactive nature, as it primarily addresses problems after they have occurred. Additionally, it may lead to a culture of blame, where individuals are held responsible for failures rather than addressing systemic issues (Hollnagel, 2018).

In contrast, *Safety-II* emphasizes ensuring that *things go right* by enhancing the system's ability to succeed under varying conditions, the *synesis*, as Hollnagel named it. This proactive approach focuses on understanding and supporting everyday operations, recognizing that complex systems operate in unpredictable environments. *Safety-II* is about creating conditions that enable successful performance even when faced with unexpected challenges and variability. It aims to improve overall system resilience by fostering an environment where success is the norm and failures are rare. This approach involves continuous learning and adaptation, encouraging systems to function effectively even in the face of disruptions. *Safety-II* promotes the idea that safety is not just the absence of accidents but the presence of adaptive capacity. Practices under this model include building strong communication channels, promoting flexibility, and ensuring that all system components can work together seamlessly under different conditions. Continuous monitoring, real-time feedback, and iterative learning processes are integral to *Safety-II*, helping organizations anticipate potential issues and adapt swiftly to changes (Hollnagel, 2018).

Reason's Swiss Cheese Model

James Reason's *Swiss Cheese Model* is a widely recognized framework for understanding how accidents occur in complex systems. This model illustrates how accidents happen when multiple layers of defense, or barriers, fail simultaneously. Each layer of defense is likened to a slice of Swiss cheese, which has holes representing weaknesses or failures. An accident occurs when the holes in each layer align, allowing a hazard to pass through all defenses and result in harm (Reason, 1997).

Incorporating *resilient safety* into this model involves reinforcing these barriers and adding adaptive measures to address unforeseen gaps. This requires regular updating and improving defenses to keep them effective against emerging threats. Conducting thorough risk assessments helps identify potential vulnerabilities in each layer of defense, ensuring that all aspects of the system are robust and capable of adapting to new challenges. By focusing on both proactive and reactive measures, organizations can prevent the alignment of holes and reduce the likelihood of accidents (Reason, 1997).

For example, in the healthcare industry, this could involve implementing multiple checks and balances at different stages of patient care to prevent errors. If one check fails, others are in place to capture the mistake. Similarly, in aviation, various redundant safety protocols are established, so if one system fails, another can compensate, thereby preventing catastrophic failures.

Woods' Resilience Engineering

David D. Woods' concept of *resilience engineering* focuses on the capacity of systems to adapt and maintain operations under stress. Resilience engineering is concerned with understanding how systems can continue to operate safely and effectively in the face of unexpected disruptions and changing conditions. This approach recognizes that failures are inevitable in complex systems, but the key to resilience is the ability to anticipate, monitor, respond to, and learn from these disruptions (Woods, 2017).

Resilience engineering involves several core principles. First, it emphasizes anticipating potential disruptions by identifying and preparing for possible challenges before they occur. This proactive stance allows organizations to develop strategies and allocate resources to mitigate risks. Continuous monitoring of system performance is also crucial, as it enables the detection of early signs of strain or failure. By observing and assessing operations in real-time, organizations can intervene before issues escalate.

Effective response to incidents is another vital aspect of *resilience engineering*. Organizations must have robust plans and capabilities in place to manage and mitigate the impact of disruptions, which includes having clear communication channels, predefined roles, and resources readily available to address emergencies. The ability to respond quickly and efficiently helps minimize damage and maintain critical functions.

Furthermore, learning from incidents is perhaps the most dynamic component of *resilience engineering*. This principle involves using experiences from past events to improve future performance and resilience. Organizations must engage in an iterative process of reflection and adaptation, integrating lessons learned into their systems and practices. This continuous learning cycle helps organizations adapt and enhance their ability to withstand and recover from adverse conditions (Woods, 2017).

For instance, in the field of healthcare, *resilience engineering* can be applied by developing flexible protocols that allow medical teams to adapt to sudden changes in patient volume or severity of cases. Hospitals can implement real-time monitoring systems to track patient vital signs and resource availability, ensuring swift responses to emerging crises. Additionally, healthcare organizations can conduct regular debriefings and simulations to learn from actual and simulated incidents, fostering a culture of continuous improvement and preparedness.

In the aviation industry, *resilience engineering* might involve the use of advanced monitoring technologies to detect anomalies in aircraft systems and the implementation of comprehensive emergency response plans. In addition, pilots and crew undergo extensive training to handle various in-flight emergencies, ensuring that they can respond effectively to unexpected situations, while lessons learned from past incidents are systematically analyzed and integrated into training programs and operational procedures to enhance overall safety and resilience.

REFERENCES

Adjekum, K., & Fernandez-Tous, M. (2020). Assessing cultural drivers of safety resilience in a collegiate aviation program. *Collegiate Aviation Review, 38*, 122. https://doi.org/10.22488/okstate.20.100208

Air Force Research Laboratory (AFRL). (n.d.). Automatic ground collision avoidance. Retrieved July 31, 2024, from https://afresearchlab.com/technology/aerospace/automatic-ground-collision-avoidance

Akselsson, R., Ek, Å., Koornneef, F., Stewart, S., & Ward, M. (2009). Resilience safety culture. In *17th World Congress on ergonomics*. IEA.

Aviation Week. (2016, September 13). *Auto-GCAS saves unconscious F-16 pilot—Declassified USAF footage* [Video]. YouTube. www.youtube.com/watch?v=WkZGL7RQBVw

Dekker, S. (2011). *Drift into failure: From hunting broken components to understanding complex systems*. Ashgate Publishing.

Dion-Schwarz, C., Evans, S. E., Geist, E., Harold, S. W., Koym, V. R., Savitz, S., & Thrall, L. (2016). *Technological lessons from the Fukushima Dai-Ichi accident*. RAND.

Galea, S., Merchant, R. M., & Lurie, N. (2020). The mental health consequences of COVID-19 and physical distancing: The need for prevention and early intervention. *JAMA Internal Medicine, 180*(6), 817–818. https://doi.org/10.1001/jamainternmed.2020.1562

Hirano, M., Yonomoto, T., Ishigaki, M., Watanabe, N., Maruyama, Y., Sibamoto, Y., …, & Moriyama, K. (2012). Insights from review and analysis of the Fukushima Dai-ichi accident: Fukushima NPP accident related. *Journal of Nuclear Science and Technology, 49*(1), 1–17.

Hollnagel, E. (2011). Prologue: The scope of resilience engineering. In E. Hollnagel, J. Pariès, D. D. Woods, & J. Wreathall (Eds.), *Resilience engineering in practice: A guidebook* (pp. xxix–xliii). Ashgate.

Hollnagel, E. (2018). *Safety-II in practice: Developing the resilience potentials*. Routledge.

Hollnagel, E., Woods, D. D., & Leveson, N. (Eds.). (2006). *Resilience engineering: Concepts and precepts*. Ashgate Publishing Ltd.

International Civil Aviation Organization (ICAO). (2018). *Safety management manual (SMM)* (4th ed.). ICAO.

Leveson, N. G. (2016). *Engineering a safer world: Systems thinking applied to safety*. The MIT Press.

Livingston, E., Desai, A., & Berkwits, M. (2020). Sourcing personal protective equipment during the COVID-19 pandemic. *JAMA, 323*(19), 1912–1914. https://doi.org/10.1001/jama.2020.5317

Parate, B. (2022). Science and technology of aircraft seat ejection: Advanced concepts. *Cogent Engineering, 9*. https://doi.org/10.1080/23311916.2022.2034267

Patriarca, R., Bergström, J., Gravio, G., & Costantino, F. (2018). Resilience engineering: Current status of the research and future challenges. *Safety Science, 102*, 79–100. https://doi.org/10.1016/J.SSCI.2017.10.005

Perrow, C. (1999). *Normal accidents: Living with high risk technologies*. Princeton University Press.

Ranney, M. L., Griffeth, V., & Jha, A. K. (2020). Critical supply shortages—The need for ventilators and personal protective equipment during the COVID-19 pandemic. *New England Journal of Medicine, 382*(18), e41.

Shanafelt, T., Ripp, J., & Trockel, M. (2020). Understanding and addressing sources of anxiety among health care professionals during the COVID-19 pandemic. *JAMA, 323*(21), 2133–2134.

Swihart, D., Barfield, A., Griffin, E., Lehmann, C., Whitcomb, S., Flynn, B., Martin, L., Skoog, M., & Prosser, K. (2011a). Automatic ground collision avoidance system design, integration, & flight test. *IEEE Aerospace and Electronic Systems Magazine, 26*, 4–11. https://doi.org/10.1109/MAES.2011.5871385

Swihart, D., Barfield, A., Griffin, E., Lehmann, R., Whitcomb, S., Skoog, M., Flynn, B., & Prosser, K. (2011b). Design, integration and flight test of an autonomous ground collision avoidance system. *Gyroscopy and Navigation, 2*, 84–91. https://doi.org/10.1134/S2075108711020088

Weick, K. E., & Sutcliffe, K. M. (2011). *Managing the unexpected: Resilient performance in an age of uncertainty* (Vol. 8). John Wiley & Sons.

Woods, D. D. (2017). Essential characteristics of resilience. In *Resilience engineering* (pp. 21–34). CRC Press.

NOTES

1 Active safety refers to systems to *prevent* accidents and hazardous situations from occurring, i.e., the Auto Ground Collision Avoidance System (AGCAS).

2 Passive safety refers to the features and design elements incorporated into a vehicle, structure, or system that aim to protect occupants and reduce injury once an accident or hazardous event has *already occurred*, i.e., ejection seat.

11 A New View on Air Show Safety

Manolis Karachalios

Excellence First, Safety Always:

The Hallmark of the Air Show Community.

INTRODUCTION

Safety has always been a paramount concern in the aviation industry within every aspect of flight operations, from commercial airlines to military missions. The detailed protocols and thorough training that define aviation safety have been developed over decades, most of the time as a result of lessons learned from accidents or incidents, resulting in a comprehensive system designed to minimize risk and prevent accidents. However, the high-energy, dynamic environment of air show performances presents unique challenges that demand a specialized approach.

Air shows are impressive events that showcase the pinnacle of aviation skill and technology. Display pilots perform complex aerobatic maneuvers at high speeds and in close formations, within confined airspace, and in front of large crowds. These performances push both pilots and their aircraft to the limits of their capabilities, creating a scenario where the margin for error is extremely small (Barker, 2003). Traditional safety models, while effective in many contexts, often fall short of addressing the complex and fast-paced nature of air shows.

In the majority of aviation operations, safety measures are designed to manage predictable risks and ensure smooth, routine flights. However, the inherent risks of air show performances require a different approach. The need for split-second decision-making, precise coordination, and flawless execution under extreme conditions calls for a safety model that is as dynamic and flexible as the environment it aims to protect.

Therefore, a new perspective on air show safety is vital—one that goes beyond conventional risk management and embraces *excellence*, *precision*, and *accuracy* as foundational principles. By raising the "*bar*," safety will be guaranteed in the air show community. This new Model must be *proactive*, anticipating potential issues before they arise and incorporating real-time monitoring to *adapt* to changing conditions immediately (Karachalios & Adjekum, 2023). It should be *integrated*, seamlessly combining various safety measures, both active and passive, into a cohesive system

DOI: 10.1201/9781003504801-15

that supports every aspect of the performance. Finally, it must be *resilient* and capable of *withstanding* and *recovering* from disruptions to ensure continuous safety and performance excellence.

Furthermore, introducing excellence as a core component of the new safety perspective that goes beyond conventional risk management in air shows is vital. *By focusing on excellence, safety—as an unquestionable byproduct—is given in the air show community; then, the hallmark of the air show performers becomes Excellence First, Safety Always.*

In this endeavor, two models are introduced and discussed in this chapter: the *STAR* and *Traffic Light Models*. Both models are intended to reverse the traditional approach to safety in air shows and concentrate on using excellence as the main catalyst to embrace a robust resilient safety culture.

EXCELLENCE IN AIR SHOW SAFETY MANAGEMENT SYSTEMS (ASMS)

How can excellence truly be embedded into a well-established air show safety management system (ASMS)? At first glance, it might seem superfluous or even idealistic. However, when we observe elite air show performers, their breathtaking displays are not merely the result of adhering to safety protocols. Instead, they arise from a synergy of technical mastery, relentless preparation, unwavering commitment to safety, and a profound respect for precision and discipline. These are the very elements we seek to embed in a new approach to safety management.

Drawing parallels between elite performers and the airshow SMS, we see that an effective ASMS must depend on consistent training, strict adherence to procedures, and a commitment to continuous improvement. According to the Federal Aviation Administration (FAA) (2020), an SMS is a formal, top-down, organization-wide approach to managing safety risk and assuring the effectiveness of safety risk controls. It includes systematic procedures, practices, and policies for managing safety risk, which must be managed with the same level of priority as other core business processes.

The FAA (2020) defines four pillars of SMS that are integral to its implementation:

1. Safety Policy: Establishes senior management's commitment to continually improving safety and defines the methods, processes, and organizational structure needed to meet safety goals.
2. Safety Risk Management (SRM): Involves identifying hazards, assessing, and mitigating risks.
3. Safety Assurance: Evaluates the continued effectiveness of implemented risk control strategies and identifies the need for new controls.
4. Safety Promotion: Includes training, communication, and other actions to create a positive safety culture.

In this book, the focus will be on SRM and specifically on Air Show Risk Management (ASRM) since it is the pillar of ASMS that should be applied at all times, regardless of the scale of air show performance operations. By integrating ASRM, air show performers can craft an ASMS capable of handling errors gracefully while maintaining high standards of safety and performance. Eventually, this system would reflect the excellence seen in elite air show performers, where preparation and execution combine to minimize risks and maximize safety.

Measuring Effectiveness in ASMS

To measure the effectiveness of integrating excellence into a SMS, one can analyze the frequency and quality of training sessions, including mental preparation, and compare these metrics to the number of incidents occurring. Research indicates that consistent, high-quality training is crucial for maintaining safety standards and reducing incidents.

Mohammadfam et al. (2012) highlight the importance of performance measurement tools like the Health, Safety, and Environment Excellence Instrument in evaluating safety systems. Their study emphasizes that a systematic approach to performance measurement is essential for understanding and improving safety outcomes.

Similarly, Hakami and Salah (2023) discuss the excellence model for effectively implementing SMS in the oil and gas industry. They stress the significance of advanced training programs and continuous evaluation in achieving safety excellence. These insights are applicable across industries, including air show safety management.

Moreover, Ulfvengren and Corrigan (2015) underline that resource allocation significantly impacts safety outcomes. Their research on implementing SMS in a lean airline demonstrates that well-resourced and high-quality training programs lead to fewer incidents and improved overall safety. This indicates that investing in comprehensive training and adequate resources is essential for achieving excellence in SMS.

Commitment and Passion

Fostering a culture of discipline and precision in high-risk industries necessitates a collective commitment and an unwavering passion for success from the entire organization. This shared dedication transforms the workplace, ensuring that every team member is not only focused but also genuinely invested in maintaining the highest standards of safety and performance.

According to Moussa et al. (2016) in their book "Committed Teams: Three Steps to Inspiring Passion and Performance," creating committed teams involves three essential steps: establishing a clear and compelling purpose, fostering a culture of collaboration and accountability, and maintaining continuous engagement and adaptability, principles crucial for inspiring passion and performance, particularly in high-risk environments where safety is paramount.

Learning from Air Show Performers

Other industries can learn valuable lessons from the culture of excellence embraced by air show performers. Their passion for their craft, meticulous daily routines, strong teamwork ethos, and efficient use of time create a foundation for success. Also, the sense of community and family culture among performers, coupled with a robust mentorship system, ensures that knowledge and skills are continuously passed down and refined. Furthermore, their practice of offering constructive criticism within a no-blame culture fosters a supportive and high-performing environment where safety is a shared priority.

INTEGRATING RESILIENCE IN AIR SHOW SAFETY

Resilience, defined as the capacity of a system to absorb disturbances and still retain its primary function and structure (Hollnagel, 2012; Woods, 2011), is essential for air show performers. In this context, resilience means that air show performers and their procedures can anticipate and prepare for potential disruptions and utilize redundant systems while they continuously learn and adapt to unforeseen challenges, ensuring rapid recovery.

Redundancy and Flexibility

Building redundancy into systems and processes ensures that air show performers have backup options available in case of failure. This could mean having multiple communication channels, spare parts for critical equipment, or alternative performance plans. Ultimately, flexibility in procedures allows teams to adapt quickly to changing circumstances without compromising safety.

Redundancy in Air Show Operations

Redundancy in air show operations involves incorporating extra components and procedures to maintain performance and safety in the event of a failure, which is crucial in the environment of air shows, where system failures can lead to severe consequences.

- Multiple Communication Channels
  Ensuring multiple communication channels are in place is vital for maintaining information flow during a flying display. Display pilots and air bosses primarily rely on radio communication for real-time coordination and instructions. However, to safeguard against any potential communication failures, they also implement backup systems. These backups often include secondary radio frequencies, which are used if the primary frequency experiences interference or technical issues. This redundancy helps to maintain operational continuity and enhances overall safety, ensuring that critical information is always accessible, regardless of any primary system malfunctions.

- Spare Parts for Critical Equipment
  Readily available spare parts for critical equipment in the aircraft can prevent extended downtime and potential performance interruptions. This is particularly important in air shows where the availability of essential components, especially any parts related to the uninterrupted aircraft engine operation, can significantly reduce the risk of prolonged out.
- Alternative Performance Plans
  In the dynamic environment of air shows, having alternative performance plans, e.g., alternate display profiles, allows display pilots to quickly adapt to unforeseen circumstances like sudden weather changes, allowing minimum interruption in the air show flow.

Flexibility in All Aspects of Flying Display Operations: Allowable Performance Error (APE)

Flexibility in procedures enables air show performers to adapt to varying conditions without compromising safety or performance. This adaptability is essential in the dynamic environment of air shows, where conditions can change rapidly, and a margin for error needs to be always considered.

However, there's a misconception that air show flying is a non-forgiving operating environment mainly due to the lack of flexibility in the decision-making process of the display pilots, as well as the lack of sufficient margins for error in their performances. To deflect this "*stigma*" of air show performances being a risk-taking environment with a "*zero margin for error*," the *allowable performance error (APE)* is conceptualized to emphasize that there should always be *considered* and *allowed an error to be more than zero* since to *err is human* (Fellmeth & Horwitz, 2011; Glosbe online dictionary, 2025; Merriam Webster, 2025; Reason, 1990).

The APE is based on, but not limited to, the following parameters: flying display profiles and energy management, dynamic weather conditions, performer physical condition, cognitive performance, putting excitement in a safe routine, and planning for emergencies. Its formula quantifies the necessary flexibility in air show operations to maintain safety and performance, with the sum of its factors needing to be *greater than zero*, indicating that the combined adaptability is sufficient to manage all conditions safely.

APE=(Flexibility in Flying Display Profiles and Energy Management)+(Dynamic Weather Conditions)+(Performer Physical Condition)+(Cognitive Performance)+(Putting Excitement in a Safe Routine)+(Planning for Emergencies) **> *0***

- Flexibility in Flying Display Profiles and Energy Management
  Display pilots manage their aircraft's energy states (speed, altitude, and power setting) effectively to execute maneuvers safely. Having flexible flying display profiles allows air show performers to adjust their routines in response

to actual flight parameters, ensuring maneuvers can be performed within safe energy constraints.

- Dynamic Weather Conditions
  Continuous monitoring of weather conditions is crucial for the safety and success of air show performances. Pilots and support teams must be vigilant in tracking real-time weather data, including wind speed and direction, visibility, and cloud base. This proactive approach enables performers to adjust their display profile dynamically, whether by altering flight paths, modifying maneuvers, or rescheduling performances entirely.
- Performer Physical Condition
  The physical condition of display pilots, encompassing health, fitness, and adequate rest periods, is paramount for safe and effective air show performances. Pilots must maintain peak physical condition to handle the rigorous demands of aerial maneuvers and to ensure quick reflexes and sharp cognitive functions. Therefore, implementing flexible scheduling practices is essential. This approach allows for adjustments based on pilots' individual needs, accommodating necessary rest and recovery periods, thereby optimizing their performance capabilities and ensuring safety during shows.
- Cognitive Performance
  Degraded cognitive performance that leads to impairment due to the physiological demands of display flying, i.e., G-loading, should be closely monitored. Therefore, margins in the maneuvers should be accounted for to allow sufficient margin for cognitive capacity; in other words, display pilots should have a flexible profile routine to transition in case they feel that they approach their cognitive limits.
- Putting Excitement in a Safe Routine
  Crafting display routines that are exciting yet safe is a delicate balance; with that in mind, display pilots should design display profiles that thrill audiences while adhering to stringent safety margins.
- Planning for Emergencies
  Comprehensive emergency planning is essential for display pilots who need to conduct regular drills for various scenarios, such as engine failures or sudden weather changes, and subsequently are prepared to respond swiftly and effectively. Such planning and preparation are critical to formation teams as well since they need to ensure all members of the team are equally prepared for any adversity.

Continuous Learning and Adaptation

Continuous learning from past experiences is essential for building resilience in air show operations. Therefore, it is essential to encourage a culture of openness and feedback that guarantees everyone is continuously improving. Examples of applying continuous learning principles by air show performers are the following:

- Thorough Debriefs
  Conducting thorough debriefs after each performance allows teams to review and evaluate every aspect of the event. This includes analyzing maneuvers, communication effectiveness, and responses to any unexpected situations. Debriefs should be structured and systematic, often guided by frameworks such as Crew Resource Management (CRM), which is widely used in aviation to improve safety and efficiency (Salas et al., 1995).
- Analyzing Successes and Failures
  It is important to analyze not only what went wrong but also what went well. Identifying successful strategies and best practices helps reinforce positive behaviors and procedures. Conversely, analyzing failures and near-misses helps identify areas for improvement. This dual approach ensures a balanced and comprehensive understanding of performance (Dekker, 2016).
- Integrating Lessons into Future Planning and Training
  The insights gained from debriefs and analyses should be integrated into future planning and training. This may involve updating standard operating procedures (SOPs), revising training programs, and incorporating new techniques or technologies (Reason, 2016).
- Encouraging a Culture of Openness and Feedback
  Fostering a culture of openness and feedback is crucial for continuous learning. Team members should feel comfortable sharing their observations and suggestions without fear of reprisal (Edmondson, 1999).

Air show performers exhibit remarkable resilience, characterized by their ability to adapt to unforeseen challenges and recover quickly from disruptions and distractions. Their extensive training and experience equip them with the skills needed to handle unplanned events seamlessly, ensuring both safety and the continuation of the show.

One vivid example of this adaptability can be seen when performers face disruptions from unexpected intrusions in the Showbox, such as birds or unauthorized drones. In such cases, they must quickly adapt their maneuvers to avoid collisions, which could potentially necessitate modifying their planned display sequence.

Another example of the adaptability of air show performers is their ability to handle changes in the event's timetable. Air shows are often large, complex events with many moving parts, and unforeseen delays or accelerations in the schedule are not uncommon. For instance, if a performer is suddenly required to fly earlier than expected due to a timetable change, they must quickly adjust their preparation and mental readiness to ensure a flawless performance. To assist display pilots in adapting to schedule changes, flying display directors and air bosses typically include a statement in the display order—and emphasize it in the air show safety briefing—that all participants need to be ready to fly at their previous time slot.

When faced with such a scenario, pilots must promptly modify their pre-flight routines, which typically include thorough checks of the aircraft, weather assessments, and visualization of their display profiles. Accelerating this process requires a high level of efficiency and alertness to ensure that safety is not compromised. Therefore, display pilots and their ground crews must work together seamlessly to complete all necessary checks swiftly and accurately. Moreover, adapting to an earlier slot means pilots must also recalibrate their mental and physical states to be performance-ready ahead of their original schedule.

Ultimately, display pilots rely on their ability to quickly transition from a state of waiting or relaxation to being fully alert and prepared to execute their planned aerobatic maneuvers precisely. Such an adaptability is a testament to their resilience and professionalism, ensuring that air shows proceed smoothly even in the face of unexpected challenges.

INTEGRATING EXCELLENCE IN AIR SHOW SAFETY

Excellence, defined as "*the possession chiefly of good qualities in an eminent or unusual degree*" (Oxford English Dictionary, 2024), ensures that air show performers not only meet but exceed the highest standards in performance and safety. When combined with resilience, excellence drives continuous improvement and the pursuit of peak performance, making air shows safer, more consistent, and, ultimately, more entertaining. Therefore, integrating excellence into air show safety requires a multifaceted approach that involves adhering to best practices, continuous improvement, and skill refinement.

Adhering to Best Practices

Adhering to best practices is essential for integrating excellence into air show safety. This involves following strict maintenance schedules and operational protocols to ensure that all equipment and procedures meet the highest safety standards. Furthermore, regular assessments and updates to the training programs based on the latest industry standards and technological advancements are necessary.

Continuous Improvement

Continuous improvement is another key component of integrating excellence into air show safety. This can be achieved by regularly updating and enhancing aircraft with the latest technology, such as advanced GPS systems and real-time monitoring tools. Allocating a budget for continuous upgrades and integrating these advancements into training programs are crucial steps. Ultimately, this commitment to continuous improvement ensures that air show performers are always operating at the cutting edge of technology, enhancing both safety and performance.

Skill Refinement and Performance Enhancement

Skill refinement and performance enhancement are also critical, which involve practicing maneuvers and incorporating new techniques learned from international air shows. Developing a training program that includes international collaboration and knowledge exchange can be highly beneficial, as it not only improves their own skills but also fosters a culture of excellence and innovation within the air show community.

Benchmarking and Performance Metrics

To integrate excellence into air show safety effectively, air show performers must also set specific performance metrics that align with both safety and excellence standards, with a benchmarking system that tracks performance against these metrics and identifies areas for improvement.

INTEGRATING RESILIENCE AND EXCELLENCE INTO AIRSHOW SAFETY RISK MANAGEMENT (ASRM)

Integrating resilience and excellence into air show SRM (ASRM) is a critical aspect of this new approach.

When resilience and excellence are integrated into the Airshow SRM (ASRM), the result is a robust safety system that can handle unexpected challenges successfully while consistently delivering high-quality performances. To integrate resilience and excellence into the ASRM framework, we'll build upon, as depicted in Figure 2.7 of the first book of this series (Karachalios & Adjekum, 2023), the following steps:

1. Identify Hazards: This initial step involves a thorough identification of potential hazards using a systematic approach. Utilizing a comprehensive checklist, referred to as "*SHOWCENTER*," each letter stands for a specific area or aspect to be examined for potential risks.
2. Assess Probability and Severity: Once hazards are identified, their likelihood and potential impact are assessed. Then it involves assigning a score based on the probability (ranging from very low to very high) and the severity (ranging from negligible to catastrophic) of each hazard.
3. Calculate Risk Scores: The risk score is calculated by multiplying the probability and severity scores. This step quantifies the level of risk associated with each hazard, providing a clear metric for prioritization.
4. Categorize Risks: Hazards are then categorized based on their risk scores. Using a color-coded or numerical table, risks are grouped into categories such as low, medium, and high, facilitating focused attention on the most critical areas.
5. Designate Mitigation Strategies: For each category of risk, specific mitigation strategies are developed. This entails detailing actions to reduce or eliminate the risks, including procedural changes, equipment upgrades, or enhanced training programs.
6. Implement Mitigation Strategies: The final step is the execution of the designated mitigation strategies, which necessitates meticulous planning and implementation, with detailed documentation covering potential hazards, risk levels, mitigation actions, responsible individuals, and the status of implementation.

Table 11.1 provides a clear overview of how resilience and excellence can be integrated into each step of the ASRM framework to enhance safety and performance in air shows.

TABLE 11.1
Resilience and Excellence Integration into ASRM

	ASRM Step	Resilience Integration	Excellence Integration
1	Identify Hazards	Utilize a diverse team of experts; incorporate historical data and real-time information.	Establish a rigorous process using advanced tools and technologies; employ predictive analytics and machine learning for comprehensive detection.
2	Assess Probability and Severity	Adopt a flexible assessment framework; use scenario-based assessments.	Implement a robust evaluation system with peer reviews; use detailed metrics and benchmarking against best practices.
3	Calculate Risk Scores	Design a dynamic risk scoring system; include contingency factors for real-time adjustments.	Ensure transparency and replicability; utilize sophisticated modeling techniques for precision.
4	Categorize Risks	Develop adaptable risk categories; regularly review and revise based on new insights.	Use a detailed categorization matrix; apply best-in-class standards to define and refine categories.
5	Designate Mitigation Strategies	Create robust strategies with backups and redundancies; encourage continuous learning and adaptation.	Develop innovative, evidence-based strategies; engage experts and stakeholders for co-creation, reflecting the latest advancements and best practices.
6	Implement Mitigation Strategies	Continuous monitoring and feedback system; train staff for adaptive responses.	Establish rigorous implementation protocols; use state-of-the-art technology to track execution and measure effectiveness against benchmarks.

In Table 11.2, it is presented a practical example of how resilience and excellence can be integrated into the ASRM framework to address the hazard of *birds* in the Showbox.

By incorporating these resilience and excellence strategies into each step of the ASRM framework, the hazard of birds in the Showbox can be managed more effectively, ensuring both the safety of the performers and the audience while maintaining high-performance standards.

Integrated ASRM Framework

To ensure a *resultful* ASRM (see Figure 11.1), additional considerations need to be considered in this dynamic and ever-evolving process of seeking excellence and

TABLE 11.2
Example for Resilience and Excellence Integration into ASRM

	ASRM Step	Resilience Integration	Excellence Integration
1	Identify Hazards	Conduct regular surveys and use radar technology to detect bird activity in and around the Showbox. Collaborate with ornithologists to understand migration patterns.	Utilize machine learning algorithms to analyze bird movement patterns and predict potential bird strike hotspots. Deploy drones for real-time monitoring.
2	Assess Probability and Severity	Use scenario-based assessments considering various bird species and their behaviors. Include seasonal variations in bird activity.	Implement a detailed evaluation system with input from wildlife experts and historical bird strike data. Benchmark against international best practices for bird strike risk assessment.
3	Calculate Risk Scores	Develop a dynamic scoring system that adjusts based on real-time bird activity data. Use predictive models to forecast changes in risk levels.	Ensure the risk scoring system is transparent and validated by external experts. Use advanced simulation software to model potential bird strike scenarios.
4	Categorize Risks	Adapt risk categories based on evolving data about bird activity and the effectiveness of deterrent measures. Regularly update categories with the latest findings.	Use a comprehensive categorization matrix that includes multiple risk dimensions. Apply leading standards from aviation safety organizations to define risk categories.
5	Designate Mitigation Strategies	Develop a multi-layered bird mitigation strategy that includes the use of bird deterrent technologies, habitat management, and controlled feeding programs.	Innovate by integrating the latest bird deterrent technologies, such as bioacoustic devices and automated laser systems. Collaborate with leading wildlife management experts for strategy development.
6	Implement Mitigation Strategies	Establish a real-time monitoring and feedback loop to adjust strategies as needed. Train staff to implement adaptive responses based on current bird activity.	Implement protocols with clear accountability and regular audits. Use state-of-the-art technology to track the implementation and effectiveness of mitigation measures, ensuring continuous improvement.

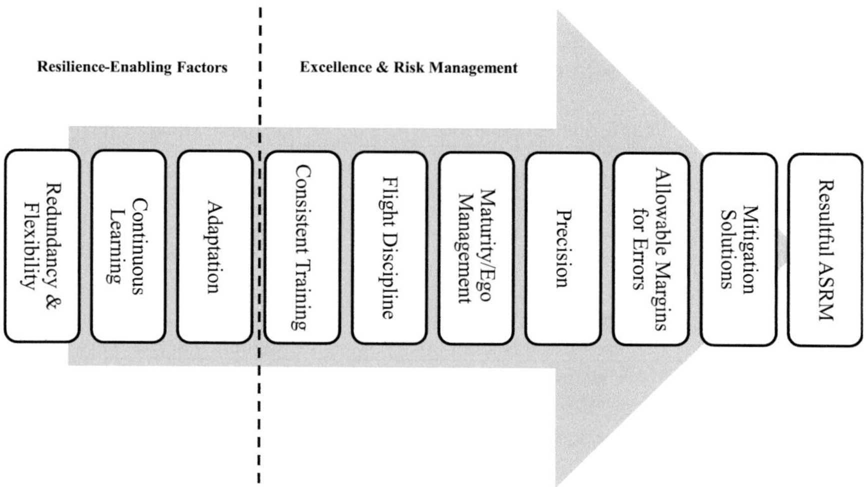

FIGURE 11.1 The path to a resultful ASRM.

safety in air show performances. The process continues, after establishing the resilience engineering steps described in the ASRM framework, with *consistent training*, which is the foundational step emphasizing the importance of regular and comprehensive training for accurately performed flights (Ericsson, 2006, 2014), particularly flying displays. This ensures that all participants are proficient in their routines and safety protocols. Following this, *flight discipline* is crucial for maintaining strict adherence to flight procedures and protocols prioritizing safety and precision. An essential aspect of this approach is maturity and ego management, which addresses the psychological factors affecting air show performance. *Managing ego* and demonstrating *maturity* is vital for making sound decisions under pressure and avoiding risky behaviors.

Precision is another critical component, focusing on the accurate execution of maneuvers and maintaining strict control over the aircraft to ensure both safety and the impressiveness of the display. Integrating *allowable margins for errors* acknowledges the inevitability of human error and ensures that safety buffers are in place to prevent minor mistakes from leading to severe consequences. This proactive measure enhances resilience by allowing for human factors without compromising safety. The development of *mitigation solutions* further strengthens ASRM by having contingency plans and safety measures ready to address any issues that arise during performances. Finally, achieving a *resultful ASRM* signifies the establishment of a robust and resilient safety framework, ensuring that air show performances are both safe and spectacular.

THE *STAR MODEL*: A PROACTIVE FRAMEWORK FOR EXCELLENCE IN AIR SHOW SAFETY

The *STAR Model* (*Safety, Training, Awareness, Resilience*) is crafted to promote a holistic approach to *preemptive* safety and *outstanding* performance, ensuring that air show performers can achieve excellence *consistently*. Unlike models that concentrate on analyzing failures (what went wrong), the *STAR Model* focuses on preventing errors (how to fly well) through continuous improvement, proactive risk management, and cultivating a culture of excellence and safety in every facet of air show operations. Ultimately, the main goal for the *STAR Model* is to keep the air show performers focused on what is important and related to the high-performance idiosyncrasies of the low-level display flying during the development of airshow ASMSs and training programs to ensure that their performances will shine bright like a "*star*" in front of their audiences.

Core Components

The *STAR Model's* four elements—*Safety*, *Training*, *Awareness*, and *Resilience*—are intricately interconnected, each reinforcing and enhancing the others. *Safety* provides the foundational protocols, *training* ensures preparedness and skill through the *EXCELS* framework, *awareness* enables proactive management of risks, and *resilience* allows for adaptability and continuous improvement to achieve performance excellence. Together, they create a robust framework that makes safety a given, fostering a culture of excellence in air show operations.

More specifically, *safety* is the bedrock of the *STAR Model*, ensuring that rigorous safety measures are ingrained in every aspect of air show design. Furthermore, air show SMSs (ASMS) could be incorporated into the operations of air show performers in order to have a systematic method of preparing and applying policies, risk management, and quality and ultimately promote safety (Karachalios & Adjekum, 2023). From the initial planning stages, strict protocols and adherence to regulatory standards are emphasized, and then routine safety briefings and thorough inspections become second nature, embedding a safety culture in every participant. For instance, comprehensive pre-flight inspections of aircraft and equipment are mandatory before each performance, ensuring that every component of the aircraft is in optimal condition.

Training under the *STAR Model* is a continuous journey aligned with the principles of the *EXCELS Model*, as discussed in Chapter 10, encompassing *Education*, *eXperience*, *Conditioning*, *Emergency preparedness*, *Leadership*, and *Safety*. This framework guarantees that display pilots receive comprehensive education on air show operations and safety protocols, gain practical experience through flying, and undergo physical and mental conditioning to maintain peak performance. Additionally, emergency preparedness drills equip them to handle unforeseen situations effectively, while leadership training fosters a sense of responsibility and initiative. Moreover, integrating safety into training reinforces its importance, creating a culture where every team member is prepared and proficient, directly supporting the overarching goal of minimizing risks.

Awareness is maintained through real-time monitoring systems, ensuring a constant state of situational alertness. Such systems use telemetry[1], which could provide continuous data on aircraft performance and pilot biometrics, while with proper communication systems, all team members, both on the ground and in the air, are kept informed about potential risks and environmental changes. For example, real-time telemetry might highlight a slight deviation in an aircraft's performance, triggering immediate audio, visual, or tactile alerts to the air show performer and the ground crew, either their maintenance team or the air boss, for proactive risk management. This *awareness* is intrinsically linked to *safety*, as it allows for the quick identification and mitigation of emerging risks, and to *training*, as it provides feedback that can be used to refine training programs.

Lastly, *resilience* is cultivated through a culture of continuous learning, innovation, and flexible response plans. Air show performers are encouraged to adapt and evolve their display profiles, using insights from past experiences to enhance future performances. Debriefing sessions, for example, provide constructive feedback, identify areas for improvement, and ensure that lessons learned are integrated into future display flights. Additionally, flexible response plans enable display pilots and their teams to swiftly adjust to unexpected disruptions, such as sudden weather changes or technical malfunctions. This resilience supports the other *STAR Models'* elements by ensuring that safety measures, training protocols, and situational awareness can be *maintained* even under challenging conditions.

Model Representation

The *STAR Model* (Figure 11.2) can be represented as a dynamic, interconnected framework, with each component forming a part of a cohesive whole. Imagine a gearbox with a central hub labeled "*Excellence*," symbolizing the ultimate goal of the *Model*. This central hub is surrounded by four interlocking gears, each representing one of the *STAR Model* components. These gears are further influenced by various performance and excellence factors that contribute to the overall effectiveness of the *Model*.

First and foremost is the *safety gear*, which is designed with intricate notches that represent the detailed and continuous attention to safety at all levels of operation. This gear ensures that safety protocols are seamlessly integrated into every phase, from planning and preparation to execution, debriefing, and analysis. Its smooth rotation signifies the effortless embedding of safety into everyday activities, reinforcing that safety is a natural and fundamental aspect of air show operations. Key performance influences for this gear include the incorporation of an ASMS, regular safety audits, updated risk assessments, and compliance with the latest safety standards and regulations.

Secondly, the *training gear* features adjustable settings and feedback loops, symbolizing the ongoing process of enhancing and maintaining skills needed for aerial excellence. It turns smoothly but is constantly fine-tuned, reflecting the use of advanced training technologies like flight simulators. This gear represents the relentless pursuit of excellence through regular training, ensuring that the team is continuously advancing its soft and human skills. Influencing factors here include access to the latest training technologies, regular skill assessments, and a feedback-rich environment that encourages constant learning and development.

Furthermore, the *awareness gear* represents precision and reliability, reflecting the high standards of air show performers. It is efficiently linked with the other gears,

FIGURE 11.2 The *STAR Model* for excellence in air show safety.

indicating that peak performance is achieved through the integration of proactive safety, continuous improvement, and resilience. Key factors include display pilot experience and the reliability of communication systems.

Lastly, the *resilience gear* is flexible yet strong, designed to adjust to different pressures and conditions without losing its integrity. This gear symbolizes the air show performer's ability to adapt to unexpected challenges and recover swiftly. Its flexibility illustrates the importance of having comprehensive contingency plans and a culture of continuous learning. Factors influencing this gear include the robustness of contingency planning, the effectiveness of communication during adversities, and the overall adaptability of the display pilot and their team to new and unexpected situations.

Interaction and Movement

As the diagram illustrates (Figure 11.2), each gear is interlocked with the others, indicating that the components do not operate in isolation. The smooth and synchronized turning of the gears demonstrates how each aspect of the *STAR Model* supports and enhances the others. For instance, improvements in the display pilot's training (*training gear*) directly enhance adaptive capabilities (*resilience gear*), while proactive safety measures reduce the risk of incidents (*safety gear*), thereby supporting enhanced performance (*awareness gear*).

Surrounding the gears can be added additional factors that further influence each component. These include, but are not limited, to the following:

- Communication
  Effective communication channels are crucial for real-time data-sharing coordination during performances and post-show debriefs.
- Innovation
  Continuous adoption of new technologies and innovative practices ensures that training and operational protocols remain cutting-edge.
- Team Cohesion
  Strong teamwork and collaboration enhance overall performance and guarantee that safety protocols are uniformly applied and respected.
- Feedback Mechanisms
  Regular feedback from all team members helps identify areas for improvement and reinforces a culture of continuous learning and adaptation.

THE *TRAFFIC LIGHT MODEL*: SYNCHRONOUS ASSESSMENT OF EXCELLENCE AND SAFETY IN AIR SHOWS

The *Swiss Cheese Model* (Reason, 2016) has been instrumental in understanding how accidents occur due to the alignment of latent failures and active errors across multiple layers of defense. However, to create a *proactive*, *dynamic*, and *synchronous* approach to safety management, a model inspired by the familiar traffic light system is conceptualized: the *Traffic Light Model*. This *Model* emphasizes *synchronous* assessment, *clear* communication, and *immediate* action based on the current level of risk, thus aiming to prevent accidents before failures and errors can align through multiple layers of defense. It consists of a sequence of traffic lights, each one representing several safety performance indicators (SPI), excellence performance indicators (EPI), or performance influencing factors (PIF). The use of traffic light colors provides an intuitive method for communicating safety status and required actions; therefore, predefined action plans for each state ensure prompt and effective responses to changing risk levels, ultimately enhancing overall safety.

MODEL REPRESENTATION

The *Traffic Light Model* (Figure 11.3) utilizes a sequence of the traffic light colors—green (dotted), orange (vertical stripes), and red (grid lines)—to represent different states of safety. Each color signals specific actions and protocols to maintain a high level of safety awareness and proactive risk management.

The bottom circle is green with dots and the word "*GO*" in bold black letters to indicate a safe and optimal condition for proceeding, while the dots represent progress, consistent with the "*GO*" instruction.

FIGURE 11.3 The *Traffic Light Model.*

The middle circle is orange with vertical stripes, and the word "*SLOW*" is in bold black letters. This orange light suggests a need for caution and reduced speed in conducting procedures to secure that it is identified as the causal factor, while the vertical stripes symbolize a gradual reduction in pace, aligning with the "*SLOW*" directive.

The top circle is red with a pattern of horizontal and vertical grid lines creating small squares. It has the word "*STOP*" in bold black letters to indicate a condition where actions must be halted immediately. Also, the use of a red grid pattern emphasizes a sense of urgency and caution.

In the context of air shows, the "*GO*" light indicates that all safety and performance criteria are met, allowing activities to proceed without restrictions. Then, the "*SLOW*" light might be used to denote minor safety concerns or areas where performance excellence needs attention, suggesting a moderated approach, and lastly, the "*STOP*" light could signal severe safety issues or non-compliance with excellence standards requiring immediate cessation of activities.

Furthermore, the *Traffic Light Model* employs a horizontal arrangement of lights. This setup differs from a traditional series of lights arranged one after the other. The horizontal alignment is designed to highlight the importance of evaluating multiple indicators at the same time (*synchronously*) rather than one after the other (*consecutively*). This framework secures that a comprehensive and simultaneous assessment of various critical performance indicators, such as SPIs, EPIs, and PFIs.

To implement the *Traffic Light Model*, continuous risk assessment protocols must be established to determine the current state (green (dotted), orange (vertical), or red (grid)); therefore, real-time monitoring systems, including sensors and data analytics, should be utilized to detect early signs of potential issues. Moreover, clear communication channels are essential to alert the crew of the current safety status. Additionally, predefined action plans for each state should be developed, allowing for standard operations, regular training, and maintenance during the green (dotted) light state; increased vigilance, immediate risk assessments, and preventive measures during the orange (vertical) light state; and halting operations, activating emergency protocols, and conducting detailed investigations during the red (grid) light state.

Examples of Air Show Performers and High-Reliability Organizations (HROs)

For air show performers, the *Traffic Light Model* can be applied to secure the highest levels of safety during performances and training.

Green (Dotted) Light State: This state signifies optimal conditions. All scheduled air show routines proceed as planned. Safety checks on the aircraft, the display pilot's fitness, and weather conditions are ideal. Briefings and debriefings provide assurance that the display pilot, along with their supporting crew, is aware of the procedures and potential risks. For example, during a performance, if all pre-flight inspections are clear, the weather is perfect, and the pilot is in peak condition, the green light is displayed, indicating that the show can proceed without concerns.

Orange (Vertical) Light State: This state is activated if there are minor technical issues with the aircraft, changes in weather conditions, or any signs of the display pilot's distraction, fatigue, or stress. Additional precautions include delaying performances, adopting a backup display profile, conducting additional technical checks, and implementing stricter pilot observation during their performance. For instance, if the aircraft shows minor technical glitches or the pilot reports feeling slightly fatigued, the orange light will be displayed. This prompts immediate risk assessments, delays in performance, and further checks to ensure safety before proceeding.

Red (Grid) Light State: This state is activated for critical issues such as severe weather conditions, substantial wildlife activity, major technical failures, or significant health concerns for display pilots. In such a state, air show performances are halted, emergency protocols are activated, and a thorough investigation is conducted to identify and rectify the issues before resuming any activities. For example, if a severe storm approaches or the aircraft experiences significant mechanical failure, the red light is displayed, stopping all activities to ensure the safety of the performers and the audience.

The Traffic Light Model could also be applicable to other high-reliability organizations (HROs), such as nuclear power plants, hospitals, and air traffic control centers, to ensure safety and operational excellence.

Green (Dotted) Light State: This state signifies optimal operational conditions where all systems and processes function correctly and staff are well-prepared and vigilant.

- Nuclear Power Plants: All reactors and safety systems are operating normally. Routine checks show no irregularities, and staff are well-trained and alert. The plant continues to produce power efficiently and safely.
- Hospitals: All medical equipment is functioning perfectly, and staff are adequately trained and rested. Patient care is delivered effectively, and there are no reported issues with supplies or personnel.
- Air Traffic Control Centers: All communication and radar systems are fully operational. Air traffic controllers are well-rested and manage traffic efficiently. There are no delays or emergencies in the airspace.

Orange (Vertical) Light State: This state indicates minor issues that require immediate attention but do not necessitate a complete halt of operations.

- Nuclear Power Plants: There might be minor technical irregularities or early signs of equipment wear that need prompt maintenance. Staff may need additional monitoring or minor adjustments in operational protocols.
- Hospitals: Some medical equipment may require servicing, or there might be minor staffing issues, such as a nurse calling in sick. Non-urgent surgeries might be rescheduled, and staff are alerted to stay vigilant.
- Air Traffic Control Centers: There might be minor technical glitches with communication systems or slightly increased air traffic requiring additional coordination. Controllers might need to implement holding patterns or alternate routing to manage the situation safely.

Red (Grid) Light State: This state signals critical issues demanding an immediate cessation of operations and activation of emergency protocols.

- Nuclear Power Plants: Severe technical failures, such as reactor malfunctions or major safety system breaches, are detected. Immediate shutdown procedures are initiated, and emergency response teams are activated to address the issues.
- Hospitals: Major emergencies, such as significant equipment failures or critical shortages of essential supplies, occur. Emergency protocols are activated, non-critical patients are transferred or discharged, and resources are focused on critical care.
- Air Traffic Control Centers: Critical failures in radar or communication systems or severe weather conditions that severely impact air traffic safety are detected. Immediate measures are taken to redirect flights, ground non-essential air traffic, and enable the safety of all airborne aircraft through emergency protocols.

NOTE

1 Telemetry is the automated process of collecting and transmitting data from remote or inaccessible sources for monitoring and analysis (Oxford English Dictionary, 2024).

REFERENCES

American Psychological Association. (2012). *The road to resilience.*

Barker, D. (2003). *Zero error margin: Airshow display flying analysed.* www.researchgate.net/publication/261136368_Zero_Error_Margin_-_Airshow_Display_Flying_Analysed

Bergström, J., van Winsen, R., & Henriqson, E. (2015). On the rationale of resilience in the domain of safety: A literature review. *Reliability Engineering & System Safety, 141*, 131–141.

Bisbey, T. M., Kilcullen, M. P., Thomas, E. J., & Salas, E. (2021). Safety culture: An integration of existing models and a framework for understanding its development. *Human Factors, 63*(1), 5–22.

Chialastri, A., & Pozzi, S. (2008). Resilience in the aviation system. In M. D. Harrison & M. A. Sujan (Eds.), *Computer Safety, Reliability, and Security* (pp. 86–98). Springer.

Dekker, S. (2016). *Just culture: Balancing safety and accountability*. CRC Press.

Edmondson, A. C. (1999). Psychological safety and learning behavior in work teams. *Administrative Science Quarterly, 44*(2), 350–383.

Ericsson, K. A. (2006). The influence of experience and deliberate practice on the development of superior expert performance. In K. A. Ericsson, N. Charness, P. J. Feltovich, & R. R. Hoffman (Eds.), *The Cambridge handbook of expertise and expert performance* (pp. 683–703). Cambridge University Press.

Ericsson, K. A. (2014). The acquisition of expert performance: An introduction to some of the issues. In *The road to excellence* (pp. 1–50). Psychology Press.

Fasey, K. J., Sarkar, M., Wagstaff, C. R. D., & Johnston, J. (2021). Understanding organizational resilience in elite sport: A Delphi study. *Psychology of Sport and Exercise, 55*, 101929.

Federal Aviation Administration (FAA). (2020, June 24). *Order 8000.369C – Safety management system.* Federal Aviation Administration (FAA). www.faa.gov/regulations_policies/orders_notices/index.cfm/go/document.current/documentNumber/8000.369

Fellmeth, A. X., & Horwitz, M. (2011). Errare humanum est, (sed) perseverare diabolicum. In *Guide to Latin in International Law*. Oxford University Press. www.oxfordreference.com/display/10.1093/acref/9780195369380.001.0001/acref-9780195369380-e-619

Flood, K., & Keegan, J. (2022). Organizational resilience in elite sport: A systematic review. *Sport Management Review, 25*(2), 256–273.

Glosbe online dictionary. (2025). *Errare humanum est.* Retrieved April 20, 2025, from https://glosbe.com/la/en/errare%20humanum%20est

Hakami, M., & Salah, M. (2023). The excellence model to implement the safety management system effectively in oil and gas industry. *Day 1 Mon, March, 13*, 2023. https://doi.org/10.2118/213980-ms

Hollnagel, E. (2006). Resilience: The challenge of the unstable. In Hollnagel, E., Woods, D. D. & Leveson, N. C. (Eds.), *Resilience engineering: Concepts and precepts* (pp. 9–18). Aldershot, UK: Ashgate.

Hollnagel, E. (2010). How resilient is your organisation? An introduction to the Resilience Analysis Grid (RAG). *Sustainable Transformation: Building a Resilient Organization.* https://hal-mines-paristech.archives-ouvertes.fr/hal-00613986

Hollnagel, E. (2014). *Safety-I and safety-II: The past and future of safety management*. CRC Press.
Hollnagel, E. (2018). *Safety-II in practice: Developing the resilience potentials*. Routledge.
Jones, L., Tanner, T., & Krishnamurthy, R. (2005). Subjective resilience: Using perceptions to quantify household resilience to climate extremes and disasters. *Regional Environmental Change, 15*(1), 123–136.
Karachalios, M., & Adjekum, D. K. (2023). *Air show performers: Safety, risk management, and psychological factors*. CRC Press. https://doi.org/10.1201/9781003431879
Merriam-Webster. (2025). *Humanum est errare*. In *Merriam-Webster.com dictionary*. Retrieved April 20, 2025, from www.merriam-webster.com/dictionary/humanum%20est%20errare
Mohammadfam, I., Mahmoudi, S., & Kianfar, A. (2012). Development of the health, safety and environment excellence instrument: A HSE-MS performance measurement tool. *Procedia Engineering, 45*, 194–198. https://doi.org/10.1016/J.PROENG.2012.08.142
Moussa, M., Boyer, M., & Newberry, D. (2016). *Committed teams: Three steps to inspiring passion and performance*. John Wiley & Sons.
Mowbray, D. (2011). *The guide to personal resilience*. The Management Advisory Service.
National Transportation Safety Board. (2010). *Loss of thrust in both engines after encountering a flock of birds and subsequent ditching on the Hudson River, US Airways Flight 1549, Airbus A320-214, N106US, Weehawken, New Jersey, January 15, 2009* (Aircraft Accident Report NTSB/AAR-10/03). www.ntsb.gov/investigations/accidentreports/reports/aar1003.pdf
Oxford English Dictionary. (2024). Telemetry. In *Oxford English dictionary*. Retrieved August 1, 2024, from www.oed.com/dictionary/telemetry_n
Reason, J. (1990). *Human error*. Cambridge University Press.
Reason, J. (1997). *Managing the risks of organizational accidents* (2nd ed.). Routledge.
Reason, J. (2016). *Managing the risks of organizational accidents*. Routledge.
Salas, E., Prince, C., Baker, D. P., & Shrestha, L. (1995). Situation awareness in team performance: Implications for measurement and training. *Human Factors, 37*(1), 123–136.
Ulfvengren, P., & Corrigan, S. (2015). Development and implementation of a safety management system in a lean airline. *Cognition, Technology & Work, 17*, 219–236.
Woods, D. D. (2011). Resilience and the ability to anticipate. In E. Hollnagel, J. Pariès, D. D. Woods, & J. Wreathall (Eds.), *Resilience engineering in practice: A guidebook* (pp. 121–126). Ashgate Publishing.

12 Study Findings on Resilient Safety in Air Show Performers

Manolis Karachalios

INTRODUCTION

Air shows captivate audiences with thrilling demonstrations of human achievements in the aerospace industry, such as the 2024 Royal International Air Tattoo air show. This event, held at Royal Air Force (RAF) Fairford, brought together 248 aircraft from 29 nations and drew in 150,000 visitors. The show featured remarkable performances, including the celebration of the iconic F-16 fast jet, a unique NATO flypast, and the RAF Typhoon's first public display using sustainable aviation fuel. Additionally, insights into the life of an Antarctic Twin Otter pilot were shared, making it an unforgettable experience (Royal International Air Tattoo (RIAT), 2024).

These exciting performances are not merely feats of skill and daring; they result from extensive planning and strict adherence to safety protocols. Behind every aerial event lies a dedicated effort to ensure the well-being of both performers and spectators. Nevertheless, achieving a balance between delivering awe-inspiring shows and maintaining stringent safety standards requires a deep understanding of *resilience* principles from both display pilots and air bosses. To explore this critical aspect further, this Chapter examines the comprehensive findings on resilient safety measures implemented by air show performers, primarily expanding on the research conducted in 2022 (Karachalios, 2022). The problem addressed in this research is the lack of empirical studies to determine how perceptions of *safety risk parameters* and *hazardous attitudes (HA)* influence *resilient safety culture (RSC)* within the international air show community when mediated by *mindfulness (MF)*. The study aimed to fill a significant gap in existing literature, which has largely focused on commercial aircraft operations, flight training programs, and air traffic control management (Adjekum & Fernandez-Tous, 2020; Akselsson et al., 2009; Hollnagel, 2018; Reason, 2016).

The findings presented in this Chapter stress the critical components that contribute to the *resilience* of safety practices in this high-risk environment, which allow them to maintain peak performance while mitigating risks. Furthermore, this Chapter provides valuable insights and practical applications to enhance safety protocols, ensuring that air show performers can continue to push the boundaries of aerial artistry while upholding the highest standards of safety and excellence.

DOI: 10.1201/9781003504801-16

INSIGHTS FROM RECENT RESEARCH ON AIR SHOW PERFORMERS

The 2022 study explores how safety risk parameters and HA influence the RSC within the international air show community, with MF acting as a mediating factor (Karachalios, 2022). It investigates the relationships between risk perception (RP), risk tolerance (RT), and HA among air show performers, examining MF as crucial for effective decision making and stress management. Techniques such as visualization and pre-performance practices are analyzed for their impact on safety and performance. Additionally, the study considers how demographic variables like flying experience, military or civilian background, age, education, and marital status influence these factors. By addressing these elements, the study provides a holistic understanding of how to cultivate a *RSC* in the air show community, filling a critical gap in the literature and offering practical insights for enhancing safety and performance standards in this high-risk arena.

Resilient Air Show Safety Model

The main research question investigated *is "What are the strengths of relationships between risk perception, risk tolerance, hazardous attitudes, and resilient safety culture when mediated by individual mindfulness among members of the international air show community?"*

Using a Structural Equation Modeling-Path Analysis (SEM-PA) (Byrne, 2010), the study tested 12 hypotheses and found a significant positive correlation between RP, RT, HA, MF, and RSC, with MF effectively mediating the impact of RP, RT, and HA on RSC.

Figure 12.1 shows the final fully mediated measurement model, which produced satisfactory goodness-of-fit indices across all fit indices.

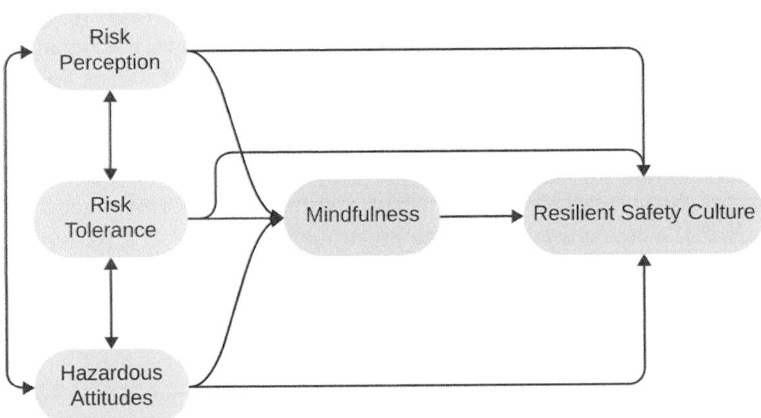

FIGURE 12.1 Resilient air show safety model.

These results, which will be analyzed in detail in the following pages, reveal the significant relationships between these variables and emphasize the mediating role of *MF* in enhancing a *RSC* within the air show community.

Findings on Resilient Safety Culture

The correlation heatmap analyzed in Chapter 3 (see Figure 3.1) illustrates the relationships between key variables in the study, derived from quantitative, qualitative, and mixed data sets. The variables examined are RP, RT, HA, MF, and RSC.

The quantitative findings depicted in the correlation heatmap of Figure 3.1 indicate that a *RSC* has a significant predictive relationship with *HA* and *MF* but a *non-significant* predictive relationship with *RP* among respondents. Notably, *MF* as a mediator significantly enhances the total effect of HA on fostering a RSC.

Furthermore, the study's qualitative findings show that the *RSC* of air show performers *negatively correlates* with *RT* and *HA*. Additionally, a *RSC* is associated with *increased risk perception* and *MF*. These results suggest that the air show community embodies the desired attributes of a RSC, as proposed by Akselsson et al. (2009), by emphasizing a learning culture and striving for resilience. The community has developed and uses *feed-forward control*[1] mechanisms, such as training and effective standard operating procedures, to keep processes within safe limits in a proactive manner.

The findings also indicate that the air show community is *mindful* of the need for high reliability, consistent with the recommendations of Weick and Sutcliffe (2011). Furthermore, the air show industry actively emulates high-reliability organizations (HROs), such as space organizations (Cantu et al., 2020), in developing a safety culture and managing risks in complex operations. The industry also emphasizes efficient change management, designs for safety, and employs a continuous improvement philosophy (Adjekum & Fernandez-Tous, 2020; Stolzer & Goglia, 2016; Teske & Adjekum, 2022).

Additionally, the findings corroborate Hollnagel's (2018) observation that even when incidents occur, the air show community's *RSC* allows the organization to adapt, recover, and operate efficiently within the margins of safety. Enhancing a solid safety culture with a proactive, RSC could assist an organization in not only improving its safety performance but also recovering from upheavals, as suggested by Shirali et al. (2016). Finally, these findings support previous research (Adjekum & Fernandez-Tous, 2020; Heese et al., 2014; Hollnagel, 2018; Hollnagel et al., 2011; Reason, 2016), which advocates for the importance of a RSC in fostering an organizational safety culture that promotes safe practices.

Relevance of Findings to Resilient Safety

The findings from this study are highly relevant to understanding and enhancing resilient safety within the air show community. Considering that *resilient safety* refers to the quality that embodies the capacity of a system, organization, or individual to not only withstand (active safety) and recover (passive safety) from adverse events but also to dynamically adapt and grow stronger in response to risks and uncertainties, the insights gained from this research shed light on several key aspects that contribute to the further understanding and development of air show resilient safety.

Predictive Relationships

The study reveals significant predictive relationships between RSC, HA, and MF. Specifically, HA *negatively* impacts RSC, while MF *positively* influences it. These relationships point out the criticality of addressing psychological factors and promoting MF practices among air show performers to foster a culture that can withstand and recover from adverse events.

Mediating Role of Mindfulness

MF emerges as a crucial mediator that enhances the overall effect of HA on a RSC. This finding highlights the value of MF training and practices in mitigating negative attitudes and promoting a proactive approach to safety. By cultivating MF, air show performers can augment their mental preparedness, situational awareness, decision making, and stress management, all of which are vital for resilient safety.

Negative Correlations with Risk Tolerance and Hazardous Attitudes

The *negative* correlations between RSC and both RT and HA indicate that a strong safety culture is characterized by lower acceptance of risk and fewer HA. This insight suggests that efforts to strengthen safety culture should focus on reducing RT and addressing HA through targeted interventions, such as training programs, mentoring, and organizational policies.

Positive Associations with Risk Perception and Mindfulness

The *positive* associations between RSC and both RP and MF suggest that individuals who perceive risks accurately and practice MF are more likely to contribute to a RSC. This finding emphasizes the need for continuous education and training to enhance RP and MF among air show performers.

Systematic Risk Management

The study highlights the air show community's systematic approach to identifying and controlling system vulnerabilities through effective risk management and controls. This proactive attitude is essential for resilient safety, as it enables organizations to anticipate potential hazards and implement measures to mitigate them before they escalate.

Learning Culture and Continuous Improvement

Resilient safety places a strong emphasis on a learning culture and continuous improvement. The results corroborate this, indicating that the air show community is dedicated to learning from the past, iteratively improving its safety procedures, and incorporating safety as a fundamental business function, ensuring that safety practices adapt to new challenges and uphold high-performance standards.

Adapting and Recovering from Incidents

Finally, the findings support the observation that a RSC enables organizations to adapt, recover, and operate efficiently within safety margins, even when incidents occur. This *adaptability* is a crucial attribute of resilient safety, ensuring that air show performers can maintain performance and safety standards despite unexpected hazards.

APPLYING FINDINGS TO ENHANCE SAFETY STANDARDS

The findings from the study have significant theoretical and practical implications for enhancing safety protocols in the air show community. By understanding the relationships between RP, RT, HA, MF, and RSC, air show organizers and performers can implement more effective safety measures (Karachalios, 2022).

Theoretical Implications

The study provides a novel theoretical foundation for assessing the strength of relationships between critical variables in the air show community. It brings to the fore the essential role of air bosses in promoting safety and ensuring effective risk management. Given their crucial position in advising performers and managing the entire air show, air bosses must adopt a proactive approach to safety, emphasizing the importance of adherence to safety standards and the promotion of a RSC.

Furthermore, the findings recognize the significance of a RSC that supports excellence, best practices, and continuous enhancements. The high standards not only applied by air show performers but also imposed by aviation authorities in recent years (in the United Kingdom, after the tragic accident in Shoreham airshow) demonstrate that a culture of excellence has been adopted by the air show community, ensuring safety is a top priority. These insights can also serve as a theoretical baseline for assessing similar variables in other high-performance, high-risk activities, such as military combat flying, aerial vehicle testing, Formula 1® racing, MotoGP®, solo climbing, and circus performances. Moreover, the study's findings provide a theoretical foundation for further research on the relationships between RSC, MF strategies, risk management, and operational performance in other high-reliability professions, such as surgeons, air traffic controllers, and nuclear plant operators.

Practical Implications

The practical implications of the study are extensive and can serve as a blueprint for new training programs and certification courses for air show performers and air bosses. Key recommendations include the following:

Risk Recognition Training

It is essential to incorporate risk recognition training into the curriculum for air show performers to improve their RT, as also suggested by Hunter (2002). This training should involve identifying potential hazards, assessing risks accurately, and making informed decisions to mitigate these risks. Scenarios and case studies from past air show incidents can be used to highlight the importance of risk recognition and the consequences of inadequate risk assessment.

Mindfulness Training

Air show associations, i.e., International Council of Air Shows (ICAS), European Airshow Council (EAC), FSA, and British Air Display Association (BADA), as well as aviation authorities, i.e., Federal Aviation Administration (FAA) and UK Civil Aviation Authority (UK CAA), could implement MF training during regular training sessions and seminars to assist both air show performers and air bosses stay mentally resilient during the air show weekend. Techniques such as meditation, breathing exercises, and visualization can help performers maintain focus, manage stress, and improve situational awareness, while such practices can be reinforced through briefings before performances and during re-certification processes.

Ego Resiliency Training

Ego resiliency training aims to help air show performers understand and manage any ego-driven behaviors. This can include exercises in self-reflection, group discussions about the impact of HA, and workshops on emotional intelligence. By recognizing and mitigating egocentric, impulsive, and macho attitudes, performers can enhance their decision making and teamwork and, with a humble, approachable, and credible attitude, entertain the audiences while representing aerial excellence.

Realistic Training Simulations

Training simulations should mimic real-life challenges that performers might face during air shows, including unexpected weather changes, mechanical failures, loss-of-sight within formation members, or other perturbations that require quick and effective responses. These simulations help build resilience and adaptability, preparing air show performers for a wide range of scenarios.

Personnel-Centric Safety Education

Developing personnel-centric safety education involves creating engaging and interactive learning experiences for air show performers. Workshops and webinars can cover topics such as risk management, safety best practices, and resilience strategies. Additionally, safety surveys and anonymous reporting systems could be developed and utilized to encourage open communication and continuous improvement. It needs to be emphasized that a personnel-centric safety education through workshops, webinars, safety surveys, and anonymous reporting systems should engage mainly air show councils but could include aviation authorities to promote best safety practices.

Role of Air Bosses

Air bosses play a crucial role in managing risk and ensuring safety during air shows. They should be trained in advanced risk management techniques and communication skills. During safety briefings, air bosses set the tone for the event, emphasizing the importance of adherence to safety protocols and fostering a culture of vigilance and responsibility. Ultimately, air bosses act as risk management intermediaries, bridging the gap between regulators and performers, ensuring the safe execution of the event while maintaining the rhythm of the air show choreography from start to finish.

Dynamic Risk Assessment

Implementing dynamic risk assessment involves using real-time physiological data, such as heart rate, to monitor performers' stress levels, as well as aircraft-performance telemetry data, to display the engine status and the aircraft flight parameters. Biodata can help air show performers identify early signs of stress or fatigue, while the aircraft telemetry information could directly alert the display pilot, the ground crew, or the air boss of a critical engine malfunction, minimizing the communication with the display pilot while they are dealing with their emergency situation. Having a dynamic risk assessment is a tool for air bosses to take proactive measures, such as adjusting the display sequence event, calling the crash and rescue services to be on alert, or giving a day off to a display pilot, with the absolute aim to ensure the safety of the performer and the audience.

Adoption of an Air Show Risk Level (ARL)

Another practical implication, as suggested in the study, is the adoption of an Air Show Risk Level (ARL) indicator, which can help standardize risk assessments in the air show community. This indicator can guide decisions on alertness, preparedness, and response strategies during air show events for both air show performers and air bosses. By providing a clear and consistent measure of risk, the ARL can facilitate better communication among organizers, participants, and regulatory authorities, ensuring that all stakeholders are aware of the current risk level and appropriate safety measures.

This concept has inspired the conceptualization and implementation of the *Traffic Light Model* discussed in Chapter 11.

Additional Thoughts

The study's findings provide a solid foundation for understanding the current safety landscape in the international air show community, but there is always room for improvement. Safety in the high-risk environment of air shows cannot rely solely on existing safety measures; it must evolve with new insights, without overwhelming display pilots, event organizers, and air bosses with additional safety measures, but rather promoting a culture of excellence that ensures safety.

Based on the study's findings, the following additional strategies could further enhance the already robust safety culture for air show performers, focusing on mentorship, collaborative safety workshops, cross-disciplinary training, and personalized safety plans:

Safety Mentorship Programs

Creating safety mentorship programs where experienced performers mentor newcomers is a crucial strategy for enhancing safety and performance in the air show community, as already covered in Chapter 9. These programs leverage the knowledge and experience of veteran performers to guide and support less experienced colleagues, fostering a culture of continuous learning and safety excellence.

Collaborative Safety Workshops

Organizing collaborative safety workshops is a vital initiative to promote excellence and safety in the air show community. These workshops, organized by prominent bodies such as the ICAS, the EAC, the BADA, and the France Spectacle Aérien (FSA), bring together performers, air bosses, safety officers, and regulators. These gatherings provide a platform for open dialogue, knowledge exchange, and collective problem-solving, which are essential for addressing safety challenges and advancing best practices.

The importance of these workshops lies in their ability to foster a culture of safety and excellence through collaboration and shared learning. By bringing together diverse stakeholders, workshops enable the air show community to pool their expertise and experiences, creating a comprehensive understanding of safety issues and potential solutions. Performers can share insights from their experiences in the field, highlighting practical safety measures and common pitfalls. Air bosses and safety officers can offer their perspectives on risk management and operational protocols, while regulators can provide updates on the latest safety regulations and industry standards.

Workshops organized by ICAS, EAC, BADA, and FSA are particularly valuable as these organizations have extensive experience and credibility in the air show industry. Their involvement ensures that the workshops are well-structured, relevant, and aligned with the highest safety standards. These organizations can also facilitate connections between different air show communities globally, promoting the exchange of best practices and innovative safety strategies across borders.

Notably, one of the key benefits of collaborative safety workshops is the opportunity to discuss and address specific safety challenges faced by the air show community. These challenges may include managing advanced display profiles, dealing with unpredictable weather conditions, ensuring crowd safety, and maintaining aircraft maintenance standards. Through structured discussions and breakout sessions, participants can analyze these challenges from multiple perspectives, identifying root causes and developing actionable solutions.

In addition to addressing immediate safety concerns, workshops provide a forum for sharing best practices that have been proven effective in different contexts. Participants present case studies, demonstrate successful safety initiatives, and share lessons learned from past incidents. This exchange of knowledge helps to disseminate effective safety practices throughout the air show community, raising the overall standard of safety performance.

The workshops also play a pivotal role in building relationships and fostering a sense of community among air show stakeholders. Networking opportunities allow participants to establish connections, share resources, and support each other in their safety efforts. This sense of camaraderie and mutual respect promotes the overall safety culture by recognizing that safety is a common shared value that everyone in this community strives to preserve.

Ultimately, air show organizations like ICAS, EAC, BADA, and FSA act as *epicenters* to the air show community, being the main source of communication for air show stakeholders. As a result, their role is more than just organizing conventions and workshops; it also includes providing systematic support and resources to help apply the established safety measures as well as the lessons learned mentioned throughout the workshops.

Cross-Disciplinary Training

Offering cross-disciplinary training sessions is an approach to enhancing the safety culture and in-flight performance of display pilots. By learning from other high-risk and high-performance professions, such as military pilots, Formula 1® drivers, and astronauts, performers can gain new perspectives on safety and risk management, which can be invaluable in their own high-stakes environment.

Military pilots, for instance, operate in extremely demanding conditions that require impeccable precision, situational awareness, cohesive teamwork, and adherence to strict safety protocols. Their training emphasizes advanced flight maneuvers, rapid decision making under pressure, and the use of cutting-edge technology (Mansikka et al., 2016; Maxim, 2019). By participating in cross-disciplinary training sessions with military pilots, air show performers can learn about these advanced techniques and incorporate them into their own routines to enhance their ability to manage risks and improve their overall performance.

Formula 1® drivers also provide a wealth of knowledge on managing high-speed, high-risk situations. Their training focuses on optimizing physical and mental performance, understanding vehicle dynamics, and maintaining peak concentration during races (Phillips, 2014). They utilize sophisticated data analysis tools to monitor and improve their performance continuously. Air show performers can benefit from understanding how Formula 1 drivers handle stress, make split-second decisions, and maintain peak physical condition.

Undeniably, astronauts operate in one of the most hazardous environments imaginable, where the margin for error is minimal. Their training involves extensive preparation for emergencies, rigorous physical conditioning, and the development of solid psychological resilience (Etheridge et al. 2016; Cripe et al. 2021; Finseth et al. 2018). Astronauts are trained to manage complex systems, execute precise operations, and maintain composure under extreme pressure. Similarly, air show performers can enhance their skills by adopting astronauts' methods for emergency preparedness, teamwork, and staying calm in high-stress situations.

Cross-disciplinary training sessions can be organized in various formats, including workshops, seminars, and practical training exercises. Workshops can feature guest speakers from different high-risk and high-performance professions who share their experiences and insights on safety and performance. Seminars can focus on specific topics such as risk assessment, stress management, and the use of technology in enhancing safety, while training exercises can simulate an actual cross-disciplinary scenario, allowing air show performers to be exposed to other high-performers' skills, techniques, and strategies in a controlled environment.

Furthermore, these cross-disciplinary training sessions can foster a culture of continuous learning and innovation. By exposing air show performers to different approaches and best practices, they can continuously refine their own techniques and procedures.

Another significant benefit of cross-disciplinary training is the opportunity to build networks and relationships with professionals from other high-risk fields, leading to ongoing collaborations, potential sponsorship opportunities, and the exchange of ideas that extend beyond the training sessions.

Personalized Safety Plans

Developing personalized safety plans for each air show performer is the ultimate goal for resilient safety. These plans should consider the performer's experience, skills, health status, and personal circumstances, providing bespoke recommendations and support to address individual risk factors.

Each performer's personalized safety plan begins with a comprehensive assessment of their background, which includes evaluating their flight experience, specific skills, recent training history, and any previous incidents or near-misses. Understanding these factors allows for the creation of a customized plan that focuses on the performer's unique needs and potential areas of vulnerability.

Health and fitness monitoring is another critical component of personalized safety plans. Regular physical exams, mental health evaluations, and fatigue assessments ensure that performers are in optimal condition to perform safely. Wearable technology, such as fitness trackers or smartwatches, can be utilized to continuously monitor sleep patterns, physical activity, and stress levels, providing real-time insights into the performer's well-being, allowing for timely interventions if any issues arise.

Personal circumstances, such as marital status, family responsibilities, and external stressors, are also taken into account. These factors can significantly impact a performer's stress levels and RT. Addressing them in the safety plan helps manage overall stress and ensures that performers are mentally and emotionally prepared for their displays.

Furthermore, training programs tailored to the individual needs of each performer are also crucial, with a focus on enhancing strengths and addressing weaknesses. However, what is essential for the sustainable peak performance of a display pilot is creating a supportive environment that continuously assists display pilots in feeling like they are part of the air show community, with shared values for passion, excellence, well-being, and safety. Such an environment will make them feel comfortable sharing their concerns and seeking help without fear of judgment or repercussions.

Ultimately, to facilitate the effective development and employment of a personalized safety plan for air show performers, the ASMS, as outlined in the first book of this series (Karachalios & Adjekum, 2023), can be integrated. The ASMS framework systematically manages safety through hazard identification, risk management, and continuous improvement processes. This is particularly beneficial for solo display pilots or small civilian teams who may not have the extensive support systems available to larger groups. The integration of ASMS into personalized safety plans also involves the use of specific tools and techniques for hazard identification and risk assessment, such as risk identification checklists and risk scoring matrices, to systematically evaluate their performance environment and identify potential hazards.

NOTE

1 Forward-feed control mechanisms are proactive control systems used to anticipate and counteract potential disturbances before they impact a process, unlike feed-back control which reacts to changes after they occur (Elliott & Sutton, 1996).

REFERENCES

Adjekum, D. K., & Fernandez Tous, M. (2020). Assessing the relationship between organizational management factors and a resilient safety culture in a collegiate aviation program with Safety Management Systems (SMS). *Safety Science, 131*, 104909. https://doi.org/10.1016/j.ssci.2020.104909

Akselsson, R., Ek, Å., Koornneef, F., Stewart, S., & Ward, M. (2009). Resilience safety culture. In *17th world congress on ergonomics*. IEA.

Byrne, B. M. (2010). *Structural equation modeling with AMOS: Basic concepts, applications, and programming* (2nd ed., pp. xix, 396). Routledge/Taylor & Francis Group.

Cantu, J., Tolk, J., Fritts, S., & Gharehyakheh, A. (2020). High Reliability Organization (HRO) systematic literature review: Discovery of culture as a foundational hallmark. *Journal of Contingencies and Crisis Management, 28*(4), 399–410. https://doi.org/10.1111/1468-5973.12293

Cripe, C., Cooper, R., & Weedn, S. (2021). Improving psychological resilience with cognitive retraining methods using EEG brain network biomarkers: Example from UND/NASA Lunar/Martian habitat. *American Journal of Aerospace Engineering, 8*(1), 14–26.

Elliott, S. J., & Sutton, T. J. (1996). Performance of feedforward and feedback systems for active control. *IEEE Transactions on Speech and Audio Processing, 4*(3), 214–223.

Etheridge, T., Szewczyk, N. J., Zuo, L., Chuang, C., Liu, Z., Ward, A. T., Dier, N. L., Rotarius, T. R., Lalande, S., Zhang, X., Zhang, M., Li, G., Dong, L., & Gao, F. (2016). Commentaries on viewpoint: A call for research to assess and promote functional resilience in astronaut crews. *Journal of Applied Physiology, 120*(4), 473–474.

Finseth, T., Keren, N., Dorneich, M., Franke, W., Anderson, C., & Shelley, M. (2018). Evaluating the effectiveness of graduated stress exposure in virtual spaceflight hazard training. *Journal of Cognitive Engineering and Decision Making, 12*, 248–268. https://doi.org/10.1177/1555343418775561

Heese, M., Kallus, W., & Kolodej, C. (2014, February). Assessing behaviour towards organizational resilience in aviation. In I. Herrera, J. M. Schraagen, J. van der Vorm, & D. D. Woods (Eds.), *Proceedings of the 5th Resilience Engineering Association Symposium: Managing trade-offs* (pp. 67–74). Resilience Engineering Association.

Hollnagel, E. (2018). *Safety-II in practice: Developing the resilience potentials*. Routledge.

Hollnagel, E., Pariès, J., Woods, D. D., & Wreathall, J. (2011). *Resilience engineering in practice: A guidebook*. Ashgate Publishing Ltd.

Hunter, D. R. (2002). Risk perception and risk tolerance in aircraft pilots (Technical Report DOT/FAA/AM-02/17; p. 26). Office of Aerospace Medicine, Federal Aviation Administration.

Karachalios, M. (2022). *An evaluation of the relationships between resilient safety culture, safety risk parameters, and mindfulness in the international air show community* [PhD dissertation, University of North Dakota]. https://commons.und.edu/theses/4267/

Karachalios, M., & Adjekum, D. K. (2023). *Air show performers: Safety, risk management, and psychological factors*. CRC Press. https://doi.org/10.1201/9781003431879

Mansikka, H., Simola, P., Virtanen, K., Harris, D., & Oksama, L. (2016). Fighter pilots' heart rate, heart rate variation and performance during instrument approaches. *Ergonomics, 59*(10), 1344–1352. https://doi.org/10.1080/00140139.2015.1136699

Maxim, I. R. (2019). Excellence achievement-striving in fighter pilots on supersonic aircraft. *Journal of Advanced Research in Social Sciences, 2*(3), Article 3. https://doi.org/10.33422/jarss.v2i3.222

Phillips, A. J. K. (2014). Uncovering Formula One driver performances from 1950 to 2013 by adjusting for team and competition effects. *Journal of Quantitative Analysis in Sports, 10*(2), 261–278. https://doi.org/10.1515/jqas-2013-0031

Reason, J. (2016). *Managing the risks of organizational accidents*. Routledge.

Royal International Air Tattoo (RIAT). (2024, July 21). Dazzling displays and memorable moments as thousands enjoy historic Royal International Air Tattoo. www.airtattoo.com/news/dazzling-displays-and-memorable-moments-as-thousands-enjoy-historic-royal-international-air-tattoo/

Shirali, G. A., Shekari, M., & Angali, K. A. (2016). Quantitative assessment of resilience safety culture using principal components analysis and numerical taxonomy: A case study in a petrochemical plant. *Journal of Loss Prevention in the Process Industries, 40*, 277–284. https://doi.org/10.1016/j.jlp.2016.01.007

Stolzer, A. J., & Goglia, J. J. (2016). *Safety management systems in aviation* (2nd ed.). Routledge.

Teske, B. E., & Adjekum, D. K. (2022). Understanding the relationship between High Reliability Theory (HRT) of mindful organizing and Safety Management Systems (SMS) within the aerospace industry: A cross-sectional quantitative assessment. *Journal of Safety Science and Resilience, 3*(2), 105–114. https://doi.org/10.1016/j.jnlssr.2022.01.002

Weick, K. E., & Sutcliffe, K. M. (2011). Managing the unexpected: Resilient performance in an age of uncertainty. *Personnel Psychology, 62*(3), 646–652. https://doi.org/10.1111/j.1744-6570.2009.01152_6.x

13 Case Studies on Resilient Safety in Air Show Performers

Manolis Karachalios

INTRODUCTION

The majority of the international air show community, which consists of air show performers from North America, Europe, and Australia, comprises professional and safety-oriented display pilots, making the long list of case studies that could be included in this chapter a book itself. However, to remain within the scope of the current publication, the following case studies are representative of the resilient safety practices followed by the majority of the elite air show performers internationally. By examining the experiences of renowned teams and individual pilots, we will attempt to uncover the strategies that have enabled them to maintain high standards of safety while delivering stunning performances, ultimately emphasizing the importance of continuous and consistent learning, proactive risk management, pathos for excellence, dynamic adaptability, and the integration of mental and physical preparedness in ensuring the safety of both themselves and spectators.

The journey of these performers is marked by a commitment to excellence and an unwavering focus on safety. From the recovery from catastrophic upheavals of demonstration teams, such as the Red Arrows in 2011, the Snowbirds in 2020, and the Thunderbirds in 2018, to the continuous learning approach of experienced civilian air show performers like The Flying Bulls, Richard Goodwin, and Skip Stewart, each case study provides valuable insights into the principles and practices of resilient safety. These stories not only inspire but also offer practical lessons that can be applied to improve safety protocols across the air show industry.

CASE STUDY 1: RED ARROWS, RAF—CONSISTENT TRAINING AND ADAPTATION

After the mid-air collision of two Hawks from the Red Arrows team in Crete in 2010 (Ranter, 2010; Royal Air Force, United Kingdom, 2010), 2011 brought unprecedented sorrow for the Red Arrows with two accidents that claimed the lives of two pilots (UK Government, 2012, 2014). Additionally, another member suffered from anxiety and had to transition to ground duty (BBC News, 2012a; Royal Air Force

DOI: 10.1201/9781003504801-17

Benevolent Fund, 2020). Everything looked pessimistic for the team and the Royal Air Force (RAF) regarding the future progress of the Reds. Yet, the leadership of the RAF decided, considering all operational factors, to continue the display schedule.

Following the accidents in 2011, the Red Arrows made adjustments to maintain their nine-aircraft formation. Red 8 was retained, and the original Red 10 was moved to the Red 5 position. However, in March 2012, the UK Ministry of Defense announced that the Red Arrows would perform with only seven aircraft for that season. To maintain formation symmetry, Red 8 also stepped down from the display team (BBC News, 2012b; European Airshows, n.d.).

Despite the reduction to seven aircraft for aerobatic displays, the Red Arrows continued to perform official flypasts with nine aircraft by utilizing Red 8 and a former Red Arrow pilot, now Red 10.

All these events before the 2012 display season meant that the team had to implement several contingency plans to adapt to the new display profile and then train consistently to once again reach the expected levels of peak performance. During the same period, the team had to deal with several important distractions to their primary task of being the flying ambassadors of the RAF, such as being involved in accident investigation boards, supporting the families of the lost pilots, and keeping closely attached to each other; yet, they managed to persevere by showing extraordinary levels of mental and psychological resilience. Moreover, they applied all principles of resilient safety, but they managed to get through this mainly by adapting their display profiles and then providing consistent flight training. The sequence of the maneuvers and formations flown were modified to demonstrate the team flying with two fewer aircraft; in some cases, the team decided to perform flypasts only instead of complex aerobatic maneuvers to compensate for the stressful times that they were facing. Such an approach to reducing the level of difficulty demonstrates an exceptional commitment to resilient safety. It acknowledges that peak performance cannot be regained quickly. Instead, in adverse situations, lowering performance demands allows peak performers to step back without stepping down. This gives them time to reassess the new prevailing conditions and progressively rebuild their performance levels. By recognizing the need for additional margins of error and approaching the process with caution, they eventually managed to reach the expected levels of performance once again. It needs to be highlighted that, according to the Red Arrows' motto, "*Éclat*"[1] is their first priority in all they do. By prioritizing excellence, safety is *inherently* achieved, reinforcing the suggestion at the beginning of Part III that "*Excellence First, Safety Always*" is the new perspective adopted by the air show community worldwide.

One significant advantage of an aerobatic team is that, with *redundancy* in resources, they can proceed with their air show performance even with a smaller number of aircraft flying in formation. This provides a distinct advantage over single-ship demonstration teams, which lack redundancy if the sole pilot is unavailable due to sickness or other operational reasons. Moreover, a multi-ship aerobatic team like the Red Arrows fosters *robust resilient safety* through an emphasis on *teamwork* and *communication*, ensuring that all pilots and support staff are fully aware of the operational status and any potential issues during flights.

The year 2011 might have been a tragic year in the long history of excellence for the Red Arrows, but 2012 marked a period of *growth*. Through the challenges faced that year, all team members affirmed that safety is not a static goal but a dynamic process requiring constant vigilance and adaptation. By 2013, the team had managed to recover from the setback and once again performed their iconic nine-ship diamond formation in front of the public (ADS Advance, 2013). This achievement was made possible by a collaborative approach from everyone involved in the Red Arrows' operations, demonstrating a commitment to maintaining excellence and upholding the highest standards of safety.

CASE STUDY 2: SNOWBIRDS, RCAF—EMPHASIS ON TEAM SPIRIT

In 2020, the Royal Canadian Air Force (RCAF) aerobatic team, The Snowbirds, faced a tragic irony: a devastating accident while they were flying across Canada to boost the public morale during the COVID-19 pandemic and pay tribute to the frontline workers in the named "*Operation Inspiration*" (see Figure 13.1) (Airshow News, 2020; Maxouris, 2020). During the accident, the passenger of the CT-114 Tutor was killed while the pilot was seriously injured. Ironically, the passenger was the public affairs officer, one of the main advocates for flying this operation. Tragically, fate had set a lethal trap for their humanitarian motivation.

The accident affected the team both operationally and emotionally. First, they had to ground all airplanes for several weeks until a thorough mechanical inspection of

FIGURE 13.1 The Snowbirds flying over Toronto, Canada as part of Operation Inspiration. (Copyright: Scott Heaney/Shutterstock.com.)

all jets took place; while emotionally, the team was affected not only by the tragic loss of the officer but also by the injury of their pilot, who stayed at the hospital for several days.

Nevertheless, what stood out from this accident was the team spirit that helped all team members remain connected, empathetic, and ultimately resilient to operate once again safely when they were cleared by the investigation authorities. Moreover, the support that the team received from the public and the assistance they provided to the families of the deceased and injured members confirmed once again that in the air show community, the feeling of family is a game changer and ensures their resilient safety.

Ultimately, it is not the procedures and safety protocols alone that help an organization remain resilient and recover; rather, it is the powerful effect of being part of a team that makes them feel like an extended family that shares the same values and culture. This is the epicenter of excellence and resilient safety. The Snowbirds exemplify a resilient safety team culture, not only for the air show community and the aviation industry but for any high-performance, high-risk organization. Such organizations recognize that hazards are ever-present, and the worst-case scenario of a tragic accident with human losses is part of the reality they face. However, with embraced and shared empathy, they can *bounce up* in their everyday operations and maintain high standards of performance.

It is worth highlighting that the Canadian Air Force conducted a meticulous accident investigation. Beyond identifying the contributory factors, they conducted a thorough risk assessment with the goal of returning the fleet to flying operations, identifying "effective preventive measures that will either prevent or reduce the risk of similar occurrences in the future" (Canadian Military Family Magazine, 2020; National Defence, Canadian Armed Forces, 2021). The investigation focused on the ejection decision making of Snowbird 11's aircraft, as well as several related upgrades to allow the then 57-year-old fleet of Tutors to continue operating until 2030. These upgrades included advancements in avionics and improvements to the aircraft escape system. Ultimately, the goal of the investigation, as emphasized by the investigation team, was a continuous dedication to learning from this accident, a key principle of resilient safety.

Even with all the related safety measures to proactively handle emergencies on takeoff, including simulator sessions for such events, the inherent risks of military flying—such as low-level operations and formation flying—require accepting a risk level as low as reasonably possible (ALARP) (Karachalios & Adjekum, 2023; Melchers, 2001; Tilman, 2011). This consideration includes various factors, such as environmental ones like wildlife (Pfeiffer et al., 2018).

In this particular incident, bird ingestion into the engine led to an engine stall, resulting in a loss of thrust at low altitude and slow speed while Snowbird 11 was taking off in formation, which caused the aircraft to lose control, ultimately forcing the crew to eject (National Defence, Canadian Armed Forces, 2021). Takeoff is one of the most critical moments, and the investigation provided insightful observations and suggestions for a comprehensive operational risk assessment and emergency training related to this phase of flight. These recommendations aim to better prepare the team for an engine failure after takeoff in a low-level environment, especially near or over populated areas. They also clarify the command to "*eject*" as part of an effective Crew Resource Management (CRM) and explore potential technical modification to stabilize the ejection seat from any tendency to pitch, roll, or yaw. Ultimately, these measures emphasize the RCAF's approach to resilient safety.

The accident, even though it did not occur during an air show, significantly affected the operational capabilities of the team. It also reinforced the increased risks associated with delayed ejection decisions in multi-crew cockpits, underlining a prevailing rule for most air shows: passengers are forbidden to fly during actual performances. This rule, which is widely accepted, recognizes the inherent risks of low-level display flying and aims to minimize the number of casualties in case of an accident. Additionally, it helps to avoid the psychological effects of potential social facilitation bias and distraction caused by the presence of a passenger in the aircraft.

CASE STUDY 3: THUNDERBIRDS, USAF—ANTICIPATION AND PREPAREDNESS

Similarly to what Hollnagel (2014) defines as safety—that it can be revealed by its absence—the same concept can be applied to resilient safety. In case parts of its aspects are not thoroughly covered, leaving gaps in the process of anticipation and preparation of adverse conditions, anomalies in the levels of performance might result. The tragic loss of Thunderbird 4 in 2018 not only highlighted the extreme risks associated with high-performance formation demonstrations but also stressed the critical importance of resilient safety practices around the ability to anticipate, prepare for, respond to, and learn from adverse events (Air Combat Command, 2018; AirshowStuff, 2018; Pawlyk, 2018). In this particular case, the unfortunate loss of Thunderbird 4 emphasized the necessity of the first two parts of resilient safety practices, i.e., to anticipate and to be prepared for adverse events by avoiding the effects of complacency and thus the insidious threat of normalization of deviance for air shows (Karachalios & Adjekum, 2023). However, the last two parts of resilient safety's practices of responding to and learning from adverse events were thoroughly investigated by the USAF by identifying the hazards and risks associated with the accident and becoming prepared for when the team resumed their training, demonstrating their ability to dynamically adapt and ultimately grow stronger after such an incident.

Examining the period before the 2018 accident, it is apparent that the Thunderbirds maintained rigorous training protocols and safety measures designed to prepare their pilots for the demanding conditions of their maneuvers. Thunderbird 4, with over 3,500 flight hours, was a testament to this rigorous preparation. However, the existing protocols did not fully account for the lethal effects of rapid G-force changes, specifically the push–pull effect[2] and its associated decrease in the typical AGSM's effectiveness, which ultimately contributed to the crash.

In the aftermath of the crash, the Thunderbirds embodied the principles of resilient safety by thoroughly analyzing the incident and identifying critical areas for improvement. The suspension of public performances for six weeks allowed the team to conduct an in-depth review of their training and safety procedures. This period was crucial for incorporating lessons learned from the incident, particularly the need to better understand and mitigate the push–pull effect, where rapid shifts from negative to positive G-forces decrease a pilot's G-tolerance, and the effectiveness of AGSM was a critical factor in the crash.

Ultimately, the Thunderbirds' response to the 2018 crash was a mix of introspection and action. They scrutinized their existing practices, acknowledged their shortcomings, and implemented significant improvements. While these changes accentuate their commitment to resilient safety, they also serve as a stark reminder of the inherent risks in high-performance aviation and the perpetual need for continuous vigilance and dynamic adaptation.

CASE STUDY 4: THE FLYING BULLS, AUSTRIA—FLEXIBILITY VIA PERFORMANCE MARGINS

The Flying Bulls have risen as an exemplary group of air show performers in the last two decades, with performances across Europe and the Middle East, with a variety of vintage aircraft (The Flying Bulls, 2025). They are one of the only few remaining fast jet display teams in Europe. They were established in the 1980s, and since they have participated in numerous high-profile flying events, they have attracted positive attention and received comments from the crowds.

Their approach to safety is pragmatic, maintaining a balanced approach to aerial entertainment and risk management, with the adoption of resilient safety's principle of *dynamic adaptability* in their operations, primarily by including performance margins and achieving high levels of safety records.

Given the limited number of pilots flying the diverse aircraft in *Hangar 7*,[3] applying a standardized set of maneuvers adaptable to all types of aircraft—with adjustments in speed, altitude, and power setting—provides essential margins for error. This approach exemplifies the pinnacle of optimal resource use while effectively applying risk management principles. The Flying Bulls display pilots who demonstrate exceptional professionalism, marked by humility and extensive knowledge, ensuring their egos are kept in check. This allows them to fly with sufficient margins to recover from potential cognitive lapses, flying errors, and external factors, even in challenging environments such as the Austrian Alps. Their approach to resilient safety in all operations is truly exquisite.

CASE STUDY 5: RICHARD GOODWIN, UNITED KINGDOM—CONTINUOUS LEARNING AND INNOVATION

Richard Goodwin is a veteran air show performer who has had no setbacks in his achievements in aviation. A retired RAF fighter pilot and then a commercial pilot for more than 25 years, Goodwin has since moved to the air show world, displaying his iconic biplanes in major European air shows (Rich Goodwin, 2025). His passion for flying and his commitment to continuous learning have led him not only to fly the two aerobatic aircraft he owns but also to develop the existing aircraft with continuous modifications and improvements during the last decade. Moreover, in an effort to offer an innovative display, he installed two jet engines into his latest biplane. Yet, this process was not made without the appropriate process of designing, regulatory approval, testing, and, finally, certification to perform in front of the crowd.

Such a detailed approach to innovation and commitment to continuous learning reflects Goodwin's resilient safety culture that promotes both proactive actions to safe flight operations and an effort to stay *active* and *non-complacent* while keeping enthusiasm at the highest levels, both for himself and also for his fans watching his display. He is only accepting calculated risks that will allow a flying display that is ALARP risky while providing an entertaining display that combines visual with the smoke, the shiny painting of the aircraft, advanced maneuvers, and audio from the contradicting sound of a propeller and jet engine at the same time!

Goodwin's passion for continuous learning expands beyond the modifications in his aircraft and entertaining enhancements to his profile to sharing his love for aviation with the new generations. His engineering accomplishments are presented to STEM[4] programs to inspire new generations of aerospace members to watch firsthand the limitless opportunities that aviation still holds for the future. Moreover, Goodwin continuously supports the air show community by participating in annual air show conventions, as he did for the European Airshow Convention in 2024 (European Airshow Council [EAC], 2025).

His approach to the essential element of resilient safety of *active* and *bidirectional* continuous learning is exemplary. An approach where Goodwin is open to receiving information, listening to constructive criticism for his aircraft modifications by the authorities, and continuously strives to provide an enthralling flying display; but more importantly, it is his unwavering passion to give back to the air show and aviation community, in general, with sharing his extensive knowledge and experiences.

CASE STUDY 6: SKIP STEWART, UNITED STATES—COMMITMENT TO SAFETY AND PRECISION

Skip Stewart, a renowned civilian air show performer, exemplifies how dedication to safety and precision can enhance performance and mitigate risks in the high-risk environment of low-level aerobatics. Stewart's career, marked by numerous accolades, including the 2013 Bill Barber Award for showmanship and the 2015 Art Scholl award, presented by International Council of Air Shows (ICAS), and several aerobatic competitions medals, and breathtaking performances, is a testament to the importance of a *comprehensive* and *disciplined* approach to resilient safety in civilian air shows (Stewart, 2024). His current mindset on safety is not a mere result of building thousands of flying hours; it is a result of a *diverse* aviation background, being an airline transport pilot, a certified flying instructor, chief pilot for a Fortune 100 company, an aerobatic competition participant, and an air show performer for more than 15 years, which as per Stewart (2011) "*air show flying makes him a better pilot, and a safer pilot*" in the airline that he currently flies for.

Stewart's displays are distinctive for their precision and their highly entertaining flying profiles, pushing the limits of his "*PROMETHEUS*[5]," as he calls his biplane based on the Greek mythological character. Such accurate flying is a result of hard daily training "*to be as safe as I can*," as Stewart said (Scott, 2005). Moreover, his flying activities as a performer extend beyond solo aerobatics in flying a well-choreographed flying profile to other acts that need detailed planning, meticulous coordination, consistent training, and ultimately precise flying. Such acts include racing cars with his *PROMETHEUS* and flying aerial battles head-to-head with other aircraft, the so-called *TinStix of Dynamite Act* (Stewart, 2024).

Risk assessment is another critical component of Stewart's safety strategy, where he is applying a proactive yet holistic approach to safety, accepting calculated risks. Before each flight, he and his team analyze the Showbox, considering factors such as weather conditions, wind speed, and potential hazards. This proactive approach allowed him to anticipate and mitigate risks, adjusting his flight profile accordingly to ensure both safety and high performance. Additionally, his commitment to safety is not limited to recognizing the inherent risks of display flying but expands to recognizing having the support of his family, who attend almost every air show. As suggested in the study (Karachalios, 2022), having family support could be a factor in moderating his risk tolerance and, thus, enhancing his resilient safety culture.

Ultimately, Stewart's career exemplifies how a *comprehensive* and *disciplined* approach to resilient safety can enhance performance in civilian air shows and is well appreciated not only by the air show community but also by the new generations of aviators who are inspired by his professionalism, his superb showmanship, and pathos for flying. All in all, his meticulous pre-flight checks, thorough risk assessments, rigorous training, and resilient safety culture underscore the importance of integrating safety and excellence into every aspect of air show operations.

NOTES

1 The term *éclat* is defined as "brilliant or conspicuous success," "dazzling display; striking effect," or "approval; acclaim," reflecting its usage in contexts denoting notable achievement or impressive display (Oxford English Dictionary, 2024a).
2 The push–pull effect is a phenomenon where a transition from negative G-forces (which cause blood to rush toward the brain) to positive G-forces (which force blood away from the brain) can lead to G-LOC (G-induced Loss of Consciousness) due to the body's inability to quickly adapt to the rapid changes in blood distribution (Metzler, 2020).
3 *Hangar 7* is the name of the main facility that hosts the aircraft for The Flying Bulls.
4 STEM programs are educational initiatives focused on Science, Technology, Engineering, and Mathematics to foster critical thinking and innovation (Oxford English Dictionary, 2024b).
5 Prometheus in Greek means "forethought" but also is related to the ancient Greek mythology who he was a titan who stole the fire from Zeus and gave it to the people of Earth (Oxford English Dictionary, 2024c).

REFERENCES

ADS Advance. (2013, March 1). Red Arrows complete first nine ship display practice for 2013 season. *ADS Advance*. www.adsadvance.co.uk/red-arrows-complete-first-nine-ship-display-practice-for-2013-season.html

Air Combat Command. (2018, October 16). *F-16CM Thunderbird accident investigation released.* Air Combat Command. www.acc.af.mil/News/Article-Display/Article/1663212/f-16cm-thunderbird-accident-investigation-released/

Airshow News. (2020, April 29). Canadian Forces Snowbirds to launch operation inspiration saturday. *Airshow News.* www.newsairshow.com/headlines/2020/4/29/canadian-forces-snowbirds-to-launch-operation-inspiration-saturday

AirshowStuff. (2018, October 16). Air Force concludes investigation into fatal crash of Thunderbird #4, releases report. *AirshowStuff.* https://airshowstuff.com/v4/2018/air-force-concludes-investigation-into-fatal-crash-of-thunderbird-4-releases-report/

BBC News. (2012a, March 4). Red Arrows' first female pilot takes ground role. *BBC News.* www.bbc.com/news/uk-england-lincolnshire-17249660

BBC News. (2012b, May 22). Red Arrows given official go-ahead for 2012 season displays. *BBC News.* www.bbc.com/news/uk-england-lincolnshire-18167918

Canadian Military Family Magazine. (2020, June 1). Snowbird 11 accident preliminary report released. *Canadian Military Family Magazine.* www.cmfmag.ca/todays_brief/snowbird-11-accident-preliminary-report-released/

European Airshows. (n.d.). *Red Arrows.* Retrieved April 20, 2025, from www.europeanairshows.co.uk/performers/red-arrows

European Airshow Council. (2025). *EAC Convention 2024.* Retrieved April 20, 2025, from www.europeanairshow.org/convention

Hollnagel, E. (2014). *Safety-I and Safety-II: The past and future of safety management.* CRC Press.

Karachalios, M. (2022). *An evaluation of the relationships between resilient safety culture, safety risk parameters, and mindfulness in the international air show community* [PhD dissertation, University of North Dakota]. https:// commons.und.edu/theses/4267/

Karachalios, M., & Adjekum, D. K. (2023). *Air show performers: Safety, risk management, and psychological factors.* CRC Press. https://doi.org/10.1201/9781003431879

Maxouris, C. (2020, May 18). Canadian Snowbirds operation meant to uplift residents' spirits ended in a deadly crash in British Columbia. *CNN.* www.cnn.com/2020/05/18/americas/canadian-snowbirds-crash-operation-inspiration/index.html

Melchers, R. E. (2001). On the ALARP approach to risk management. *Reliability Engineering & System Safety, 71*(2), 201–208.

Metzler, M. M. (2020). G-LOC due to the push-pull effect in a fatal F-16 mishap. *Aerospace Medicine and Human Performance, 91*(1), 51–55.

National Defence, Canadian Armed Forces. (2021). Flight safety investigation report for Kamloops Snowbirds accident [Flight Safety Investigation Report]. National Defence/ Canadian Armed Forces. www.canada.ca/en/department-national-defence/news/2021/03/flight-safety-investigation-report-for-kamloops-snowbirds-accident.html

Oxford English Dictionary. (2024a). Éclat. In *Oxford English Dictionary.* www.oed.com/dictionary/eclat_n

Oxford English Dictionary. (2024b). STEM. In *Oxford English Dictionary.* www.oed.com/dictionary/stem_n

Oxford English Dictionary. (2024c). Prometheus. In *Oxford English Dictionary.*

Pawlyk, O. (2018, April 18). Thunderbirds back in the air after deadly crash. *Military.com.* www.military.com/daily-news/2018/04/18/thunderbirds-back-air-after-deadly-crash.html

Pfeiffer, M., Blackwell, B., & Devault, T. (2018). Quantification of avian hazards to military aircraft and implications for wildlife management. *PLoS One, 13.* https://doi.org/10.1371/journal.pone.0206599

Ranter, H. (2010, March 23). Mid-air collision accident BAe Hawk T1A XX253, Tuesday 23 March 2010. *Aviation Safety Network.* https://asn.flightsafety.org/wikibase/73568

Rich Goodwin Airshows. (2025). *Rich Goodwin Airshows*. Retrieved April 20, 2025, from www.richgoodwinairshows.com/

Royal Air Force Benevolent Fund. (2020, May 18). Former Red Arrows pilot Kirsty Murphy on mental health. *Royal Air Force Benevolent Fund*. www.rafbf.org/news-and-stories/news/former-red-arrows-pilot-kirsty-murphy-mental-health

Royal Air Force. (2010). Service inquiry report into the Red Arrows accident in Crete on 23 March 2010 [Corporate report]. *Royal Air Force, United Kingdom*. www.gov.uk/government/publications/service-inquiry-report-into-the-red-arrows-accident-in-crete-on-23-march-2010

Scott, P. (2005, May). So you want to be an airshow pilot. *Smithsonian Magazine*. www.smithsonianmag.com/air-space-magazine/so-you-want-to-be-an-airshow-pilot-8329275/

Stewart, S. (Director). (2011, November 18). *Skip Stewart Airshows "The Art of Flight"* [Video recording]. www.youtube.com/watch?v=YbkTeoUfv54

Stewart, S. (2024). *Skip Stewart Airshows*. www.prometheusbiplane.com/

The Flying Bulls. (2025). *About us*. Retrieved April 20, 2025, from www.flyingbulls.at/en/about-us

Tillman, H. V. (2011). Risk management, military style: Especially in military operations, it's impossible to eliminate risk, but it can be minimized. Many of their risk-management techniques can apply to your flying. *Aviation Safety, 31*(12), 4–8.

UK Government. (2012). Service inquiry: Accident involving Red Arrows Hawk T Mk1 XX179 near Bournemouth [Corporate report]. *UK Ministry of Defence and Military Aviation Authority*. www.gov.uk/government/publications/service-inquiry-report-into-the-accident-involving-red-arrows-hawk-t-mk1-xx179-on-20-august-2011

UK Government. (2014). Service inquiry: Accident involving Hawk TMk1 XX177 at RAF Scampton [Corporate report]. *UK Ministry of Defence*. www.gov.uk/government/publications/service-inquiry-into-the-accident-involving-hawk-tmk1-xx177

Part IV

Top-Notch Performers

The Pinnacle of Air Show Excellence

Manolis Karachalios

As we reach the final section of this book, we've journeyed through the essential steps to achieving success in the world of air shows, as illustrated in Figure PIV.1. Our exploration began with the meticulous process of pilot selection, where we uncovered the criteria and qualities that identify the most promising candidates. From there, we examined the rigorous training programs that hone their skills, transforming raw aerobatic-flying talent into precision and mastery. We also examined the resilient safety measures that ensure both performers and audiences remain secure, understanding that safety is the bedrock upon which all successful performances are built.

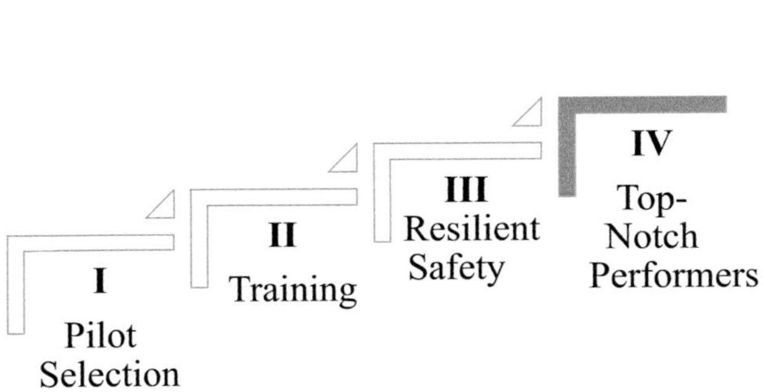

FIGURE PIV.1 Final step to air show excellence: top-notch performers.

DOI: 10.1201/9781003504801-18

Now, attention shifts to those who embody the culmination of these elements: the top-notch performers who set the standard for excellence in air shows. These elite pilots and teams represent the zenith of what can be achieved through dedication, discipline, and an unwavering commitment to their craft. They are the living embodiments of the principles and practices discussed throughout this book.

In Part IV, we honor the pinnacle of air show excellence by exploring the world of these extraordinary individuals and teams. These top-tier performers are not only masters of display flying but also exemplars of the attributes and practices that define exceptional performance. Through their profiles, valuable insights are gained into the qualities that set them apart. Their disciplined approaches, from relentless training and continuous self-improvement to their mental resilience and ability to perform under pressure, are explored. Their stories illuminate the paths they have taken to reach the top and the challenges they have overcome along the way.

14 Examples of Excellence in Air Show Performances

Manolis Karachalios

INTRODUCTION

Air shows are a remarkable blend of precision, artistry, and adrenaline, showcasing pilots and teams who have reached the pinnacle of their craft. These performers, operating within the stringent regulations, rules, training programs, and resilient safety culture that define the air show industry, consistently deliver top-notch performances. This chapter outlines some of the most exemplary individuals and teams, drawing attention to the qualities that enable them to captivate audiences around the world.

All air show performers who own the skies in front of crowds have attained a high level of proficiency and expertise. The rigorous training and safety measures in place ensure that every display pilot meets the industry's demanding standards. However, what sets the genuinely exceptional performers apart is their mastery of *showmanship*. *Showmanship* is the attribute that enables pilots to create safe, entertaining, and consistently repeatable performances. It is the ability to combine technical skills with an engaging presentation, ensuring that each show is both thrilling and safe for spectators and participants alike.

In this Chapter, we present several top performers and aerobatic teams, such as the Blue Angels, the Frecce Tricolori, and the Royal Jordanian Falcons. These profiles illustrate how these aviators exemplify the three pillars of excellence in air shows: *safety*, *entertainment*, and *consistency*.

First, *safety* is the foundation of all successful air show performances. The rigorous adherence to safety protocols, thorough pre-show planning, and continuous risk assessment ensure that every maneuver is executed with the highest regard for safety. This commitment to safety not only protects the performers and audience but also enhances the overall quality of the performance.

Next, *entertainment* is the element that transforms technical skill into a captivating spectacle. Top performers understand the importance of engaging their audience, using their aerobatic skills to create visually stunning displays that hold the spectators' attention from start to finish. The *artistry* of flight, combined with the *thrill* of aerobatic maneuvers, creates an unforgettable experience for viewers.

Last of all, *consistency* is what enables top performers to deliver high-quality performances repeatedly. Through relentless practice and a disciplined approach, these pilots ensure that each show meets the same high standards of excellence.

DOI: 10.1201/9781003504801-19

Consequently, this consistency builds trust and anticipation among audiences, who know they can expect a spectacular and safe performance every time, ultimately ensuring the air show industry's long-term relevance to their audience and industry sustainability.

PROFILES OF TOP PERFORMERS AND TEAMS

Air show performers represent aviation excellence, showcasing a blend of precision, skill, and artistry that captivates audiences worldwide. While all air show pilots exhibit exceptional attributes of showmanship, competent individuals and teams have transcended these high standards, setting new benchmarks for the industry.

Those air show performers who have demonstrated outstanding contributions to the air show community have been recognized by prestigious organizations such as the International Council of Air Shows (ICAS), the European Airshow Council (EAC), and the British Air Display Association (BADA). Firstly, the Air Show Hall of Fame, established by the ICAS Foundation, honors those who have made significant impacts on the air show industry, celebrating their careers and innovations. Similarly, the EAC recognizes excellence within the European air show community annually, presenting awards to individuals and teams who have made substantial contributions to the industry. Additionally, each year, the BADA acknowledges exceptional contributions and achievements within the UK air show community. These awards further accentuate the dedication and excellence demonstrated by top performers in the industry.

In this chapter, we profile some of the top-notch performers and teams highly recognized by the air show community, exploring the careers of notable figures such as Dave Dacy, Kent Pietsch, Bill Stein, Patty Wagstaff, the late Art Scholl, and the late Leo Loudenslager. Through their achievements, we gain a deeper understanding of the dedication, discipline, and creativity required to excel in this demanding field. These profiles illustrate how these aviators have embodied the principles of showmanship, ensuring their performances are safe, entertaining, and consistently repeatable, thereby setting the gold standard for future generations in the air show industry.

Dave Dacy: ICAS Hall of Fame Inductee, 2022

Dave Dacy stands as a paragon of excellence in the air show industry, with a career that spans over four decades. Beginning his public performances in 1974, Dacy's background in aviation and aerobatics, nurtured from an early age, laid a solid foundation for his remarkable journey. His expertise extends across various historic aircraft, notably, the Boeing Stearman, in which he performed thrilling aerobatic displays (International Council of Air Shows Foundation [ICAS], n.d.).

Dacy's commitment to safety, entertainment, and consistency has been the cornerstone of his success. As a solo performer, formation team member, and wing-walking team pilot, he captivated audiences across North America. His philosophy emphasized showmanship, ensuring performances were engaging, safe, and reliably repeatable.

A significant turning point in his career came in 1991 when he collaborated with wing walker Johnny Kazian. Their partnership, which continued with Kazian's son Tony, set new standards for wing-walking performances for nearly two decades.

Dacy's influence extended beyond his performances; as an Aerobatic Competency Evaluator, he mentored and instilled the principles of safe and entertaining airmanship in new generations of pilots.

Dave Dacy's legacy is marked by his innovative approach to air shows, his unwavering dedication to the craft, and his ability to adapt and thrive in a demanding industry. His contributions have earned him a well-deserved place in the Air Show Hall of Fame, celebrating a career that has significantly enriched the air show community.

Kent Pietsch: ICAS Hall of Fame Inductee, 2021

> *If you can't entertain, you have no business being out there.*
>
> *Kent Pietsch*

Kent Pietsch discovered his love for flying at age four, a passion that has flourished over the decades. Growing up in Minot, North Dakota, United States, he dedicated his time to being at the airport, laying the foundation for his future career (International Council of Air Shows Foundation, 2022).

For 50 years, from 1973 to 2023, Pietsch has mesmerized millions with his aerobatic routines at over 400 shows nationwide. His acts included a dead-stick routine, where he turns off the engine and descends from 6,000 feet, showcasing his precision and control. He also performs a daring rooftop landing on a moving RV, demonstrating his unmatched skill. Furthermore, his comedy act, featuring a detached aileron and a wingtip scraping pass, blends humor with the precise flying of his Interstate Cadet.

Pietsch's primary focus is audience enjoyment. He believes that the essence of his performances is to *entertain the crowd, not to showcase himself*. His *humble* demeanor and willingness to *interact* with fans make him a beloved figure wherever he performs. He appreciates the *collective effort* involved in organizing air shows and feels that the events are about the *communities* rather than himself. On receiving the Hall of Fame award, he expressed being "*very humbled, very honored, and very proud*" (Ogden, 2021).

Bill Stein: ICAS Hall of Fame Inductee, 2015

Bill Stein, a master of aerobatics and formation flight, boasts over 7,000 hours in the cockpit, perfecting his craft. His journey began as a student pilot, diving into aerobatics with unwavering dedication. Since 1995, Stein has been a standout performer at air shows across the United States, mesmerizing millions with his breathtaking displays (Stein, 2010; Bill Stein Airshows, 2025).

Stein's exceptional skills have earned him the highest honors in the industry. In 2021, he was awarded the prestigious Art Scholl Memorial Showmanship Award, a testament to his unmatched skill and showmanship. He also received the 2016 ICAS Sword of Excellence, recognizing his significant contributions to air shows. As a distinguished member of the Red Baron Pizza Squadron, Stein's legacy was further cemented with his induction into the ICAS Foundation Hall of Fame (International Council of Air Shows Foundation (ICAS), n.d.).

Every performance by Stein is a manifestation to his passion for precision flying. His air show routines are a dazzling spectacle, honed through relentless daily training to ensure every maneuver is executed with razor-sharp accuracy and exhilarating energy.

In addition to his flying skills, Stein's training regime focuses on maintaining high G-tolerance levels at all times. Even during the pre-season period, his primary focus is on physical training, ensuring he remains in peak condition to handle the intense demands of aerobatic flying.

Beyond his performances, he is dedicated to nurturing the next generation of pilots. Stein mentors new air show talent, sharing invaluable insights on enhancing both physical and mental performance in aerobatic flying. He also coaches the "*Stars of Tomorrow*" team, which has graced prestigious events like the Experimental Aircraft Association (EAA) AirVenture Air Show, showcasing the future of aerial excellence.

With his persistent pursuit of perfection and his unparalleled dedication to the art of aerobatics, Bill Stein continues to set the standard in air show performance, inspiring awe and admiration from fans and fellow pilots alike. His passion for flight and commitment to excellence make him a true icon in the world of air shows (Columbus Air Show, 2024; Stein, 2010; Bill Stein Airshows, 2025).

Patty Wagstaff: ICAS Hall of Fame Inductee, 2006

Patty Wagstaff has had a profound influence on the air show and aviation community. As one of the first female pilots to break into the male-dominated world of aerobatics, she has shattered numerous glass ceilings, serving as a trailblazer and role model for aspiring female aviators worldwide. Her fearless flying style, characterized by precision and grace, has captivated audiences and set new standards for excellence in aerobatic performance.

Wagstaff's achievements include winning the US National Aerobatic Championships three times, a feat that solidified her reputation as one of the top aerobatic pilots in the world. Her dedication to safety and meticulous attention to detail have made her performances not only thrilling but also models of how to conduct complex maneuvers with the utmost care. This focus on excellence and safety has contributed to raising awareness and setting higher standards within the air show industry (International Council of Air Shows Foundation, 2006).

Beyond her air show performances, Wagstaff has also contributed to the aviation community through her work as a flight instructor and advocate for aviation education. She has trained numerous pilots worldwide, sharing her vast knowledge and experience to help them improve their skills and adopt safer flying practices. Additionally, her involvement in organizations like the EAA and the ICAS has helped to promote and support the aviation industry.

Wagstaff's influence extends to popular culture as well. She has been featured in various media, including documentaries and interviews, bringing greater visibility to the world of aerobatics and inspiring countless individuals to pursue careers in aviation. Her story and achievements underline the possibilities within the field, encouraging both men and women to aim high and push the boundaries of what is possible.

In recognition of her contributions, Wagstaff has received numerous awards and honors, including her induction in 2006 into the Air Show Hall of Fame by the ICAS Foundation.

Art Scholl: ICAS Hall of Fame Inductee, 1996

Art Scholl remains an iconic pioneer in the world of aerobatics, celebrated for his innovative contributions and exceptional showmanship. His career began in the 1950s, and he quickly became renowned for his unique and entertaining performances that set new standards in the industry. Scholl's expertise in precision flying and his creative stunts made him a beloved performer at air shows worldwide (International Council of Air Shows Foundation (ICAS), n.d.).

To perfect his aerobatic skills, Scholl flew in competition aerobatics and was a member of the five-person team representing the United States in international competitions from 1963 through 1972. His dedication to the craft was further demonstrated when he won the US National Aerobatic Championship in 1974 with his Pitts S-2A, N13AS.

Scholl's performances, such as his "*Super Chipmunk*" routine and the thrilling inverted ribbon cut maneuver, were characterized by their precision and flair. He was known for his "*Sky Writer*" act, where he would write messages in the sky, captivating audiences with his skill and creativity.

Beyond his air show success, Scholl was a dedicated educator and mentor, influencing many aspiring pilots through his work as a flight instructor. He also made significant contributions to the film and television industry, performing stunts and flying sequences in movies like "*The Right Stuff*" and "*Blue Thunder*."

Tragically, Scholl's life was cut short in 1985 during the filming of a stunt for the movie "Top Gun" (National Transportation Safety Board, 1985; Stilwell, 2022). His loss was deeply felt in the aviation and air show community, but his legacy is remembered annually through the Art Scholl Showmanship Award, which is presented to the air show act or performer who best exemplifies the qualities of showmanship he demonstrated (International Council of Air Shows Foundation, 2025).

Leo Loudenslager: ICAS Hall of Fame Inductee, 1998

Leo Loudenslager is celebrated as one of the most influential aerobatic pilots in aviation history. Known for his exceptional precision and innovative flying techniques, Loudenslager's contributions have left an indelible mark on the air show community. His journey began with a focus on competition aerobatics, where he achieved remarkable success, including seven US National Aerobatic Championships and a World Aerobatic Championship in 1980 (International Council of Air Shows Foundation (ICAS), n.d.).

Loudenslager's dedication to the air show performance was evident in his meticulous approach to aerobatics. He modified a stock Pitts Special to create the "*Laser 200*," a custom-built aircraft that became legendary for its performance capabilities. His innovative designs and piloting skills set new standards in the industry, influencing the development of modern aerobatic aircraft.

Three critical traits characterized Loudenslager's excellence: unparalleled personal *integrity*, genuine *humility* despite his stature and reputation, and a clear *determination* to never settle for anything less than *perfection*. He served as an ambassador for the sport, instilling enthusiasm, safe practices, and showmanship and always leading by example, inspiring millions to strive for excellence.

Throughout his career, Loudenslager captivated audiences with his dynamic and precise routines. His performances were not only technically impressive but also showcased his creative flair and deep understanding of aerodynamics. His ability to push the boundaries of what was possible in aerobatic flight earned him numerous accolades, including his induction into the ICAS Foundation Hall of Fame (International Council of Air Shows Foundation (ICAS), n.d.).

Tragically, Loudenslager's life was cut short in a motorcycle accident in 1997, but his legacy continues to inspire aviators and enthusiasts. His contributions to aerobatics and his innovative spirit remain a guiding force in the world of air shows.

The Blades: The BADA Trophy Recipient, 2023

The Blades aerobatic display team has been a prominent fixture in the UK air show community, setting high standards for 17 years until its recent disbandment in January 2023. The team made its public debut in 2006 and quickly became renowned for its precision, skill, and artistry, performing at major venues across the United Kingdom and internationally. Notably, in 2017, they won the World Formation Aerobatic Challenge in China.

The Blades were recognized with the prestigious BADA Trophy, awarded to individuals or organizations that exemplify the BADA's mission to encourage, promote, and advance safety and standards in British air displays. This accolade emphasizes their significant contributions to the industry, demonstrating exceptional dedication to maintaining high safety standards and advancing the art of aerobatic display (British Air Display Association, 2024).

Beyond their thrilling performances, they supported charitable organizations such as the Royal Air Forces Association and the RAF Benevolent Fund while also providing a platform for prestigious sponsors.

Their impact extended beyond entertainment; they played a vital role in promoting the importance of safety and excellence in the air show community. By consistently delivering high-quality performances that adhered to stringent safety protocols, The Blades set a benchmark for other teams to follow. Their commitment to safety was evident in their meticulous planning, rigorous training, and adherence to best practices, which ensured not only the safety of their pilots but also that of the spectators.

Military Demonstration Teams (Single and Multi-Ship)

Military demonstration teams, either when they are operating as single-ship or in multi-ship formations, are iconic representations of their respective armed forces, showcasing top-notch performance, precision, and combat readiness. These teams not only entertain but also symbolize the excellence and professionalism of their organizations. The following examples are based on sampling recognition awards and highlight the excellence demonstrated by various teams, but it is essential to recognize that *all* military demonstration teams embody similar levels of skill and dedication.

The Blue Angels, USN

The Blue Angels, established in 1946, are the US Navy Flight Demonstration Squadron. As one of the world's premier flight demonstration teams, they exemplify excellence in air show performance. Their mission is to showcase the pride and professionalism of the US Navy and Marine Corps through flight demonstrations and community outreach (U.S. Navy, 2025).

The Blue Angels were formed after World War II to maintain public interest in naval aviation and boost Navy morale. They performed their first demonstration in June 1946, flying the F6F Hellcat. Shortly after, they transitioned to the F8F Bearcat and later to the jet-powered F9F-2 Panther in 1949, marking their first venture into jet aviation. Throughout the decades, the Blue Angels have continuously pushed the boundaries of air show performance and have flown various aircraft, including the F-11F-1 Tiger, the F-4J Phantom II, and the A-4F Skyhawk II. Since 1986, they have flown the F/A-18 Hornet A/B, and in 2021, they transitioned to the latest version of the F/A-18, the Super Hornet E/F, which they continue to fly today.

The Blue Angels' commitment to excellence is evident in all aspects of their operations, from ground training to precision flying. They practice extensively to ensure every maneuver is executed flawlessly, emphasizing accurate coordination among the team members. Their performances feature tight formations, with their trademark *Diamond Formation*, and complex coordinated aerobatics that leave audiences in awe.

In 2007, the Blue Angels were awarded the prestigious ICAS Sword of Excellence by the ICAS, which is given to individuals or organizations that have made significant contributions to the air show industry. The Blue Angels received this honor for their unparalleled contributions to aviation and their role in inspiring countless individuals worldwide.

The Blue Angels' influence extends beyond their breathtaking performances. They serve as ambassadors of goodwill, participating in community outreach programs, educational initiatives, and hospital visits. Their legacy is one of inspiration, showcasing the dedication, skill, and teamwork required to achieve excellence in aviation. The Blue Angels' history and achievements make them a shining example of excellence in air show performance.

Central to their commitment to excellence is the mantra "*Glad to be here*," which is more than just a slogan; it is a mindset that defines the team's secret formula for enduring success. According to a recent documentary by Paul Crowder (de Herrera,

2024; Prime Video, 2024), the mantra is exchanged among team members when they meet and during debriefings, reinforcing their shared gratitude and reminding them of the privilege of being part of this extraordinary organization. Furthermore, it signifies their humble acceptance of errors and unwavering commitment to continuous improvement. With this mindset, they cultivate a relentless pursuit of perfection, empowering each member to perform at their best. This family spirit, characterized by camaraderie and mutual support, ensures that their mantra creates a contagious *culture of excellence*, making each performance not only meet but exceed audience expectations.

Battle of Britain Memorial Flight, UK RAF

The RAF Battle of Britain Memorial Flight (BBMF) was recognized by the EAC in 2011 for its outstanding contribution to preserving the legacy of the RAF during World War II (European Airshow Council, n.d.). The BBMF operates from RAF Coningsby in Lincolnshire and is dedicated to maintaining and showcasing historic aircraft such as the Spitfire, Hurricane, and Lancaster bomber. These displays serve to honor those who served in the RAF and highlight the rich history of British aviation during a critical period of the 20th century (Royal Air Force, n.d.).

Established in 1957 initially as the Historic Aircraft Flight, the BBMF was created to commemorate the RAF's role in the Battle of Britain and has since expanded its scope to include all RAF campaigns during World War II. The flight operates 12 historic aircraft, including five Spitfires,[1] two Hurricanes, a Lancaster, a Dakota, and two Chipmunks used for training. These aircraft are meticulously maintained in airworthy condition, ensuring that they can be flown regularly for public displays and ceremonial events.

The BBMF performs at over 100 air shows and events annually, making around 1,000 individual aircraft appearances each year. Their displays include flypasts at significant state occasions and at major air shows across the United Kingdom and Europe. The existence of the BBMF is encapsulated in their motto, "*Lest We Forget*," reflecting their commitment to honoring the memory of those who fought and died for freedom.

The Frecce Tricolori, Italian AF

The Frecce Tricolori, officially known as the 313 degrees Gruppo Addestramento Acrobatico, Pattuglia Acrobatica Nazionale (PAN), is the Italian Air Force's aerobatic display team. Formed on March 1, 1961, the team is based at Rivolto Air Base in northern Italy. Initially flying the North American F-86 Sabre, they transitioned to the Fiat G-91 in 1964, and since 1982, they have been using the Italian-made Aermacchi MB-339 jets (Aeronautica Militare, n.d.).

The Frecce Tricolori are renowned for their intricate formations and vibrant displays, which feature the colors of the Italian flag—green, white, and red—emitted by their aircraft's smoke generators. The team consists of ten aircraft in formation, with three additional spares and a dedicated support team of approximately 60 members responsible for the maintenance and operational readiness of the aircraft.

Over the years, the Frecce Tricolori have garnered numerous accolades. In 2005, they were awarded the best exhibition at the Royal International Air Tattoo (Aeronautica Militare, n.d.). They also set a notable record in 2021 by creating the longest national flag in the sky, stretching an impressive 3.10 miles. Their commitment to excellence was further recognized when they received the EAC award in 2022 (European Airshow Council, n.d.).

The Frecce Tricolori are more than just an aerobatic team; they are a symbol of national pride and a testament to Italy's rich aviation heritage. Their performances are a highlight at major events and celebrations, including national holidays and significant international air shows. Despite facing challenges in their long history, the team has demonstrated resilience and continued to inspire millions with their breathtaking displays.

The Royal Jordanian Falcons, RJAF

The Royal Jordanian Falcons, the national aerobatic team of the Hashemite Kingdom of Jordan, were established in 1976 at the behest of His Majesty King Hussein Bin Talal. The team was formed to promote Jordan's image globally through the art and science of aviation, fostering international goodwill and showcasing the nation's commitment to peace and friendship (Royal Jordanian Falcons, n.d.).

Initially flying Pitts Special aircraft, the team later transitioned to Extra 300 aircraft in 1992 and currently operates the Extra 330LX. The RJ Falcons perform with a fleet of four aircraft, executing precise and spectacular aerobatic maneuvers that have earned them an international reputation for excellence. Their home base is the King Hussein International Airport in Aqaba, where they also contribute to local tourism by performing regular displays over the city's golden beaches.

The Royal Jordanian Falcons have performed in numerous countries, including Belgium, Egypt, France, Greece, Indonesia, Qatar, the UAE, the United Kingdom, and the United States. Over the years, the team has received multiple accolades, with the latest recognition in 2023, when they were awarded the Jock Maitland Sword by the EAC for their outstanding contribution to air displays (European Airshow Council, n.d.).

F-16 Solo Display Team, Belgian Air Force

In 2024, the Belgian Air Force F-16 Solo Display Team was honored with the EAC Chairman's Award, a prestigious recognition of their outstanding contribution to the air show community (European Airshow Council, n.d.). This accolade marks 44 years of excellence, during which the team has captivated audiences both in Belgium and internationally.

Since its inception in 1979, the Belgian F-16 Solo Display Team has featured 19 solo display pilots, each bringing their unique style and expertise to the role. These pilots underwent rigorous selection and training processes, ensuring they could perform high-intensity maneuvers with precision and safety.

Nevertheless, the team's success was not solely attributed to the skill of the pilots but also to the unwavering support of hundreds of dedicated aircraft technicians and support staff. These professionals worked tirelessly behind the scenes to maintain the aircraft in peak condition, ensuring that every display was executed flawlessly.

The Belgian F-16 Solo Display Team's impact extended beyond mere entertainment. They served as ambassadors for the Belgian Air Force, promoting its values of excellence, precision, and dedication. Additionally, their participation in international air shows strengthened diplomatic ties and enhanced the global reputation of the Belgian Air Force.

Over the years, the team performed at numerous high-profile events, earning accolades and admiration from aviation experts and enthusiasts worldwide. Their ability to consistently deliver high-quality performances established the Belgian F-16 Solo Display as a core of European air show excellence.

The 2023 season was the last for the F-16 Solo Display Team, as they departed early, marking the end of an era for one of the most iconic teams in the air show circuit (UK Flightline, 2024).

KEY ATTRIBUTES AND PRACTICES

Air show performances, whether by military demonstration teams or civilian aerobatic pilots, represent the pinnacle of aviation excellence, and therefore, there is a need to shed light on the key attributes and practices that have defined these outstanding performers. By examining the methods of teams and individuals such as the Blue Angels, the BBMF, the Frecce Tricolori, the Royal Jordanian Falcons, Dave Dacy, Kent Pietsch, Bill Stein, Patty Wagstaff, Art Scholl, Leo Loudenslager, the Blades, and the Belgian Air Force F-16 Solo Display Team, we can identify the core elements that contribute to their success.

Precision and Skill

Precision and skill are the cornerstones of exceptional air show performances. Military teams like the USN Blue Angels have demonstrated this through their precise formations and dynamic aerobatics, requiring extensive training and discipline. Similarly, civilian pilots such as Patty Wagstaff and the late Leo Loudenslager honed their craft through rigorous practice and attention to detail, performing maneuvers with exacting accuracy that captivated audiences.

Teamwork and Communication

Effective teamwork and communication are essential for both military and civilian air show performers. The Blue Angels and the Frecce Tricolori, for example, rely on flawless synchronization and clear communication to execute complex aerial routines safely. This principle was equally important for civilian teams like the Blades and performers such as Dave Dacy, who often worked in tandem with wing walkers and other pilots to deliver seamless, high-impact shows.

Dedication to Safety

Safety has always been paramount in air show performances. Military teams like the Belgian Air Force F-16 Solo Display Team adhered to strict safety protocols, including regular aircraft inspections and comprehensive pre-flight checks. Civilian performers like the late Art Scholl, known for his meticulous approach, also prioritized safety to ensure the well-being of both performers and spectators.

Continuous Improvement and Innovation

Continuous improvement and innovation drive the excellence seen in top air show performances. Teams like the Royal Jordanian Falcons evolved by adopting new aircraft and refining their routines. Performers like Leo Loudenslager, who designed and flew the revolutionary Laser 200, continually pushed the boundaries of air show performance.

Ambassadorship and Public Engagement

Air show performers serve as ambassadors for their organizations and countries. The Blue Angels and the Royal Jordanian Falcons engage in extensive public outreach, participating in community events and educational programs to promote aviation. This is also true for civilian performers like Patty Wagstaff and Bill Stein, who have been vocal advocates for aviation safety and education, inspiring future generations of pilots.

Resilience and Professionalism

Resilience and professionalism are defining traits of top air show performers. The Blue Angels, for instance, demonstrated remarkable resilience in the face of the tragic accident of 2016, rebuilding their team and continuing to perform at the highest level (Department of Navy, 2016). The Blades, who performed for 17 years, exemplified professionalism through their consistent delivery of high-quality performances until their recent disbandment. Additionally, the Belgian Air Force F-16 Solo Display Team, which recently ceased displays, left a legacy of excellence and inspiration.

Consistency in Performance

Consistency is a hallmark of exceptional air show performers, whether military or civilian. Consistently delivering high-quality performances requires solid discipline, thorough preparation, and a commitment to excellence. Teams like the Blue Angels and the Frecce Tricolori maintain rigorous training programs to ensure that each performance meets their high standards. Similarly, civilian performers like Bill Stein and Kent Pietsch focused on consistency to build and sustain their reputations.

The 10,000-Minute Rule

The notion that achieving excellence in a critical task requires a minimum level of practice is frequently advocated in studies of expertise. Gladwell, in his book Outliers (2008), suggests that 10,000 hours is the magic number for attaining true expertise. However, for air show performers, this figure seems highly impractical if it implies 10,000 flight hours. The limited duration of display flights, coupled with fuel constraints, weather conditions, and the physical and mental limits of pilots, makes accumulating such extensive practice hours practically impossible.

A quick calculation reveals that a solo display pilot, practicing twice a day for 15 minutes per session, seven days a week without weather disruptions, would need more than 56 years to reach 10,000 hours of flight time! This scenario implies that a pilot who begins training at 16 would achieve this expertise level at 72, by which time most air show performers have already retired. Therefore, the suggestion that air show performers could realistically reach 10,000 hours seems unattainable.

Even if we consider converting the benchmark from hours to minutes—10,000 minutes—it appears more plausible but still questionable. Based on optimistic assumptions, a pilot might achieve this milestone with non-stop practice of the same display profile for one year or by flying three to four times a week for approximately two years. However, this calculation excludes the necessary time for mental preparation, such as visualization and simulator practice during ground training.

The 10,000-minute rule can be a guideline for new air show performers looking to reach the levels of excellence of the top air show performers explored in this chapter. Ultimately, achieving proficiency is not simply about obtaining clearance from local aviation authorities to perform under a display authorization (DA), which ensures safe execution but does not necessarily reflect exceptional flying skills. Additionally, the idea that one can quantify expertise with an unproven number, like 10,000 hours or even minutes, is overly simplistic and fails to account for the complexities and variabilities of individual learning and performance. For air show performers, the emphasis should be on consistent dedication to excellence while avoiding the natural tendency for overconfidence and plateauing.[2]

Isorropia: The Ultimate Attribute for an Air Show Performer

Isorropia (Ισορροπία, pronounced as [ee-sho-rhoa-pee-ah]) is the common attribute found in the examples of excellence in air show performances studied in this chapter. This word, of Greek origin, derived from *ἴσος (isos)*, meaning "equal," and *ῥοπή (ropi)*, meaning "force" or "inclination," signifies a state of *balance* or *equilibrium*, where opposing forces are of equal magnitude, resulting in *stability* and *harmony*.

The word *isorropia* was chosen over the word "*balance*" because it encapsulates a deeper and more nuanced understanding of what air show performers aim to achieve. While "*balance*" generally denotes a simple state of *stability*, *isorropia* implies an intricate equilibrium of *multiple* forces and factors, resonating more profoundly with the complexities involved in air show performances.

For air show performers, achieving *isorropia* is central and is related to optimally balancing several key aspects, since they strive for excellence in their aerial acts while maintaining the highest safety standards. This balance ensures that they can push the *boundaries* of performance while maintaining sufficient *buffers for error* without compromising their *well-being* or that of the audience. Furthermore, performing intricate aerial maneuvers requires peak *physical fitness* and *mental acuity*; therefore, achieving *isorropia* in this context means being equally prepared physically and mentally to handle the demands of their performances. Also, while the primary goal of air shows is to entertain, performers must also exhibit strict discipline to execute their routines flawlessly. Ultimately, the state of *isorropia* is the *ultimate goal* of *excellence* that an air show performer reaches during their display flying—it is the Γ-shaped (*gamma*-shaped) level of peak performance that maintains with a *bouncing ball model*'s mindset.

NOTES

1 A Spitfire was lost during the tragic accident on May 25, 2024, which caused the death of the pilot (CNN, 2024).
2 Plateauing refers to the phenomenon where progress or improvement in performance levels off and becomes stagnant despite continued effort and practice (Tran et al., 2020).

REFERENCES

Aeronautica Militare. (n.d.). *Pattuglia Acrobatica Nazionale "Frecce Tricolori."* Aeronautica Militare. Retrieved July 26, 2024, from www.aeronautica.difesa.it/home/noi-siamo-l-am/personale-e-mezzi/pattuglia-acrobatica-nazionale/

British Air Display Association. (2024, January 19). *BADA awards 2023*. British Air Display Association. www.bada-uk.com/2024/01/bada-awards-2023/

CNN. (2024, May 25). *British RAF pilot dies after World War II era Spitfire crashes*. Retrieved April 20, 2025, from www.cnn.com/2024/05/25/europe/raf-spitfire-crash-intl-latam/index.html

Columbus Air Show. (2024). *Meet Bill Stein – Interview with the legendary aerobatic performer*. Columbus Air Show. https://columbusairshow.com/newsletter-2024-winter/meet-bill-stein-interview-with-the-legendary-aerobatic-performer/

de Herrera, A.. (2024, July 20). "Glad to be here": How the Blue Angels mantra inspired a film director's journey with team. *Pensacola News Journal*. www.pnj.com/story/news/military/blueangels/2024/07/20/the-blue-angels-documentary-by-paul-crowder-inspired-by-glad-to-be-here/74458235007/

Department of the Navy. (2016). *Command investigation into the circumstances surrounding the Navy Flight Demonstration Squadron Class A aviation mishap in Smyrna, Tennessee, on 2 June 2016 (Accident Investigation 5830; p. 46)*. Department of the Navy. https://news.usni.org/wp-content/uploads/2016/09/JAGMAN-and-First-Endorsement.pdf#viewer.action=download

European Airshow Council. (n.d.). *Award Winners Hall of Fame*. European Airshow Council. Retrieved July 26, 2024, from www.europeanairshow.org/award-winners-hall-of-fame

Gladwell, M. (2008). *Outliers: The story of success*. Little, Brown and Company.

International Council of Air Shows Foundation (ICAS). (n.d.). *Air Show Hall of Fame*. International Council of Air Shows Foundation. Retrieved July 26, 2024, from https://airshowfoundation.org/air-show-hall-of-fame/

International Council of Air Shows Foundation. (2006). *Patty Wagstaff*. Air Show Hall of Fame. Retrieved April 20, 2025, from https://airshowfoundation.org/patty-wagstaff/

International Council of Air Shows Foundation. (2022). *Kent Pietsch*. Retrieved April 20, 2025, from https://airshowfoundation.org/kent-pietsch/

International Council of Air Shows. (2025). *Art Scholl Showmanship Award*. Retrieved April 20, 2025, from https://airshows.aero/Page/ConvAwards/Scholl

Ogden, E. (2021, December 17). *Air show performer Kent Pietsch inducted into Hall of Fame. Minot Daily News*. Retrieved April 19, 2025, from www.minotdailynews.com/news/local-news/2021/12/air-show-performer-kent-pietsch-inducted-into-hall-of-fame/

National Transport Safety Board. (1985). *National Transportation Safety Board Aviation Accident Final Report—LAX85LA393 (LAX85LA393; p. 4)*. NTSB. https://app.ntsb.gov/pdfgenerator/ReportGeneratorFile.ashx?EventID=20001214X37820&AKey=1&RType=Final&IType=LA

Prime Video (Director). (2024, March 20). *The Blue Angels* [Video recording]. www.youtube.com/watch?v=f36UTLWPL6I

Royal Air Force. (n.d.). *Battle of Britain Memorial Flight*. Royal Air Force. Retrieved July 26, 2024, from www.raf.mod.uk/display-teams/battle-of-britain-memorial-flight/

Royal Jordanian Falcons. (n.d.). *Royal Jordanian Falcons – The Sky Falcons*. Royal Jordanian Falcons. Retrieved July 26, 2024, from https://rjfalcons.com/

Stein, B. (2010, February 18). *The physical and mental aspects of becoming great* [Lecture]. Retrieved April 20, 2025, from https://q10d2e.p3cdn1.secureserver.net/wp-content/uploads/2022/05/AerobaticCoaching.pdf

Bill Stein Airshows. (2025). *Bill Stein Airshows – Award winning pilot and coach*. Retrieved April 20, 2025, from https://billsteinairshows.com/

Stilwell, B. (2022, April 8). A stunt pilot died filming "Top Gun" in the same flat spin that almost killed Maverick. *Military.Com*. www.military.com/off-duty/stunt-pilot-died-filming-top-gun-same-flat-spin-almost-killed-maverick.html

Tran, V., Lashley, L. K., & Golden, C. J. (2020). *Career plateauing*. In R. Summers, C. J. Golden, L. K. Lashley, & E. Ailes (Eds.), *Essays in developmental psychology* (Essay 658). Nova Southeastern University. Retrieved April 20, 2025, from https://nsuworks.nova.edu/cps_facbooks/658/

UK Flightline. (2024, February 19). Belgian F-16 solo display bows out early. *Airshow Dates, News and Reviews for the UK & Europe – Flightline UK*. www.air-shows.org.uk/2024/02/airshow-news-belgian-f-16-solo-display-bows-out-early/

U.S. Navy. (2025). *History*. U.S. Navy Blue Angels. Retrieved April 20, 2025, from www.blueangels.navy.mil/history/

Epilogue
Excellence Beyond Air Shows

Manolis Karachalios

> *We are what we repeatedly do; then excellence is not an act, but a habit.*
>
> *Aristotle, Ancient Greek Philosopher*

As we conclude this exploration of "Excellence in Air Show Performers," it becomes evident that the pursuit of excellence extends far beyond the sphere of air shows and aviation as a whole. Excellence is a strategically inspiring tool for achieving high performance for individuals, organizations, and societies, offering profound vision and valuable lessons for all areas of life.

Excellence provides a clear framework for setting and achieving high standards. It involves a commitment to continuous improvement, rigorous self-assessment, and the relentless pursuit of quality.

For individuals, it involves setting high personal standards, striving for self-improvement, and consistently delivering exceptional results. This mindset fosters resilience, adaptability, and a drive to overcome challenges, enabling individuals to achieve their full potential.

For organizations, excellence serves as a foundation for sustained success. It drives the implementation of best practices, fosters innovation, and ensures a high level of performance across all operations. By cultivating a culture of excellence, organizations can enhance their competitive edge, improve customer satisfaction, and achieve long-term growth. This strategic approach encourages employees to take pride in their work, collaborate effectively, and strive for continuous improvement, leading to higher productivity and better outcomes.

In societies, the pursuit of excellence promotes overall development and progress. It inspires communities to set ambitious goals which are pragmatic, invest in education and skills development, and foster a culture of meritocracy and fairness. By prioritizing excellence, societies can drive economic growth, improve living standards, and create an environment where individuals and organizations thrive.

In air shows, we can see the epitome of the perfect blend of high performance and calculated risk, demonstrating how pushing boundaries can be accomplished safely through unwavering dedication and meticulous preparation and, ultimately, an *Isorropia* between *daring* and *caution*. Furthermore, the physiological and

DOI: 10.1201/9781003504801-20

psychological *consistency* demanded of air show performers emphasizes the importance of maintaining peak performance in all their endeavors, whether in the air or on the ground.

Beyond the breathtaking maneuvers, the essence of air shows lies in the joy and inspiration they bring. They showcase the positive aspects of life, featuring individuals who are kind, optimistic, and passionate about their craft—the art of flight. The camaraderie and shared enthusiasm within the air show community are infectious, encouraging us to approach our own lives with the same spirit of positivity and dedication.

Experiencing an air show can be transformative, sparking a desire to improve and achieve. The roar of the engines, awe-inspiring aerobatics, and pilots with seemingly "superhuman" abilities ignite dreams and aspirations. These moments remind us that greatness is within reach, inspiring us to strive for our own version of excellence.

The air show performers highlighted throughout this book exemplify the pinnacle of human potential. Their stories of *precision*, *teamwork*, *safety*, *innovation*, and *resilience* serve as beacons, illustrating that with dedication and passion, extraordinary achievements are possible. These principles transcend the air show industry and are universal truths that can guide us in any pursuit.

In our daily lives, embracing the ethos of air show performers means finding joy and fulfillment in our endeavors while inspiring those around us. It involves pushing our limits with a mindful approach to risk and a steadfast commitment to excellence. It fosters a positive environment where people are encouraged to *dream big* and *work diligently* to achieve their goals.

Ultimately, air shows celebrate what is possible when we *dare to dream* and *commit to excellence*. They remind us that the sky is not the limit; it is just the beginning.

THE BROADER IMPACT OF STRIVING FOR EXCELLENCE

Striving for excellence in any field creates a ripple effect that transcends the immediate sphere of influence. In professional sectors, this drive for excellence leads to advancements in technology, processes, and methodologies. In education, teachers and institutions that strive for excellence inspire students to achieve their best, shaping future generations.

Excellence knows no boundaries of age, nationality, or profession. From young students inspired by their first air show to seasoned professionals seeking to refine their craft, the lessons learned from air show performers can be applied universally. These lessons teach us the value of *hard work*, the importance of *resilience*, and the *joy* of achieving high standards: *not aiming to be the best, but rather to do our best.*

REFLECTIONS ON THE JOURNEY TOWARD EXCELLENCE

Reflecting on the journey toward excellence, we see that it is *not a destination* but an *ongoing process*. It requires a commitment to *lifelong* learning, a willingness to take *calculated* risks, and an *openness* to feedback and improvement. The path to excellence is paved with challenges, but it is also filled with moments of success and fulfillment.

As the author, my ultimate goal for this book is to inspire people to constantly pursue *excellence* and *eudaimonia*—a state of flourishing and well-being. The stories of air show performers presented illustrate that with *passion*, *dedication*, and a *positive outlook*, we can transform our lives and the lives of those around us. Whether in the cockpit of an aircraft, the boardroom of a company, or the classroom of a school, the *principles of excellence* can guide us toward a brighter, more fulfilling future.

Let us take the passion to do well from the air show performers and apply it to our own lives. *Let us* strive for excellence in all we do, embracing challenges and finding solutions while maintaining a resilient mindset of constantly *bouncing up*. *Let us* foster the same culture of positivity and collaborative continuous improvement, inspiring those around us to reach for the skies. Together, we can create a world where *excellence is not just a goal but a way of life*.

Glossary

Numerous definitions exist throughout international aviation authorities involving air show performers and air bosses, which may confuse the reader. However, for this book, "air show performers" will be defined as all pilots/operators flying any aircraft or unmanned air vehicle (UAV) who perform aerobatics–solo or formation, or dynamic maneuvering–solo or formation. Then, an "air boss" will be mentioned as the person in charge of the flight operations at an aviation event.

In addition, the following definitions will be referenced under the UK CAA's CAP 403[1]:

Crowd line	The line delineating the closest edge of any area, including car park(s), accessible to spectators concerning the display area/display line.
Display line	A line defining the track along which displaying aircraft may operate.
Flying display	Any flying activity deliberately performed for the purpose of providing an exhibition or entertainment at an advertised event open to the public.
Flying control committee (FCC)	A group of suitably experienced persons assembled to assist the FDD in the safety management of a flying display.
Flying display director (FDD)	The person responsible to the authorities for the safe conduct of a flying display.
Minimum separation distance	The minimum separation, in all directions, between any part of an aircraft in flight and the ground, water or any obstacle.
Show center or display datum	The display datum is the point upon which individual displays are based and is normally the center point of the crowd.
Spectator area	An area specifically designated for spectators by the event organizer or FDD and approved by the FDD for flying display safety purposes which includes all areas to which spectators have access during the flying display.
Minimum aerobatic height	The minimum height above which the aircraft must be capable of complete recovery from an aerobatic maneuver. This will be the most restrictive of: • The minimum aerobatic height specified in the Permission. • The minimum aerobatic height quoted on the relevant pilot's display authorization (in relation to the aircraft category being flown); or • The minimum aerobatic height imposed by the FDD.

NOTE

1 UK Civil Aviation Authority (UK CAA). (2025). *CAP 403: Flying displays and special events: safety and administrative requirements and guidance* (22nd ed.). UK Civil Aviation Authority.

Index

A

B

C

D

E

F

G

H

Q

R

S

T

U

V

W

Z

For Product Safety Concerns and Information please contact our EU representative GPSR@taylorandfrancis.com Taylor & Francis Verlag GmbH, Kaufingerstraße 24, 80331 München, Germany